Handbook of Research on Principles and Practices for Orchards Management

Waleed Fouad Abobatta
Horticulture Research Institute, Agriculture Research Center, Egypt

Ahmed Farag
Central Laboratory for Agricultural Climate, Agriculture Research Center, Egypt

Mohamed Abdel-Raheem
National Research Center, Egypt

IGI Global
PUBLISHER of TIMELY KNOWLEDGE

A volume in the Practice, Progress, and Proficiency in Sustainability (PPPS) Book Series

Published in the United States of America by
IGI Global
Engineering Science Reference (an imprint of IGI Global)
701 E. Chocolate Avenue
Hershey PA, USA 17033
Tel: 717-533-8845
Fax: 717-533-8661
E-mail: cust@igi-global.com
Web site: http://www.igi-global.com

Library of Congress Cataloging-in-Publication Data

Names: Abobatta, Waleed, 1968- editor. | Farag, Ahmed, 1973- editor. | Abdel-Raheem, Mohamed, editor.
Title: Handbook of research on principles and practices for orchards management / Waleed Abobatta, Ahmed Farag, and Mohamed Abdel-Raheem, editors.
Description: Hershey PA : Engineering Science Reference, [2022] | Includes bibliographical references and index. | Summary: "This book aims to improve fruit orchards' productivity through exploring the latest practical research findings in the area for professionals who want to improve their understanding of the new techniques in various agricultural management practices to increase the growth and productivity of fruit orchards under different biotic and abiotic stresses"-- Provided by publisher.
Identifiers: LCCN 2022003096 (print) | LCCN 2022003097 (ebook) | ISBN 9781668424230 (hardcover) | ISBN 9781668424254 (ebook)
Subjects: LCSH: Orchards--Management. | Fruit-culture.
Classification: LCC SB355 .H29 2022 (print) | LCC SB355 (ebook) | DDC 634--dc23/eng/20220315
LC record available at https://lccn.loc.gov/2022003096
LC ebook record available at https://lccn.loc.gov/2022003097

This book is published in the IGI Global book series Practice, Progress, and Proficiency in Sustainability (PPPS) (ISSN: 2330-3271; eISSN: 2330-328X)

British Cataloguing in Publication Data
A Cataloguing in Publication record for this book is available from the British Library.

Practice, Progress, and Proficiency in Sustainability (PPPS) Book Series

Ayman Batisha
International Sustainability Institute, Egypt

ISSN:2330-3271
EISSN:2330-328X

Mission

In a world where traditional business practices are reconsidered and economic activity is performed in a global context, new areas of economic developments are recognized as the key enablers of wealth and income production. This knowledge of information technologies provides infrastructures, systems, and services towards sustainable development.

The **Practices, Progress, and Proficiency in Sustainability (PPPS) Book Series** focuses on the local and global challenges, business opportunities, and societal needs surrounding international collaboration and sustainable development of technology. This series brings together academics, researchers, entrepreneurs, policy makers and government officers aiming to contribute to the progress and proficiency in sustainability.

Coverage

- Knowledge clusters
- E-Development
- Global Content and Knowledge Repositories
- Technological learning
- Intellectual Capital
- Innovation Networks
- Socio-Economic
- Global Business
- Sustainable Development
- Outsourcing

IGI Global is currently accepting manuscripts for publication within this series. To submit a proposal for a volume in this series, please contact our Acquisition Editors at Acquisitions@igi-global.com or visit: http://www.igi-global.com/publish/.

The Practice, Progress, and Proficiency in Sustainability (PPPS) Book Series (ISSN 2330-3271) is published by IGI Global, 701 E. Chocolate Avenue, Hershey, PA 17033-1240, USA, www.igi-global.com. This series is composed of titles available for purchase individually; each title is edited to be contextually exclusive from any other title within the series. For pricing and ordering information please visit http://www.igi-global.com/book-series/practice-progress-proficiency-sustainability/73810. Postmaster: Send all address changes to above address.

Titles in this Series

For a list of additional titles in this series, please visit: http://www.igi-global.com/book-series/practice-progress-proficiency-sustainability/73810

Climatic and Environmental Significance of Wetlands Case Studies from Eurasia and North Africa
Abdelkrim Ben Salem (Mohammed V University, Morocco) Laila Rhazi (Mohammed V University, Morocco) and Ahmed Karmaoui (University of Moulay Ismail, Morocco & Southern Center of Culture and Science, Morocco)
Engineering Science Reference • © 2022 • 310pp • H/C (ISBN: 9781799892892) • US $250.00

Eco-Friendly and Agile Energy Strategies and Policy Development
Mir Sayed Shah Danish (University of the Ryukyus, Japan) and Tomonobu Senjyu (University of the Ryukyus, Japan)
Engineering Science Reference • © 2022 • 335pp • H/C (ISBN: 9781799895022) • US $240.00

Handbook of Research on Global Institutional Roles for Inclusive Development
Neeta Baporikar (Namibia University of Science and Technology, Namibia & University of Pune, India)
Information Science Reference • © 2022 • 398pp • H/C (ISBN: 9781668424483) • US $270.00

Analyzing Sustainability in Peripheral, Ultra-Peripheral, and Low-Density Regions
Rui Alexandre Castanho (WSB University, Poland)
Engineering Science Reference • © 2022 • 334pp • H/C (ISBN: 9781668445488) • US $240.00

Frameworks for Sustainable Development Goals to Manage Economic, Social, and Environmental Shocks and Disasters
Cristina Raluca Gh. Popescu (University of Bucharest, Romania & The Bucharest University of Economic Studies, Romania)
Engineering Science Reference • © 2022 • 296pp • H/C (ISBN: 9781668467503) • US $240.00

Handbook of Research on Building Inclusive Global Knowledge Societies for Sustainable Development
Cristina Raluca Gh. Popescu (University of Bucharest, Romania & The Bucharest University of Economic Studies, Romania)
Information Science Reference • © 2022 • 405pp • H/C (ISBN: 9781668451090) • US $295.00

Handbook of Research on Algae as a Sustainable Solution for Food, Energy, and the Environment
Mostafa M. El-Sheekh (Tanta University, Egypt) Norhayati Abdullah (Universiti Teknologi Malaysia, Malaysia) and Imran Ahmad (Universiti Teknologi Malaysia, Malaysia)
Engineering Science Reference • © 2022 • 795pp • H/C (ISBN: 9781668424384) • US $325.00

IGI Global
PUBLISHER of TIMELY KNOWLEDGE

701 East Chocolate Avenue, Hershey, PA 17033, USA
Tel: 717-533-8845 x100 • Fax: 717-533-8661
E-Mail: cust@igi-global.com • www.igi-global.com

List of Contributors

Table of Contents

Detailed Table of Contents

Citrus is a valuable fruit that can be grown in varying climatic conditions. The Mediterranean climate is considered the favorable climate condition for citrus production and presents exceptional fruit quality than in native regions. There are negative impacts of climate change in the main areas of production, which have negative effects on the future of citriculture. Under semi-arid conditions, rising temperature and high solar radiation might reduce growth rate and yield. Shading is considered a common agriculture practice in nurseries and has recently been used in various commercial fruit orchards. This practice has several advantages for citrus trees particularly in arid and semi-arid environments, as it affects orchard micro-climate reducing the high temperature, light intensity, solar radiation, etc. Tree shading improves leaf photosynthetic efficacy, reduces environmental stress, enhances tree growth, increases yield, and improves fruit quality, reducing disease and pest infection. Shading is considered a principal agriculture practice as a part of integrated farm management.

The use of different techniques for recycling the orchard's biomass provided more flexibility and efficiency under the different environmental and production conditions. Aerobic composting depending on the decomposition activities of different micro-organisms and the environmental conditions (moisture, temperature, additives, machinery, etc.) was for a long time the main method for recycling the orchard's biomass. Vermicomposting introduced a promising technique for recycling the orchard's biomass depending on vermicomposting earthworm types, which led to not just recycling the orchards biomass but also producing different products (vermicompost, vermi-liquid, and earthworm biomass, feeders, etc.). Anaerobic composting mainly produces energy, as well as composting, and still needs more knowledge transfer. The management of organic soil content is a vital agricultural operation that takes into more consideration increasing the sustainability and the production of orchards as well as sequestrating the carbon dioxide in orchards' soils.

Elsayed Gamal Zaki, Egyptian Petroleum Research Institute, Egypt
Shimaa Mohamed Elsaeed, Egyptian Petroleum Research Institute, Egypt
Faissal Aziz, Laboratory of Water, Biodiversity, and Climate Change, Faculty of Sciences Semlalia, Cadi Ayyad University, Morocco
Athir Mahmood Haddad, Chemistry Department, College of Science, University of Basrah, Iraq
Aula Ali Alwattar, The University of Manchester, UK
Khalid Aziz, Laboratory of Biotechnology, Materials, and Environment, Faculty of Sciences, Ibn Zohr University, Morocco

Water plays an essential purpose in agricultural yield. Unfortunately, water shortages have led to desertification and salination of soils, threatening the durable progress of agriculture, horticulture, and food security, especially in dried regions, where water resources are a severe problem. The hydrogel absorbed the water, released it to the plants as required while maintaining the soil moist during long periods. Biopolymer as guar gum-based hydrogel is becoming one of the most green ways to hydrogel materials for water retention. Artificial intelligence (AI) is used in hydrogel technology as a result of high energy consumption to make process control and optimization of synthesis and development. Sustainable environment for hydrogel technology was integrated by AI model.

Mohamed Abul-Soud, Central Laboratory of Agricultural Climate (CLAC), Agriculture Research Center, Agriculture Ministry, Egypt
Mohamed Saad Aly Emam, Central Laboratory of Agricultural Climate (CLAC), Agriculture Research Center, Agriculture Ministry, Egypt
Ahmed Awny Farag, Central Laboratory of Agricultural Climate (CLAC), Agriculture Research Center, Agriculture Ministry, Egypt

Smart agriculture motivates more farmers every day via different applications with a view of maximizing food production as well as minimizing the climate change impacts, COVID-19 crisis, and natural resources shortages. Developing the agricultural production process based on smart agriculture and IoT technology has become extremely vital to preserve natural resources, increase sustainability and ecological agricultural food production, and satisfy food security demands. Smart agriculture is mainly based on sensing and monitoring micro-climate and environmental data (collection) and analysis. Data collected from different sources (internet, remote sensing database, and sensors) can provide a wealth of information for agriculture operators, allowing them to take appropriate actions in a timely manner. Developing microcontrollers, sensors, and actuators integrated with IoT to include all agricultural procedures especially in developed countries become more necessary.

Orchards play a key role in the non-petroleum economy of many developing countries. Developing orchards with good adaptation to abiotic and biotic stress is the main point for plant researchers to improve and increase fruit yield in many parts of the world. They have established seed production zones in the best natural stands to provide seed until orchard production can satisfy all needs. Conventional breeding and genetic improvement are time-consuming and don't associate good results in orchards. Because of their great utility, genetic markers will become standard tools for orchard improvement in the future. Biotechnology creates a genetic alteration in orchards comparatively short with slight modifications to other valid characters (tolerance to abiotic and biotic stresses, enhanced post-harvest shelf life of fruit, reduced vegetative propagation phase, and fruit production with higher nutritional values). Thus, it leads to an uprising in orchard sciences, fruit tree breeding, and orchard improvement and management.

In the US, the cultivated area and production of crops benefit from insect pollination. Mango orchards contain Zebda, Hendi sinara and Ewais varieties and Figryclan, Aranaba cultivars, cultivated together and without Zebda variety. Surveyed insects found 18 species belonging to order Diptera, Coleoptera, ymenoptera, and Neuroptera found during flowering periods. Cultivation of Zebda variety alone tended to decrease the yield/tree. Zebda cultivated mixed with selected varieties and cultivars significantly increased the yield/tree (58.10 and 56.10kg against 38.6, 36.6kg). Cultivation of mango varieties and cultivars in mixed system used application farm yard manure during flowering cycle, increased insect mango pollinators to the maximum, and increased the high mango yields and high good return for the mango growers.

Climate change conditions such as drought, rising temperature, and high solar radiation cause increasing salinity levels, which are considered the main factor that limits agricultural production. Therefore, to prevent food security in arid and semi-arid areas from being subject to climatic hazards, optimal water management is essential. To surmount this problem, research is focused on innovation likely to improve efficiency water use and protect the water resources. In order to overcome the problems of contaminated water and salinization problems, the researchers used a new technique based on a magnetic field. Thus, magnetized water used for irrigation can improve water productivity and increase the production of agricultural crops, thus sustaining the water resources for the future in view of the expected global water scarcity. In this context, this chapter discusses the characteristics linked to magnetic treatments on irrigation water by studying the parameters of water quality and their effects on productivity and fruit quality of the citrus orchards.

Climate change has become one of the most relevant global challenges that impact the overall world. One way to combat climate change is to calculate and reduce the climate impacts of single products. An estimate of the total amount of greenhouse gases (GHG) emitted from a life cycle perspective of a good or service gives an overview of the contribution to climate change from this product, usually referred to as product carbon footprint (PCF). A full life cycle product carbon footprint is necessary to identify emission hotspots in the product value chains and thereby address the climate change on the product level in the most efficient way. Products' carbon footprint also make it possible to compare the climate change impact of competing horticultural products (e.g., the fruits coming from different countries but sold in the same store). However, only a limited amount of products' carbon footprints including the raw material acquisition, production, distribution, and the consumer stage have been published about horticultural products.

Climate change impacts all agricultural production in all territories. Its effects are already clearly visible on orchards. This chapter makes it possible to better understand the changes that farmers will be faced with and therefore to consider changes to their system as soon as possible in order to sustain their work. Indeed, current cultivation practices (conventional) are based on the multitude of tool passages in succession, without considering the negative effects, in the medium and long term, on the structure of the soil and on its topsoil. Thus, the excessive number of passages of agricultural tools results in increased energy consumption and a high hourly requirement per hectare. Indeed, this phenomenon of agricultural soil compaction is one of the main causes of its physical degradation. However, the rational use of agricultural land requires the introduction of new techniques of tillage as well as appropriate mechanization, which meets the requirements of crops, the improvement of production, and the preservation of the structured soil.

Emblica officinalis (Indian Gooseberry or Aonla) is a horticultural commercial crop. Among various agroforestry, a number of fruit orchard-based agroforestry systems are practiced all over parts of the country. These systems not only provide fruits but also fulfill the necessity of timber, food, fuel, and fodder. Aonla's orchard-based agri-horticultural system has immense potential for the betterment of people's livelihoods. Due to its thin canopy and leaf shedding nature in the winter season, orchards of this fruit tree species are well suited for agro-forestry. Aonla orchard-based agroforestry was discussed in the present study. Agricultural crop (wheat and paddy) grain and straw production data were recorded under Aonla orchard-based agroforestry. Though the agricultural crop production was reduced, Aonla fruits gave an additional income to the farmers. Overall, a good economic return is obtained from Aonla orchard-based agroforestry. Utilization of Aonla orchard interspace with the traditional cropping system was profitable in the Gangetic region of Uttar Pradesh, India.

Date palm (Phoenix dactylifera L.), commonly grown in the hot arid zones predominantly in the Middle East and North Africa, provides highly healthy fruits in addition to multiple uses for all parts of the tree. With the interest of many countries of the world in the direction towards planting palm trees, it was therefore significant to point out the importance of agricultural practices that are concerned with serving the palm crown, such as pollination, pruning, fruit thinning, bunch covering, and bagging fruits. The paramount importance of soil service and irrigation operations that will obtain the highest productivity of trees and achieve remunerative profits for the stakeholders and those interested in the palm sector from breeders and factory owners are shown. The latest technologies and modern programs that serve this essential agricultural sector were reviewed.

Date palm (Phoenix dactylifera L.) plants demand reclamation of new areas or renewal of the oldest trees. Different methods to produce these plants as seeds that gave only 50% of seedlings will be female. The dates from seedling plants are often smaller and of poorer quality, so the propagation of date palm plantlets will occur with different methods. In vitro culture produced true-to-type huge numbers of plantlets under control conditions that led to many differences in morphology, physiology, and anatomy characteristics. The acclimatization was done to modify these abnormalities, and the plantlets after acclimatization need to improve growth by different treatments. Offshoots are the other method for propagation. These offshoots that grow from the base of the tree need to improve new rooting at the trunk base by growth regulators substances. This work shed light on the different methods of date palm propagation, the crucial role of the acclimatization greenhouse stage, and improved plantlets planted in the open field.

The entire world has realized the danger of chemical pesticides because of their harmful impact on environment, biodiversity, and human health as a result of the accumulation of pesticides in the food and water chain and the outbreak of epidemic diseases such as cancer, kidney failure, and liver epidemic,

Many orchids species were threatened because of environmental changes and the appearance of microbial diseases. Reduction in orchard yield and productivity depends on biotic factors like bacteria, viruses and nematodes, and pests. The yield losses range from 20 to 40% in global orchard production. Diseases causing crop orchid losses were epidemics with long term and massive effects. Many species of microbes were associated with orchids that are pathogenically resulting in symptoms of many diseases like brown spot, root rot, soft rot, brown rot, wilt, and anthracnose. This chapter will discuss diseases associated with Orchidaceae species that resulted from pests and other microbes and different biological managements for these diseases.

Foreword

This book presents a set of in-depth scientific studies and modern applied practices that deal with many aspects related to the management of fruit orchards in the conditions of dry and arid lands and in light of climatic changes facing the world. These practices include shading as a common agriculture practice in nurseries and has recently been used in various commercial fruit orchards. In addition, vermicomposting is a promising technique for recycling the orchard's biomass. One of the good practices reviewed in this book is the cultivation of mango varieties and cultivars in mixed system, used application farmyard manure during flowering cycle, which increase insect mango pollinators to the maximum, resulted in an increase in mango yield. The interspacing of seasonal crops with fruits proved to be good practice, which increase total return of the orchard. Developing the agricultural production process based on smart agriculture and IoT technology has become extremely vital to preserve natural resources, increase sustainability and ecological agricultural food production, and satisfy food security demand.

This book sheds light on many new technological practices, which include Product carbon footprint, which, is necessary to identify emission hotspots in the product value chains, smart agriculture, and IoT technology, which has become extremely vital to preserve natural resources. Using Biological control methods as alternatives to pesticides through applying bio-intensive integrated pest management strategies encourage natural enemies of pests through the creation of orchard habitat that encourages biodiversity and then Creation of natural enemies, and diseases associated with Orchidaceae species that resulted from pests and other microbes and different biological managements for these diseases.

Two chapters of this book has been devoted to date palm, including all agricultural practices that are concerned with serving the palm crown, such as pollination, Pruning, Fruit Thinning, …etc. This work shed light on the different methods of date palm propagation, the crucial role of the acclimatization greenhouse stage, and improved plantlets planted in the open field.

Ibrahim Siddik
Menoufia University, Egypt

Preface

Climate change affects all agricultural production, in all territories. Its effects are already clearly visible in fruit orchards. Therefore, modification of different agricultural practices has been more interesting in the last decades, shading net of fruit is one of these techniques, which improve orchards productivity, in addition to adapting regular practice such as fertilizing, irrigating, pruning, and pest control. Furthermore, due to climate change conditions, there is a water shortage in different countries particularly in arid and semi-arid regions, while, supplemental irrigation is required for economic production, also, Climate change causes increasing salinity levels, which are considered the main factor that limits agricultural production. Pruning is very important for different orchards and should be carried out regularly in fruit orchards to improve tree structure, minimize wind damage, and increase fruit-bearing areas. Fruit trees are infested with a number of insect pests, diseases, and weeds, which need to be controlled. Balancing nutrition has a significant impact on orchards production, new flushes, flowers, and fruit ripening are dependent on both nutrient reserves in the tree and the fertilizer applied during the growing season.

There are various impacts of climate change on main fruit crops such as citrus, which is considered a valuable fruit that can be grown in varying climatic conditions, while, under semi-arid conditions, rising temperature and high solar radiation might reduce growth rate and yield. Shading is considered a common agriculture practice in nurseries and has recently been used in various commercial fruit orchards. This practice has several advantages for citrus trees particularly in arid and semi-arid environments, as it affects orchard microclimate reducing the high temperature, light intensity, solar radiation…etc. Tree shading improves leaf photosynthetic efficacy, reduces environmental stress, enhances tree growth, increases yield, and improves fruit quality, reducing disease and pest infection, shading is considered principal agriculture practice as a part of integrated farm management.

The management of organic matter is a vital agricultural operation that takes into more consideration for increasing the sustainability and production of the orchid field as well as sequestrating the carbon dioxide in the orchid soils. Organic matter management is mainly investigating both issues, the recycling technique, and the organic soil content. The management of organic matter aimed to reduce the cost of waste management, production procedures, and operations as well as reduce the greenhouse gas (GHG) emissions, carbon and water footprint, enhance productivity, improve natural resources use efficiency, and increase the net profit. Aerobic composting depending on the decomposition activities of different microorganisms and the environmental conditions (moisture, temperature, additives, machinery, etc.) was for a long time the main method for recycling the orchid biomass. Vermicomposting introduced a promising technique for recycling the orchid biomass depending on vermicomposting earthworm types, which led to recycling the orchid biomass and producing different products (vermicompost, vermi-liquid, earthworm biomass, feeders, etc.).

Increasing salinity levels is considered the main factor that limits agricultural production, therefore, using magnetic technology to treat irrigation water that is affected by salinity can improve water productivity and increase the production of agricultural crops, thus sustaining the water resources for the future in view of the expected global water scarcity. In this context, this chapter discusses the characteristics linked to magnetic treatments on irrigation water by studying the parameters of water quality and their effects on the productivity and fruit quality of the citrus orchards.

Water plays an essential purpose in agricultural yield. Unfortunately, water shortages have led to desertification and salination of soils, threatening the durable progress of agriculture, horticulture, and food security, especially in dried regions, where water resources are a severe problem, the hydrogel absorbed the water, released it into the plants as required while maintaining the soil moist during long periods. Biopolymer as Guar gum-based hydrogel has become one of the greenest ways to hydrogel materials for water retention. Artificial intelligence is used in hydrogel technology because of its high-energy consumption to make process control and optimization of synthesis and development.

Furthermore, using new technology tools could be important to adapt fruit orchards to climate change conditions and minimize the hazards. Smart agriculture motivates more farmers every day via different applications with a view of maximizing food production as well as minimizing the ecological footprint under the climate change impacts, Covid 19 crisis, and natural resources shortages. Developing the agricultural production process based on smart agriculture and IoT technology has become extremely vital to preserve natural resources, increase sustainability and ecological agricultural food production and satisfy food security demands.

Developing Orchards with good adaptation to abiotic and biotic stress is the main point for plant researchers to improve and increase fruit yield in many parts all over the world. They have established seed production zones in the best natural stands to provide seed until orchard production can satisfy all needs. Conventional breeding and genetic improvement are time-consuming and do not associate good results in orchards. Because of their great utility, genetic markers will become standard tools for orchard improvement in the future. Biotechnology creates a genetic alteration in orchards comparatively short with slight modifications to other valid characters. Thus, it leads to an uprising in orchards sciences, fruit tree breeding, and orchards improvement and management.

In experimental work, it is often necessary to determine the parameters of water treatment and their effects on fruit orchards. One of the best ways to do this is to use regression analysis. Regression analysis is one of the most widely used methods for modeling and interpreting data, and perceptual neural networks are equivalent to linear or least squares regression models. It is therefore necessary to have a basic understanding of these types of models. The chapter presents the assumptions of the regression models with the best linear unbiased estimators. Correlation analysis measures the strength between two variables, while regression estimates the mean value of the dependent variable, based on known values of one or more predictor variables. In regression analysis, therefore, a distinction is made between dependent and independent variables.

The carbon and water footprint concepts were introduced about a decade ago, simultaneously, the water footprint has become a popular term. There is still immense potential for the management of water footprint in orchards with good agricultural practices. Carbon footprint (CF)and water footprint are being used to indicate the impacts of the carbon (C) and water (W) used by production systems. The CF can be decreased in a good sustainably managed orchard. The water and carbon footprint for orchards and the importance of them for the farmer and national levels and how to manage them and how to reduce them. The water footprint analysis shows that the sustainable practices contributed to the reduction of the blue water component use by irrigation management and the application of calculation of water requirements according.

Current cultivation practices (conventional) are based on the use of various machinery, without considering the negative effects, in the medium and long term, on the structure of the soil. Thus, the excessive using of various agricultural equipment increased the energy consumption requirements per land unit. Indeed, this phenomenon of agricultural soil compaction is one of the main causes of its physical degradation. However, the proper use of agricultural land requires the introduction of new techniques of tillage as well as appropriate mechanization which meets the requirements of crops, the improvement of production, and the preservation of the structured soil.

Emblica officinalis (Indian Gooseberry or Aonla) is a horticultural commercial crop. It is a very important fruit tree species and gives good economic returns. The fruit is having medicinal value, also, it is considered a good source of vitamin C. It has acrid, cooling, diuretic, and laxative properties, in addition, dried fruits are useful in haemorrhages, diarrhea, dysentery, anemia, jaundice, dyspepsia, and cough. Aonla fruits are commonly used for preparing candy, sweets, pickles, jelly, and jam. Besides fruits, leaves, bark, and even seeds are being used for various purposes.

Date palm commonly grown in the hot arid zones predominantly in the Middle East and North Africa provides highly healthy fruit in addition to multiple uses of all three parts. In view of the importance of this type of fruit globally and in the Arab and Islamic world, and the interest of many countries of the world in the direction of planting palm trees, it was therefore significant to point out the importance of agricultural practices. as well as explain the paramount importance of soil service and irrigation operations that will obtain the highest productivity of trees and achieve remunerative profits for the stakeholders and those interested in the palm sector from breeders and factory owners.

The propagation of date palm plantlets will be occurred by other different methods, in vitro culture that produced true- to- type huge numbers of plantlets under control conditions that led to many differences in morphology, physiology, and anatomy characteristics which need to acclimatization in the greenhouse that important steps for successfully in vitro date palm, this stage was done to modified these abnormalities, the plantlets after acclimatized need to improving growth by different treatments for speedy culture in the open field to obtain useful fruits, offshoots are the other method to produce important date palm plants, these offshoots which grow from the base of the tree need to improve new rooting at the trunk base by growth regulators substances. On the whole, this review shed light on the different methods of date palm propagation, the crucial role of the acclimatization greenhouse stage and improved plantlets to be planted in the open field.

Due to rising temperature, the insects have more activation, in addition, more use of chemical pesticides to control pests and diseases of economic crops. While, there are harmful impacts of pesticides on the environment, biodiversity, and human health As a result of the accumulation of chemicals in the food and water chain and the outbreak of epidemic diseases such as cancer, kidney failure, and liver epidemic. So start changing the trend of the world by using Biological control methods as alternatives to pesticides in order to eat healthy and safe food, raise soil fertility, increase biodiversity and maintain the period of postharvest in the long term. In addition to the booming economic aspect of the individual and society, the international organic agriculture practice is concerned with the biological control in controlling any diseases on plants and prohibited using any pesticides during the production, harvesting, processing, handling, and trading of organic crops. Furthermore, there is more attention to using Biointensive integrated pest management strategies to combat the pest incidence in fruit orchards. The hostile habit of insects resulted in the elevated pesticide treatments and affected the destruction of the agroecosystem, which will be indicated by extremely resistant insect species with the elimination of entomophages. The idea of bio-intensive management tactics viz., preliminary data collection through surveillance, accurate diagnosis, sampling and field scouting, pest forewarning threshold level assessment.

Various species of pests in our orchards have natural enemies such as predators and parasites, which play an important role in regulating pest populations. Pesticides will kill the beneficial species as well as the pests, which means the first line of defense is removed or weakened the next time the pest species attacks your fruit trees or crop. Agriculture is more sustainable and resilient to change when it mimics biological systems. The Codling moth is a widespread and common pest of apple fruit. The larvae burrow through the flesh into the core. They feed in the fruit for a number of weeks, excreting wet dark-brown frass and spoiling the fruit. Many pests invest in orchards crops such as Apple maggot, Gypsy moth, Peachtree borer, Pecan weevil, Psyllid, Sawfly, and Scale insect.

Under fluctuation in climate conditions, many Orchids species were threatened cause of environmental changes with the appearance of microbial diseases. Diseases causing crop orchid losses were epidemics with long terms and massive effects. Many species of microbes were associated with orchids are pathogenically resulting in symptoms of many diseases. This chapter will discuss diseases associated with Orchidaceae species that resulted from pests and other microbes and different biological managements for these diseases.

Finally, this book explains the most effective information for the growers of orchards in the world and techniques used in orchards management until harvesting crops, therefore, this book affects all contributors in the field of orchard crops and all the people working in the field.

The target audiences are farmers, students in university, postgraduates, and all the staff in scientific research.

Waleed Fouad Abobatta
Horticulture Research Institute, Agriculture Research Center, Egypt

Ahmed Farag
Central Laboratory for Agricultural Climate, Agriculture Research Center, Egypt

Mohamed Abdel-Raheem
National Research Center, Egypt

Acknowledgment

The world is a better place thanks to people who want to develop and teach others. What makes it even better are people who share their ideas and time to mentor future brains. Thank you to everyone who strives to grow and help others grow and development.

Editors of the book, Prof. Waleed Fouad Abobatta, Prof. Mohamed Abdel- Raheem, and Prof. Ahmed Awny Farag, thank Allah for completing this book.

We would thank and give gratitude to all professors who supported us to issue this book. In addition, we thank all colleagues, both contributors, and reviewers who spent their time and experience to produce this book in a way that outfits the readers and to provide useful and advanced scientific information.

We extend our special thanks to Prof. Dr. Ibrahim Seddik, who gave part of his time to share this book and helped us give this book a scientific value.

June 2022

Waleed Fouad Abobatta
Horticulture Research Institute, Agriculture Research Center, Egypt

Chapter 1
Shading of Citrus Orchards: Under Fluctuation of Climate Conditions

Waleed Fouad Abobatta
Horticulture Research Institute, Agriculture Research Institute, Egypt

ABSTRACT

Citrus is a valuable fruit that can be grown in varying climatic conditions. The Mediterranean climate is considered the favorable climate condition for citrus production and presents exceptional fruit quality than in native regions. There are negative impacts of climate change in the main areas of production, which have negative effects on the future of citriculture. Under semi-arid conditions, rising temperature and high solar radiation might reduce growth rate and yield. Shading is considered a common agriculture practice in nurseries and has recently been used in various commercial fruit orchards. This practice has several advantages for citrus trees particularly in arid and semi-arid environments, as it affects orchard micro-climate reducing the high temperature, light intensity, solar radiation, etc. Tree shading improves leaf photosynthetic efficacy, reduces environmental stress, enhances tree growth, increases yield, and improves fruit quality, reducing disease and pest infection. Shading is considered a principal agriculture practice as a part of integrated farm management.

INTRODUCTION

There are different challenges facing fruit production, climate change is considered one of the main threats to the sustainability of the productivity of fruit crops, due to influential fluctuations in the behavior of trees, particularly in flowering and fruit set stages, which causes negative effects on tree productivity and fruit quality (Abobatta 2021a).

Multiple phenomena of climate change affect fruit orchards such as rising temperatures, heatwaves, warmer nights in winter, elevated carbon dioxide (CO_2) levels, frequent droughts, extreme winds, and increased solar radiation (Bisbis, et al, 2019).

It is well documented that, rising temperature increases evapotranspiration reduces CO_2 assimilation and minimizes stomatal conductance through closing stomata, which reduces carbohydrates formation rate, thus reducing tree yield and affecting fruit quality (Jokar, et al, 2021). In addition, high temperature

DOI: 10.4018/978-1-6684-2423-0.ch001

increases the abscission of flowers and fruitlets of citrus trees (Mditshwa, et al, 2019; Syvertsen, et al, 2003).

Currently, there is more attention to using various techniques to alleviate the negative effects of climate change on various crops particularly fruit trees. Shade netting is one of the modern techniques that are used to reduce impacts of climate change on the physiological performance of fruit trees particularly in dry and semi-dried regions (Manja and Aoun 2019; Mahmood, et al, 2018).

Citrus trees are subjected to various environmental stress especially heat stress in the main production (warm subtropical and semi-dried) regions, particularly during summer when temperatures are above 31 ^{0}C which affects negatively yield and reduce fruit quality (Tsai, et al, 2013).

Therefore, using protective cultivation or net shading in citrus cultivation have more interesting particularly in dry and semi-dried regions (Figure 1), due to fluctuation in climate conditions, particularly high temperature, increasing solar radiation rate, and hail injuries (Abobatta, 2021b).

Figure 1. Image field to cultivation mandarin under shading net, under Egyptian desert conditions (Photo by Dr. Waleed Abobatta 2010)

Shading system is used to prevent trees from fluctuation in temperature, reduce physiological disorders like sunburn and cracking, reducing birds' damage, insect infection threats, strong wind, and sandy storms. Furthermore, shading protects newly establish citrus orchards from infection with Asian citrus psyllid and *Candidatus Liberibacter asiaticus* (Gaire, et al, 2021).

Complete net houses or top screens are considered a partial solution for mitigating impacts of climate change through modification of the local climate around the trees (Prins, 2018). Partial shading can be used to protect the fruits from excessive environmental conditions like excessive solar radiation, strong wind. In addition, could reduce water loss from the fruits, enhance fruit quality like firmness, and may prolong fruit shelf life (Botes 2018).

This work discussed the impact of climate change conditions on citrus orchards and explored the effects of net shading on growth and productivity, various types of net shading in use, the benefits of shading on citrus production. In addition to the effects of net shading on reducing solar radiation in citrus orchards. Furthermore, discussing the most important gaps that must be avoided in future researches.

CLIMATE CHANGE CONDITIONS

Under global warming and increase environmental stresses which affect negatively the growth and productivity of various fruit orchards worldwide, there is a need to use new practice management like net shading to enhance environmental conditions which improve tree vigor and yield (Sadka, et al, 2019).

The warm nights in winter and spring are one of the main factors that inhibit flower induction, reduce flowering buds, and delay bud-break particularly in deciduous fruits, which depend on the accumulation of chilling requirements during the dormancy period (Melke, 2015; Wilkie, et al, 2008).

Heatwaves are considered a serious threat to citrus production in arid and semiarid regions, as per expecting scenario, there are more and continuous intensive heatwaves for long periods that will occur in the near future. Heatwaves affect negatively particularly when it comes in crucial stages of trees like flowering, fruit set, and during June drop, it increases flower abscission and dropped new fruit set and fruitlet, also, increase adverse impacts at June drop in Northern hemisphere (Abobatta, 2021b).

Excessive solar radiation causes sunburn in fruit and increases physiological disorders, which affect drastically producer profits (Manja and Aoun, 2019).

Therefore, protecting fruit orchards from the fluctuation of climate conditions is considered a serious challenge for fruit producers worldwide. Moreover, extremely hot summer adversely affects fruit properties and the total yield of evergreen fruits like citrus fruits particularly in arid and semi-arid regions (Jamshidi, et al, 2020).

1. **Heat stress**

High temperature increases evapotranspiration and reduces stomatal conductance, there are inhibitory effects of high temperature on photosynthesis and accumulation of dry matter in the whole plant, due to closing stomata to minimize water loss (Nebauer, et al, 2011). Therefore, using shading under dry and semi-dry regions could be practical management to improve tree growth, increase yield, and enhance fruit quality.

2. **Drought**

Drought is considered one of the most important environmental stress that restrict the growth of plants, particularly those grown in semi-arid and tropical regions (Hussain et al. 2018).

Due to rising temperature and low soil moisture, citrus trees are subjected to the shortage in water supply, this conditions various metabolism processing in the plant cells are affected, and closing stomata to reduce evaporation and evapotranspiration, and reduce photosynthesis, which affect negatively tree yield (Shafqat, et al, 2021; Zhang, et al, 2018). Thus, subjecting citrus tree to drought, affect tree growth, productivity, and fruit quality (Abobatta, 2021b).

3. **Salinity**

Due to the fluctuation of climate conditions particularly drought, salinity stress become a global problem that affected the growth and productivity of glycophytes crops particularly fruit trees (Kataria, and Verma, 2018).

Citrus is considered a sensitive crop for salinity, therefore, under high temperature and low precipitation, citrus orchards in dry and semi-dried areas are subjected to salinity stress, which affects negatively on citrus production such as reduction of productivity and low fruit quality.

4. **Light intensity**

There is an important role of light intensity in regulating various processes in plants, so, there are numerous responses of light intensity include hormone regulation, flowering, shade avoidance, stomatal movement, and photosynthetic efficiency (Jiao, et al, 2007).

Why Do We Need the Shading Net in Fruit Orchards?

Due to fluctuation in climate change conditions, particularly rising temperature and drought, there is a need to use various tools to sustain fruit orchards productivity under harsh environmental conditions, such as shading, to reduce transpiration rate and enhance water use efficiency of trees (Nicolas, et al, 2008).

Shading net is a new technique to alleviate impacts of climate change conditions, shading could cause positive changes in plants such as increasing leaf area, reducing evapotranspiration, decrease radiation heat load particularly in warm regions (Mahmood, et al, 2018; Raveh, et al. 2003).

Previous reports indicated that, shading adapted orchard environmental conditions particularly light intensive, canopy temperature, humidity ratio, and soil temperature (Manja & Aoun, 2019). There are positive effects of shading net on the environment of fruit orchards, it is modifying macroclimate around the trees, for instance reducing solar radiation, light intensity, which reflects on reduction of canopy temperature, relative humidity, and soil temperature (Bastías, et al, 2012; Iglesias and Alegre, 2006; Kalcsits et al., 2017). In addition, the reduction of solar radiation is considered the main goal for using net shading in fruit orchards (Dovjik, et al, 2021). Therefore, the use of shading net technology is considered a practical solution to mitigate the adverse impacts of climate change on fruit orchards, as it acts to modify the local climate and protect trees from environmental risks.

Benefits of Net Shading in Fruit Orchards

There are positive effects of all types of shading on citrus orchards particularly photo-selective nets that improve tree performance, by increasing photosynthesis rate and net assimilation of CO_2, consequently increasing vegetative growth and improving fruit quality (Zhou, et al, 2018).

Previous work (Bosančić, et al, 2018; Mupambi, et al, 2018; Zhou, et al, 2018) indicated several benefits for shading in orchards through modifying the microclimate and enhancing plant growth as follow:

1. Decrease air velocity and reduce radiation heat-load.
2. Protect trees from cold weather.
3. Adjustment canopy temperature.
4. Increased relative humidity ratio.
5. Reducing evapotranspiration.
6. Increases net CO_2 assimilation rate.
7. Improve fruit quality.
8. Increase farmers' profits.

Types of Net Shading

Different types of shading that include closed net, screen houses, top netting, or partial shading represent a key solution for mitigating the fluctuation of climate through modifying and improving the microclimate around the trees (Prins, 2018).

There are many shade materials are in use as shown in table (1), depending on the ability to modification of the solar radiation, reduce high-temperature effects, and enhance microclimate around the crops.

Photo-selective nets protect plants from excessive solar radiation by scattering waves 8and modification of spectrum (Arthurs, et al, 2013; Rajapakse, et al, 2008), such as reflective aluminized nets (Raveh, et al, 2003), aluminized-plastic net (Alarcon, et al, 2006), 20% white shade nets (Cronjé, et al, 2020) on mandarin, photo-selective nets (Zhou, et al, 2018) on Valencia trees.

Table 1. Types of net shading using in citrus orchards

Shading type	Crop	Effect	Reference
20% white shade nets	Nadorcott mandarin	Improving fruit quality	Cronjé, et al, (2019)
Reflective aluminized nets	Citrus trees	Increase vegetative growth	Raveh, et al, (2003)
Black net	Murcott' Tangor	Improve fruit quality	Tsai, et al, (2013)
Aluminised-plastic nets	lemon	Improving water-use efficiency	Alarcon, et al, (2006)
Photo-selective nets	Valencia trees	Increased vegetative growth and photosynthetic	Zhou, et al, (2018)
Woven aluminized-plastic nets	Young citrus trees	Increase dry matter in plant tissue	Raveh, et al, (2003)
20% white shade nets	Nadorcott mandarin	Improving fruit quality	Cronjé, et al, (2019)
Gray and black nets,	′Primosole′ mandarins	improve flowering	Germanà, et al, (2002)
White net	Ponkan mandarin	Decrease sunscald and granulation in fruits	Lee et al, (2015)
White net	Murcott mandarin	Accelerate new flushing and increase growth	Abobatta, (2022) unpublished work

Furthermore, shading with photo-selective nets could be a key solution to limiting adverse effects of environmental stress through modification of the microclimate, which improves tree performance, consequently, increasing yield crop and fruit quality (Dovjik, et al, 2021). Furthermore, shading net enhances the performance of citrus tree, implement growth, increase tree crop, and improve fruit quality (Wachsmann, et al, 2012).

Raveh, et al, (2003) reported that, shading grapefruit trees with reflective aluminized nets improved average sunlight leaf conductance and increase photosynthesis, consequently increasing dry matter.

Shading Net and Citrus Trees

There are various phenomena for extreme weather conditions, like fluctuation in temperature during the growing season, warm nights, reducing relative humidity, high solar radiation, and intensive heat waves.

Citrus trees are widely cultivated in dry and semi-dried regions, especially under the Mediterranean climate. Therefore, increase solar radiation and high-temperature increase evapotranspiration particularly in the wide canopy varieties such as lemon and grapefruit (Cohen and Fuchs, 1987), which are subjected to water deficit under these conditions, consequently affecting the whole growth parameters and causing a reduction in total yield.

The adverse effects of extreme weather conditions may be due to increase evapotranspiration and cause injuries of fruit trees particularly seedless varieties like navel orange and Thompson seedless grape. However, fluctuation in temperature during the growing season, warm nights during the winter, which affect the flower bud induction, in addition to intensive heat waves during the critical growth stages of fruits in the spring and summer, increase fruitlets abortion and increase the physiological disorder of fruits. Furthermore, high solar radiation affects negatively fruit quality particularly under clear sky conditions like the Mediterranean climate.

Using shading techniques on fruit orchards like *Citrus spp.* is very important under arid and semi-arid regions to reduce the adverse effects of drastic weather events. Shading modifies the environmental conditions around the tree, which increases productivity and improves fruit quality such as reducing sunburn in mandarin (Cronjé, et al, 2019).

Shading net increase the growth of young citrus trees (Reveh, et al, 2003), increases fruit quality and reduces sunburn of Murcott tangour fruit (Tsai, et al, 2013), increases CO_2 assimilation rate in citrus leaves (Jifon and Syvertsen, 2003). Improve canopy temperature, stimulate stomatal conductance, and improve water use efficiency (Alarcon, et al, 2006). Shading of Navel orange trees increased net assimilation of CO_2 and stomatal conductance (Syvertsen, et al, 2003). Furthermore, shading-net reduced water consumption in 'Ori' mandarin orchard and improve fruit quality (Wachsmann, et al, 2012).

Shading increase the shoot/root ratio, changes the distribution of carbohydrates, and different types of sugars. The total carbohydrates increase in leaves and stem while reducing in the roots (Ebadi, et al, 2018; Zhou, et al, 2018). Furthermore, shading affects trees' morphology like shape and canopy volume than non-shading trees (Raveh, et al, 2003).

Mditshwa, et al, (2019) reported that, the shading net reduces both leaves and soil temperature, light intensity and wind speed, and notably increases relative humidity that improves WUE.

Environmental Conditions under Protective Netting

1. **Effect of solar radiation on citrus**

Solar radiation is an important factor that affects crop productivity, particularly, under the current situation of climate change conditions that cause a fluctuation in the distribution of solar radiation waves, which affect negatively the productivity of different orchards (Campillo, et al, 2012). Under dry and semi-dried regions, the clear sky increase radiation loads lead to a higher temperature in orchards and increasing degradation of chlorophyll, and act as an inhibitor factor of photosynthesis (Ribeiro and Machado, 2007). Furthermore, high solar radiation causes adverse effects on the growth and productivity of citrus orchards like accelerating transpiration average, affecting water balance in the tree (Cohen, et al, 2005; Jifon and Syvertsen, 2003), and affect negatively fruit quality and increase physiological disorders like sunscald (Figure 2).

Figure 2. Image field sunscald in Valencia orange fruit (Photo by Dr. Abobatta 2021)

In addition, Nicolás, et al, (2008) reported that, under shading conditions, there is a reduction in the transpiration rate of the lemon trees.
Fluctuation of solar radiation influences fruit crops in a number of ways as follows:

1. Limitation to leaf net photosynthetic
2. Affect carbon assimilation (A) particularly at high photon flux density.

Main Benefits of Reducing Solar Radiation by Shading in Citrus Orchards

Net shading reduces exposure of citrus trees to solar radiation, which provides many positive benefits as follows:

1. Dropping leaf temperature about 8-10 degrees less than the open field (Syvertsen and Lloyd, 1994).
2. Increase CO_2 assimilation average (Garcia-Sanchez, et al, 2015).
3. Improve leaf water potential (Alarcon, et al, 2006).
4. Increase photosynthesis efficiency, which leads to an increase in all growth parameters (Ribeiro and Machado, 2007).

On another side, reducing flux density of solar radiation limits CO_2 assimilation and reduces photosynthesis (Park and Runkle, 2018; Taiz and Zeiger, 1991).

2. **Water Use Efficiency**

Using shading regulates leaves temperature and reduces evaporation from leaves and orchard soil, which improves water use efficiency in fruit orchards (Shahak, et al. 2004). Previous reports indicated that, shading net improves water use efficiency in citrus orchards (Dovjik, et al, 2021; Ebadi, et al, 2018), García-Sánchez, et al, (2015) on lemon by suing aluminet-50 screen, also, (Alarcón, et al, 2006) using aluminized-plastic in lemon orchards to improve water use efficiency. Furthermore, Nicolás, et al, (2008) revealed that, using the shading net in lemon grove reduces transpiration rate and improves water use efficiency. Consequently, improve photosynthesis efficiency and enhances water use efficiency, which reflects in improve tree growth and productivity. Eventually, the improvement in water-use results from the adjusted microclimate under the shading net.

3. **Light intensity**

Light intensity play important role in maintaining fruit quality, it affects sugar content and peels color. Shading net regulates the quantity of solar radiation reaching the tree canopy and the soil. The reduction in the intensity of the light varies rates according to the type of net, which improves the microclimate around the trees (Shahak, 2012).
There are various advantages of reducing light intensity include:

- Increase tree vigor (e.g. leaf area, shoot length).
- Enhancing flower bud induction.
- Improve fruit set.
- Protect fruits from sunburn.

Effect of Shading on Citrus Orchards

Different types of shading that include closed net, screen houses, top netting, or partial shading represent a key solution for mitigating the fluctuation of climate through modifying and improving the microclimate around the trees (Prins, 2018).

There are adverse effects of extreme weather conditions like high solar radiation, heat waves, and fluctuation in temperature on the growth, yield, and fruit properties of citrus particularly in dry and semi-dried regions.

There are significant roles of light intensity in determining the fruit quality of citrus, it affects sugar content and peels color. Hence, using shading techniques on citrus orchards is very important particularly under arid and semi-arid regions to reduce adverse effects of drastic weather events, and improve productivity, enhance fruit quality, and reduce sunburn (Cronjé, et al, 2019).

1. Vegetative growth and leaf morphology

Shading net regulates solar radiation and changes leaf gas exchange, which improves vegetative growth in terms of shoot number and length, and leaf morphology (leaves more spreading). Furthermore, shading affects tree morphology like shape and canopy volume than non-shading trees (Raveh, et al, 2003). Shading increase the shoot/root ratio, changes the distribution of carbohydrates, and different types of sugars. The total carbohydrates increase in leaves and stem while reducing in the roots (Ebadi, et al, 2018; Zhou, et al, 2018).

Furthermore, using a shading net for newly established citrus orchards accelerate new flushing and increase growth, Murcott mandarin compared with those under open field conditions (Figure 3).

Figure 3. Image field of young Murcott mandarin trees under shading net (Photo by Dr. Abobatta 2012)

2. **Fruit set, flower induction**

Shading net application reduces tree temperature, consequently, improve flower bud induction and increasing fruit set of citrus trees. For instance, shading net enhancing flowering of ´Primosole´ mandarins trees (Germanà, et al, 2002), and increase fruit set of navel orange trees (Awad, et al, 2020).

3. **Fruit quality**

There is a positive effect of shading net on fruit size, due to reducing stem water stem of the trees and enhancing water availability in orchards soil (Shahak, et al, 2004).

Using white net improves fruit quality of ponkan mandarin especially juice percentage and reduce sunscald and granulation injuries (Lee, et al, 2015), in addition, Tsai, et al, (2013) reported that, white shade net improves the quality of 'Murcott' tangor fruit and minimizing sunscald disorder. Furthermore, using the shading net in citrus orchards ameliorate the impact of high temperature and solar radiation, and improves fruit quality in navel orange (Awad, et al, 2020).

Cultivation of navel orange trees under shading net improve fruit quality, particularly peel color than fruits of unshaded trees (Syvertsen, et al, 2003).

4. **Effect of Nets on tree Yield**

The use of shading nets is a practical way to protect trees from stresses caused by weather fluctuations and increasing yield. Shading nets treatment increased tree yield and total yield of Navel orange (Awad, et al, 2020).

Shading net improves fruit quality of Murcott" tangor" fruits, reduces sunscald, increases the size and weight of fruits (Tsai, et al, 2013).

CONCLUSION

Citrus, represent a global valuable fruit crop that can be grown in varying climatic conditions. Due to climate change, there are negative influences within the main areas of citrus production (warm subtropical and Mediterranean regions), which have considerable effects on the future of citriculture. Under warm conditions, high solar radiation levels might inhibition of photosynthesis, leading to a reduction in growth rate and yield. In addition, rising temperatures reducing stomatal conductance and diminutions in the net CO_2 assimilation rate.

Shading net is considered a new technique to alleviate impacts of climate change conditions, shading is a common agriculture practice in nurseries and has recently been used in commercial fruit orchards. This practice has numerous desirable beneficial impacts for fruit trees particularly in warm regions. Using shading net reduce environmental stress, enhancing tree growth, flowering, improve fruit quality consequently increasing yield, and reducing disease infection. With this modern in-farm technology, citrus could be produced in areas that were previously unsuitable for citrus production, as well as reducing negative impacts from climate change on citrus productivity and quality.

REFERENCES

Abobatta, W. F. (2021a). Managing citrus orchards under climate change. *MOJ Eco Environ Sci*, *6*(2), 43–44.

Abobatta, W. F. (2021b). Fruit orchards under climate change conditions: Adaptation strategies and management. *J Appl Biotechnol Bioeng*, *8*(3), 99–102. doi:10.15406/jabb.2021.08.00260

Abobatta, W.F. (2022). *Impact of shading practice on growth of Murcott mandarin seedlings.* Unpublished work.

Alarcón, J. J., Ortuno, M. F., Nicolás, E., Navarro, A., & Torrecillas, A. (2006). Improving water-use efficiency of young lemon trees by shading with aluminised-plastic nets. *Agricultural Water Management*, *82*(3), 387–398. doi:10.1016/j.agwat.2005.08.003

Awad, R. M., El-Sayed, H. A., & El-Razk, A. (2020). Vegetative Growth and Quality of Washington Navel Orange as Affected by Shading Nets and Potassium Silicate Spraying. *Egyptian Academic Journal of Biological Sciences. H, Botany*, *11*(1), 81–90. doi:10.21608/eajbsh.2020.119321

Bastías, R. M., Manfrini, L., & Grappadelli, L. C. (2012). Exploring the potential use of photo-selective nets for fruit growth regulation in apple. *Chilean Journal of Agricultural Research*, *72*(2), 224–231. doi:10.4067/S0718-58392012000200010

Bisbis, M. B., Gruda, N. S., & Blanke, M. M. (2019). Securing horticulture in a changing climate- A mini review. *Horticulturae, 5*(3), 56. doi:10.3390/horticulturae5030056

Bosančić, B., Mićić, N., Blanke, M., & Pecina, M. (2018). A main effects meta principal components analysis of netting effects on fruit: Using apple as a model crop. *Plant Growth Regulation*, *86*(3), 455–464. doi:10.100710725-018-0443-z

Botes, J. (2018). *Impact of shade netting on internal and external quality of 'Nadorcott' mandarin fruit* (Doctoral dissertation). Stellenbosch University.

Campillo, C., Fortes, R., Prieto, M. D. H., & Babatunde, E. B. (2012). Solar radiation effect on crop production.pp.167-194 *Solar Radiation, 1,* 494. Available from: https://www.intechopen.com/books/solar-radiation/solar-radiation-effect-on-crop-production

Cohen, S., & Fuchs, M. (1987). The distribution of leaf area, radiation, photosynthesis and transpiration in a Shamouti orange hedgerow orchard. Part I. Leaf area and radiation. *Agricultural and Forest Meteorology, 40*(2), 123-144. doi:10.1016/0168-1923(87)90002-5

Cronjé, P. J. R., Botes, J., Prins, D. M., Brown, R., North, J., Stander, O. P. J., Hoffman, E. W., Zacarìas, L., & Barry, G. H. (2020). The influence of 20% white shade nets on fruit quality of 'Nadorcott' mandarin. *Acta Horticulturae*, (1268), 279–284. doi:10.17660/ActaHortic.2020.1268.36

Dovjik, I., Nemera, D. B., Cohen, S., Shahak, Y., Shlizerman, L., Kamara, I., Florentin, A., Ratner, K., McWilliam, S.C., Puddephat, I.J., FitzSimons, T.R., Charuvi, D. & Sadka, A. (2021). Top Photoselective Netting in Combination with Reduced Fertigation Results in Multi-Annual Yield Increase in Valencia Oranges (Citrus sinensis). *Agronomy*, *11*(10), 2034. doi:10.3390/agronomy11102034

Ebadi, H., Raeini-Sarjaz, M., & Gholami Sefidkoohi, M. A. (2018). Vegetative growth and water use efficiency of citrus seedlings in simultaneous application of shading and partial root zone drying. *Iranian Journal of Horticultural Science*, *49*(2), 417-428. doi:10.22059/IJHS.2017.230309.1218

Gaire, S., Albrecht, U., Batuman, O., Qureshi, J., Zekri, M., & Alferez, F. (2021). Individual protective covers (IPCs) to prevent Asian citrus psyllid and Candidatus Liberibacter asiaticus from establishing in newly planted citrus trees. *Crop Protection,* 105862. doi:10.1016/j.cropro.2021.105862

García-Sánchez, F., Simón, I., Lidón, V., Manera, F. J., Simón-Grao, S., Pérez-Pérez, J. G., & Gimeno, V. (2015). Shade screen increases the vegetative growth but not the production in 'Fino 49' lemon trees grafted on Citrus macrophylla and Citrus aurantium L. *Scientia Horticulturae*, *194*, 175-180. doi:10.1016/j.scienta.2015.08.005

Germanà, C., Continella, A., & Tribulato, E. (2002, March). Net shading influence on floral induction on citrus trees. *VI International Symposium on Protected Cultivation in Mild Winter Climate: Product and Process Innovation, 614*, 527-533. 10.17660/ActaHortic.2003.614.78

Hussain, M., Farooq, S., Hasan, W., Ul-Allah, S., Tanveer, M., Farooq, M., & Nawaz, A. (2018). Drought stress in sunflower: Physiological effects and its management through breeding and agronomic alternatives. *Agricultural Water Management*, *201*, 152-166. doi:10.1016/j.agwat.2018.01.028

Iglesias, I., & Alegre, S. (2006). The effect of anti-hail nets on fruit protection, radiation, temperature, quality and profitability of 'Mondial Gala' apples. *Journal of Applied Horticulture*, *8*(2), 91–100. doi:10.37855/jah.2006.v08i02.22

Jamshidi, S., Zand-Parsa, S., Kamgar-Haghighi, A. A., Shahsavar, A. R., & Niyogi, D. (2020). Evapotranspiration, crop coefficients, and physiological responses of citrus trees in semi-arid climatic conditions. *Agricultural Water Management, 227*, 105838. doi:10.1016/j.agwat.2019.105838

Jiao, Y., Lau, O. S., & Deng, X. W. (2007). Light-regulated transcriptional networks in higher plants. *Nature Reviews Genetics*, *8*(3), 217-230. doi:10.1038/nrg2049

Jifon, J. L., & Syvertsen, J. P. (2003). Moderate shade can increase net gas exchange and reduce photoinhibition in citrus leaves. *Tree Physiology*, *23*(2), 119-127. doi:10.1093/treephys/23.2.119

Jokar, A., Zare, H., Zakerin, A., & Jahromi, A. A. (2021). The Influence of Photo-selective Netting on Tree Physiology and Fruit Quality of Fig (*Ficus carica* L.) Under Rain-fed Conditions. *International Journal of Fruit Science, 21*(1), 896-910. doi:10.1080/15538362.2021.1936345

Kalcsits, L., Musacchi, S., Layne, D. R., Schmidt, T., Mupambi, G., Serra, S., Mendoza, M., Asteggiano, L., Jaralmasjed, S., Sankaran, S., Khot, L.R. & Espinoza, C. Z. (2017). Above and below-ground environmental changes associated with the use of photoselective protective netting to reduce sunburn in apple. *Agricultural and Forest Meteorology, 237*, 9-17. doi:10.1016/j.agrformet.2017.01.016

Kataria, S., & Verma, S. K. (2018). Salinity stress responses and adaptive mechanisms in major glycophytic crops: the story so far. In *Salinity Responses and Tolerance in Plants* (Vol. 1, pp. 1–39). Springer. doi:10.1007/978-3-319-75671-4_1

Lee, T. C., Zhong, P. J., & Chang, P. T. (2015). The effects of preharvest shading and postharvest storage temperatures on the quality of 'Ponkan' (Citrus reticulata Blanco) mandarin fruits. *Scientia Horticulturae*, *188*, 57–65. doi:10.1016/j.scienta.2015.03.016

Mahmood, A., Hu, Y., Tanny, J., & Asante, E. A. (2018). Effects of shading and insect-proof screens on crop microclimate and production: A review of recent advances. *Scientia Horticulturae*, *241*, 241–251. doi:10.1016/j.scienta.2018.06.078

Manja, K., & Aoun, M. (2019). The use of nets for tree fruit crops and their impact on the production: A review. *Scientia Horticulturae*, *246*, 110–122. doi:10.1016/j.scienta.2018.10.050

Mditshwa, A., Magwaza, L. S., & Tesfay, S. Z. (2019). Shade netting on subtropical fruit: Effect on environmental conditions, tree physiology and fruit quality. *Scientia Horticulturae, 256*, 108556. doi:10.1016/j.scienta.2019.108556

Melke, A. (2015). The physiology of chilling temperature requirements for dormancy release and bud-break in temperate fruit trees grown at mild winter tropical climate. *Journal of Plant Studies*, *4*(2), 110-156. doi:10.5539/jps.v4n2p110

Mupambi, G., Anthony, B., Layne, D. R., Musacchi, S., Serra, S., Schmidt, T., & Kalcsits, L. A. (2018). The influence of protective netting on tree physiology and fruit quality of apple: A review. *Scientia Horticulturae*, *236*, 60–72. doi:10.1016/j.scienta.2018.03.014

Nebauer, S. G., Renau-Morata, B., Guardiola, J. L., & Molina, R. V. (2011). Photosynthesis down-regulation precedes carbohydrate accumulation under sink limitation in Citrus. *Tree Physiology, 31*(2), 169-177. doi:10.1093/treephys/tpq103

Nicolás, E., Barradas, V. L., Ortuño, M. F., Navarro, A., Torrecillas, A., & Alarcón, J. J. (2008). Environmental and stomatal control of transpiration, canopy conductance and decoupling coefficient in young lemon trees under shading net. *Environmental and Experimental Botany*, *63*(1-3), 200-206. doi:10.1016/j.envexpbot.2007.11.007

Park, Y., & Runkle, E. S. (2018). Far-red radiation and photosynthetic photon flux density independently regulate seedling growth but interactively regulate flowering. *Environmental and Experimental Botany, 155*, 206-216. doi:10.1016/j.envexpbot.2018.06.033

Prins, M. D. T. (2018). *The impact of shade netting on the microclimate of a citrus orchard and the tree's physiology* (Doctoral dissertation). Stellenbosch University.

Rajapakse, N. C., & Shahak, Y. (2008). 12 Light-quality manipulation by horticulture industry. Annual Plant Reviews. *Light and Plant Development*, *30*, 290.

Raveh, E., Cohen, S., Raz, T., Yakir, D., Grava, A., & Goldschmidt, E. E. (2003). Increased growth of young citrus trees under reduced radiation load in a semi-arid climate1. *Journal of Experimental Botany*, *54*(381), 365-373. doi:10.1093/jxb/erg009

Ribeiro, R. V., & Machado, E. C. (2007). Some aspects of citrus ecophysiology in subtropical climates: Re-visiting photosynthesis under natural conditions. *Brazilian Journal of Plant Physiology*, *19*(4), 393–411. doi:10.1590/S1677-04202007000400009

Sadka, A., Shahak, Y., Cohen, S., Dovjek, I., Nemera, D. B., Wachsmann, Y., Shlizerman, L., Ratner, K., Kamara, I., Morozov, M., & Charuvi, D. (2019). Top netting as a practical tool to mitigate the effect of climate change and induce productivity in citrus: summary of experiments using photo-selective nets. *XI International Symposium on Protected Cultivation in Mild Winter Climates and I International Symposium on Nettings, 1268*, 265-270.

Shafqat, W., Jaskani, M. J., Maqbool, R., Chattha, W. S., Ali, Z., Naqvi, S. A., Haider, M.S., Khan, I., & Vincent, C. I. (2021). Heat shock protein and aquaporin expression enhance water conserving behavior of citrus under water deficits and high temperature conditions. *Environmental and Experimental Botany, 181*, 104270. doi:10.1016/j.envexpbot.2020.104270

Shahak, Y. (2012). Photoselective netting: an overview of the concept, research and development and practical implementation in agriculture. *International CIPA Conference 2012 on Plasticulture for a Green Planet, 1015*, 155-162. 10.17660/ActaHortic.2014.1015.17

Syvertsen, J.P., & Lloyd, D. J. (1994). Citrus. In *Handbook of environmental physiology of fruit crops. Vol. II. Subtropical and tropical crops*. CRC Press.

Syvertsen, J. P., Goñi, C., & Otero, A. (2003). Fruit load and canopy shading affect leaf characteristics and net gas exchange of 'Spring' navel orange trees. *Tree Physiology, 23*(13), 899-906. doi:10.1093/treephys/23.13.899

Taiz, L., & Zeiger, E. (1991). *Plant Physiology*. Benjamin Cummings.

Tsai, M. S., Lee, T. C., & Chang, P. T. (2013). Comparison of paper bags, calcium carbonate, and shade nets for sunscald protection in 'Murcott'tangor fruit. *Hort Technology, 23*(5), 659-667. doi:10.21273/HORTTECH.23.5.659

Wachsmann, Y., Zur, N., Shahak, Y., Ratner, K., Giler, Y., Schlizerman, L., Sadka, A., Cohen, S., Garbinshikof, V., Giladi, B., & Faintzak, M. (2012). Photoselective anti-hail netting for improved citrus productivity and quality. *International CIPA Conference 2012 on Plasticulture for a Green Planet, 1015*, 169-176.

Wilkie, J. D., Sedgley, M., & Olesen, T. (2008). Regulation of floral initiation in horticultural trees. *Journal of Experimental Botany, 59*(12), 3215–3228. doi:10.1093/jxb/ern188 PMID:18653697

Zhang, X., Fan, Y., Jia, Y., Cui, N., Zhao, L., Hu, X., & Gong, D. (2018). Effect of water deficit on photosynthetic characteristics, yield and water use efficiency in Shiranui citrus under drip irrigation. *Nongye Gongcheng Xuebao (Beijing), 34*(3), 143–150.

Zhou, K., Jerszurki, D., Sadka, A., Shlizerman, L., Rachmilevitch, S., & Ephrath, J. (2018). Effects of photoselective netting on root growth and development of young grafted orange trees under semi-arid climate. *Scientia Horticulturae, 238*, 272–280. doi:10.1016/j.scienta.2018.04.054

Chapter 2
Orchard Biomass Management

Mohamed Saad Aly Emam
Agricultural Research Center, Egypt

Mohamed Abul-soud Mohamed
Agricultural Research Center, Egypt

ABSTRACT

The use of different techniques for recycling the orchard's biomass provided more flexibility and efficiency under the different environmental and production conditions. Aerobic composting depending on the decomposition activities of different micro-organisms and the environmental conditions (moisture, temperature, additives, machinery, etc.) was for a long time the main method for recycling the orchard's biomass. Vermicomposting introduced a promising technique for recycling the orchard's biomass depending on vermicomposting earthworm types, which led to not just recycling the orchards biomass but also producing different products (vermicompost, vermi-liquid, and earthworm biomass, feeders, etc.). Anaerobic composting mainly produces energy, as well as composting, and still needs more knowledge transfer. The management of organic soil content is a vital agricultural operation that takes into more consideration increasing the sustainability and the production of orchards as well as sequestrating the carbon dioxide in orchards' soils.

PHYSIOLOGY BASICS EARTHWORMS

Earthworms are terrestrial invertebrate animals with a similar physical structure in all species. They belong to the phylum Annelida (meaning "ringd" or "segmented") and the class Oligochaeta (small bristled worm). All of the vital organs are contained in ront third of an earthworm's body. The rings around the body delineate, and each segment has muscles and bristles, called setae, to aid in movement. The number of segments varies among earthworm species- for example, night crawlers have about 150 segments, and red wigglers have around 95. The setae anchor a segment of the worm to whatever surface it is on, while muscles and setae in other segments move the earthworm in whichever direction it wants to go. The segments lengthen or contract independently to move the worm.

DOI: 10.4018/978-1-6684-2423-0.ch002

Earthworm bodies are like tubes with no protruding sense organs or appendages, and this enables them to move easily through soil. They do not have eyes or ears, but their brain connects to nerves attached to their skin and muscles. There nerves sense light and vibrations, and their muscles move in response to the stimuli. Earthworms have well developed circulatory, digestive, nervous, excretory, muscular, and reproductive systems. The anterior end (head) has a tongue like prostomium lobe, and the posterior end has an anus.

Earthworms have simple circulatory systems. Instead of a heart, they have five pairs of aortic arches that pump blood through blood vessels. Dorsal blood vessels transport blood to the front of the earthworm's body, and ventral blood vessels carry blood to the back of the body.

The ancient Egyptians were the first to recognize the beneficial status of the earthworm. The Egyptian Pharaoh, Cleopatra (69–30 B.C.) said, "Earthworms are sacred." She recognized the important role the worms played in fertilizing the Nile Valley croplands after annual floods. Removal of earthworms from Egypt was punishable by death. Egyptian farmers were not allowed to even touch an earthworm for fear of offending the God of fertility. The Ancient Greeks considered the earthworm to have an important role in improving the quality of the soil. The Greek philosopher Aristotle (384–322 B.C.) referred to worms as the intestines of the earth (Medany, 2011).

EARTHWORM SPECIES

Earthworm species vary in how they get food, and thus inhabit different parts of the soil, and have somewhat different effects on the soil environment. They fall into three distinct ecological groups based on feeding and burrowing habits.

1 - Epigeic (litter dwelling) earthworms live and feed in surface litter. They move horizontally through leaf litter or compost with little ingestion of or burrowing into the soil. These worms are characteristically small and are not found in low organic matter soils. *Lumbricus rubellus* is an example of epigeic species. Epigeic forms of earthworms can hasten the composting process to a significant extent (Senapathi, 1988; Kale et al., 1982; Tomati et al., 1983), with production of better quality of composts, compared with those prepared through traditional methods (Tripathi and Bhardwaj, 2004).

2 - Endogeic (shallow dwelling) earthworms are active in mineral topsoil layers and associated organic matter. They create a three-dimensional maze of burrows while consuming large quantities of soil. The genuses *Diplocardia* and *Aporrectodea* have endogeic life habits.

3 - Anecic (deep burrowing) earthworms live in permanent, nearly vertical burrows that may extend several feet into the soil. They feed on surface residues and pull them into their burrows. *Lumbricus terrestris* is an example of an anecic species (Coleman and Crossley, 1996).

The most common types of earthworms used for vermicomposting are brandling worms (*Eisenia foetida*) and red worms or red wigglers (*Lumbricus rubellus*). Often found in aged manure piles, they generally have alternating red and buff-colored stripes. They are not to be confused with the common garden or field earthworm (*Allolobophora caliginosa* and other species).

Earthworms prefer a neutral to slightly alkaline pH. Liming in vermicompost bin generally enhance microbial population as well as earthworm activities. Therefore, changes in macronutrient content and some enzymatic activities of vermicompost due to lime addition to organic wastes could be an interesting study (Reinecke et al., 1992).

Figure 1. Earthworm species

EARTHWORM REPRODUCTION

The presence of a band (called a clitellum) around a worm's body indicates the ability to reproduce. The eggs and sperm of each earthworm are located in separate parts of the body to prevent self- fertilization. Two worms will mate by facing in opposite direction and connecting their clitlla, which secrete mucus that encircles both worms. While the worms are locked together, they exchange sperm. Once they have completed the exchange, the worms disengage and their movements push their mucous bands toward their heads. As the band moves over the earthworm's body, it pick up sperm and mature eggs that were deposited by the other worm. The mucous tube will slip over the earthworm's head, and both ends will close into what is called a cocoon. Inside the cocoon, the sperm and eggs unite to create worm embryos that develop into baby worms. The cocoon is golden brown, shiny, and looks like a tiny lemon the size of a match head.

Earthworms reproduce quickly when food and water are available. After the babies hatch, it takes53 to 76 days for them to reach sexual maturity.

Figure 2. These two earthworms are showed together to mate

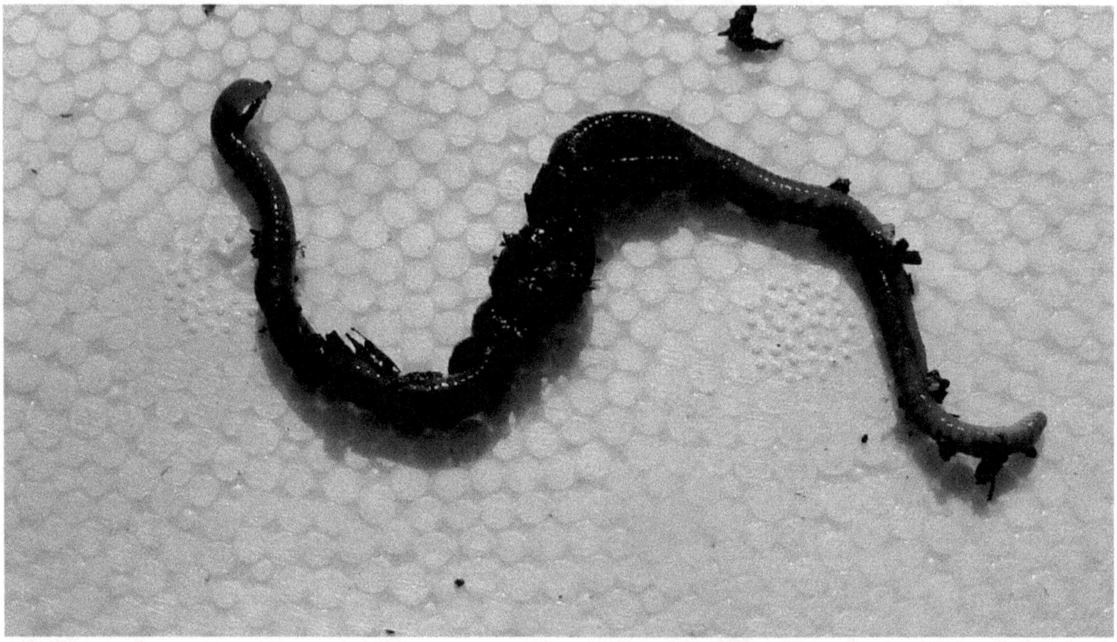

Figure 3. The cocoon is golden brown, shiny

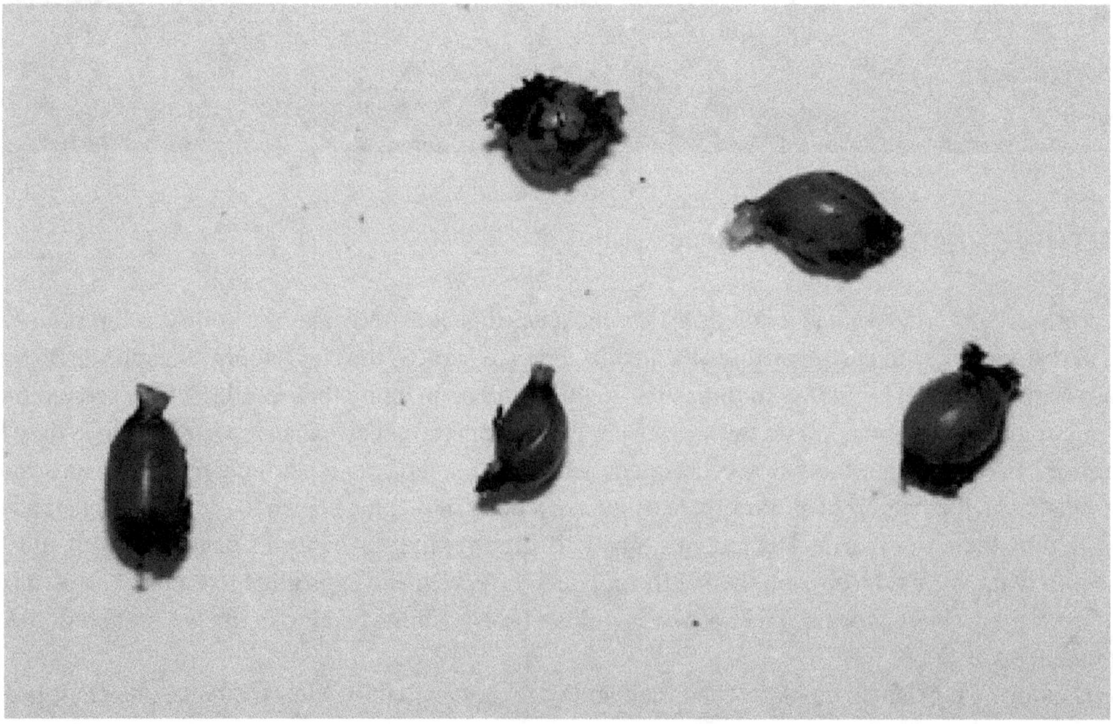

EARTHWORM FEEDS

Worms will eat compost, leaves, manure, food scarps, shredded paper, coffee grounds, spent mushroom waste, agricultural crop residues and food processing waste.

Sometimes a single type of material, such as paper, doesn't provide the ideal types of nutrients in the proper amounts for worms, so you may need to mix different kinds of materials together. One important thing for feedstock quality is its ratio of carbon to nitrogen. C: N is the ratio of total mass of elemental carbon (C) to total mass of elemental nitrogen (N). A feedstock's C: N ratio indicates how much more carbon than nitrogen exists in the organic material.

Manure, bio solids and food waste are all dense and high in nitrogen. To increase their porosity and allow airflow, it is important to thoroughly mix with another carbonaceous material.

Figure 4. Vegetable's waste for earthworm feeds

ENZYME ACTIVITY

Research has shown that some enzyme activities are correlated with overall microbial activity, soil fertility, plant growth, and plant disease resistance. During vermicomposting, earthworms enhance selectively the activities of enzymes such as invertase, urease, and alkaline phosphatases, which are of microbial origin.

VERMICULTURE

Vermiculture is the culture of earthworms. The goal is to continually increase the number of worms in order to obtain a sustainable harvest. The worms are either used to expand a vermicomposting operation or sold to customers who use them for the same or other purposes.

The huge amounts of biodegradable organic wastes that generate every day in urban and agriculture areas creating disposal problems on the environmental, gas emessions, public health, economic and social levels especially in under developed countries. These huge amounts of organic wastes could be a renewable source for many sectors (industrial & agricultural). In big cities espicially in a city like Cairo (about 18 million capita) produced around 12000 m3 organic wastes/day, at the same time of the Egyptian agriculture faces a huge lack of their needs from organic fertilizer, substrates and feeder that lead to increase the prices of food generally.

This waste can be converted into valuable compost by applying vermi-composting technology. This approach reduces pollution and provides a valuable substitute for chemical fertilizers. This process is profitable at any scale of operation, provided proper process parameters are maintained.

The feeding of earthworm in the breeding unit is depending on rabbit manure + vegetable and fruit wastes. The main objective of this room not just for multiply and producing earthworm but also to conserve an earthworm stock in safe and secure conditions. The feeding of earthworm done every 2 – 3 days. The organic wastes mixed well together and soaked in water for 30 minutes before feeding. The feeding allocated in strips to avoid any increase of temperature (thermophilic). Calcium carbonate added weekly around 5 g per each plastic box to adjust the pH to favorable pH 7.5. The plastic boxes were watering weekly to keep the moisture in range of 60 – 70%. Turning up down the plastic boxes contents weekly for offering good condition of ventilation to prevent any blocking and to allow air conditions, the turning was done by hand to conserve the earthworm and to minimize the earthworm loose.

The growth cycle of earthworm in the breeding unit is varied from summer to winter. This variation was not regarding to the breeding system but mainly to the growing unit conditions that didn't have similar control condition like the breeding room. The growth cycle during the hot days in breeding room take about 3 months to avoid the high temperature in the growing systems (vermicomposting units). While it takes 1.5 months during the cool or cold days.

The harvest of earthworm and ending the breeding cycle by remove all contents to new vermicomposting system or growing bed. Approximately, each plastic box contained about 300 to 500 g of earthworm depending on the breeding cycle.

Figure 5. Vermiculture room to culture earthworms

HARVESTING

Manual Harvesting Earthworm

Table harvesting and light separation are two different names for a common earthworm harvesting technique. To use this method, cover the table with waterproof plastic sheet, use a digging fork or shovel to remove the top 10-15 cm of material. Where most of the worms reside, from your bed and place the material on top of the plastic. Shine a bright light on the pile of harvested worms to encourage them to move deeper.

Remove the organic material and put it back in the bin for further processing, as the worms move away from the light, they will eventually from a solid mass, which you can then place in the containers.

Vertical separation is a method that lures worms to move upward toward food. It helps if you do not feed the worm for a week or two before using this technique. Place one of the following on top of the worm bed: a piece of mesh, a plastic bakery tray, or a box with wooden sides and a mesh bottom. Put enticing food, such as melon, watermelon on the mesh. The worms will crawl through it to get to the food. Then you can lift the tray or mesh of the bed, and you got your worms.

Figure 6. Manual harvesting vermicast

Vermicast Harvesting

To harvest vermicast from worm windrows or bins, growers use manual or mechanical methods, and both involve using a trammel (cylindrical) or flat shaking screen. Since most of the worms will be in the top 10-15 cm of material, if workers remove the top 15-20 cm of the windrow or bin, they will capture most of the worms. They place this worm-rich material in a new windrow. Then they use a shovel or front-end loader to pick up the rest of the organic material and empty it onto a screening device.

Harvesting Continuous Flow through Bins

With continuous flow through bins, you can harvest the vermicast without disturbing the worms or the bed. Most of the worms will be in the top 10-15 cm of the bed, and the rest of your continuous flow through will be almost full of organic material (feedstocks, bedding and vemicast). There are various methods for harvesting these beds that involve prompting a layer of finished vermicast (no more than a 2.5 cm thickness) to fall through the grates in the bottom of the bin.

To harvest from this type of continuous flow through bin, folks will usually use a hand operated crank or pulley, or a hydraulic system activated by small motors. These mechanisms pull a breaker bar or blad over the surface of the grate.

Figure 7. Harvesting machine vermicomposting

GENERAL APPLICATION OF COMPOST AND VERMICOMPOST

Vermicomposting and composting are very different processes, and it is important not to use the terms interchangeably. Composting is the controlled process of converting organic materials into a valuable soil amendment under aerobic conditions using biologically generated heat. In contrast, a vermicomposting pile or worm bin should be maintained so that it does not heat up. Too much heat can kill worms.

Now-a-days, people are more conscious about their health and consume less fruits due to more use of chemicals (fertilizers and pesticides). To overcome harmful effects of these chemicals on human health, an attempt was made to use organic manure in place of inorganic fertilizers for orchard.

Vermicomposting (Worm composting) is defined as a process in which earthworms play a major role with microbes in the conversion of organic solid waste into more stabilized dark, earth-smelling soil conditioner and nutrient-rich compost that is rich in major and micronutrients (Berritezetal., 2002). During vermicomposting, organic matter is stabilized by the enhanced decomposition (humification) in presence of earthworms (Atiyeh et al., 2001), but by a non-thermophilic process (Elvira et al., 1996, 1998).

The great advantage of worm composting is that this can be done indoors and outdoors, thus allowing year round composting. It also provides apartment dwellers with a means of composting. Vermicomposting allows obtaining organic sources of nutrients for the crops in relatively less time, which are physically, nutritionally and biochemically improved over composts. Vermicomposting is defined as a low cost technology system for processing or treatment of organic waste (Hand et al., 1988). Moreover, it does not require skilled personnel and mechanization.

Bio fertilization is now a very important method for providing the plants with their nutritional requirements without having an undesirable impact on the environment, also improve fixation of nutrients in the rhizosphere, produce growth stimulants for plants, improve soil stability and provide biological control. They also biodegrade substances, recycle nutrients, promote mycorrhiza symbiosis and develop bioremediation processes in soils contaminated with toxic, xenobiotic and recalcitrant substances (Rivera, et al., 2008).

APPLICATION OF VERMICOMPOST

Vermicompost in Soil Fertility

Vermicomposting is a process that relies on earthworms and microorganisms to break down organic matter and transform its biological, physical and chemical characteristics into a stable product that can be used as a valuable soil amendment and source of plant nutrients. Vermicomposting turns organic materials into vermicast, which is a nutrient rich, microbially active soil amendment or growth media for plants.

Organic culture is claimed to be the most benign alternative. Use of organic materials such as farmyard manure, cakes of plant origin and Vermicomposts, are important components of the bio- organic concept of orange cultivation, (Chellachamy and Dinakaran 2015).

Although microorganisms are responsible for the biochemical degradation of organic matter, earthworms are crucial drivers of the process, by fragmenting and conditioning the substrate and dramatically altering its biological activity. Earthworms act as mechanical blenders and by comminuting the organic matter they modify its physical and chemical status, gradually reducing its C: N ratio, increasing the surface area exposed to micro-organisms and making it much more favourable for microbial activity and further decomposition. Greatly during passage through the earthworm gut, they move fragments and bacteria rich excrements, thus homogenizing the organic material. The end-product, or vermicompost, is a finely divided peat-like material with high porosity and water holding capacity that contains most nutrients in forms that are readily taken up by the plants. These earthworm casts are rich in organic matter and have high rates of mineralization that implicates a greatly enhanced plant availability of nutrients, particularly ammonium and nitrate (Dominguez and Edwars, 2004).

The vermicompost technique could help directly in solving the problem of food security by offering the substrates and organic fertilizers that could be used for producing the needs from different vegetables and fruits or even from medicinal plants by use the house roofs as a green roof.

The use of vermicomposting (earthworm's farm) in urban, food industries and farm organic wastes management has received increasing attention over the last 20 years where research programs and commercial projects have been developed in many countries on all continents.

Vermicompost contains high percentages of humic substances (humic acids, fulvic acids) that contribute to numerous chemical reactions. Also this matter is full of microbial components that enhance plant growth and disease suppression by bacterial activity (Bacillus), the yeasts (Sporobolomyces and Cryptococcus), and the fungus (Trichoderma), and chemical antagonists such as phenols and amino acids (Nagavallemma, et al., 2004).

Using vermicompost in horticulture is a robust perspective that could be useful for the efficacy of organic fertilizers that attempt to achieve an increased production. Moreover, according to the nowadays trend of consuming organic and ecological produced fruits, the use of chemical fertilizers are more and more questioned by the consumers and health practitioners. In this light, we considered that the subject is one of interest especially regarding the effect of the vermicompost when used as a fertilizer or pesticide on the fruit trees (Lazar, et al., 2014).

The benefits that plants receive from vermicompost depend on the plant ability to extract from the fertilizing substrate the substances needed for the growth and development (Bohlen and Edwards 1995). Plant reaction to fertilizers is directly proportional to the quality of fertilizers used in substrate of culture and therefore the utilization of vermicompost in horticulture is highly recommended.

It can be concluded that vermicompost dose of 1.5 - 3 kg /trees applied repeatedly for three months at initial stage of tree is helpful in terms of maximum fruit yield which also improves soil health. The use of vermicompost for production of fruits appears to be an efficient plant growth medium that can ensure sustainable productivity. These findings will have long term practical implication in increasing production without any loss to soil health in drought (Kumar et al., 2018).

Vemicompost for Seed Germination and Seedling Growth

Vermicompost greatly improves seed germination, seedling growth and increased productivity well above by simply converting soil minerals to easy accessible substances for the crops. There are studies that reported high contents of vermicompost on growth hormones such as auxins, cytokinins and gibberellins.

Vermicompost is stable and fin granular organic manure, which enriches soil quality by improving its physicochemical and biological properties. It is a highly useful in raising seedlings and for crop production. Vermicompost is becoming popular as a major component of organic farming system. The level of nutrients in compost depends upon the source of the raw material and the species of earthworm. A fine worm cast is rich in N, P and K besides other nutrients. Nutrients in vermicompost are in readily available form and are released within a month of application.

Atiyeh et al. (2001) postulated that use of vermicompost as soil additives or as components of bedding media increases seed germination, growth and development of seedling and overall plant productivity.

Vermicomposting is described as "bioxidation and stabilization of organic material involved by the joint action of earthworms and mesophilic micro-organisms". Vermicompost produced by the activity of earthworms is rich in macro and micronutrients, vitamins, growth hormones, enzymes such as proteases, amylases, lipase, cellulase and chitinase and immobilized microflora. The enzymes continue to disintegrate organic matter even after they have been ejected from the worms (Barik et al., 2011). Bio fertilizers containing azotobacter produce many growth regulators such as IAA and GA2 which positively influence plant growth (Sharma and Kumar 2008).

Vermicompost in Plant Growth Promotion

Use of vermicompost as bio-fertilizers has been increasing recently due to its extraordinary nutrient status, and enhanced microbial and antagonistic activity. Vermicompost produced from different parent material such as food waste, cattle manure, pig manure, etc., when used as a media supplement, enhanced seedling growth and development, and increased productivity of a wide variety of crops (Subler et al. 1998; Atiyeh et al. 2000). Earthworms produce plant growth regulators (Gavrilov 1963). Since earthworms increase the microbial activity by several folds they are considered as important agents which enhance the production of plant growth regulators.

Vermicompost application resulted in a significant and consistent increase in plant growth in both field and greenhouse conditions (Edwards et al. 2004), thus providing a substantial evidence that biological growth promoting factors play a key role in seed germination and plant growth (Edwards 1998). Investigations revealed that plant hormones and plant-growth regulating substances (PGRs) such as auxins, gibberellins, cytokinins, ethylene and abscisic acid are produced by microorganisms (Arshad and Frankenberger 1993).

Vermicompost in Plant Disease Management

Soils with low organic matter and microbial activity are conducive to plant root diseases and addition of organic amendments can effectively suppress plant disease (Stone et al. 2004). Several researchers reported the disease suppressive properties of thermophilic compost on a wide range of phytopathogens viz., *Rhizoctonia*, *Phytopthora*, *Plasmidiophora brassicae* and *Gaeumannomyces graminis* and *Fusarium* (Kannangowa et al. 2000; Cotxarrera et al. 2002).

Earthworm feeding reduces the survival of plant pathogens such as *Fusarium* sp. and *Verticillium dahlia* and increases the densities of antagonistic fluorescent pseudomonads and filamentous actinomycetes while population densities of *Bacilli* and *Trichoderma* spp. remains unaltered (Moody et al. 1996; Elmer 2009). Earthworm activities reduce root diseases of cereals caused by *Rhizoctonia* (Doube et al. 1994). It has been proved that earthworms decreased the incidence of field diseases of clover, grains, and grapes incited by *Rhizoctonia* spp. and *Gaeumannomyces* spp. (Clapperton et al. 2001). Amendment of vermicompost at low rates (10–30%) in horticulture bedding media resulted in significant suppression of *Pythium* and *Rhizoctonia* under greenhouse conditions (Edwards et al. 2004).

Research findings proved that vermicompost when added to container media signify cantly reduced the infection of tomato plants by *P. nicotianae* var. *nicotianae* and *F. oxysporum* sp. *lycopersici* (Szczech 1999). Potato plants treated with vermicompost were less susceptible to *P. infestans* than plants treated with inorganic fertilizers (Kostecka et al. 1996).

Vermicompost Bacteria in Biomedical Waste Management

The importance of sewage sludge, biosolids and biomedical waste management by safe, cheap and easy methods need no further emphasis. All these wastes are infectious and have to be disinfected before being disposed into the environment. Biosolids also contain an array of pathogenic microorganisms (Hassen et al. 2001). Vermicomposting does not involves a thermophilic phase which might increase the risk of using this technology for management of infectious wastes, but surprisingly vermicomposting resulted into a noticeable reduction in the pathogen indicators such as fecal coliform, Salmonella sp., enteric virus and helminth ova in biosolids (Sidhu et al. 2001).

The reduction or removal of these enteric bacterial populations at the end of vermicomposting period, correlates with the findings that earthworm's diet include microorganisms and earthworms ability to selectively digest them (Edwards and Bohlen 1996). Apart from solid waste management, earthworms are also used in sewage water treatment. Vermicomposting plays a vital role for safe management of biomedical wastes and solid wastes generated from wastewater treatment plants and its bioconversion into valuable composts free from enteric bacterial populations. Depending on the earthworm species, vermicomposting was known to reduce the level of different pathogens such as Salmonella enteriditis, Escherichia coli, total and faecal coliforms, helminth ova and human viruses in different types of waste. Direct means of reduction in these microbial numbers during gut passage might be due to the digestive enzymes and mechanical grinding, while indirect means of pathogen removal might be due to promotion of aerobic conditions which could bring down the load of coliforms (Aira et al. 2011).

Vermicomposting involves the composting of organic wastes through earthworm activity. It has proven successful in processing sewage sludge and solids from wastewater, materials from breweries, paper waste, urban residues, food and animal wastes, as well as horticultural residues from processed potatoes, dead plants and the mushroom industry (Dominguez and Edwars, 2004).

Vermicompost in Nematode Control

It has been well documented that addition of organic amendments decreases the populations of plant parasitic nematodes (Akhtar and Malik 2000). Vermicompost amendments appreciably suppress plant parasitic nematodes under field conditions. Vermicomposts also suppressed the attack of Meloidogyne incognita on tobacco, pepper, strawberry and tomato and decreased the numbers of galls and egg masses of Meloidogyne javanica (Ribeiro et al. 1998; Arancon et al. 2002; Edwards et al. 2007).

Vermicompost amendment promoted fungi capable of trapping nematode and destroying nematode cysts and increased the population of plant growth-promoting rhizobacteria which produce enzymes toxic to plant parasitic nematodes (Kerry 1988; Siddiqui and Mahmood 1999). Vermicompost addition to soils planted with tomatoes, peppers, strawberry and grapes showed a significant reduction of plant parasitic nematodes and increased the population of fungivorous and bacterivorous nematodes compared to inorganic fertilizer treated plots (Arancon et al. 2002).

DERIVATIVES OF VERMICOMPOST

Vermicompost Tea

Vermicompost tea is the water extracts of solid vermicomposts from which microorganisms, soluble nutrients, and plant-beneficial substances are converted into a liquid form. It can be used in a wide range of horticultural and agricultural systems to elicit plant growth and pest and disease management responses through a variety of mechanisms. It can be applied directly to plant foliage. It is also used as a soil drench and has been shown to be effective in relatively small quantities (Edwards et al. 2011).

Worms tea can be applied on leafs or roots by sprinkler or irrigation method, diluted or not. Dilution does not affect the qualities of the extract in the preparation. There of the ''worm tea'' can be diluted up to 1:10. The extract is not toxic, does not burn the plants, no restrictions on use, can be used in any culture in greenhouses or in the field especially on organic cultures. Another positive aspect of the worm tea dilute is the fact that does not provide living conditions for Escherichia coli (Gutierrez-Miceli et al., 2008).

Organic extracts became a common agriculture practice in sustainable farming (Gross et al., 2008) as a balanced source of nutrients in available form in the rizosphere, growth stimulant and disease suppressor. In addition to its beneficial impacts on soil physical and chemical attributes as well as soil biodynamic activity.

Vermi-Liquid

This technology used the assistance of earthworms to stabilize the organic waste material and produces microorganism rich medium that enhances the process of composting (Ismail, 2005). The principle of the vermicomposting leachate collecting unit is designed to allow the percolation of water through these passages and hence collecting the nutrient and microorganisms along with it. Vermicompost leachate (VCL) as a bio-fertilizer is a liquid phase that comes from vermicompost and it has a very positive impact on soil improvement and plant growth (Chinsamy et al., 2013). Bio-fertilizers are the most importance for plant production and soil as they play an important role in increasing vegetative growth, yield and fruit quality (Hasan et al., 2013). Several epithets such as vermiwash, vermicomposting leachate, vermi-leachate, worm bed leachate and worm tea have been used to describe the liquid derived from the vermi-composting process (Quaik and Ibrahim 2013). Use of VCL as a liquid fertilizer provides the advantage of homogeneity, when applied to growth media as compared to application of solid fertilizer (Quaik et al., 2012 and Shlrene et al., 2012). Vermicomposting leachate acts as plant tonic, because it contains humic acid, fulvic acid, amino acids, vitamins, enzymes, microorganisms, actinomycetes, nutrients like nitrogen, potassium, magnesium, zinc, calcium, iron, and copper and some growth hormones like auxins and cytokines (Suthar, 2010). These characteristics increased vegetative growth and productivity of fruit trees by about 15%. Vermicomposting leachate increased disease resistance capacity in many agricultural crop plants against various bacterial, vital and fungal diseases. VCL considers as a biotic aqua fertilizer, which is applied as pesticide that also contains plant essential nutrients. VCL had anti-spawning effects on insects and increased disease resistance capacity in many agricultural crop plants against various bacterial, vital and fungal diseases (Zhu et al., 2001). Vermicomposting leachate is very good foliar spray, which prevents detachment of flowers and helps in fruit setting (Tejada et al., 2008). In addition, Sathe and Patil (2014) recorded that VCL is a good bio-fertilizer and tonic to mango (Mangifera indica), which increased the fruit production of mango and fruit quality. It was very interesting that, during the experimental period there were no insect pests and bacterial or fungal diseases attacked mango indicating the controlling capacity of VCL against pests and diseases. Furthermore, foliar spray of VCL increase salt tolerance by reducing the accumulation of Na+ in pomegranate tree, increased leaf area and total Chlorophyll (Siamak et al., 2017).

This derivate is used mainly for spraying plants, but recently was introduced in fertirigation circuit systems due to the high concentration of nutrients (Jarecki et al., 2005). Regardless the high content of nutrients, the liquid is remarkable by the presence in its composition of humic acids with an important role in the plant growth. The humic acids are also helping to a good absorption of micro and macroelements (Ordoñez et al., 2006). Significantly, vermicompost works as a 'soil conditioner' and its continued application over the years lead to total improvement in the quality of soil and farmland, even the degraded and sodic soils.

REFERENCES

Aira, M., Gómez-Brandón, M., González-Porto, P., & Domínguez, J. (2011). Selective reduction of the pathogenic load of cow manure in an industrial-scale continuous-feeding vermireactor. *Bioresource Technology*, *102*(20), 9633–9637. doi:10.1016/j.biortech.2011.07.115 PMID:21875788

Akhtar, M., & Malik, A. (2000). Role of organic amendments and soil organisms in the biological control of plant parasitic nematodes: A review. *Bioresource Technology*, *74*(1), 35–47. doi:10.1016/S0960-8524(99)00154-6

Arancon, N. Q., Edwards, C. A., & Lee, S. 2002. Management of plant parasitic nematode populations by use of vermicomposts. *Proc Brighton Crop Prot Conf Pests Dis 8B-2,* 705–716.

Arshad, M., & Frankenberger, W. T. Jr. (1993). Microbial production of plant growth regulators. In F. B. Metting Jr., (Ed.), *Soil microbial ecology: applications in agricultural and environmental management* (pp. 307–347). Marcell Dekker.

Arunkumar, J. (2000). Effect of vermicomposted sludge on growth of Amaranths dubius. *Ecotoxicol Environ Monit*, *14*(2), 157–160.

Atiyeh, R. M., Arancon, N. Q., Edwards, C. A., & Metzger, J. D. (2000). Infl uence of earthworm- processed pig manure on the growth and yield of greenhouse tomatoes. *Bioresource Technology*, *75*(3), 175–180. doi:10.1016/S0960-8524(00)00064-X

Atiyeh, R. M., Edwards, C. A., Subler, S., & Metzger, J. D. (2001). Pig manures vermicompost as a component of a horticultural bedding plant medium: Effects on physiochemical properties and plant growth. *Bioresource Technology*, *78*(1), 11–20. doi:10.1016/S0960-8524(00)00172-3 PMID:11265782

Barik, T., Gulati, J. M. L., Garnayak, L. M., & Bastia, D. K. (2011). Production of vermicompost from agricultural wastes. *Agricultural Reviews (Karnal)*, *31*(3), 172–183.

Bhawalkar, V. U. (1993). The living soil. *Extended Abstracts of congress on Traditional Science & Technologies of India*, 10.5-10.8.

Bohlen, P. J., & Edwards, C. A. (1995). Earthworm effects on N dynamics and soil respiration in microcosms receiving organic and inorganic nutrients. *Soil Biology & Biochemistry*, *27*(3), 341–348. doi:10.1016/0038-0717(94)00184-3

Chellachamy, V., & Dinakaran, S. (2015). A comparative study on ver- micomposting of epicarp of fruits (pomegranate and Sathukudi) using earthworm *Eisenia foetida. International Journal of Recent Scientific Research, 6*(3), 3125-3129.

Chinsamy, M., Kulkarni, M. G., & Staden, J. V. (2013). Garden-waste-vermicompost leachate alleviates salinity stress in tomatoseedlings by mobilizing salt tolerance mechanisms. *Plant Growth Regulation*, *71*(1), 41–47. doi:10.100710725-013-9807-6

Clapperton, M. J., Lee, N. O., Binet, F., & Conner, R. L. (2001). Earthworms indirectly reduce the effect of take-all (Gaeumannomyces graminis var. tritici) on soft white spring wheat (Triticium aestivum cv. Fielder). *Soil Biology & Biochemistry*, *33*(11), 1531–1538. doi:10.1016/S0038-0717(01)00071-2

Cotxarrera, L., Trillas-Gayl, M. I., Steinberg, C., & Alabouvette, C. (2002). Use of sewage sludge compost and Trichoderma asperellum isolates to suppress Fusarium wilt of tomato. *Soil Biology & Biochemistry*, *34*(4), 467–476. doi:10.1016/S0038-0717(01)00205-X

Dominguez, J., & Edwars, C. A. (2004). Vermicomposting organic wastes: A review. In S. H. Shakir Hanna & W. Z. A. Mikhatl (Eds.), Soil Zoology for Sustainable Development in the 21st Century (pp. 369–395). Academic Press.

Doube, B. M., Stephens, P. M., Davorena, C. W., & Ryderb, M. H. (1994). Interactions between earthworms, benefi cial soil microorganisms and root pathogens. *Applied Soil Ecology*, *1*(1), 3–10. doi:10.1016/0929-1393(94)90018-3

Edwards, C. A. (1998). The use of earthworms in processing organic wastes into plant growth media and animal feed protein. In C. A. Edwards (Ed.), *Earthworm ecology* (pp. 327–354). CRC Press.

Edwards, C. A., Arancon, N. Q., Emerson, E., & Pulliam, R. (2007). Supressing plant parasitic nematodes and arthropod pests with vermicompost teas. *BioCycle*, *48*, 38–39.

Edwards, C. A., Arancon, N. Q., & Sherman, R. (2011). *Vermiculture technology*. Taylor and Francis Group, LLC.

Edwards, C. A., & Bohlen, P. J. (1996). *Biology and ecology of earthworms* (3rd ed.). Chapman & Hall.

Edwards, C. A., Dominguez, J., & Arancon, N. Q. (2004). The infl uence of vermicomposts on pest and diseases. In S. H. Shakir Hanna & W. Z. A. Mikhail (Eds.), Soil zoology for sustainable development in the 21st centuary (pp. 397–418). Academic Press.

Elmer, W. H. (2009). Influence of earthworm activity on soil microbes and soilborne diseases of vegetables. *Plant Disease*, *93*, 175–179.

Gavrilov, K. (1963). Earthworms, producers of biologically active substances. *Zh Obshch Biol*, *24*, 149–154.

Gross, A., Arusi, R., Fine, P., & Nejidat, A. (2008). Assessment of extraction methods with fowl manure for the production of liquid organic fertilizers. *Bioresource Technology*, *99*, 327–334. https://doi.org/10.1016/j.biortech.2006.12.016

Gutierrez-miceli, Moguel-Zamudio, AbudArchila, Gutierrez-Oliva, & Dendooven. (2008). Sheep manure vermicompost supplemented with a native diazotrophic bacteria and mycorrhizas for maize cultivation. *Bioresource Technology, 99*(15), 7020-7026.

Hasan, M. A., Manna, M., Dutta, P., Bhattacharya, K., Mandal, S., Banerjee, H., Ray, S. K., & Jha, S. (2013). Foliar nutrient content in mango as influenced by organic and inorganic nutrients and their correlative relationship with yield and quality. *Acta Horticulturae*, (992), 201–206.

Hassen, A., Belguith, K., Jedidi, N., Cherif, A., Cherif, M., & Boudabous, A. (2001). Microbial characterization during composting of municipal solid waste. *Bioresource Technology*, *80*, 217–225.

Kale, R. D., Bano, K., & Krishnamoorthy, R. V. (1982). Potential of Perionyx excavatus for utilization of organic wastes. *Pedobiologia*, *23*, 419–425.

Kannangowa, T., Utkhede, R. S., Paul, J. W., & Punja, Z. K. (2000). Effect of mesophilic and thermophilic composts on suppression of Fusarium root and stem rot of greenhouse cucumber. *Canadian Journal of Microbiology*, *46*, 1021–1022.

Kerry, B. (1988). Fungal parasites of cyst nematodes. In C. A. Edwards, B. R. Stinner, D. Stinner, & S. Rabatin (Eds.), *Biological interactions in soil* (pp. 293–306). Elsevier.

Kostecka, J., Blazej, J.B., & Kolodziej, M. (1996). Investigations on application of vermicompost in potatoes farming in second year of experiment. *Zeszyty Naukowe Akademii Rolniczej W Krakowie, 310*, 69–77.

Kumar, S., Prasad, R., Shukla, A., & Kumar, A. (2018). Management of organic orchard: Effect of vermicompost on fruit yield of ber (Zizyphus mauritiana Lamk.) cv. Seo and soil health in Central India. *Indian J. of Agroforestry, 20*(2), 58-62.

Laza, M., Miter, V., Tripon, F. A., & Badiu, D. E. (2014). The Potential use of vermicompost in orchards. *Bulletin UASVM Horticulture, 71*(1), 56–58.

Medany, M. (2011). Vermiculture in Egypt: Current development and future potential Food and Agriculture Organization of the United Nations Regional Office for the Near East. Academic Press.

Moody, S. A., Piearce, T. G., & Dighton, J. (1996). Fate of some fungal spores associated with wheat straw decomposition on passage through the guts of Lumbricus terrestris and Aporrectodea longa. *Soil Biology & Biochemistry, 28*, 533–537.

Nagavallemma, K. P., Wani, S. P., Lacroix, S., Padmaja, V. V., Vineela, C., Rao, M. B., & Sahrawat, K. L. (2004). *Vermicomposting: Recycling Wastes into Valuable Organic Fertilizer.* Global Theme on Agroecosystems Report no. 8. International Crops Research Institute for the Semi-Arid Tropics, Patancheru.

Ordoñez, C., Tejada, M., Benitez, C., & Gonzalez, J. L. (2006). Characterization of a phosphorus–potassium solution obtained during a protein concentrate process from sunflower flour. Application on rye-grass. *Bioresource Technology, 97*(3), 522–528.

Quaik, S., Embrandiri, A., Rupani, P. F., & Ibrahim, M. H. (2012) Potential of vermicomposting leachate as organic foliar fertilizer and nutrient solution in hydroponic culture: a review. *2nd International Conference on Environment and BioScience IPCBEE*, 43–47.

Quaik, S., & Ibrahim, M. H. (2013). A review on potential of vermicomposting derived liquids in agricultural use. *Int J Sci Res Pub, 3*(3), 1–6.

Ribeiro, C. F., Mizobutsi, E. H., Silva, D. G., Pereira, J. C. R., & Zambolim, L. (1998). Control of Meloidognye javanica on lettuce with organic amendments. *Fitopatologia Brasileira, 23*, 42–44.

Rivera-Cruz, M., Trujillo, A., Córdova, G., Kohler, J., Caravaca, F., & Roldán, A. (2008). Poultry Manure and Banana Waste Are Effective Bio-Fertilizer Carriers for Promoting Plant Growth and Soil Sustainability in Banana Crops. *Soil Biology & Biochemistry, 40*, 3092–3095. https://dx.doi.org/10.1016/j.soilbio.2008.09.003

Roy, S. K., Trehan, S. P., & Sharma, R. C. (2000). Longterm nutrient management in potato-sun" ower-rice system for sustainable productivity. *Intl. Conference on Managing Natural Resources, New Delhi. Extended Summaries, 3*, 920-921.

Sharma, S. D., & Kumar, P. (2008). Relationship of Arbuscular Mycorrhizal Fungi and Azotobacter with Plant Growth, Fruit Yield, Soil and Leaf Nutrient Status of Mango Orchards in Northwestern Himalayan Region of India. *Journal of Applied Horticulture*, *10*, 172–178.

Siddiqui, Z. A., & Mahmood, I. (1999). Role of bacteria in the management of plant parasitic nematodes: A review. *Bioresource Technology*, *69*, 167–179.

Sidhu, J., Gibbs, R. A., Ho, G. E., & Unkovich, I. (2001). The role of indigenous microorganisms in suppression of Salmonella regrowth in composted biosolids. *Water Research*, *35*, 913–920.

Sinha, Heart, Agarwal, Asadi, & Carretero. (2005). Vermiculture and waste management: study of action of earthworms Eisinia fetida, Eudrilus enginae and Perionyx excavatus on biodegradation of some community wastes in India and Australia. *The Environmentalist, 22*(3), 261 – 268.

Sinha, R. K., Agarwal, S., Chauhan, K., & Valani, D. (2010). The wonders of earthworms & its vermicompost in farm production: Charles Darwin's 'friends of farmers', with potential to replace destructive chemical fertilizers from agriculture. *Agricultural Sciences*, *1*, 76–94.

Stone, A. G., Scheurell, S. J., & Darby, H. M. (2004). Suppression of soilborne diseases in fi eld agricultural systems: organic matter management, cover cropping and other cultural practices. In F. Magdoff & R. Weil (Eds.), *Soil organic matter in sustainable agriculture* (pp. 131–177). CRC Press LLC.

Subler, S., Edwards, C.A., & Metzger, P.J. (1998). Comparing vermicomposts and composts. *BioCycle, 39*, 63–66.

Szczech, M. M. (1999). Suppressiveness of vermicomposts against fusarium wilt of tomato. *Journal of Phytopathology*, *147*, 155–161.

Tejada, M., Gonzalez, J. L., Hernendez, M. T., & Garcia, C. (2008). Agricultural use of leachates obtained from two different vermicomposting processes. *Bioresource Technology*, *99*(14), 6228–6232. doi:10.1016/j.biortech.2007.12.031 PMID:18215517

Tomati, U., Grappelli, A., & Galli, E. (1983). Fattori difertilita nell'humus di lombrico. *Proceedings of international symposium on agricultural and environmental prospects in earthworm farming*, 49–56.

Tripathi, G., & Bhardwaj, P. (2004). Comparative studies on biomass production, life cycles and composting efficiency of Eisenia fetida (Savigny) and Lampito mauritii (Kinberg). *Bioresource Technology*, *92*, 275–283.

Uthar, S. (2010). Evidence of plant hormone like substances in vermiwash: an ecologically safe option of synthetic chemicals for sustainable farming. *Ecol Eng, 36*, 1089–1092.

Chapter 3
Optimization of Green Hydrogel in Agriculture Based on Guar Gum by Response Surface Methodology

Elsayed Gamal Zaki
https://orcid.org/0000-0001-7053-3169
Egyptian Petroleum Research Institute, Egypt

Shimaa Mohamed Elsaeed
Egyptian Petroleum Research Institute, Egypt

Faissal Aziz
Laboratory of Water, Biodiversity, and Climate Change, Faculty of Sciences Semlalia, Cadi Ayyad University, Morocco

Athir Mahmood Haddad
Chemistry Department, College of Science, University of Basrah, Iraq

Aula Ali Alwattar
The University of Manchester, UK

Khalid Aziz
Laboratory of Biotechnology, Materials, and Environment, Faculty of Sciences, Ibn Zohr University, Morocco

ABSTRACT

Water plays an essential purpose in agricultural yield. Unfortunately, water shortages have led to desertification and salination of soils, threatening the durable progress of agriculture, horticulture, and food security, especially in dried regions, where water resources are a severe problem. The hydrogel absorbed the water, released it to the plants as required while maintaining the soil moist during long periods. Biopolymer as guar gum-based hydrogel is becoming one of the most green ways to hydrogel materials for water retention. Artificial intelligence (AI) is used in hydrogel technology as a result of high energy consumption to make process control and optimization of synthesis and development. Sustainable environment for hydrogel technology was integrated by AI model.

DOI: 10.4018/978-1-6684-2423-0.ch003

HYDROGEL POLYMERS INTRODUCTION

Hydrogel Polymers

Hydrogel is a three-dimensional network of hydrophilic polymers that can hold a large amount of water while keeping its structure due to crosslinks. Hydrogel has the ability to absorb at least 20% of the total weight of water, (Duan & Jiang, 2017; Sen et al., 2020; Wang et al., 2019) and it is called "superabsorbent" (Chen et al., 1999) when it absorbs more than 95% of the total weight. Hydrogels swell in water and shrink in the absence of water (Budianto & Amalia, 2020) as this is the most characteristic property of hydrogels. Xerogels (dried hydrogel) become much smaller in size than the swollen hydrogel.

The crosslinking of polymer chains in hydrogels usually involves chemical or physical processes. Covalent bonds connect the polymer chains in the chemical crosslinked hydrogel. Thus, it is difficult to change the shape of such gels while polymer chain, which is physically entangled, are connected through non-covalent bonds. (Maitra & Shukla, 2014) The connection points of two chains are known as "crosslinks". These points could be small chemical bridges of molecular weight much smaller than the crosslinked polymer chains. Crosslinks may also be an association of macromolecular chains due to Van der Waals forces or an aggregate formed via hydrogen bonds. The crosslinked hydrogel can be formed by:

1. **Radiation**: electron beams, (Hietala et al., 1997) gamma rays, (Kademani et al., 2006) x-rays, (Sundholm et al., 1996) or U.V. light (Hu & Xia, 2004) to activate a polymer site for reaction with another polymeric region.
2. **Chemical Crosslinking**: This type of crosslinking requires a bifunctional, low molecular weight crosslinking agent. (Abed et al., 2006) Figure 1 shows the main types of crosslinking reactions: (Arias et al., 1993)

Figure 1. Schematic representation for the common crosslinking reactions (addition type)

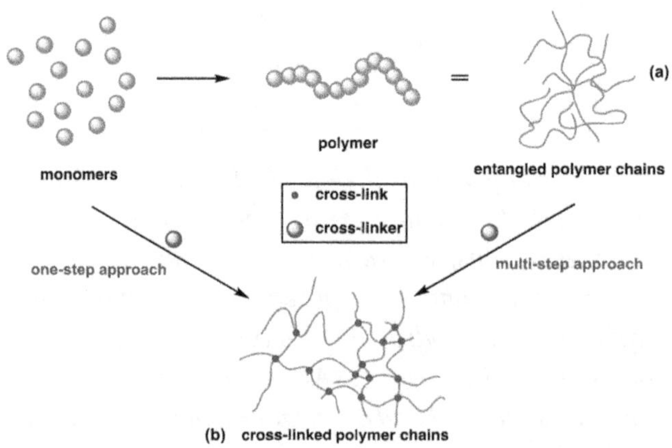

Other chemical reactions can also crosslink polymers with other functional groups. For example, Albumin and Gelatin can be crosslinked with aldehyde. (Bolto, 1995; Di Silvio et al., 1994) In addition, some hydrogels are formed by physical interaction between polymer chains. These interactions include hydrogen bonding, Van der Waals, and ionic and hydrophobic interactions. (Maitra & Shukla, 2014) Figure 2 shows the main interaction:

Figure 2. The major four types of interaction forces which control the behaviour of most responsive gels

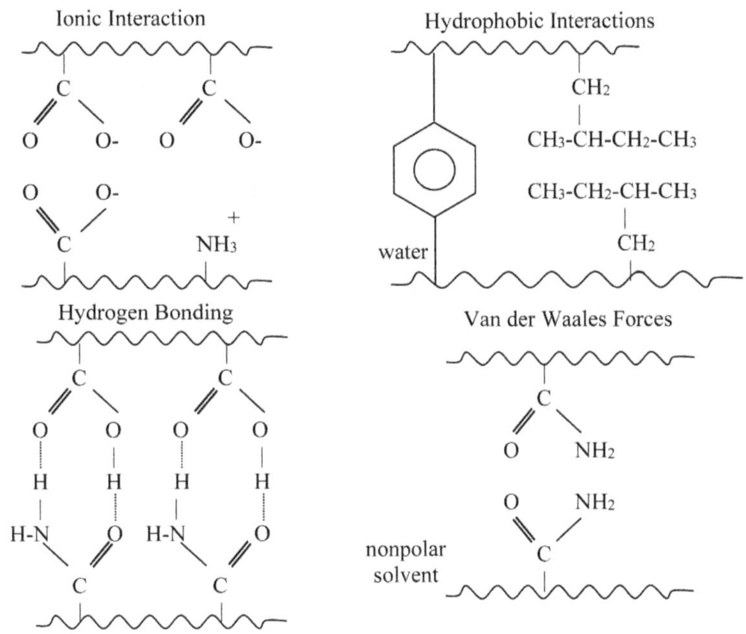

Classification of Physical Gels

1. Blend hydrogels.
2. IPN and Semi-IPN hydrogels.
3. Block copolymer hydrogels.
4. Polyelectrolyte complex hydrogels.
5. Counter ion-induced hydrogels.
6. Thermally-induced hydrogels.
7. Specific interaction induced hydrogels.

Blend Hydrogels

Blend hydrogels are usually obtained by solvent casting or precipitation from two different polymer solutions or more. The blend hydrogels normally have different properties from those made from individual polymers. The obtained hydrogels, this way, should contain at least one hydrophilic polymer to provide the water-absorbing property. For example, poly (ethylene oxide)-poly (propylene oxide)-poly (ethylene oxide) [PEO-PPO-PEO] block copolymers were blended with poly lactic acid (PLA) to make hydrogels. (Park et al., 1992) Other hydrogels were prepared by blending, chitosan with poly (ethylene oxide), (Amiji, 1995) and chitosan and poly (vinyl alcohol) blend hydrogel was also prepared by solvent casting. (Lee et al., 2019) Hyaluronic acid was blended with poly (Acrylic acid) or poly (vinyl alcohol) to make hydrogels. (Cascone et al., 1995)

IPN and Semi IPN Gels

Other types of physical gels are IPN or semi-IPN. IPNs are any materials containing two different types of polymers in a network form. In semi-IPNs only one of the components in a network form and is usually formed in the presence of the other components. (Adam et al., 1975; Fyvie et al., 1987; Yeo et al., 1981) IPNs and semi-IPNs of poly (hydroxyethylmethacrylate) (PHEMA) and poly(caprolactone) (PCL) were also prepared. (Eschbach & Huang, 1994; Eschbach et al., 1994) Carboxymethyl cellulose was mixed with gelatin which was then crosslinked by glutaraldehyde to make semi-IPNs. (Rathna et al., 1994) IPNs were prepared by crosslinking gelatin with glutaraldehyde and acrylamide with N, N¢-methylene bis acrylamide. (Kaur & Chatterji, 1990) Other IPNs were obtained by crosslinking gelatin with glutaraldehyde and (PHEMA) with N, N¢-methylene bis acrylamide. (Ma et al., 1995)

Figure 3.

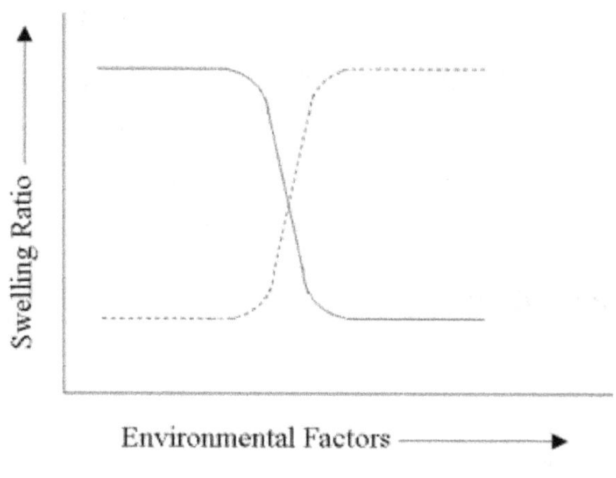

Block Copolymer Gels

The physical gels can also be made from block copolymers; thus, poly (ethylene oxide)-poly (lactic acid) block copolymer, poly (ethylene oxide)-poly (glycolic acid) block copolymers and poly (ethylene oxide)-poly(caprolactone) block copolymer are examples for this type of hydrogels. (Pérez-Luna, 2020; Sawhney et al., 1993; Sawhney et al., 1994)

Polyelectrolytic Gels

Some hydrogels are made from poly salt or polyelectrolyte complexes. Ionic interactions between two oppositely charged polyelectrolytes form poly electrolyte complexes swell in water. Examples of this type of gels are polyelectrolyte film of amino group-containing chitosan with sodium alginate, (Zhao et al., 2012) Poly (meth acryloyl ethyl trimethylammonium methyl sulfate) (PMETMMS) (Sundholm et al., 1996) and poly (vinylidene fluoride)-graft-poly (styrene sulfonic acid) as polymeric electrolyte membranes. (Hietala et al., 1999)

Counter Ion Induced Hydrogels

Certain polyelectrolytes can form hydrogels in the presence of counter ions. Sodium alginate, an atypical example, forms gels in the presence of Ca^{2+}. (Matthew et al., 1995; Miller, 2000) The electrolyte is composed of dissolving lithium perchlorate in ethylene carbonate solution immobilized with poly (methyl methacrylate). (Svanberg et al., 1999)

Thermally Induced Hydrogels

This type of gel is formed when the structural change of polymer in solution is induced by thermal energy or changes the balance between hydrogen bonding and hydrophobic interaction. Good examples for this class are the gelation of gelatin solution as the temperature is lowered, (Djabourov et al., 1985) and agarose gels can also be prepared by this method. (Bellamkonda et al., 1995)

Specific Interaction Induced Gels

Polymer networks can be formed by specific interactions, such as interactions between glucose and concanavalin A. (Lee & Park, 1996; Obaidat & Park, 1996)

Environment: Sensitive Hydrogels

"Smart hydrogels" (Kim & Park, 1998) are hydrogels with the ability to swell or shrink in response to a signal. This response, which can be swell, shrink, bend or degrade, is due to the change in the environmental conditions. These hydrogels are known as environment-sensitive as a result of this response. The swelling ratio, which is the volume of the swollen hydrogel divided by the volume of the dried hydrogel, changes rather than abruptly upon small changes in the environmental parameters. Many environmental factors are affected by gel characteristics listed in Table 1. These show a dramatic volume change called volume collapse (or phase transition) as shown in Figure 4. (Vashist et al., 2014)

Table 1. Factors that cause volume collapse for gel characteristics

Factor	Reference
1. pH	(Brandl et al., 2010; Ramanan et al., 2006)
2. Temperature	(Feil et al., 1991; Yoshida et al., 1995)
3. Electric field	(Homma et al., 2000; Kaetsu et al., 1992; Seida & Nakano, 1995)
4. Ionic strength	(Hooper et al., 1990)
5. Salt type	(Inomata et al., 1992; Seida & Nakano, 1996)
6. Solvent	(Hu et al., 1993)
7. Stress	(Sawahata et al., 1990; Tsujii et al., 1997)
8. Light	(Peppas, 2010)
9. Pressure	(Marchetti et al., 1990)

Figure 4. The volume collapse of smart hydrogels. The dotted line refers to dramatic increase in the swelling ratio and decrease the solid line) refers to the decrease

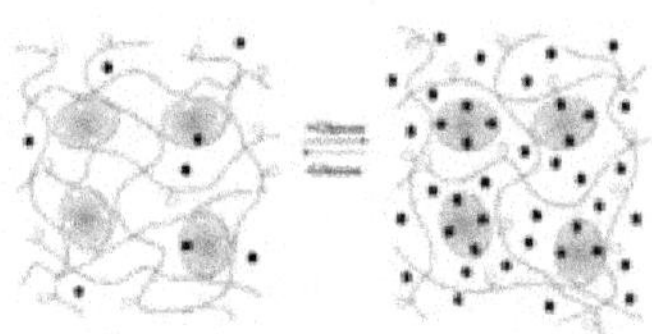

Thermoplastic Hydrogels

Thermoplastic hydrogels are based on linear copolymers of hydrophilic and hydrophobic monomers. The hydrophilic monomers are called "soft block" and the hydrophobic monomers are called "hard block" and physical gel is formed by hydrophobic interactions between hydrophobic chains of the copolymers. (Park & Robinson, 1984)

Thermoplastic hydrogels dissolve in the organic solvent, which only swell without dissolving in water, and this property provides an advantage of easy processibility. Vinyl-2-Pyrrolidone and methyl methacrylate are known to form thermoplastic hydrogels with useful properties such as melt-processibility. (Liu et al., 1994)

Figure 5.

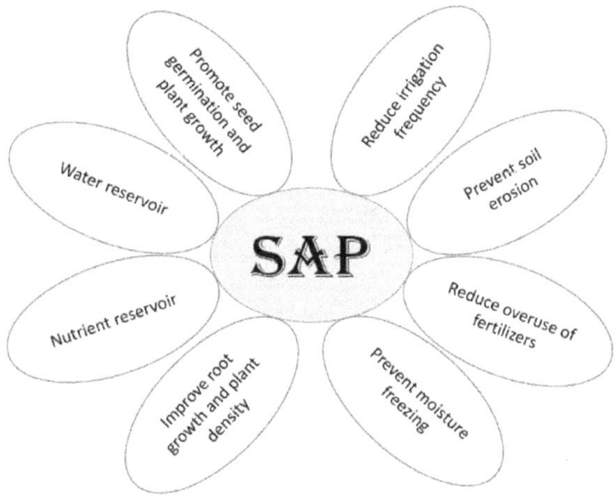

Hydrogel Foams

While hydrogels swell to a large extent in water, the equilibrium swelling usually takes a long time, from several hours to days, depending on the size and shape of the hydrogels. Hydrogel foams were recently developed (Park & Park, 1994; Park & Park, 1996) by synthesizing the hydrogels in the presence of a blowing agent. The hydrogel foams prepared with macroscopic gas cells are different from hydrogel sponges (Chirila et al., 1993) or macroporous hydrogels. (Chen & Park, 1999; Oxley et al., 1993) The size of pores in the hydrogel foams is orders of magnitude larger than the pore size (typically a few micrometers) in hydrogel sponges or macroporous hydrogels. In addition, the kinetics and the extent of swelling of hydrogel foams are much faster and larger than others. (Chen et al., 1999)

Ligand-Specific Sol-Gel Phase-Reversible Hydrogels

Physical gels can undergo Sol-Gel phase transition due to non-covalent crosslinking of polymer chains. For example, hydrogels that become sol in the presence of glucose were developed. (Lee & Park, 1996) High specific interaction between glucose and concanavalin A (con A) was used to form physical crosslinks between glucose-containing polymer chains. The glucose molecules attached to the polymer backbone react with (Con A), and since the non-covalent interactions between glucose and (Con A), the crosslinks formed are reversible.

The gel is formed by mixing glucose-containing polymers with Con A. Upon the addition of free glucose molecules, and the hydrogel dissolves to become a sol owing to the detachment of polymer chains from Con A as a result of completive binding of the free glucose to Con A. The sol can become a gel again upon removal of free glucose. (Lee & Park, 1996; Maitra & Shukla, 2014; Obaidat & Park, 1996). Many Ligands have specific interactions, such as antigen-antibody (Miyata et al., 1999).

Improving water use efficiency is important to the agricultural sector (Wen et al., 2012). Superabsorbent polymers (SAP), or hydrogels, can deny these problems because of their high water-retention and absorption capability and can be used to improve agricultural water use successfully while maintaining soil moisture and reducing irrigation water consumption. Because of the versatile number of hydrophilic

Figure 6. Allylglucose-Acrylamide copolymer. (Obaidat & Park, 1996)

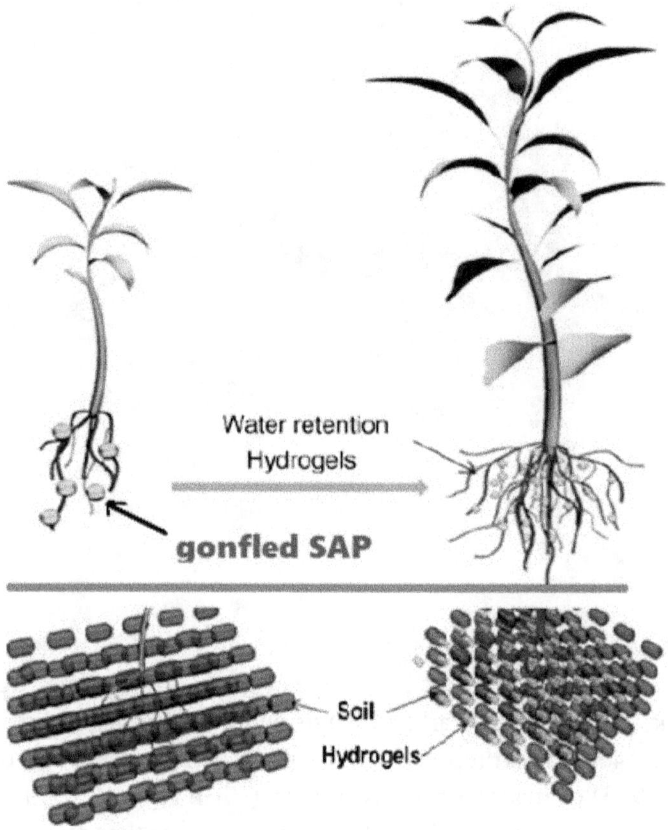

Figure 7. Allylglucose-Vinylpyrolidone copolymer. (Lee & Park, 1996)

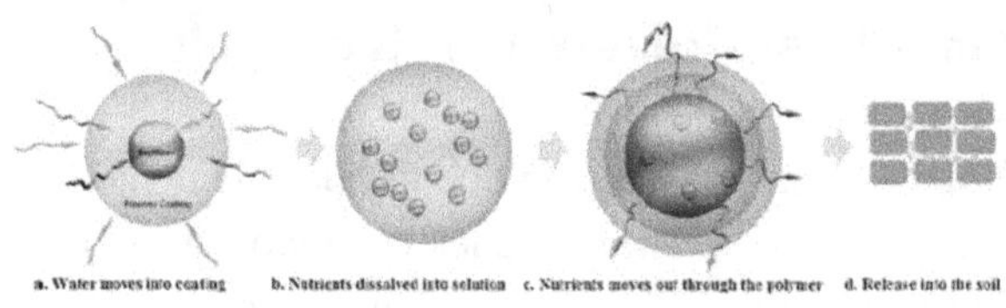

groups on their chains, such as –O.H., $-SO_3H$, -COOH, -CONH-, and $-CONH_2$, they help absorb water until thousands of times their mass without dissolution. During the swelling process of an SAP, this material changes from solid to a gel-like substance to reserve a huge volume of water even within significant compression. Therefore, the swollen SAP can gradually release the uptaken water (or nutrients) across a diffusion-driven process due to a humidity gradient inside and outside the hydrogel.

In the agricultural field, SAPs are valuable in many ways. For example, they can act as a water reservoir, root transplantation and a soil conditioner through their biodegradation or fertilizer coatings for controlled release fertilizers, among others, in order to decrease water use and optimize water resources in agriculture. Moreover, these materials enhance soil productivity, ensuring the safety of the natural environment. (Abobatta, 2018) Moreover, the additional interest of using SAPs in cultivation is attached to their swelling soil action soil, i.e, the dried gel granules can expand in size after its swelling step, which increases the soil porosity and leads to improved soil oxygenation.

The dry SAP, in powder, granules, or beads, is spread near plant roots on the application of hydrogels in the agricultural field. Furthermore, the SAP can be loaded with nutrients (like urea). (Lin et al., 2021) Then, the hydrogel absorbed the water and released it to the plants while keeping the soil moist for long periods. Moreover, the large-granule SAPs yield better results than the tine ones due to the improved aeration mechanism. However, the over dosage of these materials is eventually risky and should be avoided, so it is important to determine the optimum quantity before application.

Various synthetic hydrogels, based on synthetic polymers, like polyacrylamide, acrylic acid, and polyvinyl acetate, have been widely used for agricultural and horticultural usage because of their high swelling capacity natural hydrogels.

These synthetic materials are commonly expensive, non-degradable, and non-eco-friendly. Rather, natural polymers, including polysaccharides, proteins, polynucleotides, and lignin, exhibit significant advantages, including availability, nontoxicity, biocompatibility, biodegradability, and low production cost. Accordingly, natural polymers appear as an interesting sustainable alternative to synthetic polymers. For instance, polysaccharides can simply form hydrogels by chemical (covalent bonds) or physical (H-bonds, ionic interaction) crosslinking or an association of both, making the crosslinking of polysaccharides a polyvalent and promising procedure for superabsorbent polymers production. For that reason, numerous polysaccharide-based SAPs have been used in SAP fabrication for agricultural applications, including hydrogels based on cellulose, (Elsaeed et al., 2021) chitosan, (Essawy et al., 2016) starch, (Xiao et al., 2017) alginate, (Thakur et al., 2018) guar gum, (Thombare et al., 2018) among others.

Water Reservoir

Much attention is given to biobased SAPs because of their high absorption capacity and water retaining capability in their crosslinked structure without dissolving. Moreover, these materials can release almost 90% of absorbed water into their surroundings. However, they can be used more than one time after dying.

A lignin/sodium alginate-based hydrogel was applied to tobacco plants. The SAPs showed good absorption capacity and helped the plants grow after stopping water supply and releasing nutrients (N, P, and K). (Song et al., 2020)

Based on cellulose derivatives, a carboxymethylcellulose and starch-based hydrogel showed good water retention capacity and irrigation rate (Fidelia & Chris, 2011). Furthermore, the results revealed that adding these SAPs led to a 50-70% increase in soil water storage capacity according to the soil/hydrogel ratio.

Another carboxymethylcellulose-based SAPs was prepared via chemical crosslinking for agricultural application. These SAPs lead to saving water for tomato cultivation. (DeRosa et al., 2010) The SAP was applied as a dry powder in red soil close to the plant's roots, with several concentrations, and the results showed that the amount of absorbed water increased with increasing SAP concentration in soil. The absorbed water is later gradually released by SAP when the moisture content in the soil decreases, so there is no need for additional watering. (DeRosa et al., 2010)

Clay montmorillonite was used to reinforce a carboxymethylcellulose-based SAP produced by gamma radiation. The synthesized SAP showed high water swelling capacity, which led to using it as a water reservoir and water managing material in agriculture. (Salmawi et al., 2018)

Chitosan, the most abundant animal polysaccharide, was largely used to make hydrogels applicable in agriculture. For instance, chitosan graft acrylic acid hydrogels were synthesized via free radical polymerization and have the ability to act as a water-retaining agent. (Fang et al., 2018) The chitosan was transformed into amino ethyl chitosan using a 2-chloroethylamine hydrochloride solution as a solvent before the grafting polymerization with acrylic acid.

The prepared hydrogel thermogravimetric analysis indicated the formation of chitosan derivate and showed that the prepared SAPs had a higher thermal stability network and improved mechanical strength. However, the hydrogels showed higher water absorption and salt resistance after test swelling than chitosan-based hydrogels. The results revealed that the prepared SAP could retain 71% water after 24 h, with a maximum of 550 g/g of water absorbency at a pH 8. (Ghobashy et al., 2020)

A starch-based hydrogel was used as a soil conditioner to act as a water reservoir and was prepared by gamma-ray crosslinking with polyvinylpyrrolidone and acrylamide. (Elbarbary et al., 2017) After the applications test, the hydrogels showed a highly significant influence on the growth of sunflowers because of their ability to act as a water reservoir for the plants. (Gharekhani et al., 2017)

Finally, chitosan and sodium alginate were copolymerized with polyacrylamide to make superabsorbent polymers via gamma rays for possible use in agricultural applications. (Elbarbary et al., 2017) The maize plants treated by these SAP showed a 50% increase in the grain yield because chitosan and alginate had acted successfully as a plant growth promoter, activator, and accelerator of some biological activities (metabolic, enzymatic and photosynthetic capacity) of plants, which suggest their potential use like a water reservoir and a soil conditioner for the soil and plant.

Controlled-Release Fertilizers

As we know, fertilizers play a major role in preserving soil fertility, improving harvest quality, and increasing yield. However, most commercial uncoated fertilizers are well-known for their weak nutrient uptake efficiency by plants owing to losses in agricultural land for several reasons such as runoff, leaching and volatilization. So, in order to satisfy the demands of enhancing yields without affecting the environment, biobased SAP can be impregnated with fertilizers components, including urea, potassium ions, and phosphate or act as a coating to form environmentally friendly fertilizers. These coated fertilizers reduce environmental pollution from nutrient (N, P, K) losses through retarding or controlling the release of nutrients into the soil (the root zone).

However, besides enhancing biobased SAP as a coating, they could be charged with nutrients through two different modes, in situ loading or post-loading. In fact, in situ loading is the preferred approach because it provides higher loading potential, which leads to higher releases.

Nevertheless, hydrogels' swelling mechanism during irrigation or rain after loading stimulates nutrient discharge. First, the diffused water dissolves the nutrients to diffuse slowly to the external medium. Then the release is inhibited when the SAP dries and is auxiliary activated ahead of the watering process, offering sustained liberation of nutrients in the soil.

For instance, an alginate-based SAP formed a superabsorbent nanocomposite gel by free radical graft polymerization of sodium alginate, two acrylates, and rice husk ash. It was used for water management and plant feeding in horticultural uses. (Gharekhani et al., 2017) These nanocomposites SAPs reached 1070 g/g on equilibrium swelling capacity because of their porous structure and harder electrostatic repulsive forces of rice husk with the COOH functional groups. However, experimental results reveal that nutrient liberation via tortuosity enhancement results in a slow-release profile.

Chitosan is one of the polysaccharides used as coating fertilizers after its gelation. Chitosan was crosslinked by citric acid for making biobased hydrogel used on the coating of fertilizers. (Ahmad et al., 2015) The prepared SAP coating leads to a slow release of nutrients (phosphate ions), and the release rate becomes slower as the number of coatings and pH of the medium increase. (Ahmad et al., 2015) Otherwise, CHCAUR, (Narayanan et al., 2018) a superabsorbent hydrogel prepared by hydrothermal synthesis from chitosan, citric acid, and urea in the weight ratio 1:2:2, in the presence of water and pH 5-6 at 100 °C for 650 min then at room temperature. The prepared SAPs were highly porous owing to the evolution of NH_3 and CO_2 gases as urea is noted to react with citric acid resulting in urea citrate adduct. The SEM results revealed that the average pore diameter of the CHCAUR was about 0.359 mm and the standard deviation was determined as 0.134, where porosity was measured to be 27.61%. CHCAUR absorbed 1250 g/g of distilled water (maximum) and 600 g/g after extraction with NaOH. So, CHCAUR could be used in agriculture because chitosan is its major component. However, with a nitrogen content of 11%, CHCAUR can be used as a versatile material in the agricultural field and, particularly, a controlled release agent of micro and macronutrients to the plant and soil (Narayanan et al., 2018).

Furthermore, starch (maize starch) could be converted into carboxymethyl starch (CMS), by mixing starch (4 g) with 20 ml of an aqueous solution containing sodium hydroxide (3.2 g) and monochloroacetic acid (4 g) at 60°C for 3h, to synthesized a controlled-release fertilizer system (P-CMS-g-PAM) based on carboxymethyl starch-graft-polyacrylamide, in order to provide the phosphate fertilizer to the plant at a constant rate thereby. (Alharbi et al., 2018) Before grafting acrylamide onto CMS, the converted starch was phosphorylated with different amounts of two fertilizers: MAP (mono-ammonium phosphate) and DAP (di-ammonium phosphate). The SAPs with a high phosphate ratio to CSM (1/0.66) exhibited the maximum swelling rate and reached 87% on the percent cumulative release of phosphorous on day 30th. (Alharbi et al., 2018)

A crosslinking reaction was induced via gamma radiation between carboxymethylcellulose and polyvinylpyrrolidone has been proposed to prepare a cellulose-based SAP for agricultural application (Raafat et al., 2012). Then, urea, as an agrochemical model, urea was then loaded onto the obtained SAP to provide nitrogen nutrients. As a result, the prepared SAP, with a high swelling degree (144 g/g) and good water retention capacity (retains about 50 wt% of water after 24h at 25 °C), was found able to form a slow urea release system, due to its economic and environment-friendly properties, where the amount of released urea increased with increasing the loading percent of the SAP. (Raafat et al., 2012)

Moreover, copolymerization of starch and acrylonitrile was used to prepare a crosslinked hydrogel using methylene bisacrylamide as a crosslinker in order to form a coating for urea before application in soil. (Jyothi et al., 2018) The prepared slow-release fertilizers system showed a slow release rate of urea which improved the growth of plants without the need to make new fertilizer granules in the soil.

In conclusion, the principal advantage of biobased SAPs is managed by the slow release of water, the soil porosity, the slow-releasing of fertilizers, the longtime preserving soil humidity, and the better oxygenation of soil and plant roots. However, these natural SAPs are biodegradable, biocompatible, eco-friendly, and low-cost, with high swelling and retention capabilities. So its application in agricultural fields helps reduce irrigation water demand, increases the plant's growth rate, reduces the plant death rate and the rate of evapotranspiration, and enhances fertilizer retention in soil. They may also moderate the effect of salinity and promote better microbial activities within soil due to their biodegradability. The following figure can help understand the importance of using these materials in agricultural fields.

Response Surface Methodology of Guar Gum-based Hydrogels

(Kaith et al., 2014) study maximization of the water absorption capacity of the synthesized hydrogel was achieved through sequential experimental design-based optimization. A fractional factorial screening (Resolution-IV) approach was used to screen significant process variables to maximize percentage swelling in phase-1. Studied reaction parameters were: (i) monomer concentration, (ii) initiator concentration, (iii) cross-linker concentration, (iv) polymerization time, (v) reaction temperature, (vi) vacuum level, and (vii) pH of reaction mixture.

Jiang et al. (Jiang et al., 2019) show a thermo-sensitive hydroxybutyl chitosan (HBC) hydrogel was prepared using 1,2-butene oxide as an etherification modifying agent. In addition, response surface methodology (RSM) was applied to optimize its preparation conditions to obtain the maximum yield of HBC.

The swelling capacity of the gels in salt water (formation water) is one of the most crucial criteria in selecting hydrogel types (Heidari et al., n.d.). In this study, an efficient series of gels of acrylamide and acrylic acid, as P(AAm/A.A.) copolymers, were synthesized by a free radical method. A central-composite Design of Experiments (DOE) approach was employed to design the experiments and optimize the copolymers' desired properties. In addition, the mole ratio of AAm/A.A., the mole percentage of the crosslinker agent (N, N'-methylenebisacrylamide, MBA) and swelling time were examined as key factors affecting the swelling behavior of the hydrogels.

CONCLUSION

Nowadays, extensive information is required to prepare a hydrogel-based biopolymer, especially guar gum. Because of the typical characteristics of hydrogel-based biopolymer, it can act as an advantage in certain fields. The trials to design nonconventional methods for smart agriculture have become vital for many farmers in our countries. Artificial intelligence has recently been very important in saving time and energy.

REFERENCES

Abed, M., Haddad, A., Hassen, A., & Sultan, S. (2006). Preparation and evaluation of new hydrogels as new fertilizer delivery system. *Basrah J. Sci.(C)*, *24*(1), 103–114.

Abobatta, W. (2018). Impact of hydrogel polymer in agricultural sector. *Adv. Agric. Environ. Sci. Open Access*, *1*(2), 59–64.

Adam, G. A., Cross, A., & Haward, R. N. (1975). The effect of thermal pretreatment on the mechanical properties of polycarbonate. *Journal of Materials Science*, *10*(9), 1582–1590.

Ahmad, N. N. R., Fernando, W. J. N., & Uzir, M. H. (2015). Parametric evaluation using mechanistic model for release rate of phosphate ions from chitosan-coated phosphorus fertiliser pellets. *Biosystems Engineering*, *129*, 78–86.

Alharbi, K., Ghoneim, A., Ebid, A., El-Hamshary, H., & El-Newehy, M. H. (2018). Controlled release of phosphorous fertilizer bound to carboxymethyl starch-g-polyacrylamide and maintaining a hydration level for the plant. *International Journal of Biological Macromolecules*, *116*, 224–231.

Amiji, M. M. (1995). Permeability and blood compatibility properties of chitosan-poly (ethylene oxide) blend membranes for haemodialysis. *Biomaterials*, *16*(8), 593–599.

Arias, C., López-González, M. M. C., Fernandez-Garcia, M., Barrales-Rienda, J. M., & Madruga, E. L. (1993). Free-radical copolymerization of methyl acrylate with methyl methacrylate in benzene solution. *Polymer*, *34*(8), 1786–1789.

Bellamkonda, R., Ranieri, J. P., Bouche, N., & Aebischer, P. (1995). Hydrogel-based three-dimensional matrix for neural cells. *Journal of Biomedical Materials Research*, *29*(5), 663–671.

Bolto, B. A. (1995). Soluble polymers in water purification. *Progress in Polymer Science*, *20*(6), 987–1041.

Brandl, F., Kastner, F., Gschwind, R. M., Blunk, T., Teßmar, J., & Göpferich, A. (2010). Hydrogel-based drug delivery systems: Comparison of drug diffusivity and release kinetics. *Journal of Controlled Release*, *142*(2), 221–228.

Budianto, E., & Amalia, A. (2020). Swelling behavior and mechanical properties of Chitosan-Poly (N-vinyl-pyrrolidone) hydrogels. *Journal of Polymer Engineering*, *40*(7), 551–560.

Cascone, M. G., Sim, B., & Sandra, D. (1995). Blends of synthetic and natural polymers as drug delivery systems for growth hormone. *Biomaterials*, *16*(7), 569–574.

Chen, J., Park, H., & Park, K. (1999). Synthesis of superporous hydrogels: Hydrogels with fast swelling and superabsorbent properties. *Journal of Biomedical Materials Research: An Official Journal of The Society for Biomaterials, 44*(1), 53-62.]

Chen, J. U. N., & Park, K. (1999). Superporous hydrogels: Fast responsive hydrogel systems. *Journal of Macromolecular Science— Pure and Applied Chemistry*, *36*(7-8), 917–930.

Chirila, T. VConstable, I. JCrawford, G. JVijayasekaran, SThompson, D. EChen, Y. CGriffin, B. J. (1993). Poly (2-hydroxyethyl methacrylate) sponges as implant materials: In vivo and in vitro evaluation of cellular invasion. *Biomaterials*, *14*(1), 26–38.

DeRosa, M. C., Monreal, C., Schnitzer, M., Walsh, R., & Sultan, Y. (2010). Nanotechnology in fertilizers. *Nature Nanotechnology*, *5*(2), 91–91.

Di Silvio, L., Gurav, N., Kayser, M. V., Braden, M., & Downes, S. (1994). Biodegradable microspheres: A new delivery system for growth hormone. *Biomaterials*, *15*(11), 931–936.

Djabourov, M., Maquet, J., Theveneau, H., Leblond, J., & Papon, P. (1985). Kinetics of gelation of aqueous gelatin solutions. *British Polymer Journal*, *17*(2), 169–174.

Duan, J., & Jiang, J. (2017). Structure and properties of hydrophobic aggregation hydrogel with chemical sensitive switch. *International Journal of Polymer Science*, 2017.

Elbarbary, A. M., Abd El-Rehim, H. A., El-Sawy, N. M., Hegazy, E. S. A., & Soliman, E. S. A. (2017). Radiation induced crosslinking of polyacrylamide incorporated low molecular weights natural polymers for possible use in the agricultural applications. *Carbohydrate Polymers*, *176*, 19–28.

Elsaeed, S. M., Zaki, E. G., Ibrahim, T. M., Ibrahim Talha, N., Saad, H. A., Gobouri, A. A., ... Mohamed El-Kousy, S. (2021). Biochar grafted on cmc-terpolymer by green microwave route for sustainable agriculture. *Agriculture*, *11*(4), 350.

Eschbach, F. O., & Huang, S. J. (1994). Hydrophilic-hydrophobic binary systems of poly (2-hydroxyethyl methacrylate) and polycaprolactone. Part I: Synthesis and characterization. *Journal of Bioactive and Compatible Polymers*, *9*(1), 29–54.

Eschbach, F. O., Huang, S. J., & Cameron, J. A. (1994). Hydrophilic-hydrophobic binary systems of poly (2-hydroxyethyl methacrylate) and polycaprolactone. Part II: Degradation. *Journal of Bioactive and Compatible Polymers*, *9*(2), 210–221.

Essawy, H. A., Ghazy, M. B., Abd El-Hai, F., & Mohamed, M. F. (2016). Superabsorbent hydrogels via graft polymerization of acrylic acid from chitosan-cellulose hybrid and their potential in controlled release of soil nutrients. *International Journal of Biological Macromolecules*, *89*, 144–151.

Fang, S., Wang, G., Li, P., Xing, R., Liu, S., Qin, Y., ... Li, K. (2018). Synthesis of chitosan derivative graft acrylic acid superabsorbent polymers and its application as water retaining agent. *International Journal of Biological Macromolecules*, *115*, 754–761.

Feil, H., Bae, Y. H., Feijen, J., & Kim, S. W. (1991). Molecular separation by thermosensitive hydrogel membranes. *Journal of Membrane Science*, *64*(3), 283–294.

Fidelia, N., & Chris, B. (2011). Environmentally friendly superabsorbent polymers for water conservation in agricultural lands. *Journal of Soil Science and Environmental Management*, *2*(7), 206–211.

Fyvie, T. J., Frisch, H. L., Semlyen, J. A., Clarson, S. J., & Mark, J. E. (1987). Polymeric catenanes from crosslinked poly (2, 6-dimethyl-1, 4-phenylene oxide) and cyclic poly (dimethylsilozane). *Journal of Polymer Science. Part A, Polymer Chemistry*, *25*(9), 2503–2509.

Gharekhani, H., Olad, A., Mirmohseni, A., & Bybordi, A. (2017). Superabsorbent hydrogel made of NaAlg-g-poly (AA-co-AAm) and rice husk ash: Synthesis, characterization, and swelling kinetic studies. *Carbohydrate Polymers*, *168*, 1–13.

Ghobashy, M. M., Abd El-Wahab, H., Ismail, M. A., Naser, A. M., Abdelhai, F., El-Damhougy, B. K., ... Alkhursani, S. A. (2020). Characterization of Starch-based three components of gamma-ray cross-linked hydrogels to be used as a soil conditioner. *Materials Science and Engineering B*, *260*, 114645.

Heidari, S., Esmaeilzadeh, F., Mowla, D., Jokar, H., Cortés, F. B., Nassar, N. N., & Franco Ariza, C. A. (n.d.). *Acrylamide-Acrylic Acid as Copolymer Gel for Water Shut-Off at High Pressure and Temperature Conditions*. Available at SSRN 4127813]

Hietala, S., Holmberg, S., Karjalainen, M., Näsman, J., Paronen, M., Serimaa, R., ... Vahvaselkä, S. (1997). Structural investigation of radiation grafted and sulfonated poly (vinylidene fluoride), PVDF, membranes. *Journal of Materials Chemistry*, *7*(5), 721–726.

Hietala, S., Maunu, S. L., & Sundholm, F. (1999). Structure of styrene grafted poly (vinylidene fluoride) membranes investigated by solid-state NMR. *Macromolecules*, *32*(3), 788–791.

Homma, M., Seida, Y., & Nakano, Y. (2000). Evaluation of optimum condition for designing high-performance electro-driven polymer hydrogel systems. *Journal of Applied Polymer Science*, *75*(1), 111–118.

Hooper, H. H., Baker, J. P., Blanch, H. W., & Prausnitz, J. M. (1990). Swelling equilibria for positively ionized polyacrylamide hydrogels. *Macromolecules*, *23*(4), 1096–1104.

Hu, Y., Horie, K., Ushiki, H., Yamashita, T., & Tsunomori, F. (1993). Fluorescence studies of volume phase transition in polyacrylamide gels with a pyrenyl probe in acetone/water mixed solvent. *Macromolecules*, *26*(7), 1761–1766.

Hu, Z., & Xia, X. (2004). Hydrogel nanoparticle dispersions with inverse thermoreversible gelation. *Advanced Materials*, *16*(4), 305–309.

Inomata, H., Goto, S., Otake, K., & Saito, S. (1992). Effect of additives on phase transition of N-isopropylacrylamide gels. *Langmuir*, *8*(2), 687–690.

Jiang, C., Sun, G., Zhou, Z., Bao, Z., Lang, X., Pang, J., ... Chen, X. (2019). Optimization of the preparation conditions of thermo-sensitive chitosan hydrogel in heterogeneous reaction using response surface methodology. *International Journal of Biological Macromolecules*, *121*, 293–300.

Jyothi, A. N., Pillai, S. S., Aravind, M., Salim, S. A., & Kuzhivilayil, S. J. (2018). Cassava starch-graft-poly (acrylonitrile)-coated urea fertilizer with sustained release and water retention properties. *Advances in Polymer Technology*, *37*(7), 2687–2694.

Kademani, B. S., Kumar, V., Sagar, A., & Kumar, A. (2006). World literature on thorium research: A scientometric study based on Science Citation Index. *Scientometrics*, *69*(2), 347–364.

Kaetsu, I., Uchida, K., Morita, Y., & Okubo, M. (1992). Synthesis of electro-responsive hydrogels by radiation polymerization of sodium acrylate. *International Journal of Radiation Applications and Instrumentation Part C Radiation Physics and Chemistry*, *40*(2), 157–160.

Kaith, B. S., Sharma, R., Kalia, S., & Bhatti, M. S. (2014). Response surface methodology and optimized synthesis of guar gum-based hydrogels with enhanced swelling capacity. *RSC Advances*, *4*(76), 40339–40344.

Kaur, H., & Chatterji, P. R. (1990). Interpenetrating hydrogel networks. 2. Swelling and mechanical properties of the (gelatin-polyacrylamide) interpenetrating networks. *Macromolecules*, *23*(22), 4868–4871.

Kim, J. J., & Park, K. (1998). Smart hydrogels for bioseparation. *Bioseparation*, *7*(4), 177–184.

Lee, J. M., Noh, G. Y., Kim, B. G., Yoo, Y., Choi, W. J., Kim, D. G., ... Kim, Y. S. (2019). Synthesis of poly (phenylene polysulfide) networks from elemental sulfur and p-diiodobenzene for stretchable, healable, and reprocessable infrared optical applications. *ACS Macro Letters*, *8*(8), 912–916.

Lee, S. J., & Park, K. (1996, January). Glucose-sensitive phase-reversible hydrogels. In *ACS Symposium Series* (Vol. 627, pp. 11-16). Washington, DC: American Chemical Society.

Lin, X., Guo, L., Shaghaleh, H., Hamoud, Y. A., Xu, X., & Liu, H. (2021). A TEMPO-oxidized cellulose nanofibers/MOFs hydrogel with temperature and pH responsiveness for fertilizers slow-release. *International Journal of Biological Macromolecules, 191*, 483–491.

Liu, Y., Huglin, M. B., & Davis, T. P. (1994). Preparation and characterization of some linear copolymers as precursors to thermoplastic hydrogels. *European Polymer Journal, 30*(4), 457–463.

Ma, J. T., Liu, L. R., Yang, X. J., & De Yao, K. (1995). Bending behavior of gelatin/poly (hydroxyethyl methacrylate) IPN hydrogel under electric stimulus. *Journal of Applied Polymer Science, 56*(1), 73–77.

Maitra, J., & Shukla, V. K. (2014). Cross-linking in hydrogels-a review. *American Journal of Political Science, 4*(2), 25–31.

Marchetti, M., Prager, S., & Cussler, E. L. (1990). Thermodynamic predictions of volume changes in temperature-sensitive gels. 2. Experiments. *Macromolecules, 23*(14), 3445–3450.

Matthew, I. R., Browne, R. M., Frame, J. W., & Millar, B. G. (1995). Subperiosteal behaviour of alginate and cellulose wound dressing materials. *Biomaterials, 16*(4), 275–278.

Miller, J. S. (2000). Organometallic-and organic-based magnets: New chemistry and new materials for the new millennium. *Inorganic Chemistry, 39*(20), 4392-4408.

Miyata, T., Asami, N., & Uragami, T. (1999). A reversibly antigen-responsive hydrogel. *Nature, 399*(6738), 766–769.

Narayanan, A., Kartik, R., Sangeetha, E., & Dhamodharan, R. (2018). Super water absorbing polymeric gel from chitosan, citric acid and urea: Synthesis and mechanism of water absorption. *Carbohydrate Polymers, 191*, 152–160.

Obaidat, A. A., & Park, K. (1996). Characterization of glucose dependent gel-sol phase transition of the polymeric glucose-concanavalin A hydrogel system. *Pharmaceutical Research, 13*(7), 989–995.

Obaidat, A. A., & Park, K. (1997). Characterization of protein release through glucose-sensitive hydrogel membranes. *Biomaterials, 18*(11), 801–806.

Oxley, H. R., Corkhill, P. H., Fitton, J. H., & Tighe, B. J. (1993). Macroporous hydrogels for biomedical applications: Methodology and morphology. *Biomaterials, 14*(14), 1064–1072.

Park, H., & Park, K. (1994). Honey, I blew up the hydrogels! *Proceedings of the International Symposium on Controlled Release of Bioactive Materials, 21*, 21–22.

Park, H., & Park, K. (1996). *Hydrogels in bioapplications.* ACS Publications.

Park, K., & Robinson, J. R. (1984). Bioadhesive polymers as platforms for oral-controlled drug delivery: Method to study bioadhesion. *International Journal of Pharmaceutics, 19*(2), 107–127.

Park, T. G., Cohen, S., & Langer, R. (1992). Poly (L-lactic acid)/pluronic blends: Characterization of phase separation behavior, degradation, and morphology and use as protein-releasing matrixes. *Macromolecules, 25*(1), 116–122.

Peppas, N. A. (2010). *Biomedical applications of hydrogels handbook*. Springer Science & Business Media.

Pérez-Luna, V. H. (2020). The Effect of Glutathione Incorporated as Chain Transfer Agent in Thermosensitive Hydrogels for Controlled Release of Therapeutic Proteins. In *Controlled Drug Delivery Systems* (pp. 53–75). CRC Press.

Raafat, A. I., Eid, M., & El-Arnaouty, M. B. (2012). Radiation synthesis of superabsorbent CMC based hydrogels for agriculture applications. *Nuclear Instruments & Methods in Physics Research. Section B, Beam Interactions with Materials and Atoms*, *283*, 71–76.

Ramanan, R. M. K., Chellamuthu, P., Tang, L., & Nguyen, K. T. (2006). Development of a temperature-sensitive composite hydrogel for drug delivery applications. *Biotechnology Progress*, *22*(1), 118–125.

Rathna, G. V. N., Rao, D. M., & Chatterji, P. R. (1994). Water-induced plasticization of solution cross-linked hydrogel networks: Energetics and mechanism. *Macromolecules*, *27*(26), 7920–7922.

Salmawi, K. M. E., El-Naggar, A. A., & Ibrahim, S. M. (2018). Gamma irradiation synthesis of carboxymethyl cellulose/acrylic acid/clay superabsorbent hydrogel. *Advances in Polymer Technology*, *37*(2), 515–512.

Sawahata, K., Hara, M., Yasunaga, H., & Osada, Y. (1990). Electrically controlled drug delivery system using polyelectrolyte gels. *Journal of Controlled Release*, *14*(3), 253–262.

Sawhney, A. S., Pathak, C. P., & Hubbell, J. A. (1993). Bioerodible hydrogels based on photopolymerized poly (ethylene glycol)-co-poly (. alpha.-hydroxy acid) diacrylate macromers. *Macromolecules*, *26*(4), 581–587.

Sawhney, A. S., Pathak, C. P., van Rensburg, J. J., Dunn, R. C., & Hubbell, J. A. (1994). Optimization of photopolymerized bioerodible hydrogel properties for adhesion prevention. *Journal of Biomedical Materials Research*, *28*(7), 831–838.

Seida, Y., & Nakano, Y. (1995). Concept to control the phase behavior of stimuli-sensitive polymer gel. *Journal of Chemical Engineering of Japan*, *28*(4), 425–428.

Seida, Y., & Nakano, Y. (1996). Effect of salt on the property of adsorption in thermosensitive polymer hydrogel. *Journal of Chemical Engineering of Japan*, *29*(5), 767–772.

Sen, N., Shaikh, T., Singh, K. K., Sirsam, R., & Shenoy, K. T. (2020). Synthesis of polyacrylamide (PAM) beads in microreactors. *Chemical Engineering and Processing-Process Intensification*, *157*, 108105.

Song, B., Liang, H., Sun, R., Peng, P., Jiang, Y., & She, D. (2020). Hydrogel synthesis based on lignin/sodium alginate and application in agriculture. *International Journal of Biological Macromolecules*, *144*, 219–230.

Sundholm, F., Serimaa, R., & Sundholm, G. (1996). Polyelectrolytes and Electrochemically Active Membranes: Synthesis, Characterisation and Applications. *National Programme on Materials and Structure Research, 267*.

Svanberg, C., Adebahr, J., Ericson, H., Börjesson, L., Torell, L. M., & Scrosati, B. (1999). Diffusive and segmental dynamics in polymer gel electrolytes. *The Journal of Chemical Physics*, *111*(24), 11216–11221.

Thakur, S., Sharma, B., Verma, A., Chaudhary, J., Tamulevicius, S., & Thakur, V. K. (2018). Recent progress in sodium alginate based sustainable hydrogels for environmental applications. *Journal of Cleaner Production*, *198*, 143–159.

Thombare, N., Mishra, S., Siddiqui, M. Z., Jha, U., Singh, D., & Mahajan, G. R. (2018). Design and development of guar gum based novel, superabsorbent and moisture retaining hydrogels for agricultural applications. *Carbohydrate Polymers*, *185*, 169–178.

Tsujii, K., Hayakawa, M., Onda, T., & Tanaka, T. (1997). A novel hybrid material of polymer gels and bilayer membranes. *Macromolecules*, *30*(24), 7397–7402.

Vashist, A., Vashist, A., Gupta, Y. K., & Ahmad, S. (2014). Recent advances in hydrogel based drug delivery systems for the human body. *Journal of Materials Chemistry. B, Materials for Biology and Medicine*, *2*(2), 147–166.

Wang, L., Cavaco-Paulo, A., Xu, B., & Martins, M. (2019). Polymeric hydrogel coating for modulating the shape of keratin fiber. *Frontiers in Chemistry*, *7*, 749.

Wen, X. x., Zhang, D. q., Liao, Y. c., Jia, Z. k., & Ji, S. q. (2012). Effects of water-collecting and -retaining techniques on photosynthetic rates, yield, and water use efficiency of millet grown in a semiarid region. *Journal of Integrative Agriculture*, *11*, 1119–1128.

Xiao, X., Yu, L., Xie, F., Bao, X., Liu, H., Ji, Z., & Chen, L. (2017). One-step method to prepare starch-based superabsorbent polymer for slow release of fertilizer. *Chemical Engineering Journal*, *309*, 607–616.

Yeo, J. K., Sperling, L. H., & Thomas, D. A. (1981). Poly (n-butyl acrylate)/polystyrene interpenetrating polymer networks and related materials. III. Effect of grafting level and molecular weight in semi-2 IPNs. *Journal of Applied Polymer Science*, *26*(10), 3283–3294.

Yoshida, R., Uchida, K., Kaneko, Y., Sakai, K., Kikuchi, A., Sakurai, Y., & Okano, T. (1995). Comb-type grafted hydrogels with rapid deswelling response to temperature changes. *Nature*, *374*(6519), 240–242.

Zhao, H., Sterner, E. S., Coughlin, E. B., & Theato, P. (2012). o-Nitrobenzyl alcohol derivatives: Opportunities in polymer and materials science. *Macromolecules*, *45*(4), 1723–1736.

Chapter 4
Smart Agriculture in Orchard Farms

Mohamed Abul-Soud
Central Laboratory of Agricultural Climate (CLAC), Agriculture Research Center, Agriculture Ministry, Egypt

Mohamed Saad Aly Emam
https://orcid.org/0000-0002-6232-0662
Central Laboratory of Agricultural Climate (CLAC), Agriculture Research Center, Agriculture Ministry, Egypt

Ahmed Awny Farag
Central Laboratory of Agricultural Climate (CLAC), Agriculture Research Center, Agriculture Ministry, Egypt

ABSTRACT

Smart agriculture motivates more farmers every day via different applications with a view of maximizing food production as well as minimizing the climate change impacts, COVID-19 crisis, and natural resources shortages. Developing the agricultural production process based on smart agriculture and IoT technology has become extremely vital to preserve natural resources, increase sustainability and ecological agricultural food production, and satisfy food security demands. Smart agriculture is mainly based on sensing and monitoring micro-climate and environmental data (collection) and analysis. Data collected from different sources (internet, remote sensing database, and sensors) can provide a wealth of information for agriculture operators, allowing them to take appropriate actions in a timely manner. Developing microcontrollers, sensors, and actuators integrated with IoT to include all agricultural procedures especially in developed countries become more necessary.

SMART AGRICULTURE

Efficiency, sustainability and profitability are the main targets of smart agriculture (also known as e-agriculture, and also precision agriculture).

DOI: 10.4018/978-1-6684-2423-0.ch004

Smart agriculture is one of the means of mitigation and adaption to climate changes impacts, as it works to save energy, irrigation water, fertilizers and chemical pesticides. Smart agriculture strongly supports sustainable agriculture to achieve the highest profitable productivity while reducing environmental, social and economic impacts. Smart agriculture relies on advanced technology in growing food in sustainable and environmental behavoir, rationalizing the use of natural resources, especially water. The most prominent feature is its reliance on information management and analysis systems to make the best possible production decisions, at the lowest costs, as well as the automation of agricultural processes such as irrigation, pest control, and soil control, and crop control (USAID 2016 and FAO 2021).

Smart agriculture motivates more farmers every day via different applications with a view of maximizing the food production as well as minimizing the ecological footprint under the climate change impacts, Coved 19 crisis and natural resources shortages. Smart hydroponic greenhouse developed rapidly during the last decade depending on the enormous development of communication technology, microcontrollers, sensors and actuators as well as simplifying the programming languages.

The need to access the real-record data of micro-climate and agricultural environmental conditions play a vital roles in make many smart decisions and affect directly on agricultural production. Needless to mention the huge advantages on different scales for detecting soil moisture and determine the evapotranspiration (ET_0) for presenting Smart irrigation. Develop smart agriculture weather station and soil moisture sensor will offer the potential for providing an alarming system for extreme weather events as well as smart control in irrigation.

Smart farms have a real potential to deliver more productive and sustainable agricultural production based on a more resource-efficient approach.

The most interesting smart agriculture application could be summarized regarding Kanumuri (2020) as follows:

1. Smart agriculture weather station based on IoT,
2. Design, control, and management of irrigation system and fertigation program
3. Smart management of different agricultural systems
4. Monitoring the growth stages of crops and animals,
5. The applications of drones and robots in monitoring, control, alarm, and process,
6. Integrated pest management,
7. Accomplish arduous agricultural work through automated mechanization,
8. Smart harvesting process management,
9. Smart Market chain management.

The rapid progress of communication, IoT, android mobile applications, developing electronic boards, and automation technology drive a new revolution in the agriculture sector to perform a new technology of smart agriculture. Many researchers (TongKe (2013), Agrawal & Kamboj (2019), Madushanki *et al.,* (2019), Kanumuri (2020), FAO (2021)) had investigated the role of smart applications in precise agriculture that presented in 4 axes as follows:

First: Information and data

1. Collecting data and information (soil, crop, climate, inputs, natural resources, market, demographic, socio-economic and etc.).
2. Analyzing data and information for decision making.

3. Information share via establish database to feed platform, program and android applications for introduce the different services to the user (researcher, agents, farmers, traders and etc.).

Second: precise the agriculture process and practices

1. Forecasting the climate and market conditions.
2. Offer precise response time for meeting the extreme natural and weather events, global epidemics and struggling needs.
3. Technology transfer and knowledge sharing.
4. Optimize the natural resources management (soil, water, crop, climate, energy, labor, etc.).
5. Better inputs management (irrigation, fertilization, crops, animals mechanization, integrated pest management and etc.).
6. Increase the agriculture production and decrease the production waste.
7. Minimize the abiotic and biotic risks on the agriculture production.
8. Increase the agriculture sustainability under global and regional extreme events.
9. Support the smart automation processing of agriculture procedures.
10. Optimize the marketing exposure.
11. Improve networking and communications among the different stakeholders the rural producers, governments, market, suppliers.
12. Enhance the harvest and post-harvest process and transportation.

Third: Assist the decision makers

Actually, the close action of information and communication technology (ICT) in smart agriculture is assist the decision makers (farmers, governance, regulators, financial sector and etc.) to take the right decision on right real-time through support the decision makers by efficient and reliable data and information while the far dream is taking decision without referring to the human. ICT assist in many issue such as:

1. Land and natural resources management.
2. Development strategy plans and monitoring progress tools for poor rural.
3. Governances and regulations for the rural regions.
4. Contingency plan for meeting extreme events and risks.
5. Plans on national, regional and international scales for fighting hungry and poverty as well as disaster prevention.
6. Helping the economic sector and commercial transactions between countries to grow and prosper.
7. The financial and insurances services for rural communities.
8. Bridge among different stakeholders.
9. Improve the capacity building and empowerment management.

Fourth: Protect life

Through increasing the food production, improve the natural resource management, reduce the risks and disasters on agriculture production and take in consider the sustainability, environmental and socio-economic in consider, Life still in risk especially human life from human himself. Protect human life from human is the fourth hidden role of smart and intellegent different systems.

Several policies are recommended to facilitate application of smart and intellegent systems in Agriculture and e-commerce as follows:

1. Investing in human (training and motivating) to create ICT innovations in agriculture.
2. Investing in information and communication infrastructure.
3. Investing in farmers' capacity building through practical training on e-agriculture applications to increase the public awareness.
4. Providing financial and credit support to farmers.
5. Investing in storage and transportation to support agricultural e-commerce.
6. Facilitating farmer cooperation.
7. Improving market regulations to provide a favorable market environment.
8. Promoting more inclusive e-commerce development in rural areas.

Orchard trees

In orchards, that known as permanent crops include different fruit tree types differ strongly in their environmental and climatic needs as follows:

1. Tropical and semitropical fruits regions include mangoes, bananas, pineapples, coconut palms, citruses, and also tea, coffee and coca.
2. Subtropical fruits regions: include date palms - figs - pomegranate - some citrus varieties.
3. Warm temperate fruits regions includes grapes - olives, some varieties of apricots - some varieties of peaches - almonds and nuts, some varieties of citrus.
4. Cold temperate fruits regions includes apples, pears, cherries, some plum varieties and some peach varieties.

Moreover, the orchard trees classified regarding to the the nature of vegetative growth:

1 - Evergreens

Fruit trees that retain their leaves and do not appear bare of leaves at any time of the year. The leaves on the trees have a certain age "from 8 months to 4 years" depending on the tree type and environmental conditions. The trees get rid of part of the leaves during periods known as cycles Growth where leaves that have ended their life and fall off in the period between two activity growth cycles. The trees apear light green color (Fresh leaves) in spring, along with aged leaves of dark green color, and these fruits are: all kinds of citrus except for three-leafed oranges - bananas - mangoes - date palms - guavas - olives - drunk - pineapple - papaya - butternut – figs and cactus.

2- Deciduous

Fruit trees characterized as having a clear resting role that appear as a bare of leaves. The leaves droping begins in the late fall and continues throughout the winter until the trees emerge in early spring after gained its chilling requirments during the winter. If the winter cold is not enough to break the rest period, the emergence of new growth in the spring is much later than usual, and the flowering date may be delayed and the crop is affected in most cases. Deciduous trees include grapes - apples - pears - quince

- peaches - apricots - plums - almonds - walnuts - pecans - persimmons - figs - pomegranate. Cream trees considered semi-deciduous, because part of the leaves fall in the spring and not in the winter, because the buds are located under the necks of the leaves and not in the axillae, and the leaves must fall so that the buds are exposed and the leaves do not fall once.

Also, there is anothor classification of orcgards depending on the production sector takes the economic value in its consideration includes the following:

1 - Citrus citriculture: It includes the study of all types of citrus fruits.
2- Grape Viticulture: It includes the study of different types of grapes.
3- Palm Phoeniculture: It includes the study of date palms
4- Olive Oleiculture: It includes the study of olives
5- Nutculture: It includes the study of almonds, walnuts and hazelnuts
6- Sweets Pomology: It includes the study of apples (apples - pears - quince) and sweets (peach, apricot, plum, almond, walnut and cherry).

The previous classifications show the great diversity in orchards that enjoys by great economic value and long investment period while suffering from climate change and environmental impacts as well as diseases and insects damages. Needless to mentioned, the yield loose because of wrong agriculture practices (irrigation, fertilization, bruning, IPM, harvesting and post harvesting process, chain market and etc… that create the driving forces to apply the smart agriculture in orchards production.

Objectives of smart agriculture in orchards

The wider objectives could be summarized in the following points:

1. Establishing a smart electronic extension about orchard farms and the use of intelligent in agriculture on different scale of commercial farms.
2. Transferring and localizing the knowledge of intelligent and smart automation of modern orchard farms to commercial sector.
3. The economic objectives are increasing the income through increase the quantity and quality of product with cost reduction.
4. The environmental objectives are including increase water, fertilizers, pesticides and energy use efficiencies with the ultimate goal of conserving the natural resources.
5. The health objective is to get rid of the biohazards of using a lot of Agricultural chemicals and avoid the pollution to the ground water.
6. The development objectives are focusing on increasing the cultivated area, production yield for water and area unit and localize the modern technology.
7. The agricultural objectives are mainly to extend the cultivation area via convert non-reclaimed lands to high production area, support sustainable agriculture and increase the yield production.

1 – Agricultural and food security outcomes

a. Provide the state of art of smart climate agriculture.
b. Increase the water and fertilizers use efficiencies.
c. Increase the total yield by increasing the intensive plant density and the cultivated area as well as minimizing the climate and environmental risks.

2 - Knowledge and development outcomes

a. Develop and localize the knowledge of smart, IoT and automation systems in orchard farms.
b. Transfer the modern knowledge to the interested NGO's and public.
c. Develop the land reclamation and increase the cultivated area.
d. Develop the use of mega electronics in agricultural purposes.

3 – Economic outcomes

a. Increase the national income.
b. Increase the net return by increasing the yield via reduce the cost of orchard production by reduce operating cost.
c. Increase the yield quality of orchard farms plus the semi-control of harvesting time and period.

4 – Environmental outcomes

a. Improve irrigation and fertigation management and reduce the pollution.
b. Save the power use.
c. Reduce the use of agricultural chemicals (pesticides and fertilizers).

5 – Health outcomes

a. Minimize the use of agricultural chemicals.
b. Produce safe food with fewer chemicals.

6 – Social outcomes

a. Offer more opportunity for youth to implement new technology in agriculture.
b. Increase the income of farmers that led to enhance the life style.

Smart agriculture applications

The smart agriculture applications varied regarding to the agriculture systems, in all agriculture systems, practices, operations and process, the smart agriculture contribute strongly. The next figure (Fig. (1)) illustrates the different application catogries of smart agriculture in general. Each catogry has many applications targeted to increase the profit yield, optimize the natural resources use efficiency, minimize the climate change impacts and conserve the environmental.

Figure 1. The application catogries of smart agriculture

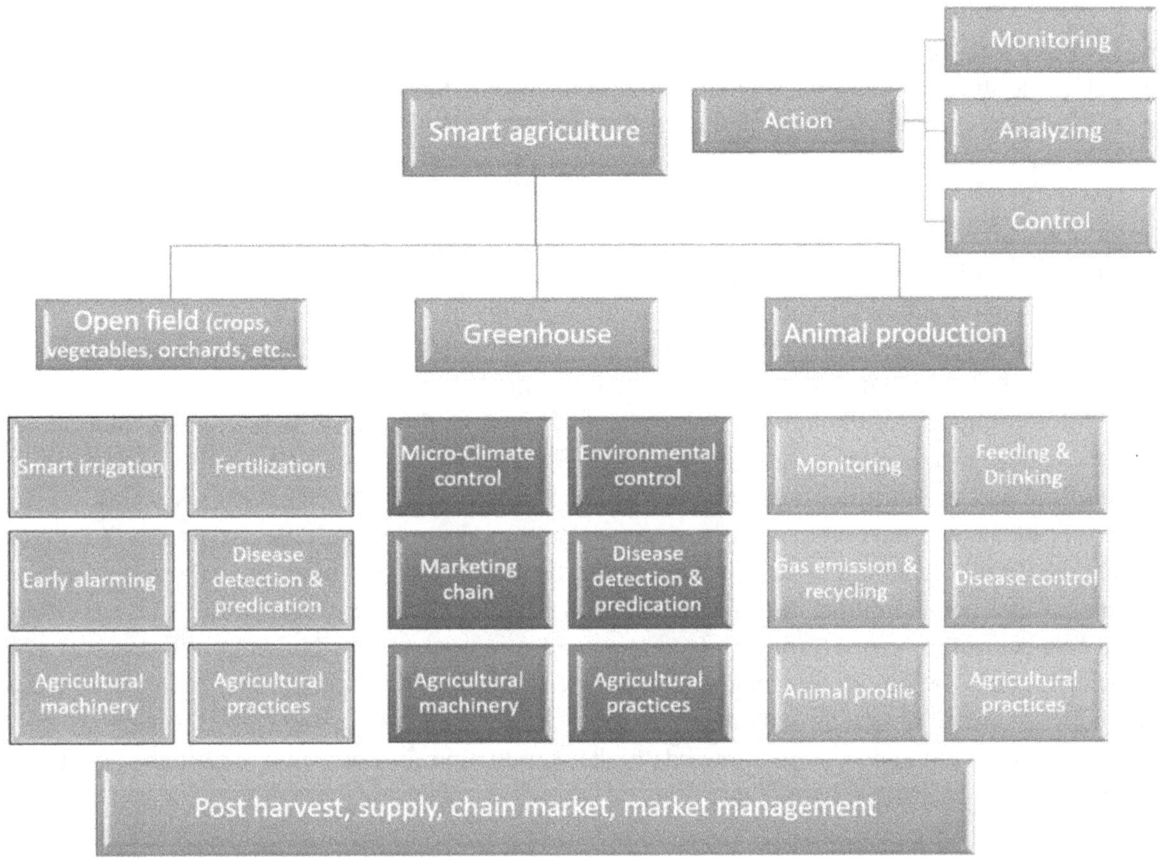

The catogry of Early alarming under open field agriculture system introduced many applications through forecasting the climatic data or detecting a real-record data.

This catogry includes many applications concern about physical factors such as extreme weather events (heat waves, storms, rainfall, frozing and etc..), irrigation statuas and calculate chilling requirments of different orchards types. While biological factors include predication and detection of diseases and insect for enhencing the IPM and reduce the biological risks.

The most important applications of smart agriculture in orchard farms could be summerized as the following agriculture practices:

- Control soil moisture via modern irrigation techniques
- Fertilization regarding the nutrional tree state (type, age, season and etc.,)
- Predication and detection climate and environmental data.
- Calculating chilling requirments
- Early alarming system
- Soil management
- Canopy volume
- Weed control

- Pest traps
- Fruits count
- Fruits size
- Disease and pest control via integrated pest management (IPM)
- Management and control agricultural machinaries
- Harvesting time and process.

Smart agriculture techologies

Smart agriculture is acting mainly in prediction, detection (monitoring), data anlysis ad act smartly (control and automation) through different technology as Fig. (2) presented.

Modern technologies play a vital role in helping to meet the growing food needs of the world's population, through the use of data management and analysis systems, and remote control techniques, in addition to the use of the most prominent technologies of the Fourth Industrial Revolution such as artificial intelligence, robotics and the Internet of Things, in order to make agriculture more productive and profitable, Less harmful to the environment and less consumption of earth's resources.

FAO (2021) reported that to achieve smart agriculture, many technologies have been developed and improved. These technolgies don't present their adventages just for smart agriculture but includes all aspects of life (home, mobile, car, hospital, learning, industral, trading, agriculture and etc., Managing farms and controlling crops through information and communication technology ("Internet of Things" (IoT) technology by linking sensors, remote control systems, and autonomous machines to the Internet, with the aim of obtaining accurate data, and investing this data in precisely directing agriculture). Automated operation and control) towards higher production at lower cost, and higher quality crops (Madushanki *et al.*, 2019).

Figure 2. The used technologies in smart agriculture

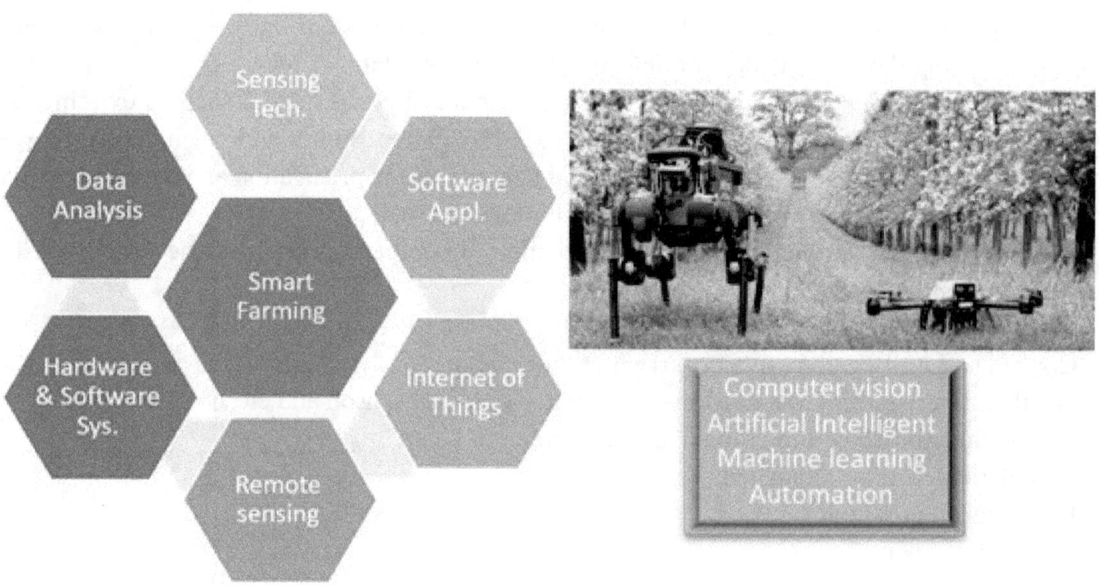

Remote Sensing

*R*emote sensing roles in precision agriculture simply observes the real-time readings of agriculture activity conditions (soil, crop or animal, climate, agriculture practices and etc.) without physically touching from the higher point (few meters to hundreds kilometers). Remote sensing basically detects interactions among reflected, absorbed, and transmitted electromagnetic energy via using satellites, aircraft, and drones. The relationship among energy types is used to determine spectral signatures of individual plants. Each plant has a unique spectral signatures according to the differences in leaf colors, textures, shapes, position and the nature of vegetative growth as well as soil moisture that determine how much energy will be reflected, absorbed or transmitted. Drones drives the developing of remote sensing in many countries regarding easy accesses and operate, low cost, multiple options and high efficiency.

Figure 3. The remote sensing basic in monitoring orchard trees (Redmond 2017 source)

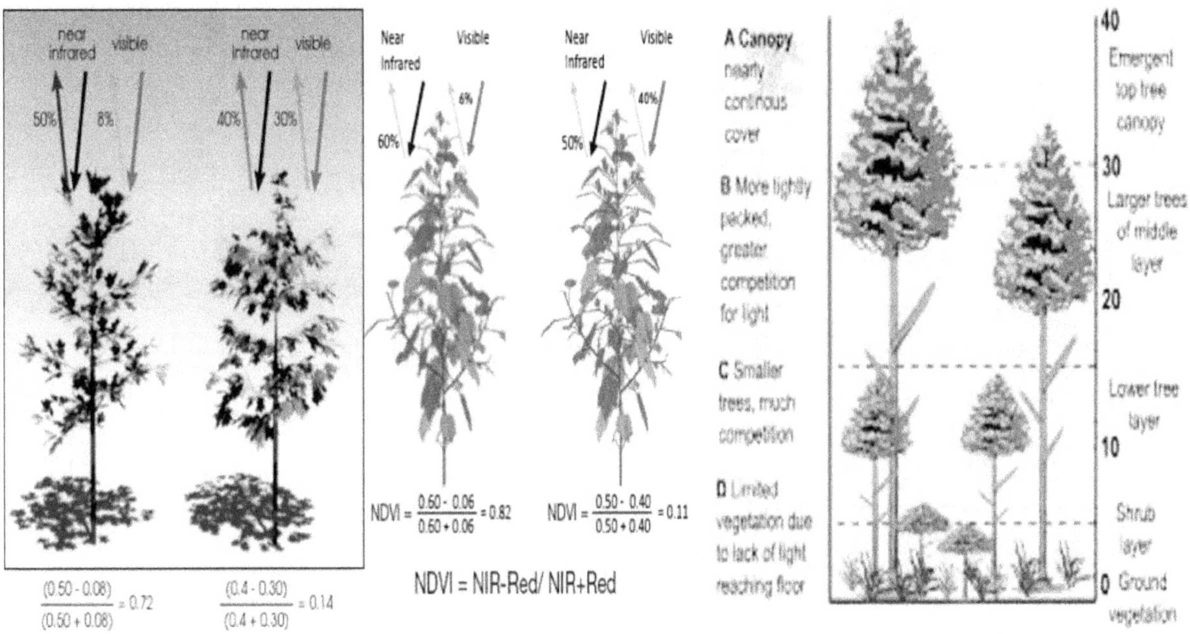

Remote sensing presented various applications in agriculture as **Anindya (2016), Raman *et al.*, 2018 and Shanmugapriya *et al.*, 2019** mentioned, examples of different agriculture applications could be developed by remote sensing as follows:

1. Identify vegetation area,
2. Agricultural land use monitoring,
3. Detect and quantify changes at the Earth's surface,
4. Crop yield forecasting,
5. Monitor and forecasting climate conditions,
6. Monitor pastures and animal herds
7. Monitor of forest fires,

8. Detect water and nutrient deficiencies,
9. Diseases, insect damage
10. Deficiency or surplus,
11. Weed infestations,
12. Hail, Wind, Herbicide, etc damage,
13. Plant populations.

*R*emote sensing information illustrated mainly in digital maps and images that help the decision makers to apply the accurate decision on time in different scales from farm up to national scale. *R*elevant technology has emerged, including numerous sensors, temporal and spectral capacities (e.g. Sentinels, Gaofen), Nano-satellites or UAV, drones and the deployment of cloud computing and machine learning techniques. These technological improvements should achieve the remote sensing applications in agriculture (Aman and Kumar 2020, Anindya 2016 and Kanumuri 2020).

Moreover, remote sensing contributes strongly in smart automated agriculture mechanization through different integrated technologies and systems including auto-steering system, high precision positioning system, variable rate technology, geomapping as well as electronic communication.

Figure 4. Typical components of remote sensing for Precision Agriculture

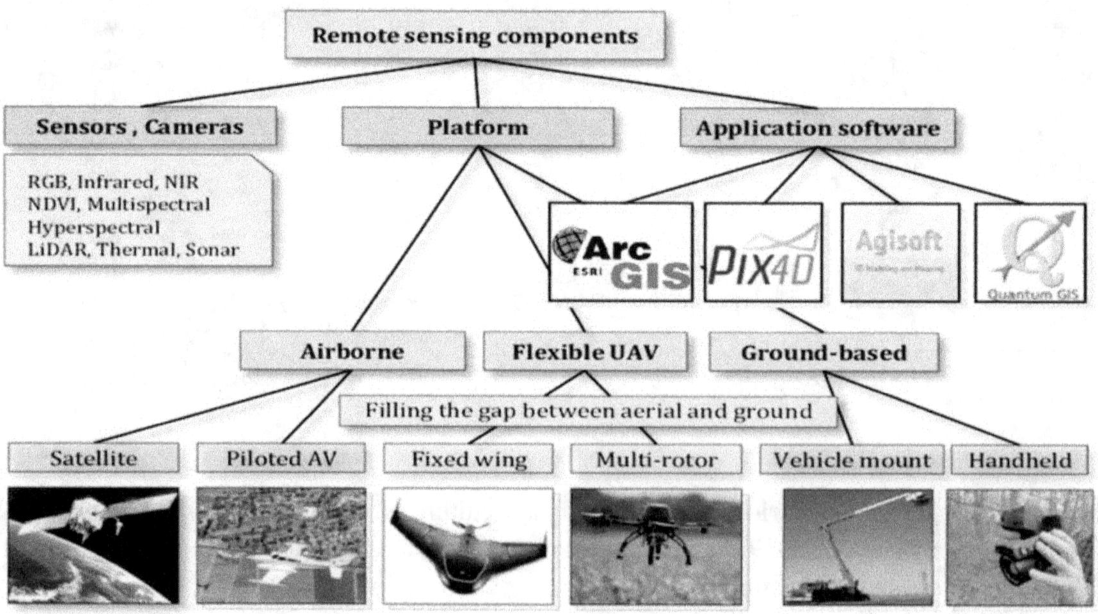

Developed Electronic boards and Sensors

The reapid development in sensor manufactory to cover the industrial sector needs for more efficient and economic factors was also the key to expend the smart agriculture applications. Low cost, accurate, efficient and durability were the most adventages of the modern sensors that match all decting and sensing requirments (climate, micro-climate and environmental conditions).

Moreover, a new generation od different sensors become available for not just to the industrial sector but also for public, students and makers communities.

The theory of the sensors' work is based on converting the required measure value into an electrical value through the transducer located inside the sensor. Simply, the sensor read the temperature for example as an electerical value through different materials that affected by increasing and decreasing temperature as an increasing or decreasing the electerical conductivity or resistance regarding the used materials that could be presented in electerical value which translated to temperature value. Analoge and digital are the two types of sensors regarding to sensor circuit and measurment type.

The costs of electronic components decreased dramatically while their availability, durability, accuracy and variety increased last few years beside the friendly application and technology transfer that courage the develop and implement of this technology and its contribution in agricultural production.

Monitoring of micro-climate and the environmental conditions plus smart automation control integrated with internet of things (IoT) gave the farmer the power and flexibility to make the right decision, to avoid the human mistakes and environmental stress impacts on food production. All micro-climate records (Air temperature, relative humidity, light intensity, carbon dioxide (CO_2), wind speed and etc..) and environmental measurements of the root zone (soil moisture, substrate and nutrient solution temperature, EC, pH, dissolved oxygen index (DOI), water level and etc...) should be sensing by different sensors to monitored for automation the soilless culture and aquaponic systems via suitable actuators. Microcontroller boards (Arduino (mini, nano, uno, Mega and etc..), microprocessor (Raspberry pi in different types and version mainly used for AI and computer vision projects). IoT allows for machine to machine interaction and controlling the soilless culture and aquaponic systems autonomously and intelligently employing deep neural networks. (Karamanis et al., (2018), Caroline et al., (2018), Mehra et al., (2018), Alipio et al., (2019), Che et al., (2019), Nianpu et al., (2019), Lizbeth (2020) and Siddiq et al., (2020)).

On the other way, the low cost of open source micro-controller (Arduino) and micro-proccessor (Rasspberry Pi) and their simillars presented a great support to make this techology available and accessable for every one. Programming these developed electronic boards become more easier and their different programming sketches are available as an open source for the public in many web sites and social media. No any more, needing for speceific smart control unit or high cost weather station, young makers and electronics enthusiasts could be programming, design, manufactory and develop low cost smart units to match the agriculture sector needs.

Table 1. Some different sensors and actuators materials used in smart system

Sensor	Job	Image
DHT 22 sensor	Sensening air temperature (°C and °F) and relative humidity (%) 5 volt power sensor	
luminsoity sensor TSL 2561	Light intensity (Lux) 5 volt power sensor	
TDS sensor	Sensing Total Dissolved Solids (TDS) measure range from 0 to 1000 ppm. 5 volt power sensor (Need to calibration).	
PH sensor kit	Water Acidity Measrments The pH scale ranges from 0 to 14. 5 volt power sensor (Need to calibration).	
Waterproof temperature Sensor	Water and nutrient solution temperature (°C and °F). 5 volt power sensor	
Float switch sensor	Water level sensing 5 volt power sensor	
Soil moisture sensors	Detecting soil moisture Capacitive and resistance value sensor 5 volt power sensor	
LCD screen (4*16)	Display the required inormation and data. 5 volt power screen	
Relay module	Opertated On/Off from 5 V provided by Arduino to 220 V for actuating different controls. 5 volt power actuator	

Internet of Things (IoT)

IoT works mainly as a mean of understanding and interconnection among different devices, sensors and applications connected to the Internet by a set of communication protocols to provide two ways bridge to collect, process, validate and analyzed different agriculture data and information to transfer to the user (Researcher, market, programmer or farmer) in GUI or transfer the information or order from the user to control system. A **GUI** (graphical user interface) is a system of interactive visual components for computer software. A GUI displays objects that convey information, and represent actions that can be taken by the user. The collection of required agriculture data and information is obtained through different techniques as follows:

1. Remote sensing
2. Electronic sensors in the field
3. Forecasting programs
4. Mathematics model programs
5. Internet (interactive platform, webs, e-commerce, etc.)

By integrate artificial intelligence techniques, the data is processing, analytical, reformulated and converted into applicable agricultural information or into intelligent automatic control in agricultural operations. Furthermore, the adoption of the microcontroller and IOT in intelligent technique to sensing the different micro-climate and environmental conditions of agriculture production could minimize the biotic and abiotic risks and human mistakes and increase the profit yield due to complicated manually monitoring and controlling process (Apolo-Apolo *et al.,* 2020, FAO 2021, Kanumuri (2020), Madushanki *et al.,* 2019 and TongKe 2013).

An IoT-based solution can provide the key answer in data collection to a Wireless Sensor Network (WSN). WSN was primarily based on IEEE 802.15.4 but recently low power Wi-Fi and Bluetooth have been gaining popularity and are the most commonly used enabling technology for IoT (Internet of Things).

Monitoring of micro-climate and the environmental conditions plus smart automation control integrated with internet of things (IOT) gave the farmer the power and flexibility to make the right decision, to avoid the human mistakes and environmental stress impacts on food production (Weissa *et al.,* 2020).

Figure 5. How IoT transform the agriculture sector (source: Arun Goyal 2019)
https://www.businessofapps.com/insights/internet-of-things-iot-agriculture-sector/

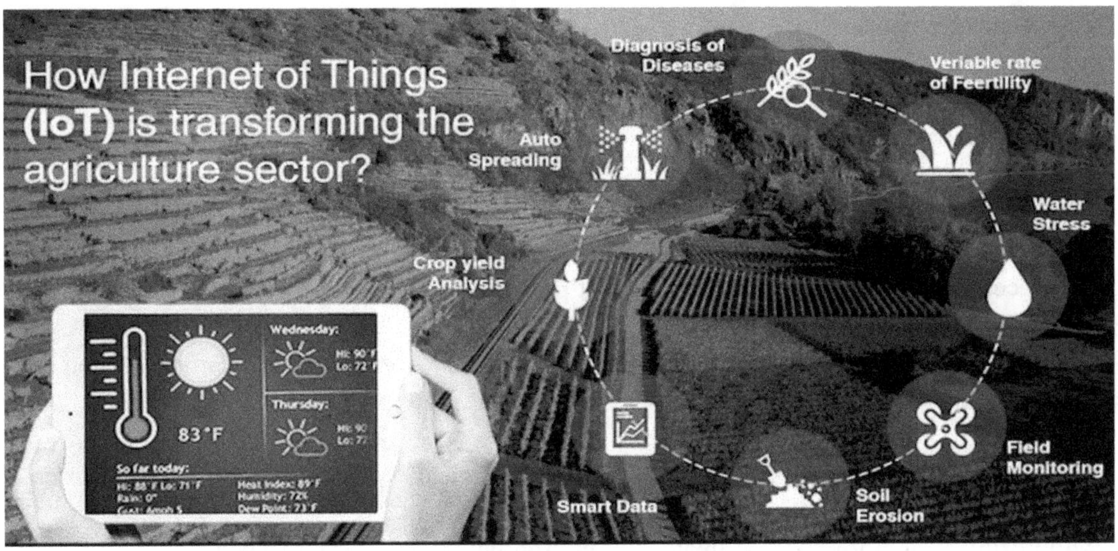

For monitoring micro-climate and the environmental conditions, microcontroller boards (i.e. Arduino (mini, nano, uno, mega and etc..), microprocessor (Raspberry pi in different types and version; and impeded system mainly used for collecting and processing agriculture data. Artificial intelligence (AI), machine learning and computer vision integrated with IOT illustrated smart and automation agriculture projects. The development of electronic industry presented many sensors for detecting micro-climate and the environmental conditions such as:

1. DHT 11and DHT 22 for measuring air temperature and relative humidity;
2. LDR module and TSL 2561 for estimating light intensity,
   Soil moisture sensor (resistive or capacitive) for detecting soil moisture.
3. waterproof temperature for recording the water temperature;
4. TDS and pH kit sensors for measuring EC and pH of fertigation respectively.
5. Digital camera for computer vision or display the agriculture systems

6. Infrared camera for remote sensing or integrated with AI and machine learning technology.

IoT allows for machine to machine interaction and controlling the agriculture production system autonomously and intelligently employing deep neural networks. Moreover, IOT

Moreover, different applications and programs could be applied via computer or mobile network, wireless with easy access, friendly use, low cost or even free download, accurate and on time.

Innovative ICT technologies for food production

Let's agree first, what's innovation today, it's traditional tomorrow. Innovative ICT technologies for food production covers all agriculture production sectors as follows:

1. Field crops, vegetables, fruits production,
2. Ornamental, medicinal, cut flower productions,
3. Greenhouse, soilless culture and indoor plant factory
4. Pasture (natural and artificial),
5. Forest management,
6. Animal production (different scales and methods),
7. Harvest and Post-harvest processes,
8. Trading, marketing,
9. Agriculture mechanization, and etc.

Innovative ICT in agriculture includes all agriculture practices, process and operation from preparation and processing (before cultivation) to not just post-harvest but actually to the client (Thomas 2018 and Tripathy and Anuradha 2018).

Michail and Thomas 2019, Thomas 2018 and Tripathy and Anuradha 2018 demonestrated the most interested ICT innovations in agriculture for farmers concern on the following items:

1. Smart agriculture weather station based on IOT (efficient, low cost, simple and update date, accuracy and etc.),
2. Monitoring system for micro-climate and environmental condition of open field and greenhouse crops as well as for different animal husbandry and pasture,
3. Intelligent control system for micro-climate and environmental condition of greenhouse, soilless culture systems, aquaculture, aquaponic and different scales of animal husbandry,
4. Design the irrigation system and network (use GPS and land coordination, water source type, reference weather data, for illustrate the recommended irrigation system and its supplies),
5. Estimate the irrigation requirement based on IoT,
6. Management of irrigation system based on the applications of weather data, or directly via soil moisture sensors or both (smart automation control for irrigation system On/Off, time, monitoring the soil moisture and alarming for irrigation system problems),
7. Administrate the fertigation program (estimate the fertilizer requirements, monitoring the EC, pH, the injection process, the fertigation efficiency),
8. Monitoring the growth stages of crops and animals (the integrated application of computer vision, remote sensing via drones and artificial intelligence),
9. The applications of drones and robots in monitoring, control, alarm and process,

10. Integrated pest management (the integrated application of computer vision, robots, drones and artificial intelligence to identify the symptoms and address the right approach and time),
11. Accomplish arduous agricultural work (integrated application of remote agriculture mechanization, robots, computer vision, machine learning and artificial intelligence),
12. Harvesting process (Recommended the harvesting date, time, method and post-treatment),
13. Market chain (management of packing, transportation, storehouse, e-trading, e-commerce),

The above innovations are for current and near future time but definitely not for future. Many internet platforms, webs, computer programs, android mobile application innovative every day to present a huge services in e-agriculture, clouding and interface the agriculture data and information from different sources (sensor network, remote sensing, webs, and etc.) as Aman and Kumar 2020 illustrated.

The agricultural sector as a very young field that is still in need ICT innovations and applications contribution in all production systems and their processes to save time, effort and cost while maximize the profit, natural use efficiency and sustainability.

Many examples of ICT innovations could be illustrated under developed countries conditions. These examples clarify how ICT could convert the conventional agriculture to smart agriculture based on IoT. A set of these examples will be presented as follows:

1 – Smart agriculture weather station

Accessing weather information daily may be is not available in many underdeveloped countries but by using simple microcontroller such as an Arduino and sensors for sensing the air temperature, relative humidity, light intensity, air speed and direction. The microcontroller and sensors are low cost with friendly access, use and apply. There are many smart agriculture weather station tutorials and its applications available on the internet. Sharing these information among farmers in the same village or rural area could be done through establishing local network by Bluetooth, WiFi or radio (nRF) modules without the need to internet access. Alarm system could be applied against the extreme weather events.

2 - Irrigation program based on detecting soil moisture and IOT

The irrigation requirement is calculated mainly based on two factors as follows:

1. The climate conditions parameters by using different applications and mathematical models that take in concern many factors like plant type and growth stage, soil type, water quality, irrigation system, etc., Many computer programs, web platforms and sites, android applications presented this service depends on actual measured weather parameters or reference climate data.
2. Detecting soil moisture based on many devices and methods such as tensiometer, digital devices, weight methods (more accurate but need more time and efforts) etc.,

Detecting soil moisture or obtained the climate condition parameters accurately to operate the modern irrigation systems is difficult process for the difficulties of update information access, cost, knowledge, huge variation of soil type, accuracy of different methods, failure to take into account extreme weather events, plant requirements, soil salinity and etc.,

ICT observes unique innovation in this vital issue through different type of soil moisture sensors (resistive and capacitive) integrated with microcontroller or microprocessor boards for operate and control the irrigation system. Robotic process automation could be integrated also to gained the micro climate condition through internet webs or smart weather station and calculate the irrigation requirements too to validate and correct the sensor detection to present precision agriculture.

Soil moisture sensors could be modified with very low cost to be more efficient, durable, and accurate to access wide soil type and depth by replacing the unreliable BCP probe by two adjustable length metal bars (steel) to select the required depth for soil moisture detection.

Figure 6. Smart weather (temperature, RH and light intensity) and soil conditions (soil moisture and temperature) station plus alarming system (Source: The Author 2021)

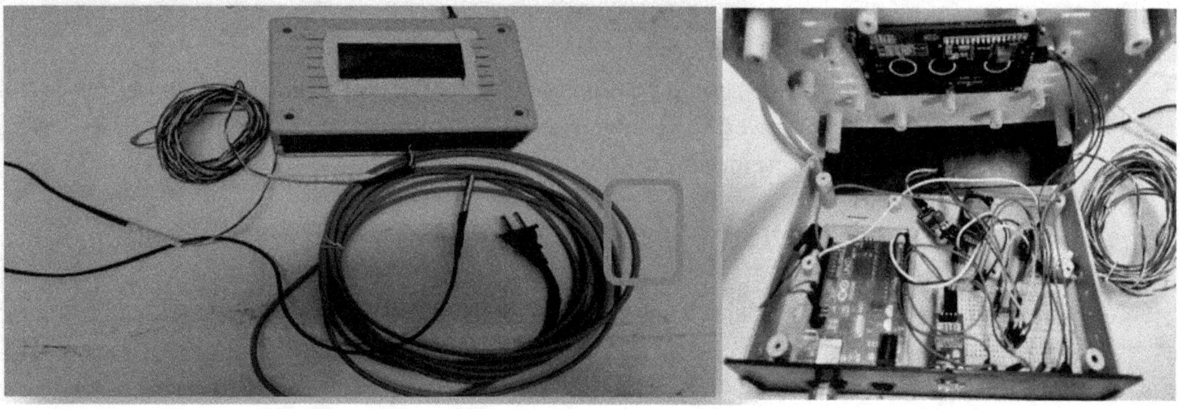

3 – Smart hydroponic greenhouse based on IOT

Smart monitoring and control systems based on Arduino Mega 2560 that programmed via Arduino IDE program. Different sensors and actuators were wired to sensing the micro-climate (DHT 22 (air temperature and relative humidity), TSL 2561 (light intensity)) and environmental conditions (TDS sensor, pH kit, float switch (tank level) and waterproof temperature (solution temperature)) plus 4 channel relay module to monitoring the real data on LCD screen (16 x 4) to implement smart automation of agricultural operations (cooling, ventilation, pumping, air supply, alarm and solution heating) based on internet of things (IOT) via Node mcu esp 8266 that programmed to transmitted the data every 30 min. via Wi-Fi through airbox to the internet web google platform (Firebase) for presenting the real time records and hosting the data.

The costs of electronic components decreased dramatically while their availability, durability, accuracy and variety increased last few years beside the friendly application and technology transfer that courage the develop and implement of this technology and its contribution in agricultural production.

The smart hydroponic greenhouse pilot illustrated good performance of monitoring the micro-climate and environmental conditions. Agricultural operations control worked according to the programming of Arduino Mega with high efficiency. Node mcu esp 8266 transmit the collected data to internet via airbox of Orange Tele. Co. with high performance. Google firebase platform display the real time records and hosting about 100 thousand of different sensors records during the lettuce season.

Moreover, the integrated Arduino and node mcu provide a wireless sensors network that offered a high capability of monitoring huge data at anywhere and anytime. Also the revealed results showed that the importance demands of smart management of nutrient solution (TDS, pH and water temperature and level), monitoring these factors without smart control were not useful enough regarding the rapid solution changes. The smart hydroponic greenhouse that provides sensing, monitoring of agricultural information, control and automation the different agricultural procedures may play a vital role in assist the farmers for offering the required information to manage the agricultural procedures for increasing sustainability and ecologically the agricultural production. The smart automation technology integrated with IOT gave the farmers the power to face the climate change risks, global pandemics, food security demands, environmental and natural resources shortage. The need to develop and improve the use of microcontroller, sensors and actuators integrated with IOT to include all agricultural procedures especially in developed country become more necessary.

Currently, the presented hydroponic greenhouse prototype is taking place in commercial greenhouse to validate the efficiency and to avoid any operating errors or sensors failures. The schematics of hydroponic greenhouse are illustrated as follows:

Figure 7. Schematic of smart hydroponic greenhouse based on Arduino and IOT. (Source: Abul-Soud et al., 2021)

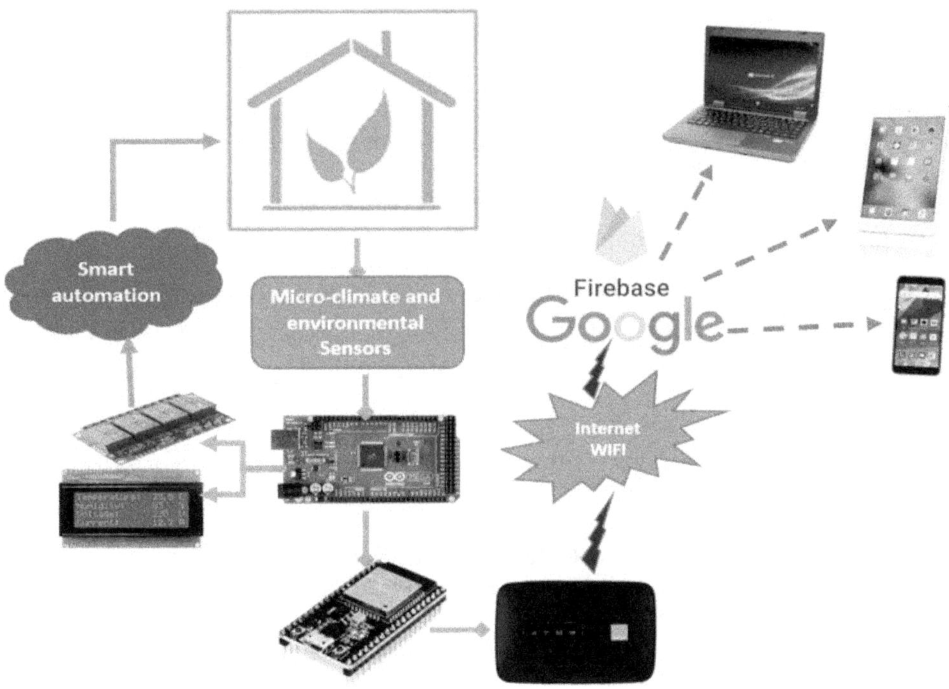

REFERENCES

Abul-Soud, M. A., Emam, M. S. A., & Mohammed, Sh. M. (2021). Smart hydroponic greenhouse (Sensing, monitoring and control) prototype based on Arduino and IoT. *International Journal of Plant and Soil Science*, *33*(4), 63–77.

Raneesha, Halgamuge, Wirasagoda, & Syed. (2019). Adoption of the Internet of Things (IoT) in Agriculture and Smart Farming towards Urban Greening: A Review. *International Journal of Advanced Computer Science and Applications*, *10*(4), 11–28.

Agrawal, K., & Kamboj, N. (2019). Smart agriculture using IOT: A futuristic approach. *International Journal of Information Dissemination and Technology*, *9*(4), 186–190.

Apolo-Apolo,, O.E., & Perez-Ruiz,, M., Martinez-Guanter, & Valente, J. (2020). A-Cloud based environment for generating yield estimation maps from apple orchard using UAV imagery and deep learning technique. *Frontiers in Plant Science*, *11*(1086), 15.

Castrignano, A., Buttafuoco, G., Khosla, R., Mouazen, A. M., Moshou, D., & Naud, O. 2020. Agricultural internet of things and decision support for precision smart farming. Academic Press.

Daum, T. (2018). *ICT Applications in Agriculture. Encyclopedia of Food Security and Sustainability*. Elsevier.

El Fiorenza, Sharma, Ranjan, & Shashank. (2018). Smart e-agriculture montoring based on Arduino using IOT. *IJSDR, 3*(10).

FAO. (2021). Climate-smart agriculture case studies, Projects from around the world. FAO.

Jain, A., & Abhay, K. (2020). Smart Agriculture Monitoring System using IoT. *International Journal for Research in Applied Science and Engineering Technology*, *8*(7), 366–372.

Kaiser, C., & Ernst, M. (2016). *Hydroponic Lettuce*. CCDCP-63. Lexington, KY: Center for Crop Diversification, University of Kentucky College of Agriculture, Food and Environment. Available: http://www.uky.edu/ccd/sites/www.uky.edu.ccd/files/hydrolettuce.pdf

Karamanis, G., Drosos, C., Papoutsidakis, M., & Tseles, D. (2018). *Implementation of an Automated System for Controlling and Monitoring a Hydroponic Greenhouse. International Journal of Engineering Science Invention*, *7*(10), 27–35.

Khanal, Kushal, Fulton, Shearer, & Ozkan. (2020). Remote Sensing in Agriculture—Accomplishments, Limitations, and Opportunities. *Remote Sensing*, *12*(3783), 1–29.

Krishna, K. R. (2018). *Agricultural Drones A Peaceful Pursuit*. Apple Academic Press Inc.

Létourneau, Guillaume, Caron, Anderson, & Cormier. (2015). Matric Potential-Based Irrigation Management of Field-Grown Strawberry: Effects on Yield and Water Use Efficiency. *Agricultural Water Management*, *161*, 102–113.

Moussaid, El Fkihi, & Zennayi. (2019). *Citrus Orchards Monitoring based on Remote Sensing and Artificial Intelligence Techniques: A Review of the Literature*. Science and Technology Publication.

Ray. (2016). Remote Sensing in Agriculture. *International Journal of Environment, Agriculture and Biotechnology (IJEAB), 1*(3), 362 – 367.

Redmond. (2017). Controller Design for an Osprey Drone to Support Precision Agriculture Research in Oil Palm Plantations. *ASABE 2017 Annual International Meeting*.

Saidu, Clarkson, Adamu, Mohammed, & Jibo. (2017). Application of ICT in Agriculture: Opportunities and Challenges in Developing Countries. *International Journal of Computer Science and Mathematical Theory, 3*(1).

Salampasis, M., & Bournaris, T. (2019). *Information and Communication Technologies in Modern Agricultural Development. 8th International Conference, HAICTA 2017*, Chania, Greece.

Shanmugapriya, P., Rathika, S., Ramesh, T., & Janaki, P. (2019). Applications of Remote Sensing in Agriculture - A Review. *International Journal of Current Microbiology and Applied Sciences*, *8*(1), 2270–2283.

Sharma, Kumar, & Malik, Abhilash, & Harender. (2018). Applications of Remote Sensing in Agriculture. Agricultural. *Allied Sciences & Biotech. for Sustainability of Agriculture, Nutrition & Food Security*, 141–146.

Siregar, Sari, & Jauhari. (2016). Automation system hydroponic using smart solar solar power plant unit. *JurnalTeknologi (Sciences & Engineering), 78*(5), 55–60.

TongKe, F. (2013). Smart Agriculture Based on Cloud Computing an d IOT. *Journal of Convergence Information Technology, 8*(2).

Tripathy, B. K., & Anuradha, J. (2018). *Internet of things (IoT). Technologies, Applications, Challenges, and Solutions*. CRC Press Taylor & Francis Group.

Weissa, M., Jacobb, F., & Duveillerc, G. (2020). Remote sensing for agricultural applications: A meta-review. *Remote Sensing of Environment, 236*. Advance online publication. doi:10.1016/j.rse.2019.111402

Yildirim, M., Dardeniz, A., Kaya, S., & Ali, B. (2016). An automated hydroponics system used in a greenhouse. *Scientific Papers, Series E, Land Reclamation, Earth Observation & Surveying Environmental Engineering, 5*, 63-66.

Zhang, D., & Wei, B. (2017). *Robotics and Mechatronics for Agriculture*. CRC Press.

Chapter 5
Genetic Resources, Breeding, and Molecular Genetic Markers for Orchard Improvement and Management

Clara R. Azzam
https://orcid.org/0000-0002-4528-5236
Cell Research Department, Field Crops Research Institute, Agricultural Research Center, Giza, Egypt

ABSTRACT

Orchards play a key role in the non-petroleum economy of many developing countries. Developing orchards with good adaptation to abiotic and biotic stress is the main point for plant researchers to improve and increase fruit yield in many parts of the world. They have established seed production zones in the best natural stands to provide seed until orchard production can satisfy all needs. Conventional breeding and genetic improvement are time-consuming and don't associate good results in orchards. Because of their great utility, genetic markers will become standard tools for orchard improvement in the future. Biotechnology creates a genetic alteration in orchards comparatively short with slight modifications to other valid characters (tolerance to abiotic and biotic stresses, enhanced post-harvest shelf life of fruit, reduced vegetative propagation phase, and fruit production with higher nutritional values). Thus, it leads to an uprising in orchard sciences, fruit tree breeding, and orchard improvement and management.

INTRODUCTION

Orchards (*Orchidaceae*) are horticultural crops with significant economic values. Its improvement has been a challenge for conventional breeding. Orchids traditional breeding is time consuming, some desired traits cannot be attained by this method due to limitations and drawbacks of the convential methods. In recent years, DNA marker and molecular breeding has been extensively employed by introducing the desired target genes into orchids.Biotechnology allows for a broader range of genetic modification in fruit trees in a relatively short period while causing minimal changes to other beneficial traits, resulting

DOI: 10.4018/978-1-6684-2423-0.ch005

in a revolution in horticultural science and fruit tree breeding. One of the main goals of fruit tree genetic modification was to reduce the size of fruit trees to increase orchard plant density (Hatzopoulos et al., 2002).The main objectives for fruit trees genetic modification and improvement in fruit trees genome are facing many problems as parthenocarpy and seedless, vigorous climbing vines (for example, kiwi fruit and rapes),dioecious, long juvenility phase in fruit trees, incompatibility issues, high heterozygosity, reticulate polyploidy structure with diploid, tetraploid, and octoploid in some species, and sterility distinguish them from agri-crops (Hatzopoulos et al., 2002).

The emerging multifaceted importance of orchards as floriculture crops has grown significantly in the last decade. Orchards are one of the world's most prominent flowering plant families, with 30,000-35,000 different species (Vendrame & Khoddamzadeh, 2017).Orchards cultivation has resulted in new cultivars and hybrids with multiple ornamental values, contributing to the rapidly expanding market and demand for exotic varieties. However, associated challenges in orchard cultivation include complex genomes, slow growth rates and poor transformation, limiting the conservation of threatened genotypes with novel characteristics (Wang et al., 2017). For example, the poplar cultivar has a development stage that lasts more than two years, from vegetative to reproductive (Lu et al., 2007). Seventy percent of *orchidaceae* species are epiphytic, accounting for two-thirds of all epiphytes worldwide (Chen et al., 2011). The remaining 25% are terrestrial and 5% require growth assistance (Atwood, 1984).Orchards have long fascinated researchers due to their specialized structures and properties. Orchards species exhibit unusual characteristics such as specialized pollination, thin non-endospermic seeds, diverse habitats, mycorrhizal-dependent germination and adaptive mechanisms (Hossain et al., 2013).

The current era has seen unprecedented advances in promoting orchard cultivation, which can be attributed to the rapid progress of biotechnological interventions. The discovery of compounds with therapeutic potential and the development/cultivation of new exotic varieties with novel characteristics such as colourful flowers and disease resistance, has broadened the horizon of orchard research and continues to attract the attention of scientific communities. Their multifaceted importance in floriculture, the food sector and medicine has significantly aided orchard commercialization, with a multi-million dollar global market growing (Hossain et al., 2013). The increasing demand for exotic varieties and the conservation of threatened species have necessitated advancements in new scientific technologies for novel flower varieties, biotic/abiotic stress tolerance and efficient propagation (Hossain et al., 2013). For various outcomes, new frontiers in plant science have focused on biotechnological interventions in orchards, such as genetic manipulations, proteome studies and functional genomics (Chen et al., 2011; Hsiao et al., 2011; daSilva et al., 2011). On the other hand, overexploitation and a complex life cycle account for critical challenges such as climatic fluctuations, threatened habitats and major bottlenecks (IUCN, 2017).

Orchards present a significant challenge for restoration/conservation due to their complicated lifestyles, highlighting the urgent need for integrated conservation measures on a broader scale. In this regard, the family Cypripedioideae (the slipper orchards) was studied and 90 percent of the species were declared threatened by the Global Red List due to over-harvesting and declining habitats (Fay & Rankou, 2016). Unsustainable harvesting of orchard species is a major issue with severe consequences in the modern era. Renanthera, Paphiopedilum, Cattleya and Phragmipedium, among others, have been severely threatened by indiscriminate/overuse and collection. Approaches to orchard conservation and the developing of new varieties should include habitat conservation(*in-situ*).

Orchards were initially cultivated as ornamentals for their multi-coloured and attractive flowers, but medicinal applications are only now gaining traction. The Chinese were the first to cultivate orchards that have been documented for their medicinal properties and uses (Jalal et al., 2008). Researchers have traced the history of orchards and their medicinal uses back 120 million years, when they were most likely cultivated for their health-promoting properties (Pant, 2013).Early documented evidence was discovered in Japanese and Chinese literature between 3000 and 4000 years ago and is regarded as a pioneer in describing various orchard species' medicinal properties (Reinikka, 1995; Bulpitt, 2005). The Indian traditional system described the pharmacological uses of orchard species such as *Orchis latifolia*, *Dendrobium macrae* and *Eulophia campestris* in Ayurveda (De et al., 2014). The following are the main genera of medicinal orchards: New orchards varieties of Ephemerantha, Eria, Galeola, Cymbidium, Cypripedium, Nevilia, Thunia, Bletilla and Anoctochilus (Szlachetko, 2001) are being discovered (Gutierrez, 2010; Pant et al., 2011). Several recent studies have isolated phytochemicals with medicinal properties from numerous species, including *jibantine orchinol*, *nidemin*, *loroglossin*, *hircinol* and *Cypripedium*. A thorough examination of crucial orchard species, plant parts, phytochemicals and ethnobotanical significance has been provided. Furthermore, the neuroprotective, anti-inflammatory, antimicrobial, hypoglycemic, anticancer, antirheumatic, wound-healing and other significant effects of orchard phytochemicals on human health have been well-known since ancient times (Gutierrez, 2010).

Therefore this chapter intends to outline the role of the genetic resources, orchards seed production, conventional and biotechnological breeding techniques, and molecular markers in orchards to develop new varieties using traditional and modern strategies for orchards Improvement and Management

GENETIC RESOURCES AND DIVERSITYFOR ORCHARDS IMPROVEMENT AND MANAGEMENT

Orchards breeding depends upon genetic diversity within and among orchards as a basis for developing new varieties with improved traits. Plant breeders using the latest methods in biotechnology and genetics into their breeding methods to more efficiently use existing differntiation but also to induce new genetic variation. Genetic diversity (GD) is the total amount of genetic variation that genotype carry within or between genetic units due to their evolutionary history, which serves as the foundation for their responses to abiotic and biotic stresses. Significant genetic variation enables natural selection to result in adaptation (Savolainen et al., 2007). When there are a lot of differences, the chances of adjusting to new environmental conditions increase (Pant et al., 2011, Khalifa et al., 2006; Azzam et al., 2007; Azzam et al., 2007; Abo-Doma & Azzam, 2007a; Azzam & Mahrous, 2010; Keshta et al., 2011; Azzam et al., 2012; Zein et al., 2012; Dobeie et al., 2017; Bosily et al., 2018; Abbas et al., 2021; Abou-Sreea et al., 2021). The development of essential plant genotypes that will aid in the restoration of wild populations of these plants and commercial cultivar breeding depends on GD research on orchard plants. As it provides the foundation for adaptability and tolerance to stressors and harsh environmental conditions, GD is critical for the long-term survival of forests (Schaberg et al., 2008). The impact of breeding on genetic diversity is depicted in Figure (1). There is no risk of genetic resource loss if orchard populations recover naturally (Isajev et al., 2009). GD, on the other hand, maybe reduced as a result of natural processes or anthropogenic intervention. Species with medium ecological amplitude, habitat fragmentation, and shortened or nonexistent gene flow, all of which are required to maintain high GD, are most vulnerable. Aside from habitat fragmentation, orchard management that involves selectively removing trees and their genes (dysgenic selection) has been shown to have an impact on orchard gene pools (Schaberg et al.,

2008; Isajev et al., 2009), decreasing the frequency and number of rare alleles and lowering estimates of future genetic potential (Adams et al., 1998; Hawle et al.,2005).

Figure 1. Scheme of the effect of breeding on genetic diversity (GD) adapted by Ivetić et al. (2016)

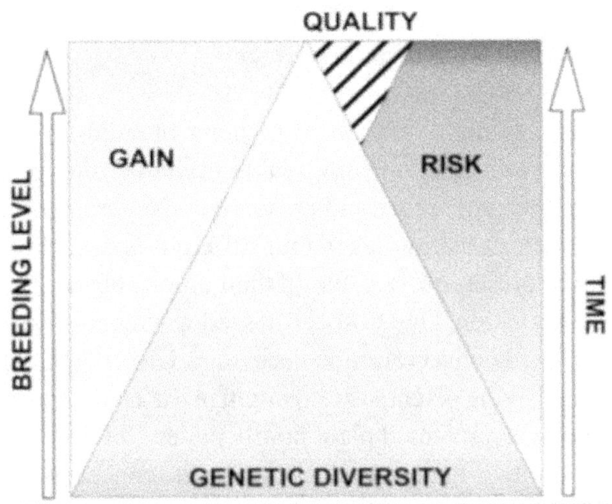

The genetic diversity of seed lots is influenced by the size of the parent population, the balance of parental reproductive success, the extent of inbreeding, and the kinship of its members (Funda et al., 2012). The genetic variability in stands was greater than in seed orchards (SO) (Kang et al., 2003). Male genetic variability outnumbers female genetic variability in SO (Kang et al., 2003), which is the polar opposite of what may be observed in natural populations (Bilir et al., 2004). Seed collection is critical in producing the vast forest reproductive material (FRM) required to keep GD alive. Many of the original GD in seed sources may be lost while collecting seeds from trees. The first is to gather at the seed stand (SS) level without considering the parents' seed lot participation.

On the other hand, the second method is to collect the same number of seeds from each parental plant (collection units), thereby preserving GD. Consequently, a seed lot contains most of the parents involved in seed production. Control over single families' participation in the seed lot is lost when seeds from different families (mother trees) are mingled (OECD, 2013;, Konnert & Hosius, 2010). Therefore, fewer seeds taken from more parent trees are more effective at preserving GD than the opposite. Phenotypic selection successfully conserves genetic variation in wild populations while also providing an accessible and low-cost source of material for future breeding efforts (El-Kassaby et al., 2011). Seeds collected over time could be mixed to maintain diversity (Kang et al., 2003). Collecting seeds from different ages of trees, on the other hand, may preserve unusual alleles for the next generation (Isajev et al., 2009). Seed harvest (systematically rejecting tiny seeds), growth rate, flowering, fruiting dates, and harvesting time should all be avoided to avoid the unintentional selection of features. Seed storability is strongly influenced by genotype, and the possibility of GD narrowing means that seeds with low vigor may lose their viability entirely after a long period of storage (Thomas et al., 2014; Stanys et al., 2012). Seed germination may significantly reduce the GD of seed lots due to differences in germination ability, speed, and seed vigor, resulting in either underestimating or overestimating the necessary population size of seed lots (Funda et al., 2012; Konnert & Hosius, 2010). Planting seeds rather than seedlings in seedling production reduces the risk of GD loss due to directional selection in seed processing and seedling production. A study of microsatellite polymorphism revealed the GD of wild cherry populations in the region and genotypes of local variations.

CONSERVATION OF GENETIC RESOURCES

Germplasm conservation is an essential component of plant breeding methods. For a long time, the regular decline in the population of economically viable orchards due to human threats and climate change has prompted an urgent need to secure and preserve endangered and threatened orchard genetic resources. Orchard germplasm conservation takes four different approaches depending on the time of preservation and the source to be preserved. Because of their unique beauty, fragility and aroma, orchards receive much attention for conservation. Due to specific seed germination and growth-related microbial and environmental factors, orchard conservation is necessary. Climate change and human intervention have exacerbated the situation, driving scientists to recognize the critical need for orchard conservation (Keel, 2005). Orchards are the most protected plant family under CITES that support its cultivation and conservation efforts through the two main approaches: *in-situ* conservation and *ex-situ* conservation (Irawati, 2013).

Most breeders use the first approach because it is the simplest way to preserve the entire plant under net house or greenhouse conditions. However, pests, insects, diseases, physiological changes caused by climate change and cultivation conditions all significantly impact this preservation (Rung et al., 2016). The second orchard preservation approach is planting tissue culture-based preservation; however, it also generally affects the genetic architecture and physio-morphological conditions stimulated by somaclonal variation resulting from *in vitro* sub-culturing. The third approach to orchard germplasm conservation is based on low temperature and dry storage (Arditti & Emst, 1993). However, these low temperature and dry storage-based methods are successful for 1 to 6 months of germplasm preservation but fail to provide high viability of germplasm after more than six months of preservation (Hirano et al., 2009).

Micropropagation is an exciting *ex-situ* conservation method and allows for the cultivation of both endangered and commercially important plant species (Shukla & Sharma, 2017).*Orchidaceae* is the most prominent family of flowering plants, with a higher market demand met either from their natural habitat (*in-situ*) or through large-scale breeding; however, a gap between demand and supply still exists, including the availability of rare endangered orchards. Furthermore, some orchards suffer from inbreeding depression, which results in a low germination rate in self-pollinated seeds (Bellusci et al., 2009).Orchards rely on mycorrhiza and other suitable environmental conditions for germination in the wild and seedlings grow faster when leaves are adapted for CO_2 support. Naturally, seed maturity is essential for seed germination and only mature seeds achieve higher levels of germination.Plant tissue culture techniques have been used for rapid propagation and *ex-situ* conservation of orchards over several decades, employing various methods and explants such as flower stalks, shoot tip nodes, stem bids and root tips rhizome segments (Pant, 2013; Reddy et al., 2016). Cryopreservation is the most viable option for long-term orchard germplasm preservation.

1.1. Cryopreservation Techniques

Cryopreservation is always the best germplasm preservation approach because all metabolic and physiological processes are halted at a cryogenic temperature of -196°C (Rung et al., 2016). However, cryopreservation of germplasm requires some pretreatments to protect the cells from the effect of instant intracellular water content freezing ice crystals at liquid nitrogen cryogenic temperatures. Typically, orchard seeds and pollen germplasm are pretreated in vitrification, desiccation and encapsulation dehydration to remove cell water content before cryopreserving in liquid nitrogen (Popova et al., 2016).

1.1.1. Vitrification

Sakai (2000) introduced the vitrification technique, which is commonly used for the longer preservation of immature and mature orchard seeds with higher than average water content. The vitrification method employs a high osmolarity vitrification solution containing cryoprotectant chemicals such as glycerol, dimethyl sulfoxide and ethyl glycol. The seeds for preservation are kept in this high osmolarity vitrification solution, which reduces the intracellular water content of the seeds and vitrifies them through osmoregulation of these cryoprotectants, lowering the freezing temperature of the cells and protecting them from ice nucleation injuries (Popova et al., 2016).

1.1.2. Desiccation

Desiccation-assisted cryopreservation is more suitable for mature orchard seeds. Desiccation is the process of slowly drying seeds at a controlled rate under constant relative humidity or drying with silica gel or a $CaCl_2.6H_2O$ salt solution to reduce the water content of the seeds before preserving them under liquid nitrogen (Rung et al., 2016). Desiccation was used to cryopreserve *Pterostylis sanguine, Paphiopedilum rothschildianum, Dendrobium candidum* and *Thelymitra macrophylla* (Rung et al., 2016; Hirano et al., 2006).

1.1.3. Encapsulation-dehydration

The assisted method for cryopreservation is encapsulation-dehydration, a technique developed to produce artificial seeds (Fabre & Dereuddre, 1990). To reduce the cellular water content, the encapsulation-dehydration method uses *in vitro* cultured for orchard plant tissues, seeds and embryos that have been partially desiccated with silica beads or the airflow of the laminar bench. These partially dried tissues are trapped and encapsulated in sodium alginate beads before being cryopreserved in liquid nitrogen (Hirano et al., 2006). Encapsulation dehydration techniques are commonly used to preserve a few orchards, namely *Cyrtopodium hatschbachii* and *Oncidium bifolium* (Surenciski et al., 2012).

2. CONVENTIONAL BREEDING APPROACHES FOR ORCHARDS IMPROVEMENT AND MANAGEMENT

A breeding system based on the seed orchards increased productivity, which will continue to be the primary method of genetic improvement. The essential element in each crop's breeding and selection processes is to bring together crucial agronomical features (Liesebach et al., 2021). Despite genetic gains and optimized management being required due to environmental issues (Scherer et al., 2020), advances in plant genomics have revealed the genetic basis of various valuable characteristics that have been limitedly used in tree improvement and conservation (Neale & Kremer, 2011; Flanagan et al., 2018). Comprehensionof genetic control and quality relationships isessential because they affect various breeding operations and decisions (White et al., 2007; Araujo et al., 2012). The lowest-intensity method fororchards improvement is a standard transplanting-like procedure called provenance testing to identify superior seed sources for reforestation. Identifying seed zones for safe seed transfer is a significant achievement of provenance testing (Konig, 2005). Creating a structured pedigree using controlled crosses (mating design) tends to be a more cost-effective approach, as genetic gain per unit of time is generally the most important measure of long-term achievement in Orchids breeding.It eventually allows for the precise estimation of genetic parameters like heritability and parent-offspring breeding values (Namkoong et al., 1988). The most common breeding framework is the recurrent selection scheme (Allard, 1960). Traditional tree improvement programs are resource and time-consuming, some desired traits, such as the mottled flowers/foliage of a single individual, cannot be attained by such approaches. These programs must establish mating designs, assign progeny tests to various sites to evaluate the parents and corresponding offspring across large agro-ecological areas, supervise those tests over overstretched periods, and finally analyze measurements to assess economic characteristics. The traditional frequent selection scheme is used in the tree breeding programs, resulting in diverse production and breeding populations (El-Kassaby et al., 2011). This process led to an abundance of reliable, affordable genetic markers, drastically altering current breeding programs (Figure 2).

2.1. Conventional Breeding Strategies

2.1.1. Crossbreeding

The orchard family, *orchidaceae*, is the most diverse flowering angiosperm family, with over 8,000 genera and 28,000 natural and artificial hybrid species (Li et al., 2021). Hybridization in agriculture refers to crossing different plants within a species. Despite its enormous potential benefits, hybridization has rarely been used as a modern breeding method in conifers due partly to the labor intensiveness of hybrid seed production by mass controlled pollination and the inefficiencies of seed production in bi-species orchards designed to produce hybrid seed. Pure species trees, on the other hand, can be added to the mix to improve specific characteristics (Park, 2014). *Phalaenopsis* and *Oncidium*, members of the *orchidaceae* family, are two of the most commonly grown orchards for commercial production. Plant flower members of orchards are both self-pollinated and cross-pollinated. However, self-pollinating species may produce fewer seeds with no embryos than cross-pollinated flowers (Li et al., 2021).

Figure 2. Scheme of Trans-disciplinary approaches such as pre-breeding and breeding strategies supported with genome-wide marker-assisted selection (MAS)combining all pedigree phenotypic information with genotypic data, adapted by Ingvarsson et al. (2016) & Myburg et al. (2019)

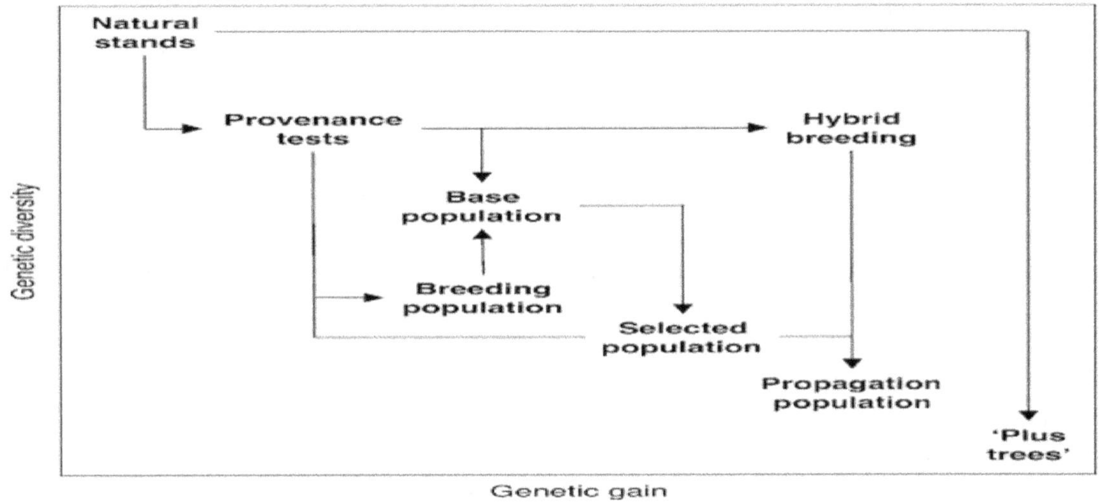

Crossbreeding or hybridization, using both natural and artificial methods, brilliantly integrates the best qualities of two parents into their hybrid offspring. *Phalaenopsis intermedia* a cross between *P. rosea* and *P. aphrodite* was described in the year 1853 as one of the oldest natural orchard hybrids. *Calanthe dominyi*, the first commercial artificial hybrid orchard, was developed *via* crossing between *C. furcate* and *C. masuca* (de Chandra et al., 2019). These commercially available natural and hybrid orchards were created through crossbreeding techniques. However, because this is a simple and effective technique, several factors such as hybrid combination fertility, targeted trait quality assessment and superior hybrid offspring selection play a significant role and must be prioritized (Su et al., 2019). The generation F_1 resulting from a cross between parents with targeted contrasting traits typically exhibits significant phenotypic differences from their parents. However, hybrids can have germination issues. For example, hybrid seeds of *Cymbidium hybrids* are difficult to germinate and culture due to intraspecific, intrageneric and intergeneric degrees of distant genetic relationship (Zhang et al., 2001). Similar problems with the hybridization process, such as parent incompatibility and post-fertilization embryo abortion, have been reported (Luo et al., 2012). Commercial orchard cultivation faces several breeding barriers, including a large and complex polyploid genome, slow growth and a long life cycle, making it challenging to generate new cultivars using the traditional breeding system.

Furthermore, the low transformation efficiency makes it difficult to develop new varieties with desired traits (Lu et al., 2007; Pan et al., 2012). For example, the high-value orchard Phalaenopsis requires more than two years to transition from the vegetative to reproductive phase (Wang et al., 2017). However, the applicability of these conditions in major commercial orchard species has yet to be investigated.Seed germination is a critical aspect of the traditional breeding system because it is directly related to the efficient success of crossbreeding; therefore, in-depth studies are required for a thorough understanding of seed germination mechanisms and plant developmental characteristics to establish an effective breeding system. Therefore, a suitable cultivation strategy is required for hybrid seeds developed through crossbreeding and the stable growth of hybrid populations.

2.1.2. Selection breeding

Selection breeding is based on natural trait variations (Osadchuk et al., 2020). Selection breeding focuses on three key genetic parameters: genetic correlations between traits, trait heritability and genotype-environment interactions (Li et al., 2021). Many important crops, including potato, banana, grape, and others, are vegetatively propagated (El-Geddawy et al., 2008). Recurrent selection breeding programs must develop and deploy new commercial clones (recombination, evaluation, and selection in successive generations). Consequently, breeding for clonal selection requires additional effort (White et al., 2007; Reis et al., 2011). Consequently, each breeding cycle typically requires two steps for evaluation and selection (initial evaluation/selection in progeny characteristics and final evaluation/selection in clonal trials) (White et al., 2007; Reis et al., 2011). Because olive trees have a relatively long juvenile phase requiring an extensive F_1 and F_2 generations study, the breeding procedure has been limited to a few cases. Plant breeding programs incorporate advances in fruit plant biological diversity, adaptability to low temperatures, disease resistance, fruit quality, and plant developmental biology by developing superior genetic lines, biological markers (Liesebach et al., 2021; Häggman et al. 2014; Baniulis et al., 2013).

Plant germplasm can be classified in a variety of ways, but the most important characteristics for ex situ germplasm management are (1) selection intensity (Table 1), (2) genetic profile (Table 2), and (3) breeding system: (a) strictly asexual; (b) mixed asexual and sexual; and (c) strictly sexual, including strictly allogamous, strictly autogamous, or mixed allogamous and autogamous (Hamrick & Godt, 1990). The classes in the preceding three features should be considered continuous rather than discrete in general. These germplasm characteristics are offered here because they dictate the types of genetic markers that are best suited for diverse plant germplasm management applications when taken together.

Table 1. Relative intensity of human selection on plant germplasm

Type of plant germplasm	Effect of human selection
Wild	Nil or weak
Weedy	Moderate
Traditional race or variety	Substantial
Elite cultivars	Strong

Table 2. Genetic profiles for various types of plant germplasm

	Genes	
Genotype	**Homozygous**	**Heterozygous**
Homogeneous	Inbred lines	Clones derived from an allogamous species
Heterogeneous	Autogamous, traditional cultivars or modern multiline cultivars	Allogamous landraces

BWB (breeding without breeding) combines the use of genotypic or phenotypic pre-selection of superior individuals, informative DNA markers for fingerprinting and pedigree reconstruction of offspring to assemble naturally created full- and half-sib families resulting from mating among selected parents (El-Kassaby et al., 2011). Having offspring without a mother breeding is fundamentally based on the use of pedigree reconstruction to form half-, and full-sib families required for standard intra-class correlation evaluates to estimate quantitative genetics parameters such as trait heritability and parental and progeny breeding values (Falconer & Mackay, 1996).

Klápště et al. (2014) compared a pedigree-free model to a marker-based pairwise relationship model in their study. Korecký et al. (2013) modified this approach to include information derived from genetic markers and traditional pedigree. The combination of historical and modern co-ancestry established through genetic markers and pedigree could not be accomplished by either technique alone, making this approach unique.

2.1.3. Mutation Breeding

Mutational breeding uses physical and chemical mutagens to improve individual traits and can shorten the breeding cycle in orchards. Several orchards have been enhanced through mutational breeding, including aroma, medicinal ingredient content, shelf life and stress resistance (de Chandra et al., 2019). Polyploidization using colchicine and other mutagen treatments is one of the greatest fundamental approaches used for mutant orchards Cymbidium, Dendrobium, Oncidium and Phalaenopsis (Cheng, 2011). Higher genomic heterozygozity can allow for a faster mutation rate. Still, the random and unpredictable nature of mutagenesis can occur throughout the genome, leading to other physio-morphological problems in orchard mutants. Consequently, more research is needed to understand the basis of mutation breeding in orchards in order to identify suitable genotypes, explants, mutagen types and optimized dose concentrations for the production of mutant orchards with desired traits.

2.2. Targeted Breeding Programs

2.2.1. Fruit Quality

Increased consumption of plant-based phytochemicals has recently prompted research into improving the phytochemical content of various fruits and vegetables. Gelvonauskis et al. (2004) investigated the biochemical composition of ripening Sorbus fruits. The combination of pigments and health-promoting chemicals found in berry fruits has recently received much attention (Dvaranauskaitė et al., 2009).The design and overall anthocyanin content are influenced by the cultivar, fruit ripeness at harvest, geographical location, and climatic conditions during plant growth (Šikšnianas et al., 2008). New cultivars with improved characteristics are critical for biological and technological progress in fruit production because they must meet the ever-changing needs of production techniques, processing sectors, and consumers. Crop productivity attributes (yield, production pattern, fruit size, suitability for manual or mechanical harvesting), fruit quality (fruit appearance, color, shipping quality, shelf life, and flavor), and resistance or tolerance to essential pests and pathogens are the selection criteria used in cultivar evaluation trials. Around the world, apple (Gelvonauskis et al., 2004; Sasnauskas et al., 2007), black currant (Šikšnianas et al., 2008), and strawberry (Rugienius et al., 2013) cultivars are being evaluated for their suitability for production practices under local climatic conditions.

2.2.2. Climatic change and Abiotic stresses

Most orchard species face elimination due to global environmental changes and overexploitation. Climate change and increasing environmental stresses may hasten dramatic temperature swings and drought. Increased demands on land resources for food and feed are highly likely in the face of future global scenarios such as continuous population growth and the issues posed by climate change. Threats from abiotic stresses, particularly how they may change due to climate change, are significant concerns for future orchards and forest trees and controlling the growing demand for wood and fruits (Lindner et al., 2010). According to Zhu et al. (2012), many species will struggle, and some tree species will likely "die back." Natural disasters are expected to become more frequent and severe due to climate change (Kurz et al., 2008; Gardiner et al., 2010).Gelvonauskis et al. (2004) investigated how winter hardiness was passed down through generations of progenies from native and exotic apple tree cultivars. Crosses produced the most resilient seedlings with the winter resistant cultivars: Kaunis, Auksis, Noris, and Tellissaare. Winter hardiness is primarily influenced by additive gene interaction (Gelvonauskis et al., 2004). Cold tolerance in garden and cherry plums was also pushed (Duchovskis et al., 2007).Furthermore, due to differing expectations about what forests need to produce other than pulp, paper, and lumber, only a limited amount of land will be available for forestry (Häggman et al., 2014).

Temperature changes have been shown to impact plum cold tolerance in the winter significantly. However, strawberry cultivars' cold tolerance ranges from –5 to –45°C, and improper cultivar selection for production could result in crop loss due to harsh winter temperatures. The buildup of oligosaccharides (Lukoševičiūtė et al., 2011) and dehydrin-like proteins during *in vitro* acclimation of strawberry and pear plants were studied to develop biochemical and molecular markers for cold resistance features (Liesebach et al., 2021).

A significant information on gene function and interconnecting regulatory networks is now available (Cramer et al., 2011). A shift in polyamine biosynthesis caused by overexpression of *ERF/AP2* in *Pinus strobus* resulted in a considerable increase in salt stress, drought, and freezing (Tang et al., 2007). Overexpression of an ethylene-responsive factor (*ERF*) like TF in *Populus alba* x *berolinensis* resulted in taller GM trees producing higher dry biomass than controls when salt (NaCl) concentrations were raised (Li et al., 2019). Other metabolite quantities, such as the osmoprotectant glycine betaine, were also found to confer salt tolerance in transgenic eucalyptus by introducing a choline oxidase gene from *Arthrobacter globiformis* (Yu et al., 2009).

2.2.3. Biotic stresses

Most orchard species face elimination due to global environmental changes and overexploitation, but microbial pathogens also threaten orchard cultivation. Disease resistance breeding requires discovering new resistance genes and analyzing their inheritance. Powdery mildew (*Sphaerotheca mors-uvae*), Septoria leaf spot (*Mycosphaerella ribis*), and anthracnose (*Pseudopeziza ribis*) are the most common fungal diseases that affect current commercial production (Sasnauskas et al., 2007).The common pathogen-fungal species are *Cladosporium cladosporioides* leaf spot, *Nigrospora oryzae* leaf spot, *Fusarium oxysporum* wilt, *Colletotrichum gloeosporioides* anthracnose, *Phytophthora capsicum* blight with root rot, *Phyllosticta capitalensis* leaf spots, *Alternaria alternata* black spot and *Phoma multi Endophytes* can also serve as conditional pathogens for orchards (Srivastava et al., 2018). Consequently, early identification of pathogen and disease are critical for plant protection. The traditional method for diagnosing orchard diseases is based on pathogen identity recognition using regional information systems and national and international databases. Because these methods are remarkably unreliable, advanced rapid disease diagnosis techniques are required. Biotechnological advancements have solved this problem *via* modern disease diagnosis techniques that allow disease diagnosis in the laboratory and at the field.

The remote hybridization method for plant species produced high-quality genetic material for plant breeding. Most black currant genetic research has focused on disease resistance gene introgression and genetic lines with cultivar breeding-relevant features. Testing for disease resistance and interspecific hybrid genetic inheritance studies revealed appropriate sources of resistance to powdery mildew, Septoria leaf spot, and anthracnose (Stanys et al., 2012; Šikšnianas et al., 2008). Crossing the interspecific hybrid *R. americanum*, *R. nigrum*, and *R. sanguineum* resulted in anthracnose-resistant F_1 progeny (Šikšnianas et al., 2008). A study of disease resistance inheritance in interspecific hybrids of *R. aureum*, *R. americanum*, *R. nigrum*, *R. Janczewskii*, *R. petraeum*, *R. pauciflorum*, *R. ussuriense*, *R. sanguineum*, and *R. uva-crispa* found that a significantfraction of the hybrids were resistant to powdery mildew, a small proportion was resistant to Septoria leaf spot, and no plants were resistant to Ribes species were found to be useful as disease resistance gene donors in this investigation (Mazeikiene et al., 2012; Sikorskaite et al., 2013). Researchers working on apple disease resistance aimed to find and develop candidate resistance genes, primarily for polygenic fungal disease resistance and hypersensitive response-based resistance to apple scab, to accelerate resistance breeding (Dvaranauskaitė et al., 2009). Also, orchard systems and their borders play an essential role in plant and diversity on multiple levels. According to Ricci et al. (2009), the populations of the codling moth (*Cydia pomonella*) in given orchards are negatively correlated with the surrounding surface zones planted with apple and treated with chemicals, which increase apple tree production within large surface zones regardless of the most effective method for codling moth pest control.

Furthermore, genetic and genomic studies on disease and insect resistance and stress response in orchard plants have been established over the last decade. The discovery and analysis of the phylogeny of plant beta-lactamase family proteins paved the way for a better comprehension of the protein's potential role in disease resistance *via* the hypersensitive response (Baniulis et al., 2013; Liobikas et al., 2006).

3. BIOTECHNOLOGICAL TECHNIQUESFOR ORCHARDS IMPROVEMENT AND MANAGEMENT

Some of the current applications of modern biotechnology fororchards and forest trees breeding programsinclude *in vitro* culture, quantitative and molecular breeding, clonal propagation, cryopreservation, and genetic engineering (Figure 3). Compared to crop plants, the introduction and rapid development of next-generation sequencing technology have played a critical role in delivering sequence information for numerous tree species with extremely high GD (Pant et al., 2011; Hamrick & Godt,1990; Abo-Doma & Azzam, 2007b; Azzam & Abo-Doma, 2007; Azzam et al., 2011; Aboelnaga et al., 2020). As forest trees have long generation intervals and life cycles and most obligate outcrosses, quantitative tree breeding is complicated. In comparison to highly domesticated crops, Orchard and tree breeding attempts have only lasted a few generations and for a short time. Therefore, breeding operations produce genetically heterogeneous seeds, which opens up enormous opportunities for tree enhancement *via* quantitative or molecular breeding. Since the late 1980s (Neale & Kremer, 2011; Flanagan et al., 2018), DNA-based molecular markers have been recognized as potential techniques for improving plant breeding.

Figure 3. Applications of modern biotechnology in orchards and forest trees breeding. Next-generation sequencing, increasing sequence information, and providing specific markers and genetic engineering possibilities are vital contributors to biotechnology. Cryopreservation of germplasm and in vitro cloning techniques that allow for the use and multiplication of specialized, potentially tailor-made cultivars in future orchards and forest treesapplications are also essential biotech features (Häggman et al., 2014)

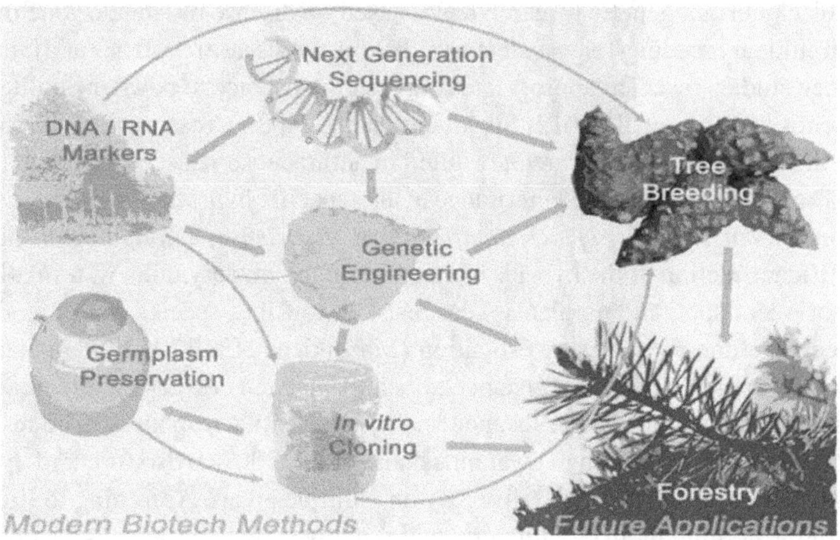

For numerous conifer species, recent breakthroughs in tree biotechnology have permitted the development of more flexible tree breeding and deployment tactics than traditional seed orchard approaches can provide (Gustavsson & Stanys, 2000)

3.1. In Vitro Breeding

Plant tissue culture techniques have been used for rapid propagation and *ex-situ* conservation of orchards over several decades, employing various methods and explants such as flower stalks, shoot tip nodes, stem bids and root tips rhizome segments (Pant, 2013; Reddy et al., 2016). Micropropagation allows the cultivation of both endangered and commercially important plant species (Shukla & Sharma, 2017). *in vitro* micropropagation is a valuable and essential breeding method for orchards, particularly for orchard seeds that are difficult to germinate and grow in the wild. It also serves as a foundation for orchard improvement and genetic engineering biotechnological interventions. *In vitro* micropropagation has been successfully applied in several species of *Paphiopedilum armeniacum*, *Bulbophyllum nipondhii, Paphiopedilum insigne* and *Anoectochilus elatus* making it a promising tool for the conservation of several threatened and endangered orchards (Zhang et al., 2015; Pakum et al., 2016; Diengdoh et al., 2017; Sherif et al., 2018).

Micropropagation and *in vitro* seed germination are greatly influenced by the composition of *in vitro* micropropagation media (Shukla & Sharma, 2007; Shekarriz et al., 2014; Dwivedi et al., 2020). MS media (Murashige & Skoog, 1962), for example, promotes higher germination in the orchard*Geodorum densiflorum*, whereas Knudson C media (Knudson, 1964) was found to be more suitable for Paphiopedilum seeds than other orchard specific culture media (Muthukrishnan et al., 2013). Furthermore, a low conc. of mineral salts in MS media (*viz.* ½ MS or ¼ MS) promotes seed germination in terrestrial orchards (Zeng et al., 2016). Along with the composition of orchard culture media, plant growth regulators, auxins and cytokinins play an essential role in the germination and development of orchard seeds and explants *in vitro* (Lindley et al., 2017). In addition to plant growth promoters, unspecified nutrient sources such as coconut water and potato extract to culture media promoted orchard*in vitro* seed germination (Pakum et al., 2016; Zhang et al., 2013).

In vitro plant tissue technologies have been employed to speed up the breeding process and screening procedures for cold hardiness in strawberry seedlings (Rugienius et al., 2013).Tissue culture propagated plants had more remarkable survival, development, and yield. Additionally, *In vitro* technology for lingonberry propagation was developed (Stanienė et al., 2002).

in vitro embryo rescue procedures,is one of the essential tissue culture techniques, usually practice with some orchards species *viz.*, sweet cherry, sour cherry, and black currant and have been devised by Stanys et al.(2006). Šikšnianas et al. (2008) used this procedure to create interspecific hybrids of *R. americanum* and *R. nigrum*. Polyploidization of *R. nigrum* was employed as an alternate to embryo rescue to offset the lower fertility during interspecific hybridization (Sasnauskas et al., 2007). Another study used an *in vitro* culture technique to create polyploid strains of Japanese quince (*Chaenomeles japonica*) (Stanys et al., 2006).Fruit plant growth and developmental management provide the means to control plant vigor and boost plant output while also providing the structure needed to satisfy the demands of production operations. Clonal material has also been used to make up for any shortages of enhanced seed from seed orchards that might occur. When germplasm conservation (Häggman et al., 2014) or the production of transgenic forest trees (Kole & Hall, 2008) has been considered, *in vitro* propagation techniques such as somatic embryogenesis (especially in conifers) and organogenesis (specifically in hardwoods) have recently been emphasized. Molecular marker technology was recently used to evaluate pear shoots' genetic and epigenetic stability developed *in vitro* and after long-term storage at low temperatures (Rugienius et al., 2013).

Multi-varietal forestry (MVF) can be implemented *via* somatic embryogenesis (SE) and cryopreservation, which is described as the use of genetically tested tree varieties in plantations (Figure 4). Generating value-added tree varieties for plantation forestry gives a more flexible tree breeding and deployment paradigm than seed orchards. Other technological technologies, such as molecular markers, can also help improve the efficacy of such a method (Park, 2014).

Figure 4. Process of conifer somatic embryogenesis (SE) in Multi-varietal forestry(MVF), adapted by Park (2014)

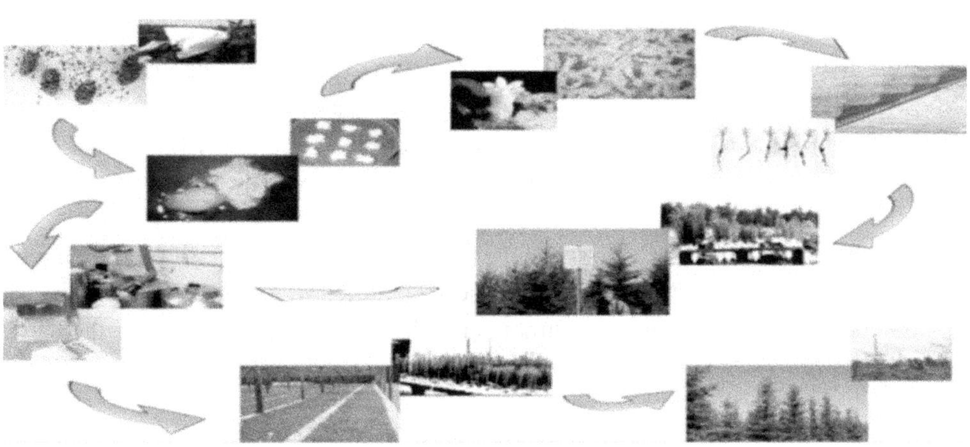

Due to the absence of an automated production system, SE's seedling production system is now slightly more expensive than traditional seedling production systems; somatic embryo (SE) can be utilized as an essential research tool for conifer development, ecophysiology, pest resistance, and functional genomics, among other aspects, in addition to the economic benefits of applying MVF. SE is an excellent method for hybrid variety development since it allows for the bulk generation of hybrid plants from a limited number of interspecific crossings (Abdrabou et al., 2017). In general, the most efficient way to initiate SE is to use immature zygotic embryos as the starting material; Embryo abortion is caused by genetic incompatibility between the developing embryo and the mother tissue. Even if embryos do not fully develop within the giant gametophyte tissue with the seed, they can generally develop if taken from seeds before abortion and raised to maturity *in vitro*, or immature embryos can be stimulated to commence SE. The latter is possible because SE frequently begins at the zygotic embryo's early cleavage polyembryony stage (Park, 2014; Rugienius et al., 2013; Lukoševičiūtė et al., 2011). Combining somatic embryogenesis with cryopreservation can greatly influence species conservation and restoration. SE allows valuable genotypes to be cryogenically kept long-term, restoring the sites with better adapted and pest-resistant genotypes. *In vitro* and thermal treatment, procedures create virus- and disease-free elite germplasm (Stanienė et al., 2002).

3.2. Molecular Genetic Markers

Molecular genetics research to develop biomarkers for early selection is not a new field of study. Random amplified polymorphic DNA (RAPD), inter simple sequence repeats (ISSRs), amplified fragment length polymorphism (AFLP), and simple sequence repeats (SSRs) were the most common of the second wave of genetic markers, which were all based on polymerase chain reaction (PCR) (Pant et al., 2011; Abbas et al., 2021; Liesebach et al., 2021; Mazeikiene et al., 2012; Azzam, 2004).However, in tree breeding populations, these markers could not overcome the low amount of linkage disequilibrium (LD) and recombination events with each generation. Therefore, QTLs could not be effectively used in marker-assisted selection (MAS) or tree breeding (Neale & Kremer, 2011). In recent years, association genetics studies have successfully identified candidate gene (SNPs) linked to anextensive range of quantitative characteristics in orchards and forest trees, including growth, stress, wood properties, and disease resistance (Neale & Kremer, 2011). Clonal propagation of orchards and forest trees has long been regarded as an essential strategy in traditional breeding for gaining additional benefits from nonadditive genetic diversity, improving orchards and forest trees by using genetically modified planting material, and increasing homogeneity.

Quantitative trait loci (QTL) mapping was once a 'moderntechnique'. It would undoubtedly improve and become more valuable with better genetic markers (microsatellite /SSRs and RFLPs). To estimate genomic breeding values, a dense SNP (single nucleotide polymorphism) marker 'chip' containing 1,000's of random SNP variants across the genome is used instead of breeding values based on field phenotypic data in repetitive trials (Zapata-Valenzuela et al., 2011; Bāders et al., 2014; Burdon & Wilcox, 2011). Grattapaglia and Resende (Grattapaglia & Resende, 2011) reviewed the evolution of the approach in orchards and forest trees.

The mismatch between science and operational application is a common problem in using molecular biology in orchards and forest tree breeding. The objective aimed to create these tools in collaboration with breeders in a real-world breeding environment (Burdon & Wilcox, 2011). The shift away from the nucleus or elite-based breeding populations, which will have repercussions for genomic selection(GS), reflects a current divide between molecular and quantitative geneticists (El-Kassaby et al., 2011; White et al., 2007; Dungey et al., 2014; Jayawickrama & Carson, 2000). Klápště et al. (2014) were dubbed "breeding without breeding". To put it another way, adopting such technology in tree breeding aids in intensifying domestication, yet domestication implies accepting higher costs in exchange for higher benefits.

Droplet polymerase chain reaction (dPCR) is a novel PCR-based orchard disease diagnosis technique, which employs the Taq DAN Polymerase to unwind a targeted DNA sequence from a complex test *via* a prevalidated primer/probe test (Taylor et al., 2017). An advanced spectroscopy method is used to diagnose disease in orchards in addition to the PCR-based nucleic acid amplification detection technique. Surface-Enhance Raman Spectroscopy (SERS) is a non-destructive, developing laser-based spectroscopy method based on Raman scrambling that employs resistant tests and atomic tests for pathogen detection in plants (Chocarro-Ruiz et al., 2017). Extensive research is also being conducted on DNA-hybridization and colourimetric biosensor-based approaches in order to develop a rapid, real-time disease diagnosis tool (Jain et al., 2021). Similarly, an advanced integrated microfluidic chip-based system for automated rapid virus detection using nucleic acid amplification is being developed.

The most common orchard virus, *Cymbidium mosaic* virus, was identified using pathogen-specific RNA purification from disease orchard samples. The isolated RNA is amplified to cDNA and optically detected using reverse transcription loop-mediated isothermal amplification (RT-LAMP) to detect the pathogen in the sample (Chang et al., 2013).

3.3. Marker-Assisted Selection (MAS)

The development of low-cost molecular genetic marker systems makes it possible to investigate the genetic basis of phenotypic features in breeding populations. Marker-assisted selection is a method of selecting desirable genotype in a breeding scheme based on DNA molecular marker that can help breeders select more efficiently for desirable crop traits. This model has significant implications for attempts to understand better the molecular genetic pathways that underpin phenotypic diversity in orchards and forest tree breeding programs and breeders interested in correctly forecasting the genetic form of marker-assisted selection (MAS) for quantitative traits merit of individuals based on genotype data. White et al. (2007) discussed association genetics in tree breeding. The ability to detect associations is influenced by the presence of population structure, the size of the test population, the degree of linkage disequilibrium in the test population, and the proportion of phenotypic variation accounted for by each causative genetic variant involved in the phenotype of interest (Neale & Kremer, 2011).

The genetic variations examined for phenotype connection may be found in known genes that influence the phenotype under investigation (the 'candidate gene' technique). They may be chosen based on allele frequencies in the population and genome distribution (the 'genome-wide approach'). Numerous population characteristics that affect power must be evaluated to determine the suitable sample size and number of genetic loci to test to achieve a certain power level (Ball, 2005; Hindorff et al., 2009; Spencer et al., 2009). Some of these factors include the extent of a locus's genetic effect, the frequency of the allele responsible for an impact in the population, and the level of LD between the causal allele and neighboring genetic markers (SNPs). The amplitude of the genetic influence is frequently stated as a ratio of illness occurs in a heterozygous individual to disease occurrence in a homozygous individual for the common allele (genotypic risk ratio (Risch & Merikangas, 1996), or relative risk per allele (Spencer et al., 2009). Power is lower for lower risk allele frequencies, lower risk per allele, and fewer genetic variant loci tested, according to simulations (Spencer et al., 2009). Individual genes with significant effects control various traits important to tree breeding programs, for instance, resistance to fusiform rust disease in *Pinus taeda* (Wilcox et al., 2007). When all SNPs were treated as random effects, Yang et al. (2010) discovered that a mixed linear model based on correlation information resulting from marker genotypes explained nearly half of the genetic variance in height in a sample human population of fewer than 4,000 individuals. Multiple genetic variants per gene have been discovered in various situations (Lango et al., 2010). Because multiple genetic variants, even when it comes to the same functional gene, can cause the same trait, a phenomenon known as allelic heterogeneity reduces the power of association analysis.When multiple genetic variations within genes affect the identical phenotype, epistatic interactions can occur, resulting in differences in heritability estimated from closely related versus distantly related individuals (Würschum et al., 2012; Zuk et al., 2012). So, analyzing association genetics data by grouping variants into functional genes, organizing genes into pathways, and integrating genetic pathways with gene expression data may provide more insights into phenotypic variation (Azzam et al., 2007; Azzam et al., 2012; Kreimer et al., 2012; Azzam et al., 2021). Only alleles with relatively large effects might be identified unless sample sizes exceed 5,000 and marker allele frequency is close to causative variant allele frequency in conventional association genetics studies (Ball, 2005; Stranger et al., 2011; Kemper et al., 2012). Yu et al. (2009) & McMullen et al.(2009) used this procedure to construct the maize Nested Association Mapping (NAM) population and strategies to cope with the population structure that occurs in populations produced through mating designs.Several SNPs that affect maize kernel composition were discovered using a combination of the NAM population and a more specialized association population of 282 inbred lines (Cook et al., 2012).Once reference genome sequences are available, and haplotype information for the parents of elite breeding populations can be easily created, similar procedures may become practical in orchards and forest tree breeding operations. In the long run, studies that focus on developing models of inheritance of complex traits in breeding populations can contribute to the development of predictive genetic models, whereas studies that focus on developing models of inheritance of complex traits in breeding populations have more immediate value in the short term (Peters & Musunuru, 2012).

Many features of importance to breeders are polygenic, influenced by many genes, each with a minor impact (Hill et al., 2008). The effectiveness of complicated trait improvement depends on these small-effect genes (Crosbie et al., 2003). Plant breeders have relied on phenotype and relative similarity to capture genetic diversity explained by these small-effect genes for decades. Breeders didn't know the underlying genetic architecture of complex features, such as the number of genes influencing the character and their position in the genome. On the contrary, rather than uncovering particular genes in breeding procedures, the focus is on forecasting the genetic value of individuals or lines. Since its inception, GS has transformed the paradigm by Meuwissen et al. (2001),owing to increased DNA sequencing technology and computational capacity efficiency. Because there is no predetermined subset of important markers utilized for selection in GS, it differs dramatically from classical MAS.

On the other hand, GS examines all markers in a population as a whole, seeking to explain the total genetic variance with dense genome-wide marker coverage by summing marker effects to predict individual breeding values (Meuwissen et al., 2001).The objective is to capture the LD between markers or marker haplotypes and causal variation by populating the genome with high-density markers.This relationship would be consistent across all families (Meuwissen et al., 2001).GS is used to choose the candidates for the next breeding cycle. As new phenotype and marker data gather, the training can be repeated iteratively (Heffner et al., 2011).

With little genetic diversity from wild populations, orchards and forest tree breeding operations are still in the early stages of breeding-testing and selection cycles. If practical, genomic selection could significantly impact orchards and forest tree breeding (Resende et al., 2012). Similarly, Isik et al. (2011) found that genomic calculated breeding values had the same level of reliability as breeding values based on relative likeness and phenotypic data in the same species. The genomic relationship matrix is then replaced for the additive genetic relationship matrix produced from pedigree to predict genomic estimated breeding values. Genomic BLUP (GBLUP) could be a helpful technique for orshards and forest tree breeding projects. Such models can capture the Mendelian segregation impact in full-sib families, which was impossible with average additive genetic links (Zapata-Valenzuela et al., 2011).

The molecular marker-assisted breeding approach uses the precision of molecular biology tools and techniques to enable fast, accurate and environmentally unaffected orchard breeding and natural and artificial hybrid selection (Jiang, 2015). RFLP (restriction fragment length polymerase), AFLP (amplified fragment length polymerase), ISSR (Inter simple sequence repeats) and SNP (single nucleotide polymorphism) markers are used by the *mmab* (Li et al., 2021). The majority of these markers are widely used in modern orchard breeding and have yielded positive results thus far. Li et al. (2015) recently developed a set of wide range genic-SSR markers in *Cymbidium ensifolium* to evaluate the orchard population's genetic relationship and trait mapping. These SSR markers are not only helpful in determining genetic relationships but they have also been used to identify root growth development mechanisms, secondary metabolites-related gene identification in many orchards and flower shape and colour-related genes in Phalaenopsis (Li et al., 2015; Wu et al., 2017; Sudarsono et al., 2017).

Similarly, SNP markers were used to create integrated genetic maps of the Dendrobium genome and identify several essential QTL sites (Lu et al., 2018). Although MMAB has played a crucial role in modern orchard breeding, its primary goal is phylogenetic analysis for genetic relationship establishment among *orchidaceae* members. Consequently, there are still opportunities for new approaches, preferably hybrid approaches that combine traditional and modern orchard breeding strategies with biotechnological interventions for improved orchard breeding. However, all orchard breeding methods, including crossbreeding, selection breeding, mutational breeding and MAAB, eventually result in the selection and identification of new, improved hybrid variety production that can be mass-produced using *in vitro* propagation methods.

3.4. Genome Analysis

Genome-wide selectionis a relatively new strategy that uses a genome-wide form of marker-assisted selection (MAS) for quantitative traits. It's a current technique in plant breeding right now, and it's being tried to see if it can include favorable alleles at a variety of low-effect loci. Individuals' genomic breeding values (GBV) in an experimental or "training" population can be predicted using prediction models built by regressing phenotypes on whole-genome marker genotypes for hundreds to thousands of individuals (Meuwissen et al., 2001).

To speed up the breeding process, recombination of individuals selected in progeny trials based on their additive values must proceed when selecting individuals from the same progeny trials based on their total genetic values are cloned for use in the clonal field trials (Kerr et al., 2004). The involvement of reactive oxygen species in the regulation of plant response to stress was shown by combining a mechanistic knowledge of the working of photosynthetic cytochrome *b6f* (Cramer et al., 2011) with its function in the formation of superoxide radicals (Baniulis et al., 2013). Furthermore, the publication of genome sequences of domestic apple and woodland strawberry plants (Velasco et al., 2010; Shulaev et al., 2011) and advancements in plant proteome research technology laid the groundwork for the development of functional genomics tools required for the discovery of molecular markers of genes involved in the regulation of disease resistance response in apple and other Rosaceae family plants (Sikorskaite et al., 2013).For screening strawberries for red stele resistance, PCR-based markers for the *Rpf1* gene were designed and employed (Sasnauskas et al., 2007; Rugienius et al., 2013).

3.5. Omics Technologies

Omics refers to a group of disciplines, including transcriptomics, genomics, proteomics and metabolomics, aiming to improve understanding of the roles and pathways of different plant molecules (Balilashaki et al., 2020). Genomics is a broad overview of the entire set of genetic instructions delivered by DNA. Orchard genomes are typically larger than those of the most model plants and genome size varies dramatically across the family, with the amount of nuclear DNA varying up to 168-fold (Leitch et al., 2014). The gene expressions associated with *in vitro* floral transition in an orchard hybrid (Dendrobium grex. Madame Thong-In) were studied (Yu et al., 2000). They studied three orchard MADS-box genes, *domads1, domads2* and *domads3* sequence, then isolated them from the TSAM cDNA library. *domads3* shares a signature amino acid with members of the independent OSMADS1 subfamily, distinct from the *agl2* sub-family. All three *domads* genes were expressed in the TSAM during the floral transition and they were also expressed later in the mature flower stage (Yu et al., 2000). (Figure5).

Figure 5. Summarized the biological roles of MADS-box genes in controlling flowering in the Orchid plant. Orthologous genes of AGL6, AP1, SOC1 and SVP have been isolated and functionally characterized either in heterologous system orchids and shown to be involved in promoting flowering. MADS-box transcription factors that function as flowering activators and suppressors are shown in green and red, respectively, whereas remaining flowering regulators are shown in black boxes. Black arrows and orange T bars indicate promoting and repressing flowering, respectively. The dashed lines with arrows indicate possible positive regulation based on the studies using heterologous systems. Double-ended diamond arrows indicate protein-protein interactions. Abbreviation:- AGL: agamous-like; AP: apetala; CO: constans; FTIP: ft-interacting protein; FUL: fruitful; LFY: leafy; SOC: suppressor of overexpression of constans; SVP: short vegetative phase, adapted by Tiwari et al. (2022)

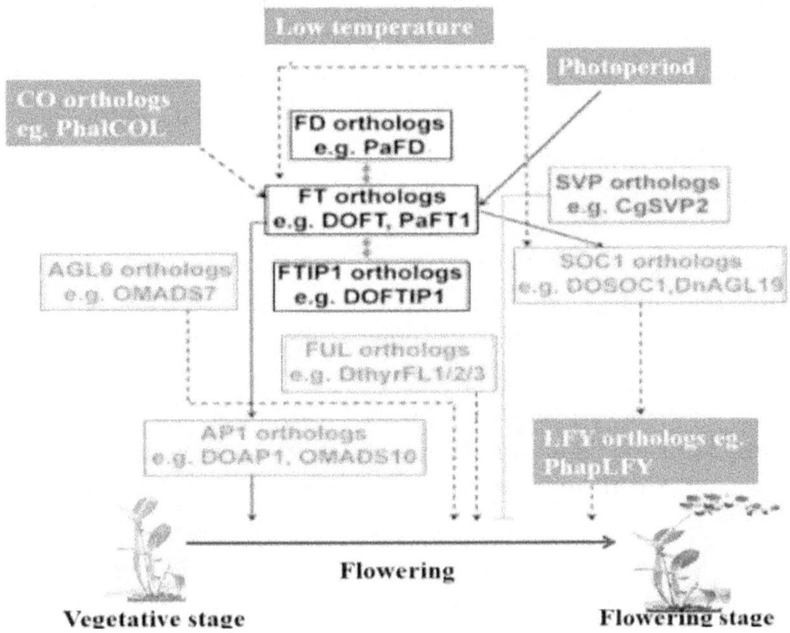

Transcriptome analysis (total clean-read pairs) was performed on *Cymbidium densifolium* pooled flower buds and mature flowers (78.1%) (Li et al., 2013). The transcriptome of the mature plant Cymbidium sinense was examined in order to identify genes involved in floral development (Zhang et al., 2013). MicroRNAs (miRNAs) are short RNA molecules that influence physiological mechanisms such as development, cell proliferation, cell death and differentiation in eukaryotes (Aceto et al., 2014). The various techniques used to study the entire "miRNome" allow for investigating these novel mechanisms of gene expression regulation (Aceto et al., 2014).

MiRNome analysis of *Orchis Italica* inflorescence revealed the presence of both conserved and novel miRNAs. In silico, searching for potential miRNA targets revealed a conserved miRNA cleavage site within the four OitaDEF-like transcripts, experimentally validated for OitaDEF2 (Aceto et al., 2014). This finding indicates that miRNAs play an essential role in the diversification of orchard perianth organs *via* the inhibitory regulation of the clade-2 DEF-like gene. Different mechanisms could regulate the expression of the other DEF-like genes, implying the existence of lineage-specific regulatory mechanisms that contribute to the functional specialization of the DEF-like clades in orchards. Advances in next-generation sequencing (NGS) technologies have resulted in developing new algorithms for the computational analysis of genome-scale RNA-seq transcriptomes (Szczesniak et al., 2016 ; Chao et al., 2017)

For species lacking reference genomes, characterizing the entire transcriptome provides valuable information about genomic features and their function, such as protein-coding/noncoding gene transcripts and alternative splicing (Sakai et al., 2016). Data from whole-transcriptome sequencing can also be used to sort through the complexities of genome-level analysis. Furthermore, because the transcriptome sequencing data platform efficiently reduces genome complexity to obtain functional information, it is frequently used as a starting point for large-scale sequencing and the development of genomics tools. On the other hand, Transcriptomics is concerned with gene expression patterns(Lowe et al., 2017). The MADS-box gene family was studied as a family with a critical role in flowering. MADS-box genes play a crucial role in flower and fruit development, while transcription factors encoding ABCDE functions control floral morphogenesis in orchards (Ng et al., 2012; Tsai et al., 2006). The Orchestra (http://orchidstra2.abrc.sinica.edu.tw) is a tool for assembling orchard transcriptomes and annotating genes. This database has been operational since 2013 (Chao et al., 2017). Orchestra 2.0 was created with a new database system to store the annotations of 510,947 protein-coding genes and 161,826 non-coding transcripts, including 18 orchard species from 12 genera in five sub-families of *orchidaceae*. The Orchestra 2.0 database reveals that RNA-Seq-based gene expression data from the *knox* genes were highly expressed in developing flower stalks and germinating seeds in *P. Aphrodite*, as well as mesocarp tissues of developing vanilla six and eight-week-old pods in *V. planifolia* (Chao et al., 2017).

Proteomics is the study of dynamic protein products, their interactions and their relationships with one another (Hsiao et al., 2011), whereas metabolomics is an intermediate step in understanding an organism's and its entire metabolic pathways (Liu & Locasale, 2017). Most orchard proteomic studies focus on flower development and tissue culture of orchards for mass production. Omics approaches can help with orchard breeding, genetic improvement, conservation and commercial production by elucidating developmental processes. This review provides an overview of the most recent developments and approaches in orchard research, focusing on developmental processes in orchards from seed formation to flower senescence using new molecular methods such as omics technologies.

Proteomics has given several insights into how gene expression products influence physical qualities or phenotype by allowing the structure-function interactions of proteins associated with genomic sequences. More recently, they have theorized on how the environment interacts with genotype to cause heritable changes in phenotype, a field known as epigenetics (Bonetta, 2008).In epigenomic studies, advances in 'omics technologies now allow epigenetic tools across entire genomes (Figure 6).DNA methylation, histone modification, and the function of non-coding RNAs are examples of epigenetic regulatory mechanisms for trees. The sites, timing, and extent of cytosine methylation in trees to generate 5-methyl cytosine are likely to impact transcription significantly. According to the findings, poplar contains a mix of three unique cytosine methylation contexts, with CG sites being the most highly methylated, followed by CHG sites and CHH sites, where H can be one of the bases of adenosine, cytosine, or thymine (Feng et al., 2010).

Cytosine can be methylated in coding regions, often known as the body of a gene, or promoter regions, which control gene production. While breeders don't have a comprehensive atlas of cytosine methylation patterns and their effects on gene expression in trees, they know that there are significant differences in cytosine methylation patterns and their impact on gene expression in different tree organs under different developmental stages or environmental stresses (Vining et al., 2012).

Small RNAs, ranging from epigenetic effects due to non-coding RNAs possibly directing DNA methylation, small interfering RNAs (siRNAs) acting to post-transcriptionally downregulate gene expression, to micro RNAs and the possibility of piRNAs interacting with Argonaute proteins, provides another promising avenue for future gene expression studies in trees (Ghildiyal & Zamore, 2009). Due to changed chromatin folding, epigenomic investigations have allowed comprehensive knowledge of such histone modifications to be mapped, resulting in up and downregulation of gene expression in large regions of chromosomes (ENCODE Project Consortium, 2012). The flood of new epigenomic data that is now starting to surface will undoubtedly lead to a better comprehension of gene regulation, stress responses, and development. All of these improvements, when combined, will ensure that orshards and forest biotechnology contributes much more to satisfying our global economic, environmental, and societal needs for many decades to come (Troggio et al., 2012).

Pollination response proteomics in the endangered orchard species Dendrobium chrysanthemum (Wang et al., 2017).The differentially expressed proteins (DEP) between the self-pollination and cross-pollination pistils of *Dendrobium chrysanthemum* were investigated using proteomic approaches to understand the pollination mechanism in *D. chrysanthemum*. The two-dimensional electrophoresis (2-DE) technique coupled with tandem mass spectrometry was used for proteomic investigation (Wang et al., 2017). He studied the comparative proteomics of pollination responses in the endangered orchard species Dendrobium chrysanthemum. Proteomic approaches were used to investigate the differentially expressed proteins (DEP) between the self-pollination and cross-pollination pistils of *Dendrobium chrysanthemum*. Two-dimensional electrophoresis (2-DE) coupled with tandem mass spectrometry was used for proteomic analysis (Wang et al., 2017).

In the two-dimensional electrophoresis (2-DE) maps between self-pollination and cross-pollination, 54 DEP spots were identified. Gene ontology analysis revealed a variety of proteins belonging to the following functional categories: metabolic process (8.94%), response to stimulus (5.69%), biosynthetic process (4.07%), protein folding (3.25%) and transport (3.25%). Identifying these DEPs during the early stages of pollination should provide new insights into the mechanism of pollination response and aid in the conservation of orchard species (3). Flower labellum tissues samples of *O. garganica*, *Ophrys exaltata* subsp. *archipelago* and *O. sphegodes* were used for the identification of candidate genes for pollinator attraction and reproductive isolation (*e.g.*, genes for hydrocarbon and anthocyanin biosynthesis and regulation and floral morphology development) using proteomics techniques (LC-MS/MS, LTQ (HPLC) (Sedeek et al., 2013)

Figure 6. The three main areas of epigenetics are DNA methylation, histone modifications (acetylation, methylation, phospholylation, etc.) and micro-RNA based mechanisms. These three processes are distinct but are interrelated and control gene expression, Adapted by Khalil (2014)

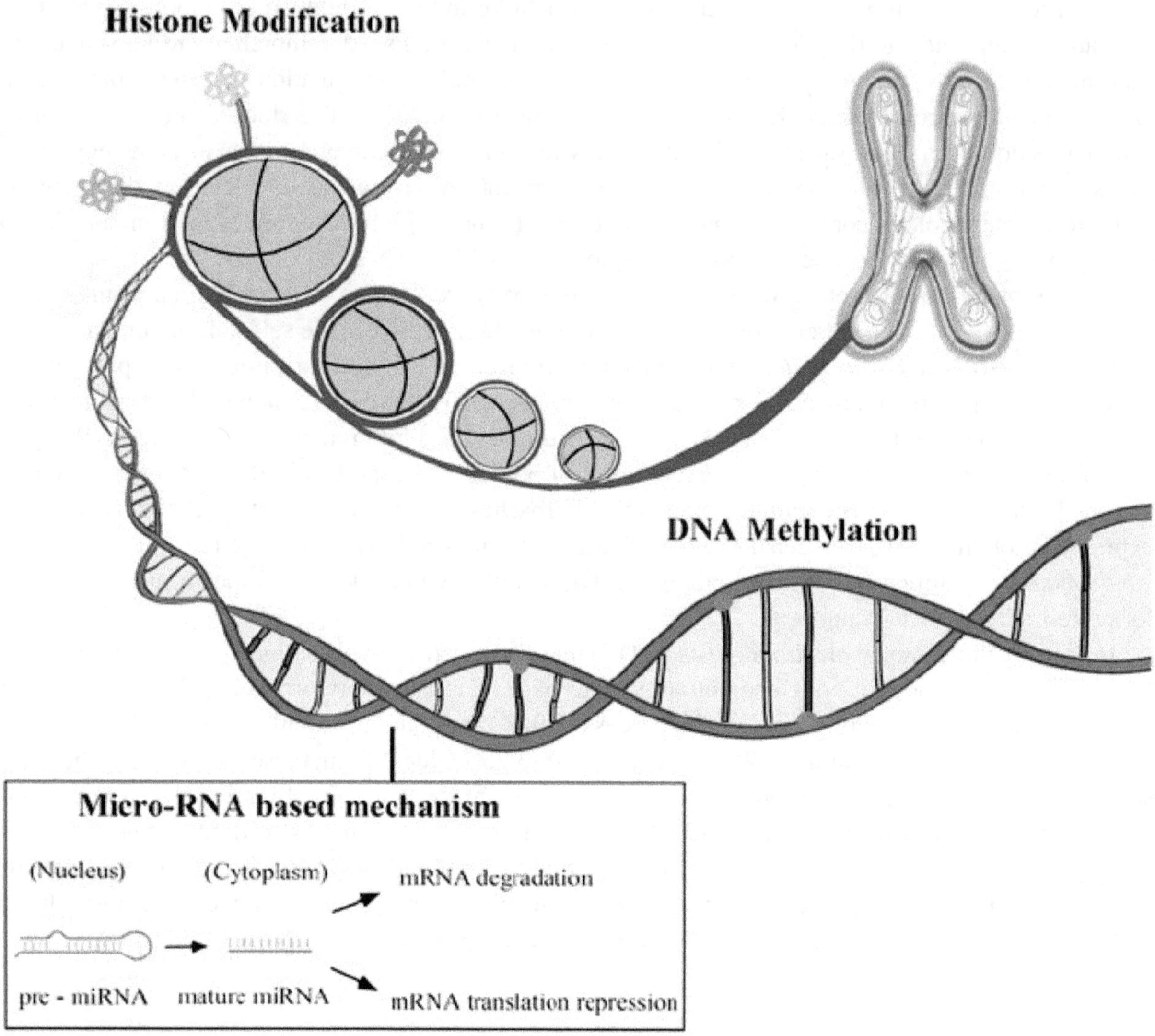

Proteins in *Cymbidium ensifolium*orchard flower structure, including inner lateral petals and labellum were analyzed using a proteomic technique (2 DE MALDI ToF/ToF) (Li et al., 2014) Yeast 2 hybrid system was used to investigate the DNA binding properties and protein-protein interactions of floral homeotic MADS-box protein complexes in *Phalaenopsis equestris* (Tsai et al., 2008).

There are few reports on the molecular mechanisms of mycorrhizal association and seed germination in orchards. So the question is whether there is a molecular difference between orchard seed development and germination and other flowering plants. According to Chen et al. (2016), some genes, such as *pamads39* and *pamads51*, belong to the Ma-subclass of type I. MADS-box genes are detected during seed development at the cellularization of developing endosperm. These genes are closely related to Arabidopsis (*agl23* & *agl62.1*) and have similar expression patterns in reproductive tissues when fertilization occurs and embryo development begins. MIKE-type genes were discovered in streptophyte lineages, revealing new information about their evolution and development. They discovered that MIKC-type genes might be involved in seed germination in *Dendrobium officinale.*

Some *MIKC* genes from *D. offcinale*, including SVP and SQUA subfamily genes, and the *MIKC* gene, showed different expressions during seed germination and these genes play the same role in other genes in flowering plants (He et al., 2019). Consequently, it was concluded that the expression pattern in orchard seed germination was the same as that of another plant, Arabidopsis. In orchards, fungi-dependent and independent germination are called symbiotic and asymbiotic germination, respectively. The first large-scale transcriptome and dataset of *Anoectochilus roxburghii* (Wall.) Lindl. seeds in both symbiotic and symbiotic seed germination were generated by Liu et al. (2015). The regulatory module was associated with 49 genes, six of which were differentially expressed in symbiotic germination vs. asymbiotic germination. It has also been proposed that fungi may induce or suppress these six genes. Valadares et al. (2014) identified 88 proteins in *Oncidium sphacelate* Lindl involved in energy metabolism, cell rescue and defense, molecular signaling and secondary metabolism. At various trophic levels of symbiotic germination. Proteomic analysis revealed that proteins involved in purine recycling, ribosome biogenesis, energy metabolism and secretion were upregulated in *O. sphacelate.*

3.6. Genetic Engineering

Breeding approaches produced many novel hybrids and cultivars with desired traits of orchards of genera such as Phalaenopsis, Cattleya, Cymbidium, Dendrobium, Oncidium, Paphiopedilum, Vanda, and others. On the other hand, traditional orchard breeding failed to address issues such as diseases, pests, and environmental stress. Therefore, orchard germplasm improvement is carried out as a supplement to the traditional orchard breeding system in order to confer new and desirable traits. Plant genetic engineering is already recognized as one of the most powerful technologies for improving plant cultivars and studying gene function (Sharma et al., 2018).Orchard genetic engineering procedures for several commercially essential orchards have been established over the last two decades.

In orchards, genetic engineering procedures deliver the desired gene *via* particle bombardment or Agrobacterium-mediated transformation systems. The first orchard genetic transformation studies focused solely on biolistic-mediated transformation systems (Yang et al., 1999; Tee et al., 2003). The first successful Agrobacterium-mediated genetic transformation was accomplished in the orchard genus *Phalaenopsis*, which expressed a *gus* gene construct (Belarmino & Mii,2000). In terms of incorporating low copy number genes at transcriptionally active chromosomal regions, Agrobacterium-mediated genetic transformation systems were more helpful than biolistic-mediated genetic transformation systems (Hiei et al., 2000). Several genetically engineered orchards have been produced in the last decade using Agrobacterium-mediated genetic transformation systems in the orchard genera Dendrobium, Phalaenopsis, Oncidium, Cymbidium and Vanda (daSilva et al., 2011; daSilva et al., 2016). Genetic engineering approaches in orchards include gene silencing studies in *Dendrobium Sonia* and *Oncidium* hybrid orchards to knock out genes and overexpression of heterologous genes to incorporate desired traits (Liu et al., 2014). With the release of the genome sequences of Dendrobium officinale and Phalaenopsis equestris, genetic engineering approaches will allow CRISPR4-Cas9 mediated genome editing in a variety of orchard species (Kui et al., 2017).

Genetic transformation permits functional genes or traits not found in even the most elite breeding stock to be incorporated into specific genotypes. Genetic engineering for insect resistance, or genes that allow trees to tolerate salt, heavy metal, or drought, for example, can be precious (Häggman et al., 2014).Breeders know a lot more about the risk associated with GE trees than they do with conventionally bred trees because of regulatory attention on genetic engineering. Nonetheless, second-generation transgenic crops (including stresses such as drought, cold, salt, heat, flood, increased yield, lower nutrient requirements, or increased tolerance to diseases and pathogens) are already being tested in field experiments worldwide (Häggman et al., 2014). *Populus alba grandidentata* was the first GE plant species, genetically transformed with *Agrobacterium tumefaciens* with neomycin phosphotransferase (*nptII*) as a marker gene and the bacterial 5-enolpyruvyl- shikimate- 3-phosphate (*epsp*) synthase gene (*aroA*) to confer herbicide tolerance (Fillatt et al., 1987). Artificial miRNAs and overexpression of siRNAs that are capable of downregulating endogenous gene function, virus-induced gene silencing (*VIGS*) that takes advantage of natural posttranscriptional silencing (*PTGS*) mechanisms, used for viral defense (Busov et al., 2010; Zhang et al., 2010).

The vertical or horizontal gene transfer of recombinant DNA into the gene pool of other plants in the vicinity of the same or different species is one of the major issues with GE trees. If practical, this method could be used to prevent plantation trees from becoming a nuisance in locations where they outperform natural species) and recommended in various cases (papaya) (Häggman et al., 2014).

Most GE crop and tree regulations are based on scientific information and data from 15–20 years ago when our comprehension of genetic engineering and its potential dangers was still in its infancy. The main goal of COST Action FP0905 is to evaluate and support scientific data on the biosafety of genetically modified trees (Fladung et al., 2012). A broad exchange of scientific knowledge has begun, providing a once-in-a-lifetime chance to establish a shared scientific foundation for biosafety research and tree development (Häggman et al., 2014).

Because of the long generation time and difficulty finding resistant genotypes in the field, disease and pest resistance by traditional selection and breeding is challenging for orchards and forest trees. Merkle et al.(2007) looked into the possibility of restoring American chestnut (*Castanea dentata*) trees that had been ravaged by the chestnut blight fungus (*Cryphonectria parasitica*) through the use of embryogenic culture to introduce antifungal genes. Most authorities worldwide limit the testing and deployment of transgenic trees due to the potential for adverse environmental effects and bio-safety concerns (Azzam, 2004; Trontin et al., 2007).

In order to identify transformed cells in the early stages of most genetic transformation methods, selectable marker genes are usually required. Antibiotic resistance genes, such as *nptII*, provide resistance to the antibiotic kanamycin in transformed tissues, are commonly used as selectable marker genes. Except for one or a few transgenic lines, transgene expression levels are generally stable, as shown in studies with the marker genes *rolC* (Fladung et al., 2012), *uidA* (Hawkins et al., 2003), the green fluorescent protein-encoding gene *gfp* (together with bar), and *nptII* (Li et al., 2019). Furthermore, in a 2-year field experiment using *gfp/bar* hybrid poplars, the stability of gene expression suppression produced by RNAi was demonstrated to be stable (Li et al., 2019). Furthermore, in research by Stefani et al. (2009), no unexpected consequences in the arbuscular or ectomycorrhizal status of transgenic aspens were found. Changes in phenotype or gene expression mediated by causes other than changes in the underlying DNA sequence are commonly referred to as epigenetics. Memory effects in the filial generation last for many years (Meehan et al., 2005).

DISCCUSSION

orchards, which are highly regarded for their aesthetic value, are among the most successful commercial crops in the global floriculture market and are widely exploited due to their ornamental and socioeconomic importance. It is one of the most prominent families of flowering plants, with approximately 28,000 species with diverse characteristics. The orchard family is the second largest family of flowering plants, with over 800 genera and a wide range of medicinal properties. With recent advances in scientific interventions and system biology, there has been a significant increase in awareness of orchard biology and its multifaceted importance. The orchard family is receiving attention in biology, evolution, classification, phytochemistry and cultivation, among others. Orchards of various varieties are cultivated and sold worldwide as cut flowers and potted plants, increasing the demand for exotic varieties in the floriculture trade (Hinsley et al., 2018).

Orchards (*Orchidaceae*) are the largest and most diverse group of flowering plants and they are gaining popularity around the world. Aside from its ornamental and aesthetic value, the orchard industry has successfully created jobs in developing countries. The orchard industry has grown by leaps and bounds in the last decade, domestically and internationally. Plant biologists and agriculturists have been drawn to new orchard species' aesthetic and ornamental qualities. For the orchard industry to gain traction and flourish to new heights, it is critical to investigate biotechnological interventions in orchards for plant trait improvement and ensure legal international trade by implementing guidelines for a multifaceted approach to orchard conservation and commercialization. Orchards have traditionally been cultivated for their cut flowers and artificially propagated varieties, with the multifaceted applications of orchards in the food sector and medicine only recently gaining global recognition (Wang et al., 2017).

Furthermore, alkaloids, polysaccharides and other essential components make these plants appealing for use in the food industry, floriculture and medicine (Wang et al., 2020). An assessment of 948 orchard species by the IUCN Global Red List revealed that 56.5 percent are threatened (IUCN,, 2017) and must be protected. Approaches in plant tissue culture have greatly aided in conserving exotic varieties in this direction. Bulbophyllum nipondhii, Anoectochilus elatus and Paphiopedilum armeniacum *in vitro* propagation (Pakum et al., 2016; Diengdoh et al., 2017; Sherif et al., 2018; Zhang et al., 2015) aided in the conservation of rare orchard varieties. Furthermore, vitrification techniques aided in preserving immature and high-water content seeds from several orchard genera, including Cymbidium, Bletilla, Dendrobium, Phaius, Encyclia and Vanda.The genetic engineering of *Oncidium* and *Odontoglossum*orchards to improve orchard growth and vase life was made using a mutant ethylene receptor gene (Raffeiner et al., 2009).

The globalization of economies has increased the demand for orchards in the floriculture trade, necessitating the development of new varieties with novel traits, quality and disease resistance (Kamboj, 2020).

Traditional plant breeding approaches formed the mainstream for orchard breeding, which was attempted through mutational and hybridization with associated limitations. In recent years, Agrobacterium-mediated and particle bombardment methods for foreign (desired) gene insertion have been widely used, contributing to significant progress in plant trait improvement (Filippo, 2017). Furthermore, recent advances in DNA molecular marker-based approaches have broadened the horizon, opening new avenues for practical orchard breeding (Das et al., 2017). Furthermore, the new era in "omics technologies," as well as the availability of next-generation sequencing technologies, have elucidated the role of essential genes involved in flower colour, resistance, flower shape, flowering time and other functions, providing a platform for 'genome-editing' in orchards, leading to improved plant traits (Li et al., 2021). The transfer of phytoene synthase-RNAi constructs into the protocorm-like bodies (PLBs) of the Oncidium hybrid resulted in the down-regulation of geranylgeranyl synthase and PSY genes, with the transgenes showing lower levels of abscisic acid and gibberellic acid than the wild type (Liu et al., 2014). Semiarti (2018) demonstrated the Agrobacterium-mediated insertion of KNAT1 (bud meristem differentiation) and AtRKD4 (embryonic differentiation) in the orchard genome to initiate bud development and embryogenesis, resulting in increased yields (Semiarti, 2018).

With unprecedented advancements and translational success in orchard biotechnology and commercialization, associated challenges continue to impact exotic variety cultivation, conservation and commercialization. The cultivation practices are the most important in this area because the efficiency of genetic transformation is low and insufficient information on gene function and breeding adds to the challenges. These constraints make it challenging to apply transgenic breeding to new varieties for trait improvement (Li et al., 2021). Another primary concern is that the false positive rate in orchard genetic manipulations is higher, implying that transgenic approaches must be improved. Other novels and exotic varieties defining socioeconomic attributes need to be further studied for trait improvement (Li et al., 2021), defining expansion of the branch of genetic engineering for crucial studies. As a possible solution, PLBs were used as transformation receptors in essential orchard species *Dendrobium* (Uddain et al., 2015). *Cymbidium* (Yang et al., 1999), *Phalaenopsis* (Chan et al., 2005) and *Oncidium* (Li et al., 2005). Other major issues in the global floriculture trade include the widespread use of online platforms, which contributes to illegal trade and unsustainable harvesting of rare varieties (Hinsley et al., 2018).

Biotechnology interventions are essential in developing orchards and ornamental plant cultivars with enhanced floral and multifaceted characteristics. Biotechnological interventions such as *in vitro* tissue culture, genetic engineering and classifying breeding methods have been routinely used to modify flower colour, appearance, fragrance, disease resistance and shelf life (Figure 7)

CONCLUSION

Orchards cultivation is rapidly expanding to meet rising market demand by developing new orchard varieties with distinct colour and appearance and quality characteristics. To fulfill this demand and supply gap, both traditional and molecular breeding approaches are employed with high effort. Traditional orchard breeding

Figure 7. Recent biotechnological interventions in orchids to improve multifaceted traits, adapted by Tiwari et al. (2022)

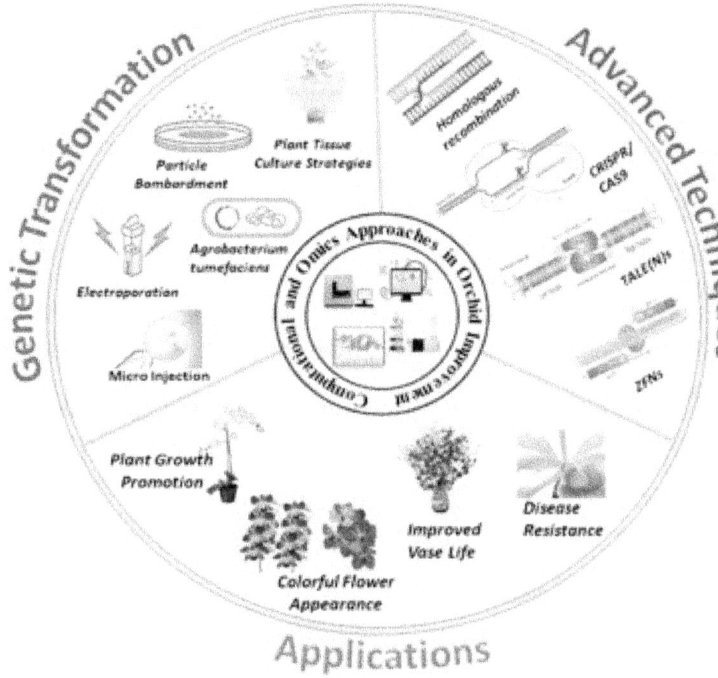

methods, while time-consuming, continue to be the dominant method for orchard breeding. Traditional orchard breeding methods include seed propagation, division of large clumps, offshoots or keikis, cutting and air layering. Conventional cultivation methods, such as crossbreeding mediated hybridization and mutation, cannot meet the increased demand for unique traits such as flower/foliage colour, morphology and enhanced shelf-life. These limitations have been actively addressed in recent years with the assistance of modern breeding approaches such as transgenic molecular breeding. Recent advances in orchard biotechnological interventions have significantly contributed to the development of exotic varieties with novel traits. Advances in biotechnological interventions, in conjunction with traditional breeding approaches and plant tissue culture, continue in order to make significant progress in developing exotic varieties with multifaceted characteristics, as well as orchard conservation and commercialization. The cost-effective genetic markers in orchards are expanding rapidly due to advances in DNA sequencing technology and investment in determining the reference genome sequences for several commercially-important species of orchards and forest trees. More sophisticated analytical methods capable of integrating the analysis of genetic variation detected by SNP assays with variation in gene expression patterns, metabolite levels, and phenotypic measurements could lead to new tools capable of more accurate genetic value prediction using molecular assays. Orchards breeding populations are likely to have very different patterns of linkage disequilibrium (LD) than crop species, and new approaches to genomic selection might be necessary. Plant tissue culture, traditional breeding approaches and biotechnological interventions have made significant contributions to introducing and improving plant traits of novel attributes in orchard varieties, resulting in remarkably improved commercialization of the plants on a global scale, generating enormous economic returns while also creating employment opportunities for a significant percentage of the population and supporting their livelihood.

REFERENCES

Abbas, M. S., Dobeie Amani, M., & Azzam Clara, R. (2021). Identification of salt tolerant Genotypes among Egyptian and Nigerian Peanut (Arachis hypogaea L.) Using Biochemical and Molecular Tools. In *Mitigating Environmental Stresses for Agricultural Sustainability in Egypt*. Springer International Publishing AG., doi:10.1007/978-3-030-64323-2_16437

Abdrabou, Fergany, Azzam, & Nahid. (2017). Devolopment of Some Canola Genotypes to Salinity Tolerance using Tissue Culture Technique. *Egypt. J. Agron., 39*(3), 431–448. doi:10.21608/agro.2017.1949.1083

Abo-Doma & Azzam. (2007). Hunting of some differentially expressed genes under salt stress in wheat. Egypt. *J. Plant Breed, 11*(3), 233–244.

Abo-Doma, A., & Azzam Clara, R. (2007). Molecular genetic relationships among some bread wheat cultivars (Triticum aestivum L.). *Egyptian Journal of Genetics and Cytology*, *36*(2), 387–400.

Aboelnaga, Abodoma, Lamyaa, & Azzam. (2020). Assessment of biodiversity among some sesame genotypes using ISSR and SRAP markers. *Arab Univ. J. Agric. Sci., 82*(3), 1–15.

Abou-Sreea, Azzam, Al-Taweel, Abdel-Aziz, & Belal. (2021). Natural Biostimulant Attenuates Salinity Stress Effects in Chili Pepper by Remodeling Antioxidant, Ion and Phytohormone Balances and Augments Gene Expression. *Plants, 10*(11), 2316. doi:10.3390/plants10112316

Aceto, N., Bardia, A., Miyamoto, D. T., Donaldson, M. C., Wittner, B. S., Spencer, J. A., Yu, M., Pely, A., Engstrom, A., Zhu, H., Brannigan, B. W., Kapur, R., Stott, S. L., Shioda, T., Ramaswamy, S., Ting, D. T., Lin, C. P., Toner, M., Haber, D. A., & Maheswaran, S. (2014). Circulating Tumor Cell Clusters Are Oligoclonal Precursors of Breast Cancer Metastasis. *Cell, 158*(5), 1110–1122. doi:10.1016/j.cell.2014.07.013 PMID:25171411

Adams, W. T., Zuo, J., Shimizu, J. Y., & Tappeiner, J. C. (1998). Impact of alternative regeneration methods on genetic diversity in coastal Douglas-fir. *Forest Science*, *44*, 390–396.

Allard, R.W. (1960). *Principles of plant breeding*. Wiley.

Araujo, J. A., Borralho, N. M. G., & Dehon, G. (2012). The importance and type of non-additive genetic effects for growth in *Eucalyptus globulus. Tree Genetics & Genomes*, *8*(2), 327–337. doi:10.100711295-011-0443-x

Arditti, J., & Emst, R. (1993). *Micropropagation of Orchid.* Wiley.

Atwood, J. T. (1984). The relationships of the slipper orchids (subfamily cypripedioideae, orchidaceae). *Selbyana*, *7*(2/4).

Azzam, Azer, Khalifa, & Abol-Ela. (2007). Characterization of peanut mutants and molecular markers associated with resistance to pod rot diseases and aflatoxin contamination by RAPD and ISSR. *Arab Journal of Biotechnology*, *10*(2), 301–320.

Azzam & Abo-Doma. (2007). Genetic relationships among some canola cultivars (*Brassica napus* L.) based on ISSR and RAPD-analyses. *Egyptian Journal of Genetics and Cytology*, *36*(2), 355–367.

Azzam, El-Rahman, & Eman. (2011). Evaluation of Genetic Relationships of some Barley Cultivars based on Phenotypic, Seed Quality and Molecular analysis. Egypt. *J. Plant Breed.*, *15*(4), 1–25.

Azzam, Al-Taweel, Abdel-Aziz, Rabe, & Abou-Sreea. (2021). Salinity Effects on Gene Expression, Morphological, and Physio-Biochemical Responses of Stevia rebaudiana Bertoni In Vitro. *Plants, 10*(4), 820. doi:10.3390/plants10040820

Azzam, C. R. (2004). Gibberellin 20-oxidase isolation and transformation its anti-sense by Agrobacterium tumfaciens to produce dwarf sunflower plants. In *The 2nd International Conf. of Biotechnology*. El-Baath University.

Azzam, C. R., Abd El Naby, Z., & Mohamed, N. (2019). Salt Tolerance Associated With Molecular Markers In Alfalfa. *Journal of Bioscience and Applied Research*, *5*(4), 416–428. doi:10.21608/jbaar.2019.110864

Azzam, C. R., Zein, S. N., & Abbas, S. M. (2007). Biochemical genetic markers for levels of resistance to Cowpea Aphid Borne Mosaic Potyvirus (CABMV) in sesame (*Sesamum indicum* L.) irradiated with gamma ray. *J. Plant Breed.*, *11*(2), 861–885.

Azzam, C. R., Zeinab, A.-E., & Kh, S. A. (2012). Influence of Agro-Ecological Conditions on Gene Expression, Yield and Yield Components of the Mono-Cut (Fahl) Type of Berseem. Egypt. *J. Plant Breed.*, *16*(2), 135–159. doi:10.12816/0003961

Azzam & Mahrous. (2010). Performance and genetic relationships among ten Egyptian wheat cultivars as revealed by RAPD-PCR analysis. *Egypt. J. Plant Breed., 14*(3), 87-102.

Băders, & Purin, Lıbiete, Nartišs, & Jansons. (2014). Fragment ăcijas ilgtermin, a dinamika meža ainav ăbez cilv¯eka saimniecisk ăs darb¯ıbas ietekmes [Long-term fragmentation dynamics in semi-natural forestlandscape]. *Mezzinatne*, *28*, 91–107.

Balilashaki, K., Moradi, S., Vahedi, M., & Khoddamzadeh, A. A. (2020). A molecular perspective on orchid development. *The Journal of Horticultural Science & Biotechnology*, *95*(5), 542–552. Advance online publication. doi:10.1080/14620316.2020.1727782

Ball, R. D. (2005). Experimental designs for reliable detection of linkage disequilibrium in unstructured random population association studies. *Genetics*, *170*(2), 859–873. doi:10.1534/genetics.103.024752 PMID:15781715

Baniulis, D., Gelvonauskienė, D., Rugienius, R., Sasnauskas, A., & Stanienė, G. (2013). Orchard Plant Breeding, Genetics, And Biotechnology Research At The Institute Of Horticulture, LRCAF. *Sodininkyste ir Darzininkyste*, *32*, 3–4.

Belarmino, M. M., & Mii, M. (2000). *Agrobacterium-mediated* genetic transformation of a Phalaenopsis orchid. *Plant Cell Reports*, *19*(5), 435–442. doi:10.1007002990050752 PMID:30754879

Bellusci, F., Pellegrino, G., & Musacchio, A. (2009). Different levels of inbreeding depression between outcrossing and selfing Serapias species. *Biologia Plantarum*, *53*(1), 175–178. doi:10.100710535-009-0029-8

Bilir, N., Kang, K. S., Zang, D., & Lindgren, D. (2004). Fertility variation and status number between a base population and a seed orchard of *Pinus brutia* Ten. *Silvae Genetica*, *53*(1-6), 161–163. doi:10.1515g-2004-0029

Bonetta, L. (2008). Epigenomics: Detailed analysis. *Nature*, *454*(7205), 795–798. doi:10.1038/454795a PMID:18685708

Bosily, M. A., Noaman, M. M., El-Banna, M. N., Azzam Clara, R., & Nassar, M. A. (2018). Breeding for barley resistance to leaf rust disease using marker-assisted selection. *Proceeding of The 7th Field Crops Research Institute Conference*, 397-437.

Bulpitt, C. (2005). The uses and misuses of orchids in medicine. *QJM*, *98*(9), 625–631. doi:10.1093/qjmed/hci094 PMID:16006500

Burdon, R. D., & Wilcox, P. L. (2011). Integration of molecular markers in breeding. In *Genetics, genomics and breeding of conifers* (pp. 276–322). CRC Press.

Busov, V. B., Strauss, S. H., & Pilate, G. (2010). *Transformation as a tool for genetic analysis in Populus. Genetics and genomics of Populus*. Springer.

Chan, Y.-L., Lin, K.-H., & Liao, S. (2005). Gene stacking in Phalaenopsis orchid enhances dual tolerance to pathogen attack. *Transgenic Research*, *14*(3), 279–288. doi:10.100711248-005-0106-5 PMID:16145836

Chang, W. H., Yang, S. Y., Lin, C. L., Wang, C.-H., Li, P.-C., Chen, T.-Y., Jan, F.-J., & Lee, G.-B. (2013). Detection of viruses directly from the fresh leaves of a Phalaenopsis orchid using a microfluidic system. *Nanomedicine; Nanotechnology, Biology, and Medicine*, *9*(8), 1274–1282. doi:10.1016/j.nano.2013.05.016 PMID:23751373

Chao, Y., Li, L., Girodat, D., Förstner, K. U., Said, N., Corcoran, C., Śmiga, M., Papenfort, K., Reinhardt, R., Wieden, H.-J., Luisi, B. F., & Vogel, J. (2017). *In Vivo* Cleavage Map Illuminates the Central Role of RNase E in Coding and Non-coding RNA Pathways. *Molecular Cell*, *65*(1), 39–51. doi:10.1016/j.molcel.2016.11.002 PMID:28061332

Chen, P., Taylor, N. J., Dueker, K. G., Keifer, I. S., Wilson, A. K., McGuffy, C. L., Novitsky, C. G., Spears, A. J., & Holbrook, W. S. (2016). pSIN: A scalable, Parallel algorithm for Seismic INterferometry of large-N ambient-noise data. *Computers & Geosciences*, *93*, 88–95. doi:10.1016/j.cageo.2016.05.003

Chen, W. H., Hsu, C. Y., Cheng, H. Y., Chang, H., Chen, H. H., & Ger, M. J. (2011). Downregulation of putative UDP-glucose: Flavonoid 3-O-glucosyltransferase gene alters flower coloring in Phalaenopsis. *Plant Cell Reports*, *30*(6), 1007–1017. doi:10.100700299-011-1006-1 PMID:21274540

Cheng, Q. Q. (2011). *Studies on leaf culture and induction of octoploid of Phalaenopsis cultivars* [M. D. Dissertation]. Shantou University.

Chocarro-Ruiz, B., Fernández-Gavela, A., Herranz, S., & Lechuga, L. M. (2017). Nanophotonic label-free biosensors for environmental monitoring. *Current Opinion in Biotechnology*, *45*, 175–183. doi:10.1016/j.copbio.2017.03.016 PMID:28458110

Cook, J. P., McMullen, M. D., Holland, J. B., Tian, F., Bradbury, P., Ross-Ibarra, J., Buckler, E. S., & Flint-Garcia, S. A. (2012). Genetic architecture of maize kernel composition in the nested association mapping and inbred association panels. *Plant Physiology*, *158*(2), 824–834. doi:10.1104/pp.111.185033 PMID:22135431

Cramer, W. A., Zakharov, S. D., Hasan, S. S., Zhang, H., Baniulis, D., & Zhalnina, M. V. (2011). Membrane proteins in four acts: Function precedes structure determination. *Methods (San Diego, Calif.)*, *55*(4), 415–420. doi:10.1016/j.ymeth.2011.11.001 PMID:22079407

Crosbie, T. M., Eathington, S. R., Johnson, G. R., Edwards, M., & Reiter, R. (2003). Plant breeding: past, present, and future. In *Plant breeding: the Arnel R. Hallauer International Symposium*. Blackwell.

Das, G., Patra, J. K., & Baek, K. H. (2017). Insight into MAS: A molecular tool for the development of stress-resistant and quality rice through gene stacking. *Front. Plant Sci.*, *8*, 1321. doi:10.3389/fpls.2017.01321 PMID:28775736

daSilva, J. A. T., Chin, D. P., Van, P. T., & Mii, M. (2011). Transgenic orchids. *Scientia Horticulturae*, *130*(4), 673–680. doi:10.1016/j.scienta.2011.08.025

daSilva, J. A. T., Dobranszki, J., Cardoso, J. C., Chandler, S. F., & Zeng, S. (2016). Methods for genetic transformation in Dendrobium. *Plant Cell Reports*, *35*(3), 483–504. doi:10.100700299-015-1917-3 PMID:26724929

De, L. C., Pathak, P., Rao, A. N., & Rajeevan, P. K. (2014). *Commercial Orchids*. Walter de Gruyter GmbH & Co KG.

de Chandra, L., Pathak, P., Rao, A. N., & Rajeevan, P. K. (2019). Breeding approaches for improved genotypes. Commercial Orchids.

de Paula, Figueiredo, & de Paula. (2014). Physiological changes in eucalyptus hybrids under different irrigation regimes. *Revista Ciencia Agronomica*, *45*(4), 805–814. doi:10.1590/S1806-66902014000400019

Diengdoh, R. V., Kumaria, S., Tandon, P., & Das, M. C. (2017). Asymbiotic germination and seed storage of Paphiopedilum insigne, an endangered lady's slipper orchid. *SAJB*, *112*, 215–224. doi:10.1016/j.sajb.2017.05.028

Dobeie, Abbas, Soliman, & Azzam. (2017). *In vitro* screening of some Egyptian and Nigerian peanut genotypes for salt tolerance. Egypt. *J. Plant Breed.*, *21*(6), 1035–1050. doi:10.12816/0046384

Duchovskis, P., Stanys, V., Sasnauskas, A., & Bobinas, C. (2007). Cold resistance of *Prunus domestica* L. and *Prunus cerasifera* Ehrh. in Lithuania. *Acta Horticulturae*, (734), 299–303. doi:10.17660/ActaHortic.2007.734.39

Dungey, Yanchuk, & Burdon. (2014). A 'Reality Check' in the Management of Tree Breeding Programs. In Challenges and Opportunities for the World's Forests in the 21st Century. Forestry Sciences. doi:10.1007/978-94-007-7076-8_19

Dvaranauskaitė, A., Venskutonis, P. R., Raynaud, C., Talou, T., Viškelis, P., & Sasnauskas, A. (2009). Variations in the essential oil composition in buds of six blackcurrant (*Ribes nigrum* L.) cultivars at various development phases. *Food Chemistry*, *11*(2), 671–679. doi:10.1016/j.foodchem.2008.10.005

Dwivedi, P., Amin, D., & Sharma, A. (2020). Effect of differential concentration of micronutrient copper and zinc on *in vitro* morphogenesis of *Foeniculum vulgare* Mill. *Plant Physiol. Rep.*, *25*(1), 178–184. doi:10.100740502-019-00478-4

El-Geddawy, Azzam, & Khalil. (2008). Somaclonal variation in sugarcane through tissue culture and subsequent screening for molecular polymorphisms. In *The 3rd International Conference "Meeting the Challenges of Sugar Crops & Integrated Industries in Developing Countries."* Sinai University.

El-Kassaby, Y. A., Cappa, E. P., Liewlaksaneeyanawin, C., Klápšte, J., & Lstiburek, M. (2011). Breeding without breeding: Is a complete pedigree necessary for efficient breeding? *PLoS One*, *6*(10), e25737. doi:10.1371/journal.pone.0025737 PMID:21991342

ENCODE Project Consortium. (2012). An integrated encyclopaedia of DNA elements in the human genome. *Nature*, *489*(7414), 57–74. doi:10.1038/nature11247 PMID:22955616

Fabre, J., & Dereuddre, J. (1990). Encapsulation-dehydration: A new approach to cryopreservation of Solanum shoots tips. *Cryo Letters*, *11*, 413–426.

Falconer, D. S., & Mackay, T. F. C. (1996). *Introduction to quantitative genetics*. Longman.

Fay, M., & Rankou, H. (2016). *Slipper orchids on the IUCN Red List. In: 2015 Annual Report to the Environment Agency—Abu Dhabi*. Framework Support for Implementing the Strategic Plan of the IUCN Species Survival Commission.

Feng, S., Cokus, S. J., Zhang, X., Chen, P.-Y., Bostick, M., Goll, M. G., Hetzel, J., Jain, J., Strauss, S. H., Halpern, M. E., Ukomadu, C., Sadler, K. C., Pradhan, S., Pellegrini, M., & Jacobsen, S. E. (2010). Conservation and divergence of methylation patterning in plants and animals. *Proceedings of the National Academy of Sciences of the United States of America*, *107*(19), 8689–8694. doi:10.1073/pnas.1002720107 PMID:20395551

Filippo, G. (2017). Molecular-assisted breeding. In P. Roberto & G. Giuseppe (Eds.), *More Food: Road to Survival* (pp. 373–398).

Fillatti, J. J., Sellmer, J., McCown, B., Haissig, B., & Comai, L. (1987). Agrobacterium mediated transformation and regeneration of Populus. *Molecular & General Genetics*, *206*(2), 192–199. doi:10.1007/BF00333574

Fladung, M., Altosaar, I., Bartsch, D., Baucher, M., Boscaleri, F., Gallardo, F., Häggman, H., Hoenicka, H., Nielsen, K., Paffetti, D., Séguin, A., Stotzky, G., & Vettori, C. (2012). European discussion forum on transgenic tree biosafety. *Nature Biotechnology*, *30*(1), 37–38. doi:10.1038/nbt.2078 PMID:22231091

Flanagan, S. P., Forester, B. R., Latch, E. K., Aitken, S. N., & Hoban, S. (2018). Guidelines for planning genomic assessment and monitoring of locally adaptive variation to inform species conservation. *Evolutionary Applications*, *11*(7), 1035–1052. doi:10.1111/eva.12569 PMID:30026796

Funda, T., Lstiburek, M., Klapšte, J., & El-Kassaby, Y. A. (2012). Optimization of genetic gain and diversity in seed orchard crops considering variation in seed germination. *Scandinavian Journal of Forest Research*, *27*(8), 787–793. doi:10.1080/02827581.2012.686627

Gardiner, B., Blennow, K., & Carnus, J.-M. (2010). *Destructive storms in European forests: past and forthcoming impacts.* Final report to EC DG environment. https://ec.europa.eu/environment/forests/fprotection.htm

Gelvonauskis, B., Shikshnianas, T., Duchovskis, P., & Gelvonauskienė, D. (2004). Investigation of apple genetic resources to select donors resistant to fungal diseases and for their winterhardiness. *Zeszyty problemowe postępów nauk rolniczych, 497*, 659–663.

Ghildiyal, M., & Zamore, P. D. (2009). Small silencing RNAs: An expanding universe. *Nature Reviews. Genetics*, *10*(2), 94–108. doi:10.1038/nrg2504 PMID:19148191

Grattapaglia, D., & Resende, M. D. V. (2011). Genomic selection in forest tree breeding. *Tree Genetics & Genomes*, *7*(2), 241–255. doi:10.100711295-010-0328-4

Gustavsson, B. A., & Stanys, V. (2000). Field performance of 'Sanna' lingonberry derived by micropropagation vs. stem cuttings. *HortScience*, *35*(4), 742–744. doi:10.21273/HORTSCI.35.4.742

Gutierrez, R. M. P. (2010). Orchids: A review of uses in traditional medicine, its phytochemistry, and pharmacology. *Journal of Medicinal Plants Research*, *4*, 592–638.

Häggman, H., Sutela, S., Walter, C., & Fladung, M. (2014). Biosafety Considerations in the Context of Deployment of GE Trees. T. Challenges and Opportunities for the World's Forests in the 21st Century. *Forestry Sciences*, *81*, 491–525. doi:10.1007/978-94-007-7076-8_21

Hamrick, J. L., & Godt, M. J. W. (1990). Allozyme diversity in plant species. In Plant population genetics, breeding, and genetic resources. Sinauer Assoc.

Hatzopoulos, P., Banilas, G., Giannoulia, K., Gazis, F., Nikoloudakis, N., Milioni, D., & Haralampidis, K. (2002). Breeding, molecular markers and molecular biology of the olive tree. *European Journal of Lipid Science and Technology*, *104*(9-10), 574–586. doi:10.1002/1438-9312(200210)104:9/10<574::AID-EJLT574>3.0.CO;2-1

Hawkins, S., Leplé, J., Cornu, D., Jouanin, L., & Pilate, G. (2003). Stability of transgene expression in poplar: A model forest tree species. *Annals of Science*, *60*, 427–438. doi:10.1051/forest:2003035

Hawley, G. J., Schaberg, P. G., DeHayes, D. H., & Brissette, J. C. (2005). Silviculture alters the genetic structure of an eastern hemlock forest in Maine,USA. *Canadian Journal of Forest Research*, *35*(1), 143–150. doi:10.1139/x04-148

He, H., Shang, Y., Yang, X., Di, Y., Lin, J., Zhu, Y., Zheng, W., Zhao, J., Ji, M., Dong, L., Deng, N., Lei, Y., & Chai, Z. (2019). Constructing an Associative Memory System Using Spiking Neural Network. *Frontiers in Neuroscience*, *13*, 650. doi:10.3389/fnins.2019.00650 PMID:31333397

Heffner, E. L., Jannink, J.-L., & Sorrells, M. E. (2011). Genomic selection accuracy using multifamily prediction models in a wheat breeding program. *The Plant Genome*, *4*(1), 65–75. doi:10.3835/plantgenome.2010.12.0029

Hiei, Y., Komari, T., Ishida, Y., & Saito, H. (2000). Development of Agrobacterium-mediated transformation method for monocotyledonous plants. *Ikushugaku Kenkyu*, *2*(4), 205–213. doi:10.1270/jsbbr.2.205

Hill, W. G., Goddard, M. E., & Visscher, P. M. (2008). Data and theory point to mainly additive genetic variance for complex traits. *PLOS Genetics*, *4*(2), e1000008. doi:10.1371/journal.pgen.1000008 PMID:18454194

Hindorff, L. A., Sethupathy, P., Junkins, H. A., Ramos, E. M., Mehta, J. P., Collins, F. S., & Manolio, T. A. (2009). Potential etiologic and functional implications of genome-wide association loci for human diseases and traits. *Proceedings of the National Academy of Sciences of the United States of America*, *106*(23), 9362–9367. doi:10.1073/pnas.0903103106 PMID:19474294

Hinsley, A., De Boer, H. J., Fay, M. F., Gale, S. W., Gardiner, L. M., Gunasekara, R. S., Kumar, P., Masters, S., Metusala, D., Roberts, D. L., Veldman, S., Wong, S., & Phelps, J. (2018). A review of the trade-in orchids and their implications for conservation. *Botanical Journal of the Linnean Society*, *186*(4), 435–455. doi:10.1093/botlinnean/box083

Hirano, T., Godo, T., Miyoshi, K., Ishikawa, K., Ishikawa, M., & Mii, M. (2009). Cryopreservation and low-temperature storage of seeds of Phaius tankervilleae. *Plant Biotechnology Reports*, *3*(1), 103–109. doi:10.100711816-008-0080-5

Hirano, T., Ishikawa, K., & Mii, M. (2006). Advances on orchid cryopreservation. In Floriculture Ornamental and Plant Biotechnology: Advances and Topical Issues (Vol. 2). Global Science Book.

Hossain, M. M., Kant, R., Van, P. T., Winarto, B., Zeng, S., & Teixeira da Silva, J. A. (2013). The application of biotechnology to orchids. *Critical Reviews in Plant Sciences*, *32*(2), 69–139. doi:10.1080/07352689.2012.715984

Hsiao, Y. Y., Pan, Z. J., Hsu, C. C., Yang, Y. P., Hsu, Y. C., Chuang, Y. C., Shih, H.-H., Chen, W.-H., Tsai, W.-C., & Chen, H.-H. (2011). Research on orchid biology and biotechnology. *Plant & Cell Physiology*, *52*(9), 1467–1486. doi:10.1093/pcp/pcr100 PMID:21791545

Ingvarsson, P. K., Hvidsten, T. R., & Street, N. R. (2016). Towards integration of population and comparative genomics in forest trees. *The New Phytologist*, *212*(2), 338–344. doi:10.1111/nph.14153 PMID:27575589

Irawati, A. (2013). Conservation of orchids the gems of the tropics. In M. Normah, H. Chin, & B. Reed (Eds.), *Conservation ofTropicalplant species* (pp. 171–187). Springer. doi:10.1007/978-1-4614-3776-5_9

Isajev, V., Ivetić, V., Lučić, A., & Rakonjac, Lj. (2009). Gene pool conservation and tree improvement in Serbia. *Genetika*, *41*(3), 309–327. doi:10.2298/GENSR0903309I

Isik, Whetten, Zapata-Valenzuela, Ogut, & McKeand. (2011). Genomic selection in loblolly pine–from lab to field. From IUFRO tree biotechnology conference 2011: From genomes to integration and delivery. *BMC, 5*(7).

IUCN. (2017). *The IUCN red list of threatened species*. https://www.iucnredlist.org/

Ivetić, Devetaković, Nonić, Stanković, & Šijačić-Nikolić. (2016). Genetic diversity and forest reproductive material – from seed source selection to planting. *iForest-Biogeosciences and Forestry, 9*(5), 801-812. doi:10.3832/ifor1577-009

Jain, A., Sarsaiya, S., Chen, J., Wu, Q., Lu, Y., & Shi, J. (2021). Changes in global Orchidaceae disease geographical research trends: Recent incidences, distributions, treatment, and challenges. *Bioengineered, 12*(1), 13–29. doi:10.1080/21655979.2020.1853447 PMID:33283604

Jalal, J. S., Kumar, P., & Pangtey, Y. (2008). Ethnomedicinal Orchids of Uttarakhand, Western Himalaya. *Ethnobotanical Leaflets., 2008*, 164.

Jayawickrama, K. J. S., & Carson, M. J. (2000). A breeding strategy for the New Zealand radiata pine breeding cooperative. *Silvae Genetica, 49*, 82–90.

Jiang, G. L. (2015). Molecular marker-assisted breeding: a plant breeder's review. In Advances in Plant Breeding Strategies: Breed, Biotechnology Molecular Tools. Springer.

Kamboj, D. (2020). *Dendrobium and Venda Orchids as Potential Cut Flower in North Indian Market.* GINMA.

Kang, K. S., Bila, A. D., Harju, A. M., & Lindgren, D. (2003). Estimation of fertility variation in forest tree populations. *Forestry, 76*(3), 329–344. doi:10.1093/forestry/76.3.329

Keel, B. G. (2005). Assisted migration. In M. Allaby (Ed.), *Dictionary of Ecology* (p. 36). Oxford University Press.

Kemper, K. E., Daetwyler, H. D., Visscher, P. M., & Goddard, M. E. (2012). Comparing linkage and association analyses in sheep points to a better way of doing GWAS. *Genetical Research, 94*(4), 191–203. doi:10.1017/S0016672312000365 PMID:22950900

Kerr, R. J., Dieters, M. J., & Tier, B. (2004). Simulation of the comparative gains from four hybrid tree breeding strategies. *Canadian Journal of Research, 34*(1), 209–220. doi:10.1139/x03-180

Keshta, & Hassan Nemat, Azzam, & Hassanin. (2011). Embryogenic callus induction of some sunflower (*Helianthus annuus* L.) genotypes under *in vitro* salt stress. *J. Plant Production. Mansoura Univ., 2*(2), 327–333. doi:10.21608/jpp.2011.85524

Khalifa, M. M. A., Azzam Clara, R., & Azer, S. A. (2006). Biochemical markers associated with disease resistance to damping-off and root-rot diseases of peanut mutants and their productivity. *Egyptian J. of Phytopathology, 34*(2), 53–74.

Khalil, C. A. (2014). The emerging role of epigenetics incardiovascular disease. *Therapeutic Advances in Chronic Disease, 5*(4), 178–187. doi:10.1177/2040622314529325 PMID:24982752

Klápště, J., Lstibůrek, M., & El-Kassaby, Y. A. (2014). Estimates of genetic parameters and breeding values from western larch open-pollinated families using marker-based relationship. *Tree Genetics & Genomes, 10*(2), 241–249. doi:10.100711295-013-0673-1

Knudson, L. (1946). A new nutrient solution for germination of orchid seed. *American Orchid Society Bulletin, 15*, 14–217.

Kole, C., & Hall, T. C. (2008). *Compendium of transgenic crop plants: transgenic forest tree species.* Wiley. doi:10.1002/9781405181099

Konig, A. O. (2005). Provenance research: evaluating the spatial pattern of genetic variation. In Conservation and management of forest genetic resources in Europe. Arbora Publishers.

Konnert, M., & Hosius, B. (2010). Contribution of forest genetics for a sustainable forest management. *Forstarchiv (Hannover)*, *4*, 170–174.

Korecký, J., Klápště, J., Lstibůrek, M., Kobliha, J., Nelson, C. D., & El-Kassaby, Y. A. (2013). Comparison of genetic parameters from marker-based relationship, sibship, and combined models in Scots pine multi-site open-pollinated tests. *Tree Genetics & Genomes*, *9*(5), 1227–1235. Advance online publication. doi:10.100711295-013-0630-z

Kreimer, A., Litvin, O., Hao, K., Molony, C., Pe'er, D., & Pe'er, I. (2012). Inference of modules associated to eQTLs. *Nucleic Acids Research*, *40*(13), e98. doi:10.1093/nar/gks269 PMID:22447449

Kui, L., Chen, H., Zhang, W., He, S., Xiong, Z., Zhang, Y., Yan, L., Zhong, C., He, F., Chen, J., Zeng, P., Zhang, G., Yang, S., Dong, Y., Wang, W., & Cai, J. (2017). Building a Genetic Manipulation Tool Box for Orchid Biology: Identification of Constitutive Promoters and Application of CRISPR/Cas9 in the Orchid, Dendrobium officinale. *Front. Plant Sci.*, *7*, 2036. doi:10.3389/fpls.2016.02036 PMID:28127299

Kurz, W. A., Dymond, C. C., Stinson, G., Rampley, G. J., Neilson, E. T., Carroll, A. L., Ebata, T., & Safranyik, L. (2008). Mountain pine beetle and forest carbon: Feedback to climate change. *Nature*, *454*(7190), 987–990. doi:10.1038/nature06777 PMID:18432244

Lango, A. H. (2010). Hundreds of variants clustered in genomic loci and biological pathways affect human height. *Nature*, *467*(7317), 832–838.

Leitch, C. C., Lodh, S., Prieto-Echagüe, V., Badano, J. L., & Zaghloul, N. A. (2014). Basal body proteins regulate Notch signaling through endosomal trafficking. *Journal of Cell Science*, *127*, 2407–2419. PMID:24681783

Li, D. M., Zhao, C., Liu, X., Liu, X., Lin, Y., Liu, J., Chen, H., & Lv, F. (2015). *De novo* assembly and characterization of the root transcriptome and development of simple sequence repeat markers in Paphiopedilum concolor. *Genetics and Molecular Research*, *14*(2), 6189–6201. doi:10.4238/2015.June.9.5 PMID:26125820

Li, J. W., Ya, J. D., Ye, D. P., Liu, C., Liu, Q., Pan, R., He, Z.-X., Pan, B., Cai, J., Lin, D.-L., & Jin, X.-H. (2021). Taxonomy notes on Vandeae (*Orchidaceae*) from China: Five new species and two new records. *Plant Diversity*, *43*(5), 379–389. doi:10.1016/j.pld.2021.01.009 PMID:34816063

Li, S. H., Kuoh, C. S., Chen, Y. H., Chen, H. H., & Chen, W. H. (2005). Osmotic sucrose enhancement of single-cell embryogenesis and transformation efficiency in Oncidium. *Plant Cell, Tissue and Organ Culture*, *81*(2), 183–192. doi:10.100711240-004-4955-z

Li, T., Kung, H. J., Mack, P. C., & Gandara, D. R. (2013). Genotyping and genomic profiling of non-small-cell lung cancer: Implications for current and future therapies. *Journal of Clinical Oncology*, *31*(8), 1039–1049. doi:10.1200/JCO.2012.45.3753 PMID:23401433

Li, X., Jin, F., Jin, L., Jackson, A., Huang, C., Li, K., & Shu, X. (2014). Development of Cymbidium ensifolium genic SSR markers and their utility in genetic diversity and population structure analysis in cymbidiums. *BMC Genetics*, *15*(1), 124. doi:10.118612863-014-0124-5 PMID:25481640

Li, Y., Klápště, J., Telfer, E., Wilcox, P., Graham, N., Macdonald, L., & Dungey, H. S. (2019). Genomic selection for non-key traits in radiata pine when the documented pedigree is corrected using DNA marker information. *BMC Genomics*, *20*(1), 1026. doi:10.118612864-019-6420-8 PMID:31881838

Liesebach, H., Liepe, K., & Baucker, C. (2021). Towards new seed orchard designs in Germany – A review. *Silvae Genetica*, *70*(1), 84–98. doi:10.2478g-2021-0007

Lindley, C. P., Arniputri, R. B., Soliah, L. A., & Cahyono, O. (2017). Effects of organic additives and naphthalene acetic acid (NAA) application on the in vitro growth of black orchid hybrid Coelogyne pandurata Lindley. *Bulgarian Journal of Agricultural Science*, *23*(6), 951–957.

Lindner, M., Maroschek, M., Netherer, S., Kremer, A., Barbati, A., Garcia-Gonzalo, J., Seidl, R., Delzon, S., Corona, P., Kolström, M., Lexer, M. J., & Marchetti, M. (2010). Climate change impacts, adaptive capacity, and vulnerability of European forest ecosystems. *Forest Ecology and Management*, *259*(4), 698–709. doi:10.1016/j.foreco.2009.09.023

Liobikas, J., Baniulis, D., Stanys, V., Toleikis, A., & Eriksson, O. (2006). Identification and analysis of a novel serine beta-lactamase-like plant protein using a bioinformatics approach. *Biologija (Vilnius, Lithuania)*, *3*, 126–129.

Liu, J., Chiou, C., Shen, C., Chen, P. J., Liu, Y. C., Jian, C. D., Shen, X. L., Shen, F. Q., & Yeh, K. W. (2014). RNA Interference-Based Gene Silencing of Phytoene Synthase Impairs Growth, Carotenoids, and Plastid Phenotype in Oncidium Hybrid Orchid. 3. *SpringerPlus*, *3*(1), 478. doi:10.1186/2193-1801-3-478 PMID:25221736

Liu, J., Zimmer, K., Rusch, D. B., Paranjape, N., Podicheti, R., Tang, H., & Calvi, B. R. (2015). DNA sequence templates adjacent nucleosome and ORC sites at gene amplification origins in Drosophila. *Nucleic Acids Research*, *43*(18), 8746–8761. doi:10.1093/nar/gkv766 PMID:26227968

Liu, X., & Locasale, J. W. (2017). Metabolomics: A Primer. *Trends in Biochemical Sciences*, *42*(4), 274–284. doi:10.1016/j.tibs.2017.01.004 PMID:28196646

Lowe, R., Wu, Y., Tamar, A., Harb, J., Abbeel, O. P., & Mordatch, I. (2017). Multi-agent actor-critic for mixed cooperative-competitive environments. Advances in Neural Information Processing Systems, 6382–6393.

Lu, H. C., Chen, H. H., Tsai, W. C., Chen, W. H., Su, H. J., Chang, C. N., & Yeh, H. H. (2007). Strategies for functional validation of genes involved in reproductive stages of orchids. *Plant Physiology*, *143*(2), 558–569. doi:10.1104/pp.106.092742 PMID:17189336

Lu, J. J., Liu, Y. Y., Xu, J., Mei, Z. W., Shi, Y. J., Liu, P. L., He, J. B., Wang, X. T., Meng, Y. J., Feng, S. G., Shen, C. J., & Wang, H. Z. (2018). High-density genetic map construction and stem total polysaccharide content related QTL exploration for Chinese endemic Dendrobium (*Orchidaceae*). *Front. Plant Sci.*, *9*, 398. doi:10.3389/fpls.2018.00398 PMID:29636767

Lukoševičiūtė, V., Stanienė, G., Blažytė, A., Sasnauskas, A., & Gelvonauskienė, D. (2011). Angliavandenių įtaka kultūrinių kriaušių mikroūglių užsigrūdinimui žemoje teigiamoje temperatūroje. *Sodininkyste ir Darzininkyste*, *30*(3–4), 17–29.

Luo, Y. H., Huang, M. L., & Wu, J. S. (2012). Progress in Oncidium breeding study. *Jiangxi Nongye Daxue Xuebao*, *24*, 15–20.

Mazeikiene, I., Bendokas, V., Stanys, V., & Siksnianas, T. (2012). Molecular markers linked to resistance to the gall mite in blackcurrant. *Plant Breeding*, *131*(6), 762–766. doi:10.1111/j.1439-0523.2012.01995.x

McMullen. (2009). Genetic properties of the maize nested association mapping population. *Science*, *325*(5941), 737–740.

Meehan, R. R., Dunican, D. S., Ruzov, A., & Pennings, S. (2005). Epigenetic silencing in embryogenesis. *Experimental Cell Research*, *309*(2), 241–249. doi:10.1016/j.yexcr.2005.06.023 PMID:16112110

Merkle, S. A., Andrade, G. M., Nairn, C. J., Powell, W. A., & Maynard, C. A. (2007). Restoration of threatened species: A noble cause for transgenic trees. *Tree Genetics & Genomes*, *3*(2), 111–118. doi:10.100711295-006-0050-4

Meuwissen, T. H. E., Goddard, M. E., & Hayes, B. J. (2001). Prediction of total genetic value using genome-wide dense marker maps. *Genetics*, *157*(4), 1819–1829. doi:10.1093/genetics/157.4.1819 PMID:11290733

Murashige, T., & Skoog, F. (1962). A revised medium for rapid growth and bioassays with tobacco tissue cultures. *Physiologia Plantarum*, *15*(3), 473–497. doi:10.1111/j.1399-3054.1962.tb08052.x

Muthukrishnan, S., Kumar, T. S., & Rao, M. V. (2013). Effects of different media and organic additives on seed germination of *Geodorum densiflorum* (Lam) Schltran endangered orchid. *International Journal of Scientific Research*, *2*(8), 2277–8179.

Myburg, A. A., Hussey, S. G., Wang, J. P., Street, N. R., & Mizrachi, E. (2019). Systems and synthetic biology of forest trees: A bioengineering paradigm for Woody biomass feedstocks. *Front. Plant Sci.*, *10*, 775. doi:10.3389/fpls.2019.00775 PMID:31281326

Namkoong, G., Kang, H. C., & Brouard, J. S. (1988). Tree breeding: Principles and strategies. Springer, New York, Monograph. *Theoretical and Applied Genetics*, 11.

Neale, D. B., & Kremer, A. (2011). Forest tree genomics: Growing resources and applications. *Nature Reviews. Genetics*, *12*(2), 111–122. doi:10.1038/nrg2931 PMID:21245829

Ng, T. B., Liu, J., Wong, J. H., Ye, X., Wing Sze, S. C., Tong, Y., & Zhang, K. Y. (2012). Review of research on Dendrobium, a prized folk medicine. *Applied Microbiology and Biotechnology*, *93*(5), 1795–1803. doi:10.100700253-011-3829-7 PMID:22322870

OECD. (2013). *OECD guidelines on the production of forest reproductive materials*. Organization for Economic Cooperation and Development Trade and Agriculture Directorate.

Osadchuk, V.D., Saranchuk, I.I., Lesyk, O.B., & Olifirovych, V.O. (2020). Selective Breeding in Plant Growing in Bukovina. *Taurian Scientific Herald*, p. 16.

Pakum, W., Watthana, S., Srimuang, K. O., & Kongbangkerd, A. (2016). Influence of medium composition on *in vitro* propagation of Thai's endangered orchid: *Bulbophyllum nipondhii* Seiden. *Plant Tissue Culture & Biotechnology*, *25*(1), 37–46. doi:10.3329/ptcb.v26i1.29765

Pan, I. C., Liao, D. C., Wu, F. H., Daniell, H., Singh, N. D., Chang, C., Shih, M. C., Chan, M. T., & Lin, C. S. (2012). Complete chloroplast genome sequence of an orchid model plant candidate: *Erycina pusilla* apply in tropical Oncidium breeding. *PLoS One*, *7*(4), 34738. doi:10.1371/journal.pone.0034738 PMID:22496851

Pant, B. (2013). Medicinal orchids and their uses: Tissue culture a potential alternative for conservation. *African Journal of Plant Science*, *7*(10), 448–467. doi:10.5897/AJPS2013.1031

Pant, B., Shrestha, S., & Pradhan, S. (2011). *In vitro* seed germination and seedling development of *Phaius tancarvilleae* (L'Her.) Blume. *TheScientificWorldJournal*, *9*, 50–52.

Park, Y.-S. (2014). Conifer Somatic Embryogenesis and Multi- Varietal Forestry. Challenges and Opportunities for the World's Forests in the 21st Century. *Forestry Sciences*, *81*, 425–440. doi:10.1007/978-94-007-7076-8_17

Peters, D. T., & Musunuru, K. (2012). Functional evaluation of genetic variation in complex human traits. *Human Molecular Genetics*, *21*(R1), R18–R23. Advance online publication. doi:10.1093/hmg/dds363 PMID:22936690

Popova, E., Kim, H. H., Saxena, P. K., Engelmann, F., & Pritchard, H. W. (2016). Frozen beauty: The cryo-biotechnology of orchid diversity. *Biotechnology Advances*, *34*(4), 380–403. doi:10.1016/j.biotechadv.2016.01.001 PMID:26792590

Raffeiner, B., Serek, M., & Winkelmann, T. (2009). *Agrobacterium tumefaciens* mediated transformation of *Oncidium* and *Odontoglossum* orchid species with the ethylene receptor mutant gene etr1-1. *Plant Cell, Tissue and Organ Culture*, *98*(2), 125–134. doi:10.100711240-009-9545-7

Reddy, M. S., Murali, T. S., Suryanarayanan, T. S., Rajulu, M. B. G., & Thirunavukkarasu, N. (2016). Pestalotiopsis species occur as generalist endophytes in trees of Western Ghats forests of southern India. *Fungal Ecology*, *24*, 70–75. doi:10.1016/j.funeco.2016.09.002

Reinikka, M. A. (1995). *A history of the Orchid*. Portland Timber Press.

Reis, C. A. F., Gonçalves, F. M. A., Rosse, L. N., Costa, R. R. G. F., & Ramalho, M. A. P. (2011). Correspondence between performance of *Eucalyptus* spp. Trees selected from family and clonal tests. *Genetics and Molecular Research*, *10*(2), 1172–1179. doi:10.4238/vol10-2gmr1078 PMID:21732281

Resende, M. F. R. Jr, Muñoz, P., Acosta, J. J., Peter, G. F., Davis, J. M., Grattapaglia, D., Resende, M. D. V., & Kirst, M. (2012). Accelerating the domestication of trees using genomic selection: Accuracy of prediction models across ages and environments. *The New Phytologist*, *193*(3), 617–624. doi:10.1111/j.1469-8137.2011.03895.x PMID:21973055

Ricci, B., Franck, P., Toubon, J.-F., Bouvier, J.-C., Sauphanor, B., & Lavigne, C. (2009). The influence of landscape on insect pest dynamics: A case study in southeastern France. *Landscape Ecology*, *24*(3), 337–349. doi:10.100710980-008-9308-6

Risch, N., & Merikangas, K. (1996). The future of genetic studies of complex human diseases. *Science, 273*(5281), 1516–1517. doi:10.1126cience.273.5281.1516 PMID:8801636

Rugienius, R., Šikšnianas, T., & Sasnauskas, A. (2013). Žemuogių veislių, hibridų ir atrinktų linijų tyrimai. *Sodininkystė ir daržininkystė, 31*(1–2), 3–13.

Rung, Y. W., Shao, Y. C., Ting, F. H., Keng, C. C., Ting, I., Yen, H. L., & Yu, S. C. 2016. Cryopreservation of Orchid Genetic Resources by Desiccation: A Case Study of Bletilla formosana. Cryopreservation in Eukaryotes. doi:10.5772/65302

Sakai, Keene, Renard, & De Backer. (2016). FBN1: The disease-causing gene for *Marfan syndrome* and other genetic disorders. *Gene, 591*(1), 279-291. doi:10.1016/j.gene.2016.07.033

Sakai, A. (2000). Development of cryopreservation techniques. In F. Engelmann & H. Takagi (Eds.), *Cryopreservation of Tropical Plant Germplasm* (pp. 1–7). IPGRI.

Sasnauskas, A., Rugienius, R., Gelvonauskienė, D., Zalunskaitė, I., & Stanienė, G. (2007). Screening of strawberries with red stele (*Phytophthora Fragariae*) resistance gene *Rpf1* using sequence specific DNA markers. *Acta Horticulturae*, (760), 165–169. doi:10.17660/ActaHortic.2007.760.21

Savolainen, O., Bokma, F., Knurr, T., Karkkainen, K., Pyhajarvi, T., & … . (2007). Adaptation of forest trees to climate change. In *Climate change and forest genetic diversity: Implications for sustainable forest management in Europe* (pp. 19–30). Bioversity International.

Schaberg, P. G., Dehayes, D. H., Hawley, G. J., & Nijensohn, S. E. (2008). Anthropogenic alterations of genetic diversity within three populations: Implications for forest ecosystem resilience. *Forest Ecology and Management*, *256*(5), 855–862. doi:10.1016/j.foreco.2008.06.038

Scherer, L., Svenning, J. C., Huang, J., Seymour, C. L., Sandel, B., Mueller, N., Kummu, M., Bekunda, M., Bruelheide, H., Hochman, Z., Siebert, S., Rueda, O., & van Bodegom, P. M. (2020). Global priorities of environmental issues to combat food insecurity and biodiversity loss. *The Science of the Total Environment*, *730*, 139096. doi:10.1016/j.scitotenv.2020.139096 PMID:32388110

Sedeek, K. E. M., Qi, W., Schauer, M. A., Gupta, A. K., Poveda, L., Xu, S., Liu, Z.-J., Grossniklaus, U., Schiestl, F. P., & Schlüter, P. M. (2013). Transcriptome and Proteome Data Reveal Candidate Genes for Pollinator Attraction in Sexually Deceptive Orchids. *PLoS One*, *8*(5), e64621. Advance online publication. doi:10.1371/journal.pone.0064621 PMID:23734209

Semiarti, E. (2018). Biotechnology for Indonesian orchid conservation and industry. *Proceeding of Inventing Prosperous Future through Biological Research and Tropical Biodiversity Management, in AIP Conference Proceedings 2002*. 10.1063/1.5050118

Sharma, A., Verma, P., Mathur, A., & Mathur, A. K. (2018). Overexpression of tryptophan decarboxylase and strictosidine synthase enhanced terpenoid indole alkaloid pathway activity and antineoplastic vinblastine biosynthesis in Catharanthus roseus. *Protoplasma*, *255*(5), 1281–1294. doi:10.100700709-018-1233-1 PMID:29508069

Shekarriz, P., Kafi, M., Deilamy, S. D., & Mirmasoumi, M. (2014). Coconut water and peptone improve seed germination and protocorm-like body formation of hybrid Phalaenopsis. *Agriculture Science Developments*, *3*(10), 317–322.

Sherif, N. A., Benjamin, J. H. F., Kumar, T. S., & Rao, M. V. (2018). Somatic embryogenesis, acclimatization, and genetic homogeneity assessment of regenerated plantlets of *Anoectochilus elatus* Lindl., an endangered terrestrial jewel orchid. *Plant Cell, Tissue and Organ Culture*, *132*(2), 303–316. doi:10.100711240-017-1330-4

Shukla, S. P., & Sharma, A. (2017). *In vitro* seed germination, proliferation, and ISSR marker-based clonal fidelity analysis of *Shorea tumbuggaia* Roxb.: An endangered and high trade medicinal tree of Eastern Ghats. *In Vitro Cellular & Developmental Biology. Plant*, *53*(3), 200–208. doi:10.100711627-017-9818-5

Shulaev, V., Sargent, D. J., Crowhurst, R. N., Mockler, T. C., Folkerts, O., Delcher, A. L., Jaiswal, P., Mockaitis, K., Liston, A., Mane, S. P., Burns, P., Davis, T. M., Slovin, J. P., Bassil, N., Hellens, R. P., Evans, C., Harkins, T., Kodira, C., Desany, B., ... Folta, K. M. (2011). The genome of woodland strawberry (*Fragaria vesca*). *Nature Genetics*, *43*(2), 109–116. doi:10.1038/ng.740 PMID:21186353

Sikorskaite, S., Rajamäki, M. L., Baniulis, D., Stanys, V., & Valkonen, J. P. T. (2013). Protocol: Optimised methodology for isolation of nuclei from leaves of species in the *Solanaceae* and *Rosaceae* families. *Plant Methods*, *9*(1), 31–39. doi:10.1186/1746-4811-9-31 PMID:23886449

Šikšnianas, T., Stanienė, G., Stanys, V., & Sasnauskas, A. (2008). *Ribes sanguineum* Pursh. as donor of leaf fungal disease resistance in blackcurrant breeding. *Biologija (Vilnius, Lithuania)*, *54*(2), 79–82. doi:10.2478/v10054-008-0015-7

Spencer, C. C., Su, Z., Donnelly, P., & Marchini, J. (2009). Designing genome-wide association studies: Sample size, power, imputation, and the choice of genotyping chip. *PLOS Genetics*, *5*(5), e1000477. doi:10.1371/journal.pgen.1000477 PMID:19492015

Srivastava, S., Kadooka, C., & Uchida, J. Y. (2018). Fusarium species as a pathogen on orchids. *Microbiological Research*, *207*, 188–195. doi:10.1016/j.micres.2017.12.002 PMID:29458853

Stanienė, G., Stanys, V., & Kawecki, Z. (2002). Peculiarities of propagation in vitro of Vaccinium vitis-idaea L. and V. praestans Lamb. *Biologija (Vilnius, Lithuania)*, *48*(1), 84–86.

Stanys, V., Baniulis, D., Morkunaite-Haimi, S., Siksnianiene, J. B., Frercks, B., Gelvonauskiene, D., Stepulaitiene, I., Staniene, G., & Siksnianas, T. (2012). Characterising the genetic diversity of Lithuanian sweet cherry (*Prunus avium* L.) cultivars using SSR markers. *Scientia Horticulturae*, *142*, 136–142. doi:10.1016/j.scienta.2012.05.011

Stanys, V., Weckman, A., Staniene, G., & Duchovskis, P. (2006). *In vitro* induction of polyploidy in japanese quince (*Chaenomeles japonica*). *Plant Cell, Tissue and Organ Culture*, *84*(3), 263–268. doi:10.100711240-005-9029-3

Stefani, F. O. P., Moncalvo, J., Séguin, A., Bérubé, J. A., & Hamelin, R. C. (2009). Impact of an 8-year-old transgenic poplar plantation on the ectomycorrhizal fungal community. *Applied and Environmental Microbiology*, *75*(23), 7527–7536. doi:10.1128/AEM.01120-09 PMID:19801471

Stranger, B. E., Stahl, E. A., & Raj, T. (2011). Progress and promise of genome-wide association studies for human complex trait genetics. *Genetics*, *187*(2), 367–383. doi:10.1534/genetics.110.120907 PMID:21115973

Su, J. S., Jiang, J. F., Zhang, F., Liu, Y., Ding, L., Chen, S. M., & Chen, F. D. (2019). Current achievements and prospects in the genetic breeding of Chrysanthemum: A review. *Horticulture Research*, *6*(1), 109. doi:10.103841438-019-0193-8 PMID:31666962

Sudarsono, S., Haristianita, M. D., Handini, A. S., & Sukma, D. (2017). Molecular marker development based on diversity of genes associated with pigment biosynthetic pathways to support breeding for novel colors in Phalaenopsis. *Acta Horticulturae*, (1167), 305–312. doi:10.17660/ActaHortic.2017.1167.44

Surenciski, M. R., Flachsland, E. A., Terada, G., Mroginski, L. A., & Rey, H. Y. (2012). Cryopreservation of *Cyrtopodium hatschbachii* Pabst (*Orchidaceae*) immature seeds by encapsulation dehydration. *Biocell*, *36*(1), 31–36. doi:10.32604/biocell.2012.36.031 PMID:23173302

Szczesniak, M. W., Rosikiewicz, W., & Makalowska, I. (2016). CANTATAdb:A Collection of Plant Long Non-Coding RNAs. *Plant & Cell Physiology*, *57*(1), e8. doi:10.1093/pcp/pcv201 PMID:26657895

Szlachetko, D. L. (2001). Genera et species Orchidalium. *Polish Botanical Journal*, *46*, 11–26.

Tang, W., Newton, R., Li, C., & Charles, T. (2007). Enhanced stress tolerance in transgenic pine expressing the pepper *CaPF1* gene is associated with the polyamine biosynthesis. *Plant Cell Reports*, *26*(1), 115–124. doi:10.100700299-006-0228-0 PMID:16937149

Taylor, S. C., Laperriere, G., & Germain, H. (2017). Droplet Digital PCR versus q-PCR for gene expression analysis with low abundant targets: From variable nonsense to publication-quality data. *Scientific Reports*, *7*(1), 2409. doi:10.103841598-017-02217-x PMID:28546538

Tee, C. S., Maziah, M., Tan, C. S., & Abdullah, M. P. (2003). Evaluation of different promoters Belarmino, driving the GFP reporter gene and selected target tissues for particle bombardment of Dendrobium Sonia 17. *Plant Cell Reports*, *21*(5), 452–458. doi:10.100700299-002-0539-8 PMID:12789448

Thomas, E., Jalonen, R., Loo, J., Boshier, D., Gallo, L., Cavers, S., Bordács, S., Smith, P., & Bozzano, M. (2014). Genetic considerations in ecosystem restoration using native tree species. *Forest Ecology and Management*, *333*, 66–75. doi:10.1016/j.foreco.2014.07.015

Tiwari, P., Sharma, A., Bose, S. K., & Gautam, A. (2022). Biotechnological interventions in Orchids: Recent updates, Translational success, and Commercial outcomes. Research Square. doi:10.21203/rs.3.rs-1382018/v1

Troggio, M., Gleave, A., Salvi, S., Chagne, D., Cestaro, A., Kumar, S., Crowhurst, R. N., & Gardiner, S. E. (2012). Apple, from genome to breeding. *Tree Genetics & Genomes*, *8*(3), 509–529. doi:10.100711295-012-0492-9

Trontin, J. F., Walter, C., Klimaszewska, K., Park, Y. S., & Lelu-Walter, M. A. (2007). Recent progress on transformation of four *Pinus* species. Invited review. *Transgenic Plant Journal*, *1*(2), 314–329.

Tsai, J., Knutson, B., & Fung, H. H. (2006). Cultural variation in affect valuation. *Journal of Personality and Social Psychology*, *90*(2), 288–307. doi:10.1037/0022-3514.90.2.288 PMID:16536652

Tsai, K. H., Chou, C., & Kuo, J. H. (2008). The curvilinear relationships between responsive and proactive market orientations and new product performance: A contingent link. *Industrial Marketing Management*, *37*(8), 884–894. doi:10.1016/j.indmarman.2007.03.005

Uddain, J., Zakaria, L., Lynn, C. B., & Subramaniam, S. (2015). Preliminary assessment on Agrobacterium-mediated transformation of Dendrobium Broga Giant orchid's PLBs. *J. Emir. J. Food Agric.*, *27*(9), 669–677. doi:10.9755/ejfa.2015.05.211

Velasco, R., Zharkikh, A., Affourtit, J., Dhingra, A., Cestaro, A., Kalyanaraman, A., Fontana, P., Bhatnagar, S. K., Troggio, M., Pruss, D., Salvi, S., Pindo, M., Baldi, P., Castelletti, S., Cavaiuolo, M., Coppola, G., Costa, F., Cova, V., Dal Ri, A., ... Viola, R. (2010). The genome of the domesticated apple (Malus × Domestica Borkh.). *Nature Genetics*, *42*(10), 833–839. doi:10.1038/ng.654 PMID:20802477

Vendrame, W. A., & Khoddamzadeh, A. A. (2017). Orchid biotechnology. *Horticultural Reviews*, *44*, 173–228.

Vining, K. J., Pomraning, K. R., Wilhelm, L. J., Priest, H. D., Pellegrini, M., Mockler, T. C., Freitag, M., & Strauss, S. H. (2012). Dynamic DNA cytosine methylation in the Populus trichocarpa genome: Tissue-level variation and relationship to gene expression. *Biomed Cent Genom*, *13*(1), 27. doi:10.1186/1471-2164-13-27 PMID:22251412

Wang, H. M., Tong, C. G., & Jang, S. (2017). Current progress in orchid flowering/flower development research. *Plant Signaling & Behavior*, *12*(5), 5. doi:10.1080/15592324.2017.1322245 PMID:28448202

Wang, J. Y., Liu, Z. J., Zhang, G. Q., Niu, S. C., Zhang, Y. Q., & Peng, C. C. (2020). Evolution of two Ubiquitin-like a system of autophagy in orchids. *Horticultural Plant Journal*, *6*(5), 321–334. doi:10.1016/j.hpj.2020.05.006

White, T. L., Adams, W. T., & Neale, D. B. (2007). *Forest genetics*. CABI. doi:10.1079/9781845932855.0000

Wilcox, P. L., Echt, C. E., & Burdon, R. D. (2007). Gene-assisted selection: applications of association genetics for forest tree breeding. In *Association mapping in plants* (Vol. 278). Springer. doi:10.1007/978-0-387-36011-9_10

Wu, W. L., Chung, Y. L., & Kuo, Y. T. (2017). Development of SSR markers in Phalaenopsis orchids, their characterization, cross-transferability, and application for identification. *Orchid Biotechnol.*, *III*, 91–107.

Würschum, T., Maurer, H. P., Dreyer, F., & Reif, J. C. (2012). Effect of inter- and intragenic epistasis on the heritability of oil content in rapeseed (*Brassica napus* L.). *Theoretical and Applied Genetics*, *126*(2), 435–441. doi:10.100700122-012-1991-7 PMID:23052025

Yang, J., Benyamin, B., McEvoy, B. P., Gordon, S., Henders, A. K., Nyholt, D. R., Madden, P. A., Heath, A. C., Martin, N. G., Montgomery, G. W., Goddard, M. E., & Visscher, P. M. (2010). Common SNPs explain a large proportion of the heritability for human height. *Nature Genetics*, *42*(7), 565–569. doi:10.1038/ng.608 PMID:20562875

Yang, J., Lee, H., Shin, D. H., Oh, S. K., Seon, J. H., Paek, K. Y., & Han, K. (1999). Genetic transformation of Cymbidium orchid by particle bombardment. *Plant Cell Reports*, *18*(12), 978–984. doi:10.1007002990050694

Yu, H., Yang, S. H., & Goh, C. J. (2000). *DOH1*, a Class 1 *knox* Gene, Is Required for Maintenance ofthe Basic Plant Architecture and Floral Transition in Orchid. *The Plant Cell*, *12*(11), 2143–2159. doi:10.1105/tpc.12.11.2143 PMID:11090215

Yu, X., Kikuchi, A., Matsunaga, E., Morishita, Y., Nanto, K., Sakurai, N., Suzuki, H., Shibata, D., Shimada, T., & Watanabe, K. N. (2009). Establishment of the evaluation system of salt tolerance on transgenic woody plants in the special netted-house. *Plant Biotechnology (Sheffield, England)*, *26*(1), 135–141. doi:10.5511/plantbiotechnology.26.135

Zein, S. N., El-khalik, A., & Khatab, S. (2012). Characterization of Tobacco mosaic Tobamovirus (TMV-S) isolated from sunflower (*Helianthus annuus* L.) in Egypt. *International Journal of Virology*, *8*(1), 27–38. doi:10.3923/ijv.2012.27.38

Zeng, S., Huang, W., Wu, K., Zhang, J., Silva, J. A. T., & Duan, J. (2016). *In vitro* propagation of *Paphiopedilum orchids. Critical Reviews in Biotechnology*, *36*(3), 521–534. PMID:25582733

Zhang, F., Maeder, M. L., Unger-Wallace, E., Hoshaw, J. P., Reyon, D., Christian, M., Li, X., Pierick, C. J., Dobbs, D., Peterson, T., Joung, J. K., & Voytas, D. F. (2010). High frequency targeted mutagenesis in Arabidopsis thaliana using zinc finger nucleases. *Proceedings of the National Academy of Sciences of the United States of America*, *107*(26), 12028–12033. doi:10.1073/pnas.0914991107 PMID:20508152

Zhang, Y., Lee, Y. I., Deng, L., & Zhao, S. (2013). Asymbiotic germination of immature seeds and the seedling development of Cypripedium macranthos Sw., an endangered lady's slipper orchid. *Scientia Horticulturae*, *164*, 130–136. doi:10.1016/j.scienta.2013.08.006

Zhang, Y. Y., Wu, K. L., Zhang, J. X., Deng, R. F., Duan, J., Silva, J. A. T., Huang, W. C., & Zeng, S. J. (2015). Embryo development in association with symbiotic seed germination in vitro of *Paphiopedilum armeniacum* S.C. Chen et F.Y. Liu. *Scientific Reports*, *12*(5), 16356. doi:10.1038rep16356 PMID:26559888

Zhang, Y. Y., Wu, K. L., Zhang, J. X., Deng, R. F., Duan, J., Silva, J. A. T., Huang, W. C., & Zeng, S. J. (2015). Embryo development in association with symbiotic seed germination in vitro of *Paphiopedilum armeniacum* S.C. Chen et F.Y. Liu. *Scientific Reports*, *12*(5), 16356. doi:10.1038rep16356 PMID:26559888

Zhang, Z. S., He, Q. Y., Fu, X. L., Ou, X. J., Lin, W. Q., & Jiang, J. Y. (2001). Studies on the wide cross of Chinese orchids and the germination of their hybrid seeds. *Journal of South China Agricultural University*, *2*, 62–65.

Zhu, K., Woodall, C. W., & Clark, J. S. (2012). Failure to migrate: Lack of tree range expansion in response to climate change. *Global Change Biology*, *18*(3), 1042–1105. doi:10.1111/j.1365-2486.2011.02571.x

Zuk, O., Hechter, E., Sunyaev, S. R., & Lander, E. S. (2012). The mystery of missing heritability: Genetic interactions create phantom heritability. *Proceedings of the National Academy of Sciences of the United States of America*, *109*(4), 1193–1198. doi:10.1073/pnas.1119675109 PMID:22223662

APPENDIX: ABBREVIATIONS

SO: Seed Orchards, **SS:** Seed stand, **GD:** Genetic diversity, **GS:** Genomic selection
GE: Genetic engineering, **GBV:** Genomic breeding values, **SE:** Somatic embryogenesis
LD: linkage disequilibrium, **BWB:** Breeding without breeding, **VF:**Multi-varietal forestry
GBLUP: Genomic BLUP, **PLBs:** Protocorm-like bodies, **siRNAs:**Small interfering RNAs
2-DE: Two-dimensional electrophoresis, **miRNAs:** Micro RNA and RNA molecules
DEP: The differentially expressed proteins, **GWS:** Genome-wide selection
NGS: Next-generation sequencing, **PCR:** Polymerase chain reaction,
dPCR: Droplet polymerase chain reaction, **ISSR:** Inter simple sequence repeats
SSR: Simple sequence repeats, **RAPD:** Random amplified polymorphic DNA
AFLP: Amplified fragment length polymorphism, **QTL:** Quantitative trait loci
RFLP: Restriction fragment length polymerase, **SNPs:** Single nucleotide polymorphism
HPLC: High-performance liquid chromatography, **MAS:** Marker-assisted selection
RT-LAMP: loop-mediated isothermal amplification, **VIGS:** Virus-induced gene silencing
SERs: Surface-Enhance Raman Spectroscopy, **PTGS:** Posttranscriptional silencing
NAM: Nested Association Mapping

Chapter 6
The Relationship of Insects and the Process of Pollination to Increase the Production of Orchards

Mohamed Abdel-Raheem Ali Abdel-Raheem
Agricultural and Biological Research Institute, National Research Center, Egypt

ABSTRACT

In the US, the cultivated area and production of crops benefit from insect pollination. Mango orchards contain Zebda, Hendi sinara and Ewais varieties and Figryclan, Aranaba cultivars, cultivated together and without Zebda variety. Surveyed insects found 18 species belonging to order Diptera, Coleoptera, ymenoptera, and Neuroptera found during flowering periods. Cultivation of Zebda variety alone tended to decrease the yield/tree. Zebda cultivated mixed with selected varieties and cultivars significantly increased the yield/tree (58.10 and 56.10kg against 38.6, 36.6kg). Cultivation of mango varieties and cultivars in mixed system used application farm yard manure during flowering cycle, increased insect mango pollinators to the maximum, and increased the high mango yields and high good return for the mango growers.

INTRODUCTION

Insect pollinators play a fundamental role in the production of many fruits, vegetables and field crops (Klein et al., 2007) and numerous studies have valued insect pollination as an ecosystem service for agricultural food production at both global (Gallai et al., 2009; Winfree et al., 2011) and national scales (Smith et al., 2011).

A pollinator is an animal that moves pollen from the male anther of a flower to the female stigma of a flower. This helps to bring about fertilization of the ovules in the flower by the male gametes from the pollen grains. Approximately, 80 percent of all flowering plants species are pollinated by animals, including vertebrates and mammals but the main pollinators are insects. Pollinators are responsible for

DOI: 10.4018/978-1-6684-2423-0.ch006

providing with a wide variety of orchard, agricultural crops, horticultural crops and forage production. More than three quarters of the world's food crops rely at least on some parts on pollination by insects and other animals.

Pollination is an essential step in the reproductive process of the world's nearly 300,000 species of flowering plants because it is usually required for the production of seeds, (Committee on the Status of Pollinators in North America NRC, 2007, Axelrod, 1960, Grimaldi and Engel, 2005, Kevan and Viana, 2003, Davies, et al., 2004, Klein, et al., 2007, Ollerton, et al., 2011).

Pollination is the transfer of pollen, bearing the male gamete, from the anther of a flower to the stigma of a flower. After landing on a receptive stigma, a pollen grain germinates and a pollen tube develops, growing through the supporting style to the ovary. Genetic material in the pollen grain travels through the pollen tube to the ovary where it unites with an egg, the female gamete, in a process called fertilization. The fertilized egg develops into a seed, and that process is often accompanied by the development of fruit from surrounding tissue (Raghavan, 2000). Depending on the species, from one to several hundred eggs must be fertilized to ensure a high quality fruit because each egg requires a separate pollen grain for fertilization. Plants with incompletely pollinated flowers have fewer seeds and reduced fitness, and they produce inferior fruit with reduced market value (Ricketts, et al., 2004, Kasina, et al., 2009).

Pollination can result from the action of abiotic forces such as wind and water, but 80% of the Angiosperms rely on animals, including bats, flies, butterflies, beetles and other insects, Committee on the Status of Pollinators in North America NRC (2007). The majority of pollinators are insects, and the majority of those are bees (*Anthophila*), (Grimaldi and Engel, 2005), of which there are approximately 17,000 described species and as many as 30,000 species worldwide, Committee on the Status of Pollinators in North America NRC, 2007, Michener, 2000), With rare exception, bees collect pollen and nectar from flowers for food, transferring pollen in the process. North America is home to nearly 4,500 species of bees (Michener, 2000). Most are solitary, but there are 49 known species of the primitively eusocial bumble bee in the US, 41 of which are also found in Canada; an additional 11 species are found in Mexico. The highly eusocial western honey bee, *Apis mellifera*, was introduced to North America from Europe and Africa beginning in 1622 (Sheppard, 1989,). It is the only species of honey bee in North America.

There is evidence that such increases in agricultural productivity are leveling off, and while new agricultural technologies must have a role, maintaining and improving ecosystem services will be crucial to future food security, (Godfray, et al., 2010). In a global economy, changes in pollination services are likely to have ramifications for geographically distant markets and human responses, such as developing new suppliers. Worldwide a variety of insects including social and solitary bees, flies, wasps, beetles, butterflies and moths provide an ecosystem service to humans by pollinating many crops. Insect pollination has been shown to increase or stabilize yields and quality of fruit, vegetable, oil, seed and nut crops (. Klein, et al.,2007, Garibaldi, et al., 2011). Global cultivation of insect-pollinated crops has expanded since the 1960s, leading to about a 300% increase in demand for pollination services (Aizen and Harder, 2009). The global economic value of this pollination service was estimated (in 2005 US$) to be $215 billion or 9.5% of global food production value (Gallai, et al., 2009).

In Egypt mango varieties are mainly pollinated by insects and are dominant. While, other methods i.e. rains, winds or other natural methods are less (rarely) effective pollinating agents as compared with insect pollinators (Singh,1968, Anderson et al., 1982, Singh, 1988, Singh,1996, Dag and Gazit, 2000, Hsin Sung et al., 2006, Fajardo et al., 2008, Kumar et al., 2016, MunjAy et al., 2017 and Mohsen, 2019. Insect species visiting mango flowers in India and indicated that highest number of species were from order Diptera (17), followed by Coleoptera (3), Hymenoptera (3), Heteroptera (3), Lepidoptera (3),

(Singh. 1996). Effective insect pollinators are essential for good fruit set and yield in mango, (Dag and Gazit, 2000). Insects visiting mango bloom were forty six species or types were found, most belonged to the orders Diptera (26), Hymenoptera (12) and Cleoptera (6). Hsin Sung et al., (2006) one hundred and twenty six individual insects belonging to 39 species in 23 families and five order were recorded as a visitor or pollinator on mango flowers from February 13 to March 17, 2005 in southern Taiwan. Most of these insects belonged to the Diptera and Hymenoptera. Kumar et al., (2016) many plants are depends on insects for pollination, mango tree are pollinated most predominantly insects, like numerous insects of the order Hymenoptera, Diptera, Lepidoptera and Coleoptera. Pollen grains have been observed adhering to the bodies of many species belonging to these orders.

SURVEY OF INSECT POLLINATORS FOUND IN MANGO ORCHARDS DURING FLOWERING PERIODS

The insects found (12) of insects species belonged to order Diptera were surveyed as follows in desending order *Musca domestica* L., *Stomoxys calcitrans* (L.), *Fannia canicularis* (L.), *Muscina stabulans* (Fallen), *Chrysomya albiceps* Wiedemann, *Lucilia sericata* Meigen, *Calliphora vieina* (Rbineau- Desvoidy), *Sacrophaga carnaria* L., Tabanus sp., *Hermetia illucens* (L.), Syrphus sp. and *Eristalis tenax* (L.), Hymenoptera contain (3) species, *Apis mellifera* L., Solenopsis sp. and Aptanteles sp, (Mohsen, 2019). Then, order Coleoptera contain (2) species, *Coccinella septempunctata* L. and *Pancoda fasicata* F., finally, order Neuroptera contain one species, *Chrysoperia carnea* Stephans. (Mohsen, 2019).

WORLD TOP POLLINATORS

Entomophily is by far the most common mean of pollen transfer and it played a vital role in the evolution of angiosperms. There are about 250000 species of flowering plants globally which are pollinated by 200000 species of animals. Out of 95 percent of the flower which are cross pollinated, more than 85 percent depend on insects for pollination.

Insect pollinators include honey bees, bumble bees, pollen wasps, ants, flies including bee flies, hoverflies and mosquitoes, butterflies and moths and flower beetles. 50 percent of the plant species propagated by seeds are dependent on insect pollination whereas one third of the food supply is either directly or indirectly depend on these insects pollinated plants. Similarly more than 65 percent of all flowering plants are insect pollinated.

World top pollinators are 73% bees, 19% flies, 6.5% bats, 5% beetles, 5% wasps, 4% birds and 4% butterflies and moths.

ORCHARD POLLINATION

Pollination involves the integration of several biological and physical factors, including cultivar compatibility, synchronous blooming, insects, and proper weather conditions.

The nearer the pollinizer to the producing tree, the better distribution by the bees of pollen to all blossoms. Crabapples are often used as pollinizers.

Pollination is the sexual portion of a tree's life cycle and involves the integration of several biological and physical factors, including cultivar compatibility, synchronous blooming, insects, and proper weather conditions. If any one of these components is missing or limiting, crop yield and quality can be affected. Pollinators are important for all tree fruit crops, whether cultivars are self-fertile or self-sterile. Bees are by far the most important group of pollinators, but some types of flies, such as syrphids, some beetles, and some types of thrips may also assist to lesser degrees. Honey bees are the most widely used bee for pollination because they are easy to manage and can be moved in and out of orchards for bloom. However, several species of mason bees (*Osmia* spp.) are also managed for tree fruit pollination, and more than 120 species of wild bees frequent tree fruit orchards and contribute significantly to pollination.

Cold periods during flowering can reduce pollination and subsequent fruit set. Pollen may fail to germinate when temperatures are below 41°F, and pollen tube growth is extremely slow below 51°F. Therefore, in some situations, temperatures could be warm enough for bees to fly (65°F for the honey bee, 5 to 10 degrees cooler for bumble bees and solitary bees), but if the weather turns cold, the pollen tubes may not grow fast enough before the embryo sac deteriorates.

EFFECT OF MANGO CULTIVATION ALONE OR MIXED WITH OTHER ON INSECT POLLINATORS, AND YIELD

Mango cultivation alone or mango cultivation mixed system with other varieties and cultivars on yield/tree of zebda cultivated alone gave in the average 38.6 and 36.0 kg/tree and insect pollinators (36 and 32 insects) in both seasons, respectively, while yield of zebda mixed with other varieties and cultivars significantly increased average yield / tree as compared with zebda cultivated alone 58.1 and 56.00 kg /tree and insect pollinators (64 and 61 insects). Mixed mango cultivation either zebda or other varieties gave high yield / tree increased and increasing insect pollinators than those zebda cultivated alone or other varieties without zebda recording 36.42, 37.6; 22.43, 20.23; 31.00, 30.00 and 34.6, 36.6kg / tree for forgiclane, Awise, Hendi senra and Aranaba varieties and cultivars for both seasons, respectively, (Mohsen, 2019).

SPECIES COMPOSITION IN APPLE ORCHARDS

Hymenopteran insects dominate apple flower visitation, particularly honeybees and bumblebees. Also, widespread honeybees, represent valuable complementary pollination vectors for apple (Pardo and Borges2020). It has been shown that wild bees are important for global crop pollination including for apples (Garibaldi, et al. 2013 and Martins, et al., 2015), and that non-bee pollinators, such as hoverflies, can play an important role in some crops as both pollinators and pest control agents (Rader, et al. 2016). Our results also indicated that introduced species dominate the insect community visiting apple flowers and that, although native or endemic species are abundant in apple orchards, they might have a lower contribution to apple pollination. These results are in accordance with previous studies showing that native pollinators are abundant and spread across the landscape in Terceira Island (Traveset, et al., 2016). However, we observed significant differences in the relative abundance and species richness of native and introduced species in relation to the management implemented. In organic and conventional orchards with herbicides there was a clear dominance of introduced species visiting apple flowers, whereas in conventional without herbicides the relative abundances were almost equivalent. Moreover, the richness of native species visiting apple flowers and sampled in pan traps was higher in conventional orchards without herbicides as compared to the other management types. This could be a result of the large quantity of beneficial flowering plant species present in the herbaceous cover, particularly white clover, Trifolium repens, which is highly valuable as pollen source for long-tongued pollinator species (Campbell, et al., 2012). Underground cover was absent in conventional orchards that applied herbicides as expected. In the selected organic orchards, the herbaceous cover was evidently present, but mostly dominated by Plantago species, which are mainly wind-pollinated and thus not expected to attract pollinator insects (García and Miñarro, 2014). This result suggests that the type of herbaceous vegetation cover is of paramount importance for the maintenance of a diverse native pollinator load.

EFFECT OF MANAGEMENT ON APPLE POLLINATION

Results indicated that management did not have a significant impact on insect visitation rate to apple flowers. These findings are in line with those from other studies in apple (Brittain, et al., 2010), and other perennial crops, which found no effect of management on pollinator visitation. However, evidence from a large-scale study found contrasting results, with higher flower visitation rates in organic apple orchards vs. integrated pest management sites. Inconsistencies among studies may be explained by the fact that perennial crops usually possess a range of permanent structural elements such as hedgerows that can potentially attract pollinators regardless of the management implemented, thus being ultimately less impacted in terms of floral resources than annual crops (Porcel, et al., 2018). Apple orchards in Terceira Island were all surrounded by diverse structural elements such as hedges and stonewalls that could serve as nesting habitats and attract pollinators even in orchards that applied herbicides. Additionally, the effect of landscape composition around orchards could modulate that of management. We found that visitation rates were negatively associated with higher amounts of agricultural land around orchards. This result is in agreement with previous studies, which have showed that a predominance of agricultural habitats surrounding apple orchards decreases wild bee richness, abundance and diversity (Sheffield, et al., 2013, Watson, et al., 2011, Marini, et al., 2012, Proesmans, et al., 2019, Wu, et al., 2019 and Mallinger, et al., 2016). Therefore, the effects of management on pollination visitation rates seem to depend to a large extent on habitat composition at the landscape level, and could be further modulated by the existence of within-field habitat heterogeneity at the local scale (Pardo and Borges2020).

By contrast, our results suggested that management, together with landscape composition, could have a significant impact on insect abundance, species richness and diversity at the orchard scale. Some studies had previously reported an effect of management on pollinator species richness and diversity in apple orchards (Samnegård, et al. 2019 and Sheffield, et al., 2013), whereas others had found a non-significant effect (Joshi, et al., 2020 and Sheffield, et al., 2008). In this work, conventional orchards without herbicides had a higher species abundance than conventional with herbicides and, unexpectedly, than organic orchards. This latter result could be explained by the more beneficial underground herbaceous cover with respect to organic orchards as discussed in the previous section. Other studies in cider-apple orchards have shown that plant species richness and the abundance of some plant species in the groundcover favored the presence of pollinators assemblages (García and Miñarro, 2014). Despite their greater insect abundance, conventional orchards without herbicides showed a lower species richness (rarefied) and diversity than conventional with herbicides sites, contrary to our initial expectations.

These orchards also showed a highly uneven species diversity profile, i.e., they host an insect community in which species differ widely in their abundance. This highly uneven distribution can lead to lower species richness after rarefaction, in contrast with the other management systems that showed a more even distribution. It has been shown that when communities are highly uneven, or there is extreme dominance by one or a few species, they could be less resistant to environmental stress (Wittebolle, et al., 2009).

In addition, populations of rare species are more likely to experience local extinctions (Matthews and Whittaker, 2015). Last, our results indicated that management may have a weak impact on pollination service provision, showing a significant effect exclusively on fruit set. In fact, the effect of management on pollination services at local scale appears to be erratic (Porcel, et al., 2018). Organic management has been shown to increase pollination in annual systems like strawberries (Andersson, et al., 2012), but evidence on its effects in perennial crops like apple is still inconsistent (Pardo and Borges, 2020). Here, management seems insufficient for the improvement of overall pollination services in apple, although further studies undertaking hand pollination and bagging experiments would be needed in order to draw more general conclusions.

Apple Yield

The fruit was affected by pollination treatment with significantly greater fruit set following hand pollination compared to open pollinated blossoms which in turn set more fruit than pollinator excluded blossoms. The same treatment effects on fruit set were hand pollinated blossoms set more fruit than open pollinated blossoms which again set more than pollinator excluded blossoms. The same relationship was found for apple number at harvest with hand pollinated greater than open pollinated which was greater than pollinator excluded, (Garratt, et al., 2014).

Apple Quality

The number of seeds per apple was significantly affected by treatment with a greater seed number per apple in hand pollinated than open pollinated fruit which in turn had greater seed numbers than pollinator excluded fruit. The same significant relationship was found with hand pollinated greater than open pollinated which was greater than pollinator excluded fruit.

Pollination treatment had a significant effect on the size, weight and firmness of apples. Open pollinated apples were significantly larger and heavier than hand pollinated apples. Furthermore open pollinated apples were significantly softer than both pollinators excluded and hand pollinated apples. Pollinator excluded apples were significantly firmer than open pollinated and hand pollinated apples. Pollination treatment significantly affected the concentration of calcium, magnesium and zinc in experimental apples. In all cases, hand pollinated apples had a significantly lower concentration than pollinator excluded apples. For calcium and magnesium, concentrations in hand pollinated apples were also significantly lower than in open pollinated fruit which were also significantly lower than concentrations found of magnesium and zinc in pollinator excluded fruit, (Garratt, et al., 2014).

The market value of insect pollination to apple production is clear for both Cox and Gala, adding >£11,000 and >£14,000 in additional output per hectare, respectively. Our data also shows, however, that there is evidence of potential economic pollination deficits for Gala where pollination affects both yield and quality. Under optimal pollination conditions, market output of Gala would be increased by

£6500/ha. Current levels of investment in pollination services, either through management of wild populations or introduction of managed species, remains largely unknown and as such the importance of insect pollinators may be undervalued. The continued decline of insect pollinators could have serious ramifications for the apple industry and the £37 million service they provide to Cox and Gala production. Furthermore, the large scope for potential improvement in market output for Gala, highlights the potential for investment in management to boost pollination services, (Garratt, et al., 2014).

Insects were carrying low numbers of avocado pollen grains when visiting female phase flowers, during a period where no overlap was evident between male and female flowers on the same tree or cultivar, (Lisa and Mark, 2011).

In Australia, in a block with no pollinizers, 68.6% of the bees that were washed were carrying pollen (mean = 4.7 pollen grains/bee). The pollen found on these bees may be the result of foraging trips on male phase 'Hass' flowers the previous day.

In New Zealand, in a block with pollinizers, 84% of the bees were carrying pollen (mean = 38.7 pollen grains/bee). The number of pollen grains found on honey bees is much lower than those found by Vithanage (1990). However, Vithanage did not specify when his bees were caught; they may have been caught at a time when female and male flowers overlapped. Furthermore, his method did not state whether the bees were caught on male or female phase flowers, (Lisa and Mark, 2011).

Bees caught on male phase 'Bacon' flowers in the New Zealand orchard were carrying up to 2500 pollen grains, indicating that honey bees are capable of carrying large numbers of avocado pollen grains. The low pollen count suggests the bees on the female phase flowers had not recently visited a male phase flower and therefore had few pollen grains available to transfer to avocado stigmas.

In the Australian orchard, the mean number of pollen grains/insect visit increased when flowers were visited at least 3.2 hours after anthesis. This corresponds to the time at which flowers on the same tree and/or cultivar have reopened in their male phase (close pollination), (Lisa and Mark, 2011).

This indicates that much of the pollination occurs in a short time frame, near the end of the female phase when the male phase flowers open. This is potentially problematic, because the timing and duration of both phases of flowering are affected by climatic conditions; thus, in unsuitable weather, this bisexual phase may not occur. This trend was not evident in the New Zealand orchard. The presence of pollinizer trees meant that crosspollination could occur throughout anthesis. However, since there were a large number of visits recorded at this orchard that did not result in pollen being transferred, we suspect that bees were not frequently crossing between cultivars. Both the low insect visitation (in New Zealand), and the low numbers of pollen grains transferred onto avocado flowers (at all sites), may be factors that are currently limiting pollination and consequently fruit set in some orchards in Australia and New Zealand. Further research in these countries should focus on determining the effectiveness of different insects at pollinating avocado flowers and investigating the overlap between male and female phase flowers (taking into account the timing of anther dehiscence), under various climatic conditions. If under current growing conditions there are times when not enough pollen is being transferred, then there is potential to better manage pollination of avocado, thus increasing fruit production in 'Off' years, (Lisa and Mark, 2011).

CONCLUSION

Honeybees, represent important pollination vectors for apple, further calling attention to the need of preserving wild pollinators in agroecosystems. In addition, results from this work suggest that management has a limited effect on apple pollination on an oceanic island, showing no significant effect on pollinator visitation rates and ultimately overall pollination services. On the contrary, a negative effect of increasing surrounding agricultural area on pollinator visitation was observed. Although conventional orchards without herbicides seem to benefit insect abundance at the orchard scale, together with favouring native species, they also showed a very uneven species distribution, leading to lower species richness and diversity than conventional orchards with herbicide application. These findings indicate that the effect of management in apple pollination may depend largely on habitat composition at the landscape level. Nevertheless, further studies are needed to gain a better understanding of the effect of these drivers on insect pollination in apple and other agroecosystems.

REFERENCES

Aizen, M. A., & Harder, L. D. (2009). The global stock of domesticated honey bees is growing slower than agricultural demand for pollination. *Current Biology*, *19*(11), 915–918. doi:10.1016/j.cub.2009.03.071 PMID:19427214

Anderson, D. L., Sedgley, H., Shortant, T. R. T., & Allwood, A. J. (1982). Insect pollination of mango in nort-hen Australia. *Aust. Agric. Res.*, *33*(3), 541–548. doi:10.1071/AR9820541

Andersson, G. K. S., Rundlöf, M., & Smith, H. G. (2012). Organic farming improves pollination success in Strawberries. *PLoS One*, *7*(2), 1–4. doi:10.1371/journal.pone.0031599 PMID:22355380

Axelrod, D. (1960). The evolution of flowering plants. In S. Tax (Ed.), *Evolution after Darwin.1* (pp. 227–305). University of Chicago Press.

Brittain, C., Bommarco, R., Vighi, M., Settele, J., & Potts, S. G. (2010). Organic farming in isolated landscapes does not benefit flower-visiting insects and pollination. *Biological Conservation*, *143*(8), 1860–1867. doi:10.1016/j.biocon.2010.04.029

Campbell, A. J., Biesmeijer, J. C., Varma, V., & Wäckers, F. L. (2012). Realising multiple ecosystem services based on the response of three beneficial insect groups to floral traits and trait diversity. *Basic and Applied Ecology*, *13*(4), 363–370. doi:10.1016/j.baae.2012.04.003

Committee on the Status of Pollinators in North America NRC. (2007). *Status of pollinators in North America*. National Academies Press.

Dag, A., Eisenstin, D., & Gazit, S. (2000). Effect of temperature regime on pollen and the effective pollination of kent mango in Israel. *Scientia Horticulturae*, *86*(1), 1–11. doi:10.1016/S0304-4238(99)00134-X

Dag, A., & Gazit, S. (2000). Mango pollinators in Israel. *Journal of Applied Horticulture*, *2*(1), 39–43. doi:10.37855/jah.2000.v02i01.12

Davies, T. J., Barraclough, T. G., Chase, M. W., Soltis, P. S., Soltis, D. E., & Savolainen, V. (2004). Darwin's abominable mystery: Insights from a supertree of the angiosperms. *Proceedings of the National Academy of Sciences of the United States of America, 101*(7), 1904–1909. doi:10.1073/pnas.0308127100 PMID:14766971

Evans & Goodwin (2011). The role of insect pollinators in avocado (Persea americana) pollination in New Zealand and Australia. *Proceedings VII World Avocado Congress 2011 (Actas VII Congreso Mundial del Aguacate 2011).*

Fajardo, A. C., Medina, J. R., Opina, O. S., & Cervancia, C. R. (2008). Insect pollinators and floral visitors of mango. *Philippine Agricultural Scientist, 91*, 372–382.

Gallai, N., Salles, J. M., Settele, J., & Vaissiere, B. E. (2009). Economic valuation of the vulnerability of world agriculture confronted with pollinator decline. *Ecological Economics, 68*(3), 810–821. doi:10.1016/j.ecolecon.2008.06.014

García, R. R., & Miñarro, M. (2014). Role of floral resources in the conservation of pollinator communities in cider-apple orchards. *Agriculture, Ecosystems & Environment, 183*, 118–126. doi:10.1016/j.agee.2013.10.017

García, R. R., & Miñarro, M. (2014). Role of floral resources in the conservation of pollinator communities in cider-apple orchards. *Agriculture, Ecosystems & Environment, 183*, 118–126. doi:10.1016/j.agee.2013.10.017

Garibaldi, L. A., Aizen, M. A., Klein, A. M., Cunningham, S. A., & Harder, L. D. (2011). Global growth and stability of agricultural yield decrease with pollinator dependence. *Proceedings of the National Academy of Sciences of the United States of America, 108*(14), 5909–5914. doi:10.1073/pnas.1012431108 PMID:21422295

Garibaldi, L. A., Steffan-Dewenter, I., Winfree, R., Aizen, M. A., Bommarco, R., Cunningham, S. A., Kremen, C., Carvalheiro, L. G., Harder, L. D., Afik, O., Bartomeus, I., Benjamin, F., Boreux, V., Cariveau, D., Chacoff, N. P., Dudenhoffer, J. H., Freitas, B. M., Ghazoul, J., Greenleaf, S., ... Klein, A. M. (2013). Wild pollinators enhance fruit set of crops regardless of honey bee abundance. *Science, 339*(6127), 1608–1611. doi:10.1126cience.1230200 PMID:23449997

Garratt, M. P. D., Breeze, T. D., Jenner, N., Polce, C., Biesmeijer, J. C., & Potts, S. G. (2014). Avoiding a bad apple: Insect pollination enhances fruit quality and economic value. *Agriculture, Ecosystems & Environment, 184*, 34–40. doi:10.1016/j.agee.2013.10.032 PMID:24748698

Godfray, H. C. J., Crute, I. R., Haddad, L., Lawrence, D., Muir, J. F., Nisbett, N., Pretty, J., Robinson, S., Toulmin, C., & Whiteley, R. (2010). The future of the global food system. *Philosophical Transactions of the Royal Society of London. Series B, Biological Sciences, 365*(1554), 2769–2777. doi:10.1098/rstb.2010.0180 PMID:20713383

Grimaldi, D., & Engel, M. (2005). *Evolution of the insects*. Cambridge University Press.

Joshi, N. K., Leslie, T., Rajotte, E. G., & Biddinger, D. J. (2020). Environmental impacts of reduced-risk and conventional pesticide programs differ in commercial apple orchards, but similarly influence pollinator community. *Chemosphere, 240*, 124926. doi:10.1016/j.chemosphere.2019.124926 PMID:31726586

Kasina, J. M., Mburu, J., Kraemer, M., & Holm-Mueller, K. (2009). Economic benefit of crop pollination by bees: A case of kakamega small-holder farming in western Kenya. *Journal of Economic Entomology*, *102*(2), 467–473. doi:10.1603/029.102.0201 PMID:19449623

Kearns, C. A., Inouye, D. W., & Waser, N. M. (1998). Endangered mutualisms: The conservation of plant-pollinator interactions. *Annual Review of Ecology and Systematics*, *29*(1), 83–112. doi:10.1146/annurev.ecolsys.29.1.83

Kevan, P. G., & Viana, B. F. (2003). The global decline of pollination services. *Biodiversity (Ottawa)*, *4*(4), 3–8. doi:10.1080/14888386.2003.9712703

Klein, A.-M., Vaissière, B. E., Cane, J. H., Steffan-Dewenter, I., Cunningham, S. A., Kremen, C., & Tscharntke, T. (2007). Importance of pollinators in changing landscapes for world crops. *Proceedings. Biological Sciences*, *274*(1608), 303–313. doi:10.1098/rspb.2006.3721 PMID:17164193

Klein, A. M., Vaissiere, B. E., Cane, J. H., Steffan-Dewenter, I., Cunningham, S. A., Kremen, C., & Tscharntke, T. (2007). Importance of pollinators in changing landscapes for world crops. *Proceedings of the Royal Society of London*, *274*, 303–313. PMID:17164193

Kumar, S., Joship, C., Pashupati, N., Singh, V. K., & Mansotra, D. K. (2016). Role of insects in pollination of mango trees. *International Research Journal of Biological Sciences*, *5*(1), 64–67.

Mallinger, R. E., Gibbs, J., & Gratton, C. (2016). Diverse landscapes have a higher abundance and species richness of spring wild bees by providing complementary floral resources over bees' foraging periods. *Landscape Ecology*, *31*(7), 1523–1535. doi:10.100710980-015-0332-z

Marini, L., Quaranta, M., Fontana, P., Biesmeijer, J. C., & Bommarco, R. (2012). Landscape context and elevation affect pollinator communities in intensive apple orchards. *Basic and Applied Ecology*, *13*(8), 681–689. doi:10.1016/j.baae.2012.09.003

Martins, K. T., Gonzalez, A., & Lechowicz, M. J. (2015). Pollination services are mediated by bee functional diversity and landscape context. *Agriculture, Ecosystems & Environment*, *200*, 12–20. doi:10.1016/j.agee.2014.10.018

Matthews, T. J., & Whittaker, R. J. (2015). On the species abundance distribution in applied ecology and biodiversity management. *Journal of Applied Ecology*, *52*(2), 443–454. doi:10.1111/1365-2664.12380

Michener, C. D. (2000). *The bees of the world*. Johns Hopkins University Press.

Mohsen, A.M.A. (2019). Survey of insect mango pollinators and the pollination occurrence of mango trees in relations to fruit yields. *Current Science International, 8*(2), 245-251.

Munj, A. Y., Zote, V. K., Rut, R. A., & Salvi, B. R. (2017). Survey and surveillance of pollinators of mango in south konkan coastal region of Maharashtra. *Journal of Entomology and Zoology Studies*, *5*(3), 190–192.

Ollerton, J., Winfree, R., & Tarrant, S. (2011). How many flowering plants are pollinated by animals? *Oikos*, *120*(3), 321–326. doi:10.1111/j.1600-0706.2010.18644.x

Pardo, A., & Borges, P. A. V. (2020). Worldwide importance of insect pollination in apple orchards: A review. *Agriculture, Ecosystems & Environment, 293*, 106839. doi:10.1016/j.agee.2020.106839

Porcel, M., Andersson, G. K. S., Pålsson, J., & Tasin, M. (2018). Organic management in apple orchards: Higher impacts on biological control than on pollination. *Journal of Applied Ecology, 55*(6), 2779–2789. doi:10.1111/1365-2664.13247

Proesmans, W., Smagghe, G., Meeus, I., Bonte, D., & Verheyen, K. (2019). The effect of mass-flowering orchards and semi-natural habitat on bumblebee colony performance. *Landscape Ecology, 34*(5), 1033–1044. doi:10.100710980-019-00836-5

Rader, R., Bartomeus, I., Garibaldi, L. A., Garratt, M. P. D., Howlett, B. G., Winfree, R., Cunningham, S. A., Mayfield, M. M., Arthur, A. D., & Andersson, G. K. S. (2016). Non-bee insects are important contributors to global crop pollination. *Proceedings of the National Academy of Sciences of the United States of America, 113*, 146–151. doi:10.1073/pnas.1517092112 PMID:26621730

Raghavan, V. (2000). Developmental biology of flowering plants. Springer-Verlag GmbH and Co. KG.

Ricketts, T. H., Daily, G. C., Ehrlich, P. R., & Michener, C. D. (2004). Economic value of tropical forest to coffee production. *Proceedings of the National Academy of Sciences of the United States of America, 101*(34), 12579–12582. doi:10.1073/pnas.0405147101 PMID:15306689

Roubik, D. W. (Ed.). (1995). *Pollination of cultivated plants in the tropics*. FAO.

Samnegård, U., Alins, G., Boreux, V., Bosch, J., García, D., Happe, A. K., Klein, A. M., Miñarro, M., Mody, K., Porcel, M., Rodrigo, A., Roquer-Beni, L., Tasin, M., & Hambäck, P. A. (2019). Management trade-offs on ecosystem services in apple orchards across Europe: Direct and indirect effects of organic production. *Journal of Applied Ecology, 56*(4), 802–811. doi:10.1111/1365-2664.13292

Sheffield, C. S., Kevan, P. G., Pindar, A., & Packer, L. (2013). Bee (Hymenoptera: Apoidea) diversity within apple orchards and old fields in the Annapolis Valley, Nova Scotia, Canada. *Canadian Entomologist, 145*(1), 94–114. doi:10.4039/tce.2012.89

Sheffield, C. S., Kevan, P. G., Westby, S. M., & Smith, R. F. (2008). Diversity of cavity-nesting bees (Hymenoptera: Apoidea) within apple orchards and wild habitats in the Annapolis Valley, Nova Scotia, Canada. *Canadian Entomologist, 140*(2), 235–249. doi:10.4039/n07-058

Sheppard, W. S. (1989). A history of the introduction of honey bee races into the USA. Part I of a two-part series. *American Bee Journal, 129*, 617–619.

Singh, G. (1988). Insect pollinators of mango and their role in fruit setting. *Acta Horticulturae*, (231), 629–632.

Singh, G. (1996). Pollination, pollinators and fruit setting in mango. *Acta Horticulturae*, (455), 116–123.

Singh, L. B. (1968). The mango: Botany, cultivation and utilization. Leonard Hill, London 263pp. Tim, A. Heard, 1999. The role of stingless bees in crop pollination. *Annual Review of Entomology, 44*, 183–206.

Smith, P., Ashmore, M., Black, H., Burgess, P., Evans, C., Hails, R., Potts, S. G., Quine, T., & Thomson, A. (2011). UK National Ecosystem Assessment – Chapter 14: Regulating Services. UNEP-WCMC.

Sung, Ying, Chang, Chen, Chen, & Ho. (2006). Pollinators and their behaviors on mango flowers in Sothern Taiwan. *Formosan Entomol.*, *26*, 161–170.

Traveset, A., Tur, C., Trøjelsgaard, K., Heleno, R., Castro-Urgal, R., Olesen, J. M., & Santos, A. (2016). Global patterns of mainland and insular pollination networks. *Global Ecology and Biogeography*, *25*(7), 880–890. doi:10.1111/geb.12362

Vithanage, V. (1990). The role of the European honeybee (Apis mellifera L.) in avocado pollination. *Journal of Horticultural Science*, *65*(1), 81–86. doi:10.1080/00221589.1990.11516033

Watson, J. C., Wolf, A. T., & Ascher, J. S. (2011). Forested landscapes promote richness and abundance of native bees (hymenoptera: Apoidea: Anthophila) in Wisconsin apple orchards. *Environmental Entomology*, *40*(3), 621–632. doi:10.1603/EN10231 PMID:22251640

Winfree, R., Gross, B. J., & Kremen, C. (2011). Valuing pollination services to agriculture. *Ecological Economics*, *71*, 80–88. doi:10.1016/j.ecolecon.2011.08.001

Wittebolle, L., Marzorati, M., Clement, L., Balloi, A., Daffonchio, D., Heylen, K., De Vos, P., Verstraete, W., & Boon, N. (2009). Initial community evenness favours functionality under selective stress. *Nature*, *458*(7238), 623–626. doi:10.1038/nature07840 PMID:19270679

Wu, P., Axmacher, J. C., Li, X., Song, X., Yu, Z., Xu, H., Tscharntke, T., Westphal, C., & Liu, Y. (2019). Contrasting effects of natural shrubland and plantation forests on bee assemblages at neighboring apple orchards in Beijing, China. *Biological Conservation*, *237*, 456–462. doi:10.1016/j.biocon.2019.07.029

Chapter 7
Impact of Magnetic Water Treatments on Citrus Cultivation:
Water Treatment and Cultivation Under Abiotic Stress

Waleed Fouad Abobatta
Horticulture Research Institute, Agriculture Research Center, Egypt

Anis Elaoud
University of Carthage, Tunisia

ABSTRACT

Climate change conditions such as drought, rising temperature, and high solar radiation cause increasing salinity levels, which are considered the main factor that limits agricultural production. Therefore, to prevent food security in arid and semi-arid areas from being subject to climatic hazards, optimal water management is essential. To surmount this problem, research is focused on innovation likely to improve efficiency water use and protect the water resources. In order to overcome the problems of contaminated water and salinization problems, the researchers used a new technique based on a magnetic field. Thus, magnetized water used for irrigation can improve water productivity and increase the production of agricultural crops, thus sustaining the water resources for the future in view of the expected global water scarcity. In this context, this chapter discusses the characteristics linked to magnetic treatments on irrigation water by studying the parameters of water quality and their effects on productivity and fruit quality of the citrus orchards.

DOI: 10.4018/978-1-6684-2423-0.ch007

INTRODUCTION

Agricultural production is one of the most basic elements contribute to the economic income and food security, despite the problems that accompanied such as lack of water, desertification, salinity and low yield. Recently, depleting water resources and the increased demand for food to meet human needs are the most critical problems in the world. Water globally is plentiful, but there is 97% saline, 2.25% is trapped in glaciers and ice, while, only 0.75% remains in aquifers, rivers, and freshwater lakes. Most of this freshwater (69%) is used for agricultural production, 23% for industry and 8% for domestic use. In the Mediterranean region, Water resources represent an important component in the field of agriculture, drinking water, and industry. It should be noted that agriculture represents the most water-consuming activity in the region. However, the scarcity of water resources particularly in south Mediterranean, in both quality and quantity, as well as its distribution in time and space, are among the main factors limiting the development of agriculture.

This problem is intensifying with recent climate change. Indeed, research is directed more and more towards strategies, which aim at the preservation, conservation and enhancement of both the quantity and the quality of these water resources, and even innovations likely to improve the efficiency and the productivity of irrigation water.

In order to meet this urgent need for water saving, several techniques and approaches are used, hydraulic techniques (irrigation system, dose and frequency of irrigations, drainage, etc.), biological techniques (varieties of plants tolerant to drought and salinity, supply of organic matter, etc.), physical techniques (soil amendment), chemical techniques (supply of gypsum, etc.). Magnetic water treatment technology is one of the solutions that can be used to remedied the water shortage resources, this technique has become the focus of researchers compared the physical and other chemicals treatments, due to it is simplecity, as ecofriendly, and safety.

The use of magnetic water leads to change or break down hydrogen bonds between molecules. This leads to change the properties of water, such as electrical systems, increasing the proportion of oxygen dissolved in water, increase the ability to dissolve salts and acids, polymerization, the surface tension, change in the speed of chemical reactions, properties of evaporation, moisture, elasticity, electrical insulation and increased permeability. Later, these changes will effect on the qualities of the material that enters the structure through its effect on the qualities physical or in chemical or physiological processes and biochemical (Al-Jbouri et al., 2006). Cai et al., (2009) who reported that magnetic field (at 500 mT) caused changes in physicochemical properties of water; these changes include decreasing water surface tension and increase viscosity. More that their results suggested an increase of activation energy and water molecule size due to extra hydrogen bond formation; this effect is linear with exposure period. The importance's of physical treatment of saline water using magnetic devices become feasible. Magnetic water (MW) is considered as environmentally friendly technique (Nimmi and Medhu, 2009). MW is produced when water passes through the magnetic field of magnetic permanent device or electromagnetic one, installed on feed pipeline, where all water and salt molecules have internal vibration (Babu, 2010). The used of magnetic water to soil have the removal of excess soluble salts. Bringing down of pH esteems and the dissolving of slightly soluble components such as phosphate, carbonates and sulphates. Moreover, the attractive strategy of magnetic method for saline water is allegedly a successful technique for soil desalinization (Mostafazadeh et al., 2011). The application of irrigation magnetic water to soil salinity was increased leaching of excess soluble salts, lowered soil alkalinity and dissolved slightly soluble salts and increasing of macro-micronutrients available in soil (El-Sonbaty 2021).

Citrus one of the main fruit crop in subtropical, arid, and semiarid regions, particularly under Mediterranean climate, citrus ranking as third crop globally following grape and apple (FAO, 2016). Due to fluctuation in climate conditions, there are increasing in soil and water salinity, which affect negatively citrus growth, total yield, and reduce fruit quality (Abobatta, 2018). Salinity and heat stresses may affect the productivity of orange trees, but high temperature and radiation have a positive effect on ripening and fruit quality (Abobatta, 2020; Colauto Stenzel, et al, 2006). In addition, citrus is one of the most important horticultural crops that can adapt well to various management practices for irrigation (Martínez-Ferri, et al, 2013).

This new approach appears to be ecological and inexpensive for the treatment and valuation of saline water and for increasing yield. In the literature, the magnetic anti-scale treatment is the most developed. However, a few research is directed towards treatment to deal with the salinity of irrigation water. In this work, the treating of saline water by this technology tested on the citrus orchards. Some results of the effect of this process on irrigation water and the impact on the crop and yield will be examined.

This work explores a new technology based on the magnetic treatment of the brackish water, which can constitute a solution of recourse allowing the use of this water in agriculture and mainly on citrus cultivation.

IRRIGATION WATER QUALITY

The quality of irrigation water is an important parameter for crop yield, maintenance of soil productivity and environmental protection. Therefore, the physical and chemical properties of the soil, such as its structure (aggregation stability) and permeability, are very sensitive to the types of potentially exchangeable ions present in irrigation water, table 1 shows the acceptable values for different elements found in irrigation water.

Table 1. Standard of irrigation water quality (Couture, 2004)

Alkalinity ($CaCO_3$)	1 to 100 ppm (not more than 200 ppm)
Aluminium (Al^{3+})	0 to 5 ppm
Bicarbonate (HCO^{3-})	30 to 50 ppm (not more than 150)
Boron (B)	0,2 to 0,5 ppm (not more than 0,8)
Calcium (Ca2+)	40 to 120 ppm
Chlorine (Cl^-)	0 to 100 ppm (not more than 140)
Copper (Cu^{2+})	0,08 to 0,15 ppm (not more than 0,2)
Fluor (F^-)	0 (not more than 1)
Iron (Fe^{3+})	1 to 2 ppm (not more than 5)
Magnesium (Mg^{2+})	6 to 25 ppm
Manganese (Mn^{2+})	0,2 to 0,7 ppm (not more than 2)
Molybdenum (Mb)	0,02 to 0,05 ppm (not more than 0.07)
pH	5 to 7
Potassium (K^+)	0,5 to 5 ppm
Sodium absorption ratio (RAS)*	0 to 4 ppm
Sodium (Na^+)	0 to 30 ppm (not more than 50 ppm)
Sulfate ($SO4^{2-}$)	24 to 240 ppm
Total dissolved salts	70 to 700 ppm (not more than 875)
Zinc (Zn^{2+})	0,1 to 0,2 ppm (not more than 2,0)
Salinity (ds)	Less than 1,0-1,5

PHYSICO-CHEMICAL PARAMETERS OF WATER

Salinity

Since soluble salt is present in all irrigation water, salt is a common problem for farmers in arid climates. In fact, high concentrations of salt in water or soil will adversely affect crop yields, leading to soil degradation and pollution of groundwater. This problem is commonly encountered in coastal areas due to overexploitation of irrigation wells and depletion of aquifers leading to ocean intrusion (Otsuka & Ozeki, 2006).

The salts causing this problem are calcium Ca^{2+}, magnesium Mg^{2+}, sodium Na^+, potassium K^+, chloride Cl^-, sulfate SO_4^{2-}, and bicarbonate HCO_3^-. It can be measured in two ways, namely total dissolved salts (TDS) or electric conductivity (EC) (ds/cm).

At high concentrations in irrigation water, sodium is one of the undesirable salts. Its presence makes the ground harder, drier and more waterlogged, in addition, excessive sodium in the water causes toxicity of crops (Zörb, et al, 2019).

Sodium Absorption Ratio (SAR)

The sodium problem is reduced if the amount of calcium and magnesium is high in relation to the amount of sodium. This relationship is expressed as the rate or ratio of sodium adsorption *'SAR'* (Food and Agriculture Organization [FAO] 2008). This parameter indicates the effect of relative cation concentration on accumulated sodium in soil.

Sodium adsorption rate is commonly used as an index of sodium threats in water and soil, and as a soil exchangeable sodium substitute (FAO, 2008).

The SAR is calculated according to the following formula (cation concentration is expressed in (meq/l)): $SAR=Na/(0.5(Ca^{2+}+Mg^{2+})^{1/2}$

Water with a SAR more than nine meq/l should not be used even if the total salt content is relatively low. Continuous use of water with high SAR causes soil degradation (Couture, 2006). Young, et al, (2003) reported that, irrigation water could divided to four classes as shown in (Table 2).

Table 2. Classification of irrigation water according to its sodium content

Class 1	Low rate of Sodium	SAR<10
Class 2	Average rate	10,1<SAR<18
Class 3	High rate	18.1<SAR<26
Class 4	Very high rate	SAR>26.1

This classification is based primarily on the effect of sodium on physical soil conditions.

pH of Irrigation Water

Hydrogen potential (expressed as pH) is a measure of the chemical activity of hydrogen ions (H^+) in a solution. In practice, the normal pH between 6.5 and 8.5, considered the proper water of life. For natural water, it mainly consists of two balanced types of carbonic acid pH (low acidity) (Yun-Zhu, et al, 2012).

Electrical Conductivity

Conductivity is converted into the capacity of an aqueous solution to conduct electricity. It is inversely proportional to resistivity. It is expressed in siemens (S/cm), micro-siemens per cm (μS/cm), or milli-siemens per cm (mS/cm). Conductivity is directly proportional to the quality of solids (mineral salts) dissolved in water.

Generally, the relationship between conductivity and ionic concentration is approximately as follows: 2μS/cm = 1 ppm (parts per million) (Yun-Zhu, et al, 2012).

Alkalinity and Hardness

Alkalinity and hardness are two concepts that cause water pipes to clog. Hardness reflects the total calcium and magnesium content of the water. While, alkalinity is the ability of a mineral solution to neutralize hydrogen ions.

Generally, when the calcium and magnesium content of the water is high, it must have an equivalent amount of bicarbonate or carbonate, which will increase its alkalinity (Holysz, et al, 2007).

MAGNETIC PROCESSING TECHNIQUE

The process consists of passing the water to be treated through a series of permanent magnets. Some manufacturers insist on the importance of the spiral.

The water flows through the magnet and there is a flow restriction, so the water flow changes along its path. Magnets are generally made of iron (51%), cobalt (24%), nickel (14%), aluminum (8%) and copper (3%).

Concept and Principle of the Magnetic Field

A shepherd named Magnes had identified the property of magnetism 2500 years ago. He was the first who observed magnetism when a nail from his shoes was pulled to a rock. This stone is a natural magnet that has the ability to pull iron. The first law of magnetism states that different poles hold together and identical poles repel each other (Joshi and Kamat, 1966). This means that there is a region of southern polarity close to the Earth's geographic North Pole.

The magnetic field is represented by lines of force that always follow the same direction, from the North Pole to the South Pole.

The flow lines are tighter at the ends of the magnet, which leads to the conclusion that the concentration of the force lines is a measure of the density of the field. The higher magnetic field density, the closer the force lines will be, the magnetic flux density unit (B) is the Tesla in the international system (Tai, et al, 2008).

According to previous researches, water magnetization is possible by passing water through a magnetic field, studies are still under development and researchers are trying to use it in different fields. The general operating scope for magnetic processing is the result of the physics of the interaction between a magnetic field and a moving electrical charge (Craig, 2009).

This principle is based on the passage of water through a magnetic softener, during which a Lorentz force is exerted on each ion that is in the opposite direction of the other. The reorientation of particles increases the frequency of collisions between ions on both sides, combining to form an insoluble precipitate (Tai, et al, 2008).

Different Types of Magnetic Devices

The magnetic water treatment equipment is environmentally friendly, reasonably priced and does not consume too much energy. In fact, they are accessible in different configurations (Gutfleisch, et al,

2011). Some manufacturers use electromagnets, while others use single or permanent magnet devices with different magnetic field directions.

The most effective arrangement, where the magnetic field is perpendicular to the flow or radial direction, while there are also fields parallel to the flow.

Effects of Magnetized Water on Seed Germination

The effects of magnetized water on seed germination are demonstrated by results from several researchers:

1. Increasing the germination rate of tomato seeds compared to control seeds (Labiod, 2010).
2. Magnetic water treatment increases the sprouting of bean seeds by 2-3 days (Elloumi, 2016).

APPLICATIONS AND EFFECTS

Today, magnetic processing has attracted a great deal of interest in many areas such as the environment, health, industry, etc., in particular, we are interested in its application in agriculture.

Health Application

The earth is considered a huge natural magnet that can transmit magnetic energy to all living organisms. The development of life is linked to electromagnetic radiation. Furthermore, this inevitable phenomenon affects the lives cycle of animals and plants.

For thousands of years, magnets have been popular among Chinese, Indians, Pharaonic, Arabs, Chaldeans, Hebrews, Greeks and Romans in medicine. Usually, use permanent magnets with a power between 500 and 4000 Gauss in water treatments. Water treated with a magnetic field can increase vitality, eliminate pain, increase blood circulation, reduce plaque and normalize bowel problems. In fact, when water is magnetized by the magnet's north pole, it prevents the activity and spread of bacteria, fungi, and tumors.

It is also effective for various forms of infection (such as typhoid fever, measles, smallpox, urethritis, and blood poisoning) and growth.

In case of influenza, it can also be used as a preventive measure. Arctic water can also be swallowed to escape bad breath and relieve various diseases, such as inflammation of the oral mucosa, gingivitis, sore throat, tonsillitis, etc. However, Antarctic water is used to treat diseases that accompany muscle pain and weakness (Sylvan, 2002).

On the other hand, research from the Leningrad clinic shows that magnetized water is beneficial for the kidneys, liver, and gallbladder. It can also solve obesity and fat (Donnet, 1988).

Industrial Application

According to Smith's research (Smith, 2003), in the 1990s, the cost of inefficient heat transfer and scales removal in the United Kingdom was estimated at £1 billion per year. Removing a 25 mm thick layer of $CaCO_3$ calcium carbonate can reduce the cost of 95% heat transfer. After properly configuring and using of the magnetic equipment, they were very efficient to reduce the amount of calcite in the pipeline.

Smith, et al, (2003) confirmed that permanent magnets reduced the formation of scale in six hot water storage tanks by up to 70%.

Agricultural Application

Magnetic treatment is an economical technology (at least 20% water saving) and does not require a significant energy input and can improve the quantity and quality of vegetation and fruit. In addition, it solves the problem of limestone in the pipeline by reducing the accumulation of limestone in the pipeline, thus increasing crop yield.

Previous works indicated that physical treatment reduces the risk of soil salinization by reducing the size of water crystals, so improve absorbed the crystals by the root hairs (Lin, et al, 1990). Using magnetically treated water in irrigating chickpea plants increased the plant height than that of chickpea plants irrigated with untreated water (Nasher, 2008), also, Taimourya, et al, (2015) showed that the height of potato plants irrigated with magnetized water increased by about 13.9% under different test condition.

Similarly, using magnetically treated water improve vegetative growth (i.e. number of main branches, dry matter, and root mass) and crop yield in different crops such as beans, chickpeas, wheat, and lentils (Qados and Hozayn, 2010).

A large number of studies have confirmed that the magnetic field can improve the physical and chemical properties of water (Wang, et al, 2018). In other words, it will remove limestone accumulated in water pipes and heating systems, increase germination and stimulate the growth of various plants.

THE INFLUENCE OF MAGNETIC FIELDS ON CALCIUM CARBONATE PRECIPITATION

The problem of hard scale in the heating or cooling system can significantly reduce its efficiency. According to research by several researchers, magnetic treatment of water reduce precipitation of calcium carbonate in pipes and minimizing carbonate crystal growth (Banejad & Abdosalehi, 2009). In addition, Young et al., (2003) found that the copper heating coil after treatment would become 2.5 times thinner due to the effect of the magnetic field.

Effect on Surface Tension

It is well documented that, the number of magnetic treatments increases, the surface tension decreases. The maximum mass and diameter of a drop of water are strongly affected by the magnetic field (Sueda, et al, 2007). In addition, Otsuka, et al, (2006) found that after magnetic treatment, no change in the properties of pure water distilled from ultra-pure vacuum extracted water was observed. However, exposed distilled water to oxygen and subjected to the same magnetic treatment, properties such as surface tension will change. The degree of influence of magnetic treatment on water is determined quantitatively by the angle of contact.

In addition, (Ben Amor, et al, 2017) investigated the influence of the number of permanent magnetization treatments on surface tension by measuring the surface tension and visualization of the behaviour of the dye in the water sample. This experiment proved that increasing the number of magnetic treatments reduces the surface tension.

Effect on pH and Conductivity

Previous reports indicated that, the flow rate and temperature of the water have a great influence on the magnetic treatment of the water. Ben Amor, et al, (2021) using four magnets of different sizes and forces to treated water, so, the pH value increased slightly over time and then returned to its original value.

Surendran, et al, (2016) stated in their work that the decrease in conductivity of the treated water is caused by the presence of fine colloidal molecules (in Brownian motion) and electrolytic substances, which are determined by their depository capacity subjected to magnetic treatment. Joshi and Kamat, (1966) reported that the pH value of distilled water varies up to 0.4 units.

Parsons, et al, (1997) reported that after the water passed the magnetic field, the pH value also dropped by 0.5 pH units.

Effect on Plant Growth and Productivity

Magnetic water treatment has a direct effect on plant growth, seed germination, and production. Irrigation with treated water promotes the development of plant metabolism at the level of plant metabolism, at the level of absorption of photosynthesis (Abobatta, 2019). There is a stimulating effect of magnetic water in nutrient uptake and increases absorption, which improves plant growth (Yano, et al, 2004).

Moreover, water magnetization considered a very important factor in improves germination, seed growth, vigour of seedlings, and increases plant growth (Nasher, et al, 2008). Magnetized water treatment promotes the absorption of NPK, improves the thickness of the stems, increases the size of the roots, increases the number of leaves, and dry matter (Alikamanolu, et al, 2007).

Several researchers such as De-Souza, et al, (2006) found that tomato irrigated with magnetized water increased fruit yield by the plant, enhancing fruit quality (i.e. average fruit weight and fruit diameter), and overall fruit yield compared to controls. In addition, Lin and Yotvat, (1990) reported that magnetic water treatment reduces soil salinity due to reduced crystal size, which increases absorbed rates by the root system. Nasher, (2008) showed that chickpea plants irrigated with magnetically treated water are longer than those irrigated with raw water.

Additionally, Taimourya, et al, (2015) observed an increase in the height of potato plants irrigated with treated water. In this context, compared to non-magnetized water irrigation, there is a remarkable increase in aerial parts mass, the number of branches, dry matter, root mass, and an increase in yield of the potato crop. In addition, Qados and Hozayn, (2010) found that magnetic water treatment increase different plants height like in the chickpea, bean, lentil, and wheat compared to the non-magnetized water irrigation.

EXPERIMENTAL SITE AND DEVICE

This work aims to investigate the impact of magnetized irrigation saline water on the growth and fruit quality of citrus varieties that include clementine, and (Maltese and Thomson) orange, that irrigated with saline water

Experimental Site

The experiment was conducted on a farm in the Cap Bon region of Tunisia. The experiment site was located in Soliman (36.69491S 10.50111N), For conducting the experiment, a plot of land was prepared and divided into two parts, the experimental test for magnetized water treatment, and a control part (untreated water).

Magnetic Devices

During this, four magnetic treatment devices with a strength of 0.35 Tesla were used in the part that was irrigated with treated water during this experiment, while, there is no devices used in the control part, which, just irrigated with normal water.

The experimental device is a random block complete with three replicates. In this work, one treated part: with magnetized water (MG) and another control: raw water (NM). The use magnetic device is 0.35 T (Fig 1.). The varieties treated are Clementine 'Cassar', Thomson 'Navel' and Maltese semi-sanguine.

Figure 1. Experimental device

RESULTS OF MAGNETIZATION PRACTICE

Vegetative Parameters

The positive effect of irrigation by magnetized water on the vigor and leaves the color of the tree (A) is observed as shown in (Figure 2) in comparison to the control where the appearance of yellowing of foliage (B).

Figure 2. Appearance of a 'Thomson' Navel orange tree irrigated by magnetized water (A) and raw water (B)

(A) (B)

Pest Attack

This experiment indicated that irrigation with magnetized water reduces infection of the mite (*Eriophyes sheldoni*) on the fruits, compared with control fruits that irrigated with normal water (Figure 3).

Figure 3. Citrus fruit irrigated by magnetized water Citrus fruit irrigated by raw water

Similarly, it is noted that the aphid attack is much more visible on the new shoots of the tree irrigated by raw water comparing that irrigated by magnetized water.

Fertilizer Absorption

According to the visual diagnosis of the tree under experiment, the nutrient deficiency symptoms are more visible on the control one.

Yield Parameters

Fruit Set

Figure 4. shows that the best fruit set is obtained for the plant irrigated with magnetic water (940 fruits of clementine, 300 fruits of Maltese, and 230 fruits of Thomson) compared to that irrigated with the normal water. The results obtained indicate that, the fruit retain ratio is higher on the trees irrigated with magnetizing water (A), also, the circumference of these fruits is higher than that from control treatments (B).

Figure 4. (A) Average number of fruits retain. (B) Average fruit circumference

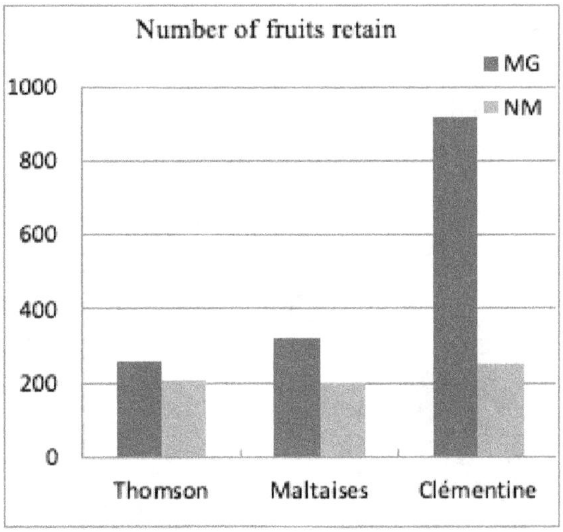

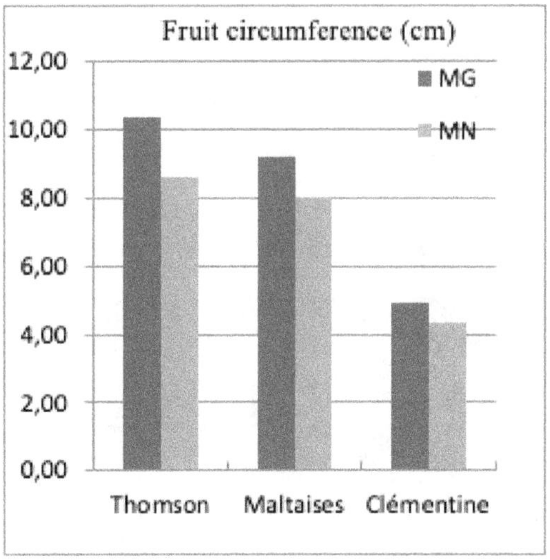

(MG: irrigation with magnetized water, MN: irrigation with raw water)

Fruit Weight

The Cassar clementine fruits were harvested in December 2020, the results showed that improving all parameters of fruit quality such as increasing fruit weight and fruit brightness, from trees that watering with magnetizing water compared to those treated with normal water (Fig 5.). The results show an average increase of 48% in yield compared to the plot irrigated by raw water.

Figure 5. A- fruits irrigated by magnetized water B- fruits irrigated by raw water

Fruit Size

The results obtained showed in (figure 6) that the fruits irrigated with magnetized water show an increase in average size reaching 7.5 cm and 5.3 cm for the samples taken by the control plants.

Figure 6. Size of fruits irrigated with raw water (1) and magnetized water (2)

1

2

SOLUTIONS AND RECOMMENDATIONS

Among the proposed solutions: use a high-intensity magnetic device and enhance the saltwater, this generates an improvement in fruits, sizes, yields.

FUTURE RESEARCH DIRECTIONS

This chapter illustrates a technique for preserving and using saltwater in irrigation. The magnetization technique can be one of the new solutions to develop citrus cultivation under arid and semiarid conditions, furthermore, this study can be carried out on other fruits.

CONCLUSION

The magnetic treatment of water has been of interest to several researchers around the world for years. There are numerous reports about this technique, but unfortunately, the results of several experiments are sometimes non-reproducible and not significant.

Indeed, the influence of the magnetic field remains enigmatic, despite numerous attempts to explain the observed phenomena. Still in this perspective, during this work, we are interested in the influence of magnetic treatment on the physicochemical properties of water, soil, and some citrus crops. In this context, this chapter explains the potential effects of magnetic water devices on different parameters of brackish water and the growth and productivity of citrus trees. Where this study clarified the positive impacts of magnetic treatment on water properties and citrus cultivars production.

The results of this experiment show that magnetic treatment increases the pH value of water, particularly when using high-tension devices, while, pH returns to the original values after a different period varies according to the treatment and soil conditions, which confirms the strength of the water memory effect.

Regarding trees production, it is remarkable that irrigation by treated water increases total yield, improves fruit quality compared to irrigation by non-magnetized water, in addition, there is a beneficial effect of magnetic treatments in the reduction of pests and diseases.

ACKNOWLEDGMENT

Our thanks to the Tunisian Ministry of Agriculture and the Ministry of Higher Education, specifically to the PEJC 18PEJC 05-01 and 20PEJC 04-03 program.

REFERENCES

Abobatta, W. F. (2018). Challenges for Citrus Production in Egypt. *Acta Scientific Agriculture*, *2*(8), 40–41.

Abobatta, W. F. (2019). Overview of Role of Magnetizing Treated Water in Agricultural Sector Development. *Adv Agri Tech Plant Sciences*, *2*(1), 180023.

Abobatta, W. F. (2020). Citriculture and Climate Change. *Modern Concepts & Developments in Agronomy*, *6*(3), 649–650. doi:10.31031/MCDA.2020.06.000639

Al-Jubouri, A.A.A. & Hamza, J.H. (2006). *Magnetically water treatment technology and its impact in the agricultural field.* Baghdad University - Faculty of Agriculture – Dept. of Field Crop Sci. Crop Pro.

Alikamanoğlu, S., Yaycılı, O., Atak, Ç., & Rzakoulieva, A. (2007). Effect of Magnetic Field and Gamma Radiation on Paulowinia Tomentosa Tissue Culture. *Biotechnology, Biotechnological Equipment, 21*(1), 49–53. doi:10.1080/13102818.2007.10817412

Babu, C. (2010). *Use of magnetic water and polymer in agriculture*. Ph.D., Agronomy D.W.S.R. Centre.

Banejad, H., & Abdosalehi, E. (2009). The effect of magnetic field on water hardness reducing. In *Thirteenth international water technology conference, IWTC* (pp. 117-128). Academic Press.

Ben Amor, H., Elaoud, A., Ben Hassen, H., Ben Amor, T., Ben Salah, N., Stuerga, D., & Elmoueddeb, K. (2021). Experimental study and data analysis of the effects of iIons in water on evaporation under static magnetic conditions. *Arabian Journal for Science and Engineering*. Advance online publication. doi:10.100713369-021-05519-5

Ben Amor, H., Elaoud, A., Ben Salah, N., & Elmoueddeb, K. (2017). Effect of Magnetic Treatment on Surface Tension and Water Evaporation. *International Journal of Advance Industrial Engineering.*, *5*(3), 119–124. Doi.Org/10.14741/Ijae/5.3.4

Cai, R., Yang, H., He, J., & Zhu, W. (2009). The effects of magnetic fields on water molecular hydrogen bonds. *Journal of Molecular Structure*, *938*(1-3), 15–19. doi:10.1016/j.molstruc.2009.08.037

Cho, Y. I., Lee, S., Kim, W., & Suh, S. (2003). *Physical water treatment for the mitigation of mineral fouling in cooling-tower water applications*. Engineering Conferences International, ECI Digital Archives. https://dc.engconfintl.org/heatexchanger/4

Colauto Stenzel, N. M., Vieira, J. N. C. S., Marur, C. J., Santos, S. M. B., & Gomes, J. C. (2006). Maturation curves and degree-days accumulation for fruits of 'folha murcha' orange trees. *Scientia Agrícola*, *63*(3), 219–225. doi:10.1590/S0103-90162006000300002

Couture, I. (2004). *Analyse d'eau pour fin d'irrigation*. MAPAQ Montérégie-Est. https://www.agrireseau.net/petitsfruits/documents/61548/analyse-d_eau-pour-fin-d_irrigation

Couture, I. (2006). *Principaux critères pour évaluer la qualité de l'eau en micro-irrigation Colloque sur l'irrigation l'eau, source de qualité et de rendement*. Boucherville.

Craig, A. M. (2009). *Investigation of the quality of water treated by magnetic fields; Bachelor of Engineering (Environmental)*. University of Southern, Queensland Faculty of Engineering and Surveying.

De Souza, A., Garcí, D., Sueiro, L., Gilart, F., Porras, E., & Licea, L. (2006). Pre-sowing magnetic treatments of tomato seeds increase the growth and yield of plants. *Bioelectromagnetics: Journal of the Bioelectromagnetics Society, The Society for Physical Regulation in Biology and Medicine. The European Bioelectromagnetics Association*, *27*(4), 247–257. doi:10.1002/bem.20206

Donnet, L. (1988). *Les aimants pour votre santé*. L'eau merveilleuse.

El-Sonbaty, A. E. (2021). Effect of organic matter combined with mineral N fertilizers different rates with or without magnetic water on soil fertility and maize productivity. *Plant Cell Biotechnology and Molecular Biology*, *22*(67 &68), 96–112.

Elloumi, M. (2016). *La gouvernance des eaux souterraines en Tunisie*. IWMI Project report. http://gwmena.iwmi.org/wpcontent/uploads/sites/3/2017/04/Rep.7-Groundwater-governance-inTunisia_final_cover.pdf

FAO. (2008). Phocaides, Manuel des techniques d'irrigation sous pression. FAO.

Food and Agriculture Organization of the United Nations (FAO). (2016). *Citrus fruit fresh and processed statistical bulletin*. FAO.

Gutfleisch, O., Willard, M. A., Brück, E., Chen, C. H., Sankar, S. G., & Liu, J. P. (2011). Magnetic materials and devices for the 21st century: Stronger, lighter, and more energy efficient. *Advanced Materials*, *23*(7), 821–842. doi:10.1002/adma.201002180 PMID:21294168

Holysz, L., Szczes, A., & Chibowski, E. (2007). Effects of a static magnetic field on water and electrolyte solutions. *Journal of Colloid and Interface Science*, *316*(2), 996–1002. doi:10.1016/j.jcis.2007.08.026 PMID:17897662

Joshi, K., & Kamat, P. (1966). Effect of magnetic field on the physical properties of water. *Journal of Evolutionary Biology Research*, *2*(1), 7–14.

Labiod, K. (2010). *Etude du caractère incrustant des eaux et son inhibition par des essais chimiques et s'électrodéposition, Master*. University of Constantine.

Lin, I., & Yotvat, T. (1990). Exposure of irrigation and drinking water to a magnetic field with controlled power and direction. *Journal of Magnetism and Magnetic Materials*, *83*(1-3), 525–526. doi:10.1016/0304-8853(90)90611-S

Martínez-Ferri, E., Muriel-Fernández, J. L., & Díaz, J. R. (2013). Soil water balance modelling using SWAP: An application for irrigation water management and climate change adaptation in citrus. *Outlook on Agriculture*, *42*(2), 93–102. doi:10.5367/oa.2013.0125

Mostafazadeh, F. B., Khoshravesh, M., Mousavi, S. F., & Kiani, A. R. (2011). Effects of magnetized water on soil sulfate ions in trickle irrigation. In *Proceedings of the second international conference on environmental engineering and applications*. IACSIT Press.

Nasher, S. H. (2008). The effect of magnetic water on growth of chick-pea seeds. *Engineering and Technology Journal*, *26*(9), 4.

Nimmi, V., & Medhu, G. (2009). Effect of pre-sowing treatment with permanent magnetic field on germination and growth of Chili. *International Agrophysics*, *23*, 195–198.

Otsuka, I., & Ozeki, S. (2006). Does Magnetic Treatment of Water Change Its Properties? *The Journal of Physical Chemistry B*, *110*(4), 1509–1512. doi:10.1021/jp056198x PMID:16471705

Parsons, S., Wang, B., Judd, J., & Stephenson, T. (1997). Magnetic treatment of calcium carbonate scale-effect of pH control. *Water Research*, *31*(2), 339–342. doi:10.1016/S0043-1354(96)00238-2

Qados, A. M. S. A., & Hozayn, M. (2010). Magnetic water technology, a novel tool to increase growth, yield and chemical constituents of lentil (Lens esculenta) under greenhouse condition. *American-Eurasian Journal of Agricultural & Environmental Sciences*, *7*(4), 457–462.

Smith, C., Coetzee, P. P., & Meyer, J. P. (2003). The effectiveness of a magnetic physical water treatment device on scaling in domestic hot-water storage tanks. *Water S.A.*, *29*(3), 231–236.

Sueda, M., Katsuki, A., Nonomura, M., Kobayashi, R., & Tanimoto, Y. (2007). Effects of High Magnetic Field on Water Surface Phenomena. *The Journal of Physical Chemistry C*, *111*(39), 14389–14393. doi:10.1021/jp072713a

Surendran, U., Sandeep, O., & Joseph, E. J. (2016). The impacts of magnetic treatment of irrigation water on plant, water and soil characteristics. *Agricultural Water Management*, *178*, 21–29. doi:10.1016/j.agwat.2016.08.016

Tai, C., Chang, M., Shieh, R., & Chen, T. (2008). Magnetic effects on crystal growth rate of calcite in a constant-composition environment. *Journal of Crystal Growth*, *310*(15), 3690–3697. doi:10.1016/j.jcrysgro.2008.05.024

Taimoury, H., Oussible, M., Bourarach, E. H., Harif, A., Hassanain, N., Masmoudi, L., & Baamal, L. (2015). Évaluation de la productivité de la pomme de terre (SolanumTuberosum L) sous l'effet de l'irrigation avec une eau traitée magnétiquement dans la région de Chaouia (Maroc). *Revue Marocaine des Sciences Agronomiques et Vétérinaires*, *3*(3), 30–36.

Wang, Y., Wei, H., & Li, Z. (2018). Effect of magnetic field on the physical properties of water. *Results in Physics*, *8*, 262–267. doi:10.1016/j.rinp.2017.12.022

Yano, A., Yoshiaki, O., Tomoyuki, H., & Kazuhiro, F. (2004). Effects of a 60 Hz magnetic field on photosynthetic CO_2 uptake and early growth of radish seedlings. *Bioelectromagnetics*, *25*(8), 572–581. doi:10.1002/bem.20036 PMID:15515039

Yun-Zhu, G., Da-Chuan, Y., Hui-Ling, C., Jian-Yu, S., Chen-Yan, Z., Yong-Ming, L., Huan-Huan, H., Yue, L., Yan, W., Wei-Hong, G., Ai-Rong, Q., & Peng, S. (2012). Evaporation Rate of Water as a Function of a Magnetic Field and Field Gradient. *International Journal of Molecular Sciences*, *13*(12), 16916–16928. doi:10.3390/ijms131216916 PMID:23443127

Zörb, C., Geilfus, C. M., & Dietz, K. J. (2019). Salinity and crop yield. *Plant Biology*, *21*(S1), 31–38. doi:10.1111/plb.12884 PMID:30059606

Chapter 8
Carbon and Water Footprint

Ahmed Awny Farag
Agriculture Research Center, Egypt

Mohamed Ahmed Abdrabbo
Central Laboratory for Agricultural Climate, Egypt

ABSTRACT

Climate change has become one of the most relevant global challenges that impact the overall world. One way to combat climate change is to calculate and reduce the climate impacts of single products. An estimate of the total amount of greenhouse gases (GHG) emitted from a life cycle perspective of a good or service gives an overview of the contribution to climate change from this product, usually referred to as product carbon footprint (PCF). A full life cycle product carbon footprint is necessary to identify emission hotspots in the product value chains and thereby address the climate change on the product level in the most efficient way. Products' carbon footprint also make it possible to compare the climate change impact of competing horticultural products (e.g., the fruits coming from different countries but sold in the same store). However, only a limited amount of products' carbon footprints including the raw material acquisition, production, distribution, and the consumer stage have been published about horticultural products.

INTRODUCTION

Climate change has become one of the most relevant global challenges that impact the overall world. One way to combat climate change is to calculate and reduce the climate impacts of single products. An estimate of the total amount of greenhouse gases (GHG) emitted from a life cycle perspective of a good or service gives an overview of the contribution to climate change from this product. Usually referred to as product carbon footprint (PCF). A full life cycle product carbon footprint is necessary to identify emission hotspots in the product value chains and thereby address the climate change on the product level in the most efficient way. products carbon footprint also make it possible to compare the climate change impact of competing horticultural products, e.g. The fruits coming from different countries but sold in the same store. However, only a limited amount of products' carbon footprint including the

DOI: 10.4018/978-1-6684-2423-0.ch008

raw material acquisition, production, distribution, and the consumer stage have been published about horticultural products.

Climate change is considered to be one of the most complex, multifaceted and most serious threats to mankind. Guided by the principles of common but differentiated responsibilities and respective capabilities of the United Nations Framework Convention on Climate Change, the actual response to the climate challenge depends on the ability of individual countries to adapt or build adaptability to the changing climate, while contributing to global greenhouse gas mitigation efforts

Through the Nationally Determined Contributions of Contracting Parties. In accordance with the provisions of the Paris Agreement, parties are invited to envision a long-term climate vision based on a half-century strategy for low greenhouse gas emissions.

At the beginning of 2007, several British retail organizations created a demand for information about the carbon footprints of agricultural products (the sum of greenhouse gas emissions that can be emitted from the product) by announcing that they would introduce a carbon footprint labeling scheme for their products. In 2007 the British Standards Institution (BSI) began developing a protocol for calculating the carbon footprints of various products and services. This protocol was published as the Publically Available Specification (PAS) 2050, in October 2008 (BSI 2008).

The expression "carbon impression" has acquired expanded ubiquity lately and is presently generally utilized in government, business, and the media. Albeit widely utilized in the general society area, further examination shows that this term has not been enough characterized in logical writing. An enormous scope of definitions exists for this term. In spite of the absence of logical underwriting, the term carbon impression has immediately turned into a generally acknowledged "trendy expression" to additionally animate shoppers' developing worry for issues identified with environment change by depicting anything from the tightest to the vastest understanding of nursery gas estimation and reduction. As a general rule, differentiation in the writing is essentially engaged on two major questions: units of estimation and extent of estimation.

Different Definitions of Carbon Footprint

The carbon footprint expression is related to and grew out of the older idea of ecological footprint, a concept invented in the early 1990s by Canadian ecologist William Rees and Swiss-born regional planner Mathis Wackernagel at the University of British Columbia. An ecological footprint is the total area of land required to sustain an activity or population. It includes environmental impacts, such as water use and the amount of land used for food production. In contrast, a carbon footprint is expressed as a measure of weight, as in tons of CO_2 or CO_2 equivalent per year (William E. Rees, 1992)

1- BP (2007): The carbon footprint is the amount of carbon dioxide emitted due to your daily activities.

2- Carbon Trust (2007): A. It is a methodology to estimate the total emission of greenhouse gases (GHG) in carbon equivalents from a product across its life cycle from the production of raw material used in its manufacture, to disposal of the finished product (excluding in-use emissions.

Carbon Trust (2007): B. a technique for identifying and measuring the individual greenhouse gas emissions from each activity within a supply chain process step and the framework for attributing these to each output product will refer to this as the product's "carbon footprint".

3- Parliamentary Office of Science and Technology (POST 2006): "A 'carbon footprint' is the total amount of CO2 and other greenhouse gases, emitted over the full life cycle of a process or product.

It is expressed as grams of CO2 equivalent per kilowatt-hour of generation (g CO_2 eq/kWh), which accounts for the different global warming effects of other greenhouse gases."

4- Global Footprint Network (GFN 2007): The demand on biocapacity required to sequester (through photosynthesis) the carbon dioxide (CO2) emissions from fossil fuel combustion.

5- Grub & Ellis (2007): A carbon footprint is a measure of the amount of carbon dioxide emitted through the combustion of fossil fuels. In the case of a business organization, it is the amount of CO2 emitted either directly or indirectly as a result of its everyday operations. It also might reflect the fossil energy represented in a product or commodity reaching the market.

6- ETAP (2007): The 'Carbon Footprint' is a measure of the impact human activities have on the environment in terms of the number of greenhouse gases produced, measured in tonnes of carbon dioxide.

7- Energetics (2007): The full extent of direct and indirect CO_2 emissions caused by your business activities.

8- Wiedmann and Minx, (2008): The carbon footprint is a measure of the exclusive total amount of carbon dioxide emissions that are directly and indirectly caused by an activity or is accumulated over the life stages of a product.

Through the above definitions, we suggest the following definition of the term "carbon footprint" for orchard farms in the field of agriculture. "The carbon footprint is a measure of the exclusive total amount of carbon dioxide and other greenhouse gases emissions that are directly and indirectly caused by horticultural operations (irrigation pumps, fertilizer used, pest control, plowing, waste burning ... etc.) during the life stages of fruit production. It is expressed as grams of CO_2 equivalent per kilograms of fruit or yield (CO_2eq/kg yield)". Carbon footprints for different crops are very difficult to compare because the calculation is dependent on local yields and environmental conditions.

The CO_2 equivalent: CO_2e (carbon dioxide equivalent) expresses the impact of each different greenhouse gas in terms of the amount of CO_2 that would create the same amount of warming.

Therefore, as Hammond (2007) suggests, a more accurate term would be 'carbon weight' as opposed to 'footprint', which intuitively makes one think of the land area. However, the term 'footprint' is likely to remain the popular and preferred word despite its unusual measurement unit, as it is much more visual and easier to understand and hence, more meaningful than the idea of carbon as weight. Another benefit of using the term footprint is the underlying connotation of a footprint representing a legacy, i.e. the long-term impact carbon has on the earth and atmosphere.

Carbon footprints are primarily measured in metric tons as they usually represent significant quantities of gas. However, they are generally converted into kilograms during the calculation process as emission factors are primarily provided in this smaller unit (DCCEE, 2011)

Table 1. The six Kyoto greenhouse gases and their global warming potential

Kyoto greenhouse gas	Chemical formula	Global warming potential	Major sources (examples)
Carbon dioxide	CO_2	1	Fossil fuel combustion, land use change, cement manufacturing
Methane	CH_4	21	Natural gas, coal mines, anaerobic decomposition of organic waste
Nitrous oxide	N_2O	310	Fertilizers, agriculture, combustion
Perfluorocarbons	(7 types)	7000	Electronics, aluminum production
Hydrofluorocarbons	(13 types)	11700	Refrigerants
Sulfur hexafluoride	SF6	23900	High voltage switchgear

Carbon Footprint Reduction

The first step in reducing emissions from orchards is to produce fruit as efficiently as possible—that is, to change how we farm. A set of proven GHG-efficient farming technologies and practices—some of which are already being deployed—could achieve about 20 percent of the sector's required emissions reduction by 2050.

The greenhouse gas (GHG) emissions from agricultural activities can be reduced through more efficient management of carbon and nitrogen flows within agricultural systems. Here are some beneficial management practices recognized for lowering greenhouse gas emissions.

Agricultural ecosystems have large reserves of carbon, especially in soil organic matter. The following agricultural practices enhance carbon sequestration by either increasing carbon storage or decreasing the loss of stored carbon:

- Avoid excessive use of fertilizer especially nitrogen
- Plow management
- Waste management especially pruning parts
- Replacement of a part of N fertilizers with biological N fixation

Renewable energy sources can displace fossil fuel use, reducing GHG emissions on and off-farm. They can also help decrease reliance on energy sources with volatile prices, and create new economic diversification opportunities for agricultural producers. Renewable energy technologies suitable for on-farm use include:

- Anaerobic digestion
- Electrification
- Geothermal
- Gasification
- Wind
- Solar electric (photovoltaic)
- Biofuel production from crops or crop residues

Ecological Footprint

The Ecological Footprint quantifies the amount of biologically productive land and water area necessary to produce all of the resources used by an individual, a population, or an activity, taking into account the absorption of wastes generated. This may be compared to bio-capacity, which is the amount of productive land available to create these resources and absorb the wastes (Ewing et al., 2008). Nonetheless, the display of the Ecological Footprint might be much better. The Ecological Footprint is a static metric that overlooks technological advancements, subterranean resources, and biodiversity concerns. It is a stock metric that does not examine flows and does not include equity measures.

Greenhouse Gas Emissions from Fruit Production

The most plenty GHGs are carbon dioxide (CO2), methane (CH4), and nitrous oxide (N2O) which all absorbs some of the thermal radiation that leaves the earth's surface, which then is re-emitted in all direction including at the earth's surface (Houghton, 2015).

The GHG emissions from using fossil fuels in the agriculture sector worldwide contributed with 400-600 Mt CO2-eq/yr in 2010; however, those emissions are not categorized under the AFOLU sector, which means that they are not included in the Intergovernmental Panel Climate Change (IPCC) statistics (IPCC, 2013).

The GHG emissions from using fossil fuels in agriculture worldwide contributed with 400-600 Mt CO2-eq/yr. in 2010; however, those emissions are not categorized under the AFOLU sector, which means that they are not included in the Intergovernmental Panel Climate Change (IPCC) statistics (IPCC, 2013). Carbon dioxide many of the processes and inputs in fruit production contribute to emissions of CO_2. Inputs, such as fertilizers, lime, pesticides, water, fossil fuels, and electricity, emit CO_2 when produced, stored, and/or transported (IPCC, 2014; FAO, 2015). These inputs are utilized in different cultivation phases, which is causes additional emissions of CO_2 gas. Irrigation system pumps water and the application of fertilizers, pesticides, and lime are done with the use of machinery (Mithraratne et al. 2010; Keyes et al. 2015). All these processes need some form of equipment that uses diesel fuel or electricity (Mithraratne et al. 2010; Keyes et al. 2015). Fruit production like other agricultural practices can be considered both as a sink for CO2, through soil carbon sequestration, and as a source of it by soil management such as tillage (IPCC, 2014). Methane Anthropogenic sources of CH4 that can be assigned to fruit production are energy production, from coal and gas, burning of biomass from e.g. pruning, and from manure storage (Swedish Environmental Protection Agency, 2006; Cederberg, 2010; Aguilera et al. 2015; Houghton, 2015). Atmospheric concentrations of the CH4 have more than doubled since the 1800s and the concentration is higher than it has ever been during the last 800 000 years, according to ice core records (Houghton, 2015; IPCC, 2013). Out of the yearly emissions of CH4, about 50% comes from agriculture, and if food waste handling is added, the percentage goes up to 60% (Cederberg, 2010). Nitrous oxide The anthropogenic emissions of N2O are dominated by the production and use of ammonia as fertilizer, storage of manure, cultivation of legumes, combustion of fossil fuels, and burning biomass (Cederberg, 2010; IPCC, 2014). All of these are processes used within fruit production, depending on the production method. Conventional fruit production uses large amounts of synthetic fertilizers, and the emissions of N2O from it was in 2015, 660 Mt/yr. worldwide (FAOSTAT). The emissions from manure, which is an important part of organic fruit production, were much smaller, 147 Mt/yr., (FAOSTAT). The formation of N2O in the soil is an unavoidable byproduct of denitrification, which means that a high input will probably lead to increased emissions (Cederberg, 2010).

Figure 1. Higher carbon footprint of fruit and vegetables kg CO_2e per kg of product

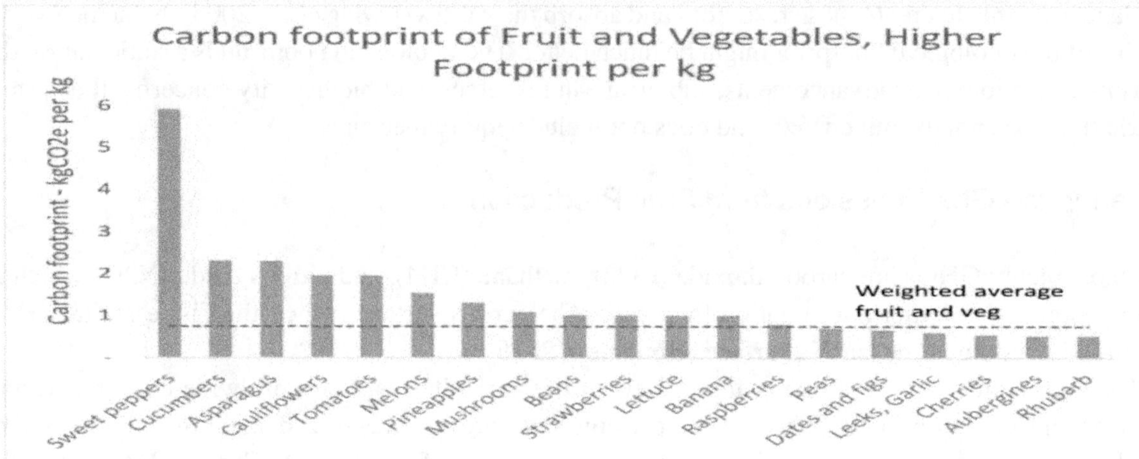

Figure 2. Average carbon footprint of fruit and vegetables kg CO_2e per kg of product

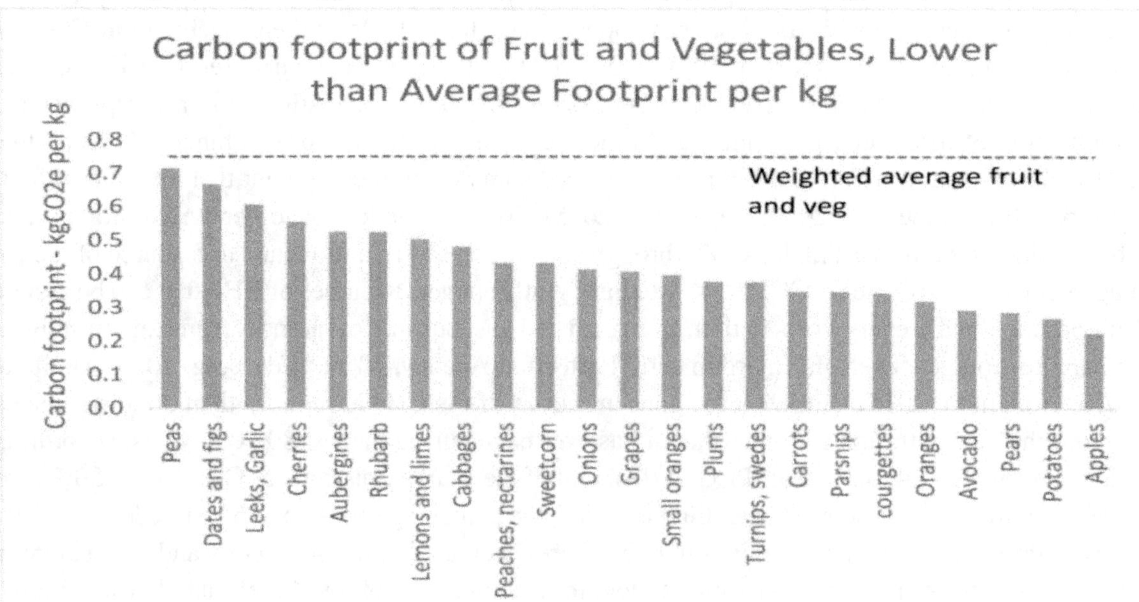

WATER FOOTPRINT

The water footprint started to gain broad interest from about 2008, the year in which the Water Footprint Network (WFN) was established – a network of academic institutions, governments, non-governmental organizations, companies, investors, and UN institutions. The world's freshwater resources throughout the world have become scarcer during the past decades due to an increase in population and economic activity and a subsequent increase in water appropriation. The 'water footprint' concept was developed in order to have an indicator of 'water use' in relation to its consumption (Hoekstra, 2003). The water footprint can help to identify products that can be risky to produce at a certain location due to water scarcity and to achieve more sustainable and equitable use of freshwater. The water footprint of any product or service is the volume of freshwater employed to produce the product, measured over the various steps of the production chain. 'Water use' is measured in terms of water volumes consumed (evaporated) and/or polluted. The water footprint is a geographically explicit indicator, not only showing volumes of water use and pollution but also the locations and timing of water use. The total water footprint of a product swear into three components:

1- The blue water footprint is the volume of freshwater that evaporated from the global blue water resources surface water and groundwater to produce the product and services consumed by the individual or community.
2- The green water footprint is the volume of water evaporated from the global green water resources rainwater stored in the soil as soil moisture.
3- The gray water footprint is the volume of polluted water that associates with the production of all products and services for the individual or community.

The total water footprint per year water footprint of an Indian individual is 1089 m^3 compared to the global average of 1385 m^3, whereas for an individual of the US, it is 2842 m^3. Consumption of agricultural products determines the global water footprint related to consumption (Mekonnen and Hoekstra, 2011). Among agricultural commodities, the global water footprint of vegetables and fruits is 322 m^3/ t and 967 m^3 /t, respectively. In India, the total water footprint for vegetables and fruits is 29 billion cubic meters and 13 billion cubic meters, respectively (Jayaram, 2016). Water footprints of some vegetables and fruits as well as their derived products are quite high, which is a matter of great concern.

Figure 3. The differences between the green, blue and grey-water in the water foot print definity

Figure 4. Of a water footprint [Hoekstra, 2008]

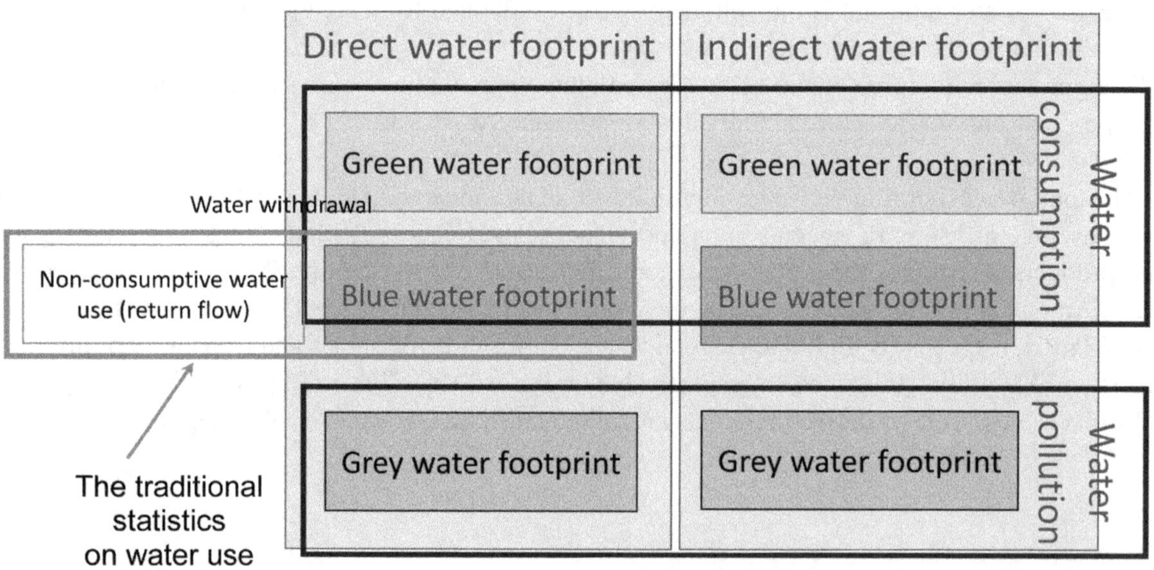

CF and WF Similarities and Differ

The water footprint and Carbon Footprint concepts have similarities; however, their assets and intended purposes differ. The Carbon Footprint was edited to quantify the contribution of various activities to climate change. The history of the water footprint is in the exploration of water use along supply chains and in the search for a tool to understand the global dimension of water as a natural resource. Although each footprint has different assets and characteristics and addresses different research and policy questions, there is a direction among practitioners in the fields of environmental policy and corporate social responsibility to treat the water footprint in a similar way as the carbon footprint. For example, terms such as 'carbon neutral' and 'carbon offsetting' are immediately adapted to 'water neutral' and 'water offsetting' without any particular attention to the appropriateness and applicability of these thinner to water. Similarly, initiatives are taken to expand water labels for products in similarity to carbon labels and to incorporate the water footprint into Life Cycle Assessment (LCA) for products in the same way as was done with the carbon footprint. In particular, people have a tendency to interpret the numbers of the water footprint without considering their spatial and temporal characteristics as is commonly done in carbon footprint analysis. Each footprint needs to be seen within its appropriate context and interpreted with attention as it is built around different research questions and tells a different story.

Table 2. Comparison of carbon and water footprints

	CF	WF
Measurement	The mankind emission of greenhouse gases (GHG).	The human personalization of freshwater resources in nomenclature of volumes of water consumed and polluted.
Unit	Carbon dioxide equivalents (CO_2eq) per unit of time or per unit of product.	Volume of Water per unit of time or per unit of product.
Footprint Ingredients	GHG types: CO_2, CH_4, N_2O, HFC, PFC, and SF6. Emissions per type of gas are weighted by their global warming potential. see table (1)	Blue footprint, green footprint and grey footprint.
Entities that footprint calculate	Processes, products, companies, industry sectors, individual consumers, groups of consumers.	Processes, products, companies, industry sectors, individual consumers, groups of consumers.
Calculation methods	*Small entities The method of Life Cycle Assessment (LCA) *National and global studies The method of Environmentally Extended Input-Output Analysis (EE-IOA) * Hybrid LCA and EE-IOA for products, nations, organizations	*Processes, products and businesses, but also for sector, national and global studies 1- Water Footprint Assessment (WFA) products, the accounting along supply chains in WFA is similar to the 2- accounting in the Life Cycle Inventory stage of LCA studies sector, national and global studies 1- accounting in WFA, which is based on drawing national virtual water trade balances 2- method of EE-IOA is used as an alternative
Scope	1. Direct emissions 2. Indirect emissions from electricity used 3. Other indirect emissions	Always includes direct and indirect WF.
Sustainability of the footprint	More information is needed to determine the CF's long-term viability. A maximum permissible GHG concentration for the whole world must be estimated, which must then be converted into a CF cap. CF benchmarks can be utilized for certain processes and goods	More information is needed to determine the WF's long-term viability. Freshwater availability and waste assimilation capacity per catchment area must be determined, resulting in a WF cap for the catchment. WF benchmarks can be used for certain processes and goods.

Carbon Dioxide Equivalent Emissions for Each Life Cycle Phase

Footprints due to the cultivation of almond, hazelnut and pistachio are respectively 2.30, 1.29 and 2.53 kg CO_{2eq}/kg of fruit, from which we can respectively deduct 0.23, 0.40 and 0.53 kg CO_{2eq}/kg of raw products due to the recovery of thermal energy from the shells and 0.14, 0.26 and 0.37 kg CO_{2eq}/kg of fruit for temporary carbon dioxide sequestration by the plants during their life. Thus, net emissions per kg of raw product considered in this study are:

- Almond: 1.92 kg CO2eq/kg;
- Hazelnut: 0.52 kg CO2eq/kg;
- Pistachio: 1.74 kg CO2eq/kg.

WATER FOOTPRINT METHODOLOGIES AND ASSOCIATED METHODS ARE DESCRIBED

There are three main ways to calculating a water footprint that are currently being used (RPA & Cranfield University 2011):

1. The volumetric approach, which is based on an assessment of the volume of water associated with a specific production activity;
2. The stress weighted approach, which is based on an assessment of the amount of freshwater consumed in an activity combined with an assessment of the implications of that consumption in terms of water stress; and 3. The impact assessment approaches, which are based on water consumption using an inventory analysis similar to the volumetric approach.

Sustainability of Agricultural and Water Management

Both the scientific and political communities are becoming increasingly interested in the development and application of footprint approaches for environmental assessment. The primary concept behind all "footprints" is to assess human impact on the environment in terms of production and/or consumption, on a micro, or macro scale. Several other "footprints," both environmental (e.g. carbon footprint, CF, and water footprint, WF), economic (e.g. economic footprint), and social (e.g. social and poverty footprint), were defined as means of assessing and communicating sustainability elements, inspired by the ecological footprint (EF) concept developed in the early 1990s (Wackernagel and Rees 1996). Existing footprints, in most cases, incorporate life cycle thinking (Čuček et al 2012), emphasizing problematic environmental aspects that are critical for assessing the sustainability of production and consumption patterns, such as resource consumption and CO_2 emissions leading to global warming.

A lot of water is consumed and polluted as a result of human activities. Current water footprint methodologies are based on different interpretations of what "footprint" means: on the one hand, there are methods that simply make an inventory of a resource's consumption (in the case of water, for example, entailing different typologies of freshwater/groundwater, etc.); on the other hand, other methods attempt to compare consumption with availability, incorporating somehow the system's carrying capacity in terms of the capability of providing the resource. The term "water footprint" (WF) has been described as the "total of domestic water usage and net virtual water import" as a measure of a country's actual appropriation of global water resources. (Hoekstra and Hung, 2002). Indeed, it is commonly acknowledged that humanity is currently devouring more resources than our earth can create and replenish. Interventions are required not just to educate and increase awareness about the environmental effects of our consumption habits but also to produce actual actions that will lead to more sustainable consumption patterns and styles. Relevant research needs for better water sustainability assessment are reported in the document. Due to seasonal precipitation changes, the area hydrologic status may change throughout the year. If the rainy seasons cannot fully compensate for the dry seasons due to a lack of storage capacity in the local watershed or greater evaporation of stored water, this seasonal variance may produce additional water stress.

By distinguishing between blue, (possibly) green, and grey water, as well as subtypes as needed, the product's water footprint should reflect the source of the water used, as well as water quantity and quality impacts. The existing water footprint approaches are not well adapted to be utilized as a thorough and robust recommended method by the European Commission, but rather as a starting point for accumulating preliminary experience as a contribution to future method development (EC-JRC, 2011).

Freshwater flows are non-exhaustible in principle, but because they offer a life-sustaining element to the biosphere, unsustainable removal has significant negative consequences on ecosystems. Depletion of freshwater resources occurs when replenishment capacity is surpassed by substantial withdrawal, or freshwater flows are lowered by a reduced regeneration rate, both of which have implications for future resource availability (Bauer and Zapp 2005). However, other important qualitative features such as heat emissions and microbiological contaminations are yet unknown, with the latter being a major cause of human sickness in Asia and Africa. Similarly, the additional reduction in freshwater availability as a result of deteriorated reservoir quality should be addressed.

ORCHARD PRODUCTION UNDER CLIMATE CHANGE

An orchard's life begins with the preparation of the soil through tillage and fertilization, thus establishing favorable concision for plant growth, and installation of infrastructure, such as irrigation systems and frost protection (Goossens et al. 2017). When the trees grow and become orchards, they are two to three years old or more depending on the nature of fruit trees (Iriarte et al. 2014; Goossens et al. 2017). The climate impact from the trees starts already in the nursery phase (Photo1), which is not always included in environmental assessment studies. Many fruit nursery already protected by shading nets at the commercial level. That protection allows plants to grow in optimized conditions. When those plants are transplanted to the open field condition the growth rate decline, if not protect the plant via natural protection (cultivate tropical fruits between date palm trees or artificial protection via transplant the fruit trees with greenhouse covered with screen net or plastic), which take a longer period to start flowering and production (Medany et al., 2009). Figures (4 and 5) illustrates the maximum and minimum air temperature in the open field as well as under the white screen net cultivated with mango trees. The extreme weather events become more frequent and more severe cause a decline in the production of the orchards especially tropical fruit trees such as mango and citrus then protection of such orchard become more essential (Saleh et al., 2017). The flowering and fruit set phonological phases are the most critical phases of the production of the fruit orchard, hot sand storms, fluctuation of air temperature, and frost can decline the seasonal production significantly (El-Gayar et al., 2019). During the last few years, growers of tropical fruit start to cope with the extreme weather events which affected the production of most of the old orchards, covering tropical fruit trees especially mango becomes a normal practice with using cheap greenhouse construction (woody greenhouse) and cover this structure with screen net (Photo 2). The economic study investigated under Egyptian conditions illustrated the production cover the annual cost of wooden greenhouse covered with white screen net in the second year of cultivation of Keit mango trees. The water use efficiency of orchard fruits is higher than open field conditions due to lower solar radiation, lower air temperature and high relative humidity under net houses then lower evapotranspiration, and lower water consumption.

Figure 5. Maximum and minimum air temperature under white screen net and open field condition for mango trees (medany et al., 2009)

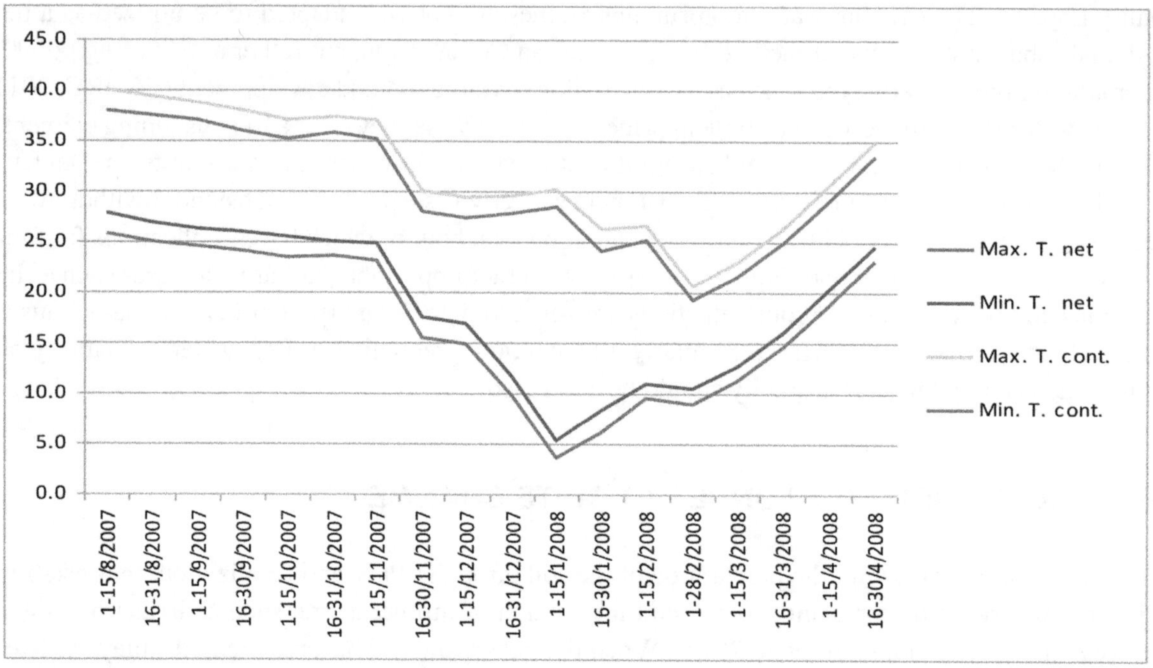

Figure 6. Average relative humidity under white screen net and open field condition for mango trees (medany et al., 2009)

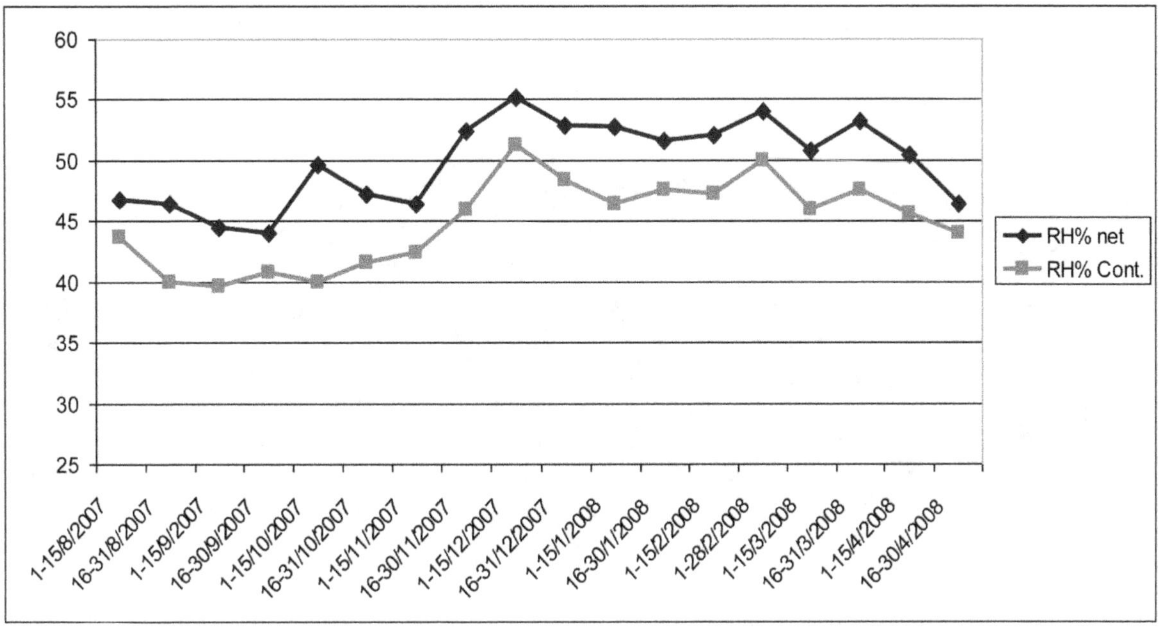

Figure 7. Mango nursery under greenhouse (Authors)

Figure 8. Mango grow under the net house during fruiting theological stage (Authors)

The Potential for Improving the Carbon and Water Footprint

Smart agricultural practices such as low or no-tillage, irrigation management based on climatic data, using certificate seeds for higher irrigation water efficiency, using the proper quantity of fertilizers based on soil analysis and crop nutrients requirement, protected agriculture are also good practices to improve water use efficiency which led to improving the water footprint for the horticulture products. Many efforts had been conducted in developing countries such as Egypt to enhance on-farm agronomy practices to meet the challenges of climate change impacts in the agriculture sector such as replacing the clayey canals with cement canals for most surface irrigation networks, replacing flood irrigation with drip irrigation system for orchards and vegetable production to improve the efficiency of the irrigation at the national level, the national project for cultivating 100 thousand acres of greenhouse (less water consumption), adopt the weather prediction models for agriculture management via giving a recommendation for the farmers before extreme weather events to reduce the harmful of the cultivated crops (Saleh et al., 2017). In recent years, a number of water footprint concepts have been created in attempt to provide a quantitative and systematic way to measuring and better managing important concerns associated to water consumption. The awareness of people about climate change, carbon and water footprint start to be better because of fluctuation of weather conditions, newspaper and media become more concerned about the climate change issues. The research of carbon and water footprint in the developing countries such as Egypt is still limited in comparison with the developed country, maybe during the next few years, the researches in this field will be better due to the availability of scientific researchers who work in this field.

REFERENCES

Bauer, C., & Zapp, P. (2005). Towards generic factors for land use and water consumption. In A. Dubreuil (Ed.), *Life cycle assessment of metals: issues and research directions*. SETAC—USA.

BP. (2007). *What is a Carbon Footprint?* http://www.bp.com/liveassets/bp_internet/globalbp/STAGING/global_assets/downloads/A/ABP_ADV_ what_on_earth_is_a_carbon_footprint.pdf

Trust, C. (2007). *Carbon Footprint Measurement Methodology, Version 1.1*. Carbon Trust. http://www.carbontrust.co.uk

Čuček, L., Kravanja, Z., & Klemeš, J. J. (2012). A review of footprint analysis tools for monitoring impacts on sustainability. *Journal of Cleaner Production, 34*, 9–20. doi:10.1016/j.jclepro.2012.02.036

Department of Climate Change and Energy Efficiency (DCCEE). (2011). National greenhouse gas account factors. Commonwealth of Australia.

EC-JRC. (2011). International Reference Life Cycle Data System (ILCD) Handbook- Recommendations for Life Cycle Impact Assessment in the European context. European Commission-Joint Research Centre - Institute for Environment and Sustainability.

El-Gayar, S., Negm, A., & Abdrabbo, M. (2019). Greenhouse Operation and Management in Egypt. Handbook of Environmental Chemistry, 74, 489–560.

Energetics. (2007). *The Reality of Carbon Neutrality*. www.energetics.com.au/file?node_id=21228

ETAP. (2007). *The Carbon Trust Helps UK Businesses Reduce their Environmental Impact*. Press Release. https://ec.europa.eu/environment/etap/pdfs/jan07_carbon_trust_initiative.pdf

Ewing, B., Goldfinger, S., Wackernagel, M., Stechbart, M., Stechbart, S., Rizk, M., Reed, A. & Kitzes A. (2008). *The Ecological Footprint Atlas*. Global Footprint Network.

GFN. (2007). *Ecological Footprint Glossary*. Global Footprint Network. Accessed July 2007 from http://www.footprintnetwork.org/gfn_sub.php?content=glossary.

Goossens, Y., Annaert, B. J., De Tavernier, J., Mathijs, E., Keulemans, W., & Geeraerd, A. (2017). Life cycle assessment (LCA) for apple orchard production systems including low and high productive years in conventional, integrated and organic farms. *Agricultural Systems, 153*, 81–93. doi:10.1016/j.agsy.2017.01.007

Grubb, E., & Ellis, C. (2007). *Meeting the Carbon Challenge: The Role of Commercial Real Estate Owners*. Users & Managers.

Hammond, G. (2007). Time to give due weight to the 'carbon footprint' issue. *Nature, 445*(7125), 256. doi:10.1038/445256b PMID:17230169

Hoekstra, A. Y., & Hung, P. Q. (2002). *Virtual Water Trade: A Quantification of Virtual Water Flows Between Nations in Relation to International Crop Trade*. Value of Water Research Report Series No 11. IHE.

Hoekstra, A. Y. (2008). *Water neutral: Reducing and offsetting the impacts of water footprints*. Value of Water Research Report Series No 28, UNESCO-IHE. www.waterfootprint.org/Reports/Report28-WaterNeutral.pdf

Hoekstra, A. Y. (2003). *Virtual water trade: Proceedings of the International Expert Meeting on Virtual Water Trade*. Value of Water Research Report Series No. 12, UNESCO-IHE Institute for Water Education.

Houghton, J. (2015). *Global warming – the complete briefing* (5th ed.). Cambridge university press. doi:10.1017/CBO9781316134245

IPCC. (2013). The Physical Science Basis – Anthropogenic and natural radiative forcing. Working group I. *Climatic Change*.

Iriarte, A., Almeida, M. G., & Villalobos, P. (2014). Carbon footprint of premium quality export bananas: Case study in Ecuador, the world's largest exporter. *The Science of the Total Environment*, *472*, 1082–1088. doi:10.1016/j.scitotenv.2013.11.072 PMID:24361571

Medany, M. A., Abdrabbo, M. A. A., Awny, A. A., Hassanien, M. K., & Abou-Hadid, A. F. (2009). Growth and productivity of mango grown under greenhouse conditions. *Egyptian Journal of Horticulture*, *36*, 373–382.

Mekonnen, M. M., & Hoekstra, A. Y. (2011). The green, blue and grey water footprint of crops and derived crop products. *Hydrology and Earth System Sciences*, *15*(5), 1577–1600. doi:10.5194/hess-15-1577-2011

POST. (2006). *Carbon footprint of electricity generation*. POST note 268, October 2006, Parliamentary Office of Science and Technology. http://www.parliament.uk/documents/upload/postp n268.pdf

Saleh, S. M., Heggi, M. A. M., Abdrabbo, M. A. A. & Farag, A. A. (2017). Heat Waves Investigation During Last Decades in Some Climatic Regions in Egypt. *Egypt. J. Agric. Res., 95*(2).

Wackernagel, M., & Rees, W. E. (1996). *Our Ecological Footprint: Reducing Human Impact on the Earth*. New Society Publishers.

Wiedmann, T., & Minx, J. (2008). A Definition of 'Carbon Footprint. In C. C. Pertsova (Ed.), *Ecological Economics Research Trends* (pp. 1–11). Nova Science Publishers. https://www.novapublishers.com/catalog/product_info.php?products_id=5999

Rees, W. E. (1992). Ecological footprints and appropriated carrying capacity: What urban economics leaves out. *Environment and Urbanization*, *4*(2), 121–130. doi:10.1177/095624789200400212

Chapter 9
New Tillage Technique for Better Orchard Management:
Soil Management for Orchards

Anis Elaoud
University of Carthage, Tunisia

Rim Jalel
AVFA, Tunisia

ABSTRACT

Climate change impacts all agricultural production in all territories. Its effects are already clearly visible on orchards. This chapter makes it possible to better understand the changes that farmers will be faced with and therefore to consider changes to their system as soon as possible in order to sustain their work. Indeed, current cultivation practices (conventional) are based on the multitude of tool passages in succession, without considering the negative effects, in the medium and long term, on the structure of the soil and on its topsoil. Thus, the excessive number of passages of agricultural tools results in increased energy consumption and a high hourly requirement per hectare. Indeed, this phenomenon of agricultural soil compaction is one of the main causes of its physical degradation. However, the rational use of agricultural land requires the introduction of new techniques of tillage as well as appropriate mechanization, which meets the requirements of crops, the improvement of production, and the preservation of the structured soil.

INTRODUCTION

GENERALITY

The upgrading and modernization of the agricultural sector and the intensification of agricultural production have been based on a better use of agronomic innovations and technological advances, as well as an increase in the degree of mechanization on farms agricultural. Mainly the intensification of the vine sector

DOI: 10.4018/978-1-6684-2423-0.ch009

is facing several constraints, including those related to the interaction between machine-soil-plant. This interaction has a direct and immediate impact on the practical interventions necessary for the preparation of the soil, the planting and the management of the vines. Some interventions may have negative effects on soil structure and soil development, which can affect crop development and production.

Problematic

Climate change is impacting all agricultural production in all regions. Its effects are already visible on the cultivation of the vine. This chapter provides a better understanding of the changes that farmers will be facing and thus, as soon as possible, consider changes to their system in order to sustain their work. Indeed, current (conventional) cultural practices are based on the multitude of successive tool passes, without considering the negative effects, in the medium and long term, on the soil structure and its topsoil. Thus, the excessive number of passes through agricultural tools results in increased energy consumption and a high hourly need per hectare. This often translates into increased farm mechanization costs. On the other hand, the impact of the multitude of mechanized operations and inadequate tools, in view of the setting up and running of various cropping operations for the cultivation of vines, can be the cause of problems of compaction of soils function time and space. Such problems generate restrictive effects on the root development of crops and their production.

Indeed, this phenomenon of agricultural soil compaction is one of the main causes of its physical and biological degradation. However, the rational use of agricultural land requires the introduction of new tillage techniques (planting on teenagers or permanent bed) as well as appropriate mechanization to meet the requirements of crops, improvement of production, and preservation of soil structure and optimization of energy and economic loads. Indeed, by acting on the cultivation mode and tractor/tool suitability, potential savings of the order of 20% can be achieved (Schrock et al., 1986; Grogan et al., 1987). As a result, the correct energy and dimensional adequacy of the tractor-tool assembly helps to improve operational and crop efficiency.

On the other hand, the impact of the multitude of mechanized operations and inappropriate tools, for the installation and conduct of different cropping operations for a given crop, may be the cause of soil settlement problems, variables in time and space. Such problems have binding effects on the root development of the crops, their development and their final production. Indeed, this phenomenon of settling of agricultural soils is one of the main causes of its physical and biological degradation.

Many agricultural practices are carried out over short periods and frequently when the soil is wet. The practicability of a soil reflects its ability to accept the passage of gear (Elaoud and Chehaibi, 2011). Thus, the soil settling factors applied to the technique and soil tillage tools are related to the initial soil condition (water and structural condition), the weight of the agricultural tractor, the number of passes, the speed of circulation, type of tires and inflation pressure, etc. These factors are also related to the technique and tillage tools applied (Vitlox and Loyen, 2002). A poor match between these different parameters lead to a greater susceptibility of the soil to compaction (Pouya et al., 2013). In Tunisia, as in some developing countries, agricultural mechanization is faced with a problem of choice and suitability tractor/ tool to achieve a soil preparation that meets agronomic, climatic and energy requirements.

However, the rational exploitation of agricultural land requires the introduction of new tillage techniques and appropriate mechanization to meet the requirements of crops, to improve production, and the preservation of the soil structure and the optimization of energy and economic loads. Therefore, the good energy and dimensional adequacy of the tractor/tool unit contributes to the improvement of the operational efficiency of the different cropping operations. In addition, the choice of an appropriate tillage tool, such as a heavy cultivator instead of a disc plough, can reduce traction effort and, depending on the type and state of the soil, generate an energy reduction of up to 40% during the intervention period (Michel et al., 1985).

In recent years, the cost of agricultural equipment has risen sharply, which has resulted in difficulties in acquiring, renewing and modernizing tools. This has slowed down the adoption of some relatively costly technological innovations, such as the use of high value economic and technological combined tools. As a result, cultivation techniques have changed little over the last long period.

SOIL-TRACTOR INTERACTION

The conditions of agricultural soils are monitored by physical, mechanical and applied techniques. The factors that influence the condition of agricultural soils are related to the gear (gear load, tool, etc.) and other soil-related factors such as texture, organic matter, structure, water content, etc. (Lebert et al., 1989).

Weight of Gear

The load of the tractor plays an important role in the settling of the ground. This weight varies from one machine to another depending on its power and the work to be performed usually from 1 to 12 tons. The total mass of the tractor is distributed proportionally on the two front and rear axles which transmit the latter to the wheels.

According to Asselin (2010), to reduce the effect of soil compaction, soil pressure must be reduced. The weight of the machine used with the weight of the tool plays a role in characterizing this phenomenon. In fact, the greater the weight of the machine, the more intense the forces applied by anthropogenic activity, the tighter the ground is and the more massive and structural the extreme (Weill, 2009). The settling affects first the topsoil layer then the ploughing sole and it can exceed the sole and reach 60 cm and even more.

Gear Speed

The speed of advance is a parameter that essentially characterizes the machine; it is evaluated in km per hour. In modern agriculture, the trend is to increase the speed at which work is carried out thanks to the increase in the power of the tractors but also to the improvement of the design of the equipment (reinforced chassis, active suspensions, electronic system, etc.). Several studies have shown the effects of working speed (slow, medium and fast), on physical soil parameters (penetration resistance, density, depth of tillage and crop profile) and on crop yield.

According to Lefebvre et al. (2010), settlement becomes less severe when the speed of the tractor's advance increases and the longer the duration of the machine's circulation increases, the more the propagation of settlement increases in depth. Similarly, Chehaibi et al. (2008) reported that on a tuber plot, working at slow speeds resulted in better yields. The latter was 28% and 25% higher than the average and fast-speed plots.

Gear Slip

The slip is the common property that depends on the design of the wheels and the condition of the agricultural land. The variation of the distance tilled by the gear wheel during a single complete turn. The spike of the driving wheel, anchored in the furnished earth, comes to rest on a layer of earth which it compresses. As long as the earth resists, the machine is propelled in the direction of advance. When the resistance of the earth decreases, it gives way to the effort exerted by the crampon: it is the appearance of the slip. Settlement, a scourge that opposes the development of crops, has the effect of increasing the cohesion of the earth's structure; however, it reinforces the point of reaction against which the wheel spike is supported. The slip is usually expressed in %, and varies according to the condition of the tyres and the physical state of the ground.

Dwyer et al. (1976) reported that the wheel slip rate, up to the reduction in travel, was set at 0.2 for the traction tool and 0.1 for animated tools. As a general rule, slip rates below 20% will be accompanied by better traction efficiency, but the traction produced is low. On the other hand, slippage rates above 20% should be avoided because they cause considerable energy losses (Perdok et al., 1983).

In addition, Smith et al. (1993) reported that excessive loosening of soil during cropping operations may further promote wheel slippage when evaluating a tool in clay soil. They concluded that at 1800 rpm engine speed, the change from fourth to fifth speed resulted in an increase in wheel slip from 17.1% to 20%. A similar change in engine speed at 2100 rpm resulted in an increase in wheel slip from 18.2% to 21.5%. As with the increase in wheel slip, the energy loss increased at the wheel/ground interface and fuel consumption was noted. The slippage of the 12 wheels during the cultivation operation varies from 9.5 to 12.5% due to a change in speed at a particular engine speed.

To reduce slippage, several solutions are proposed such as increasing the load on the driving wheels (ballast) or improving the grip between wheel and ground by using twin wheels, low pressure wheels within the limits of use.

Pull Force

The tractive effort developed by a tractor is directly related to the energy required by the working tools in a given soil. For a working speed and a type of tool, the tensile effort varies with the density and humidity of the soil and with the working depth. These factors have been the focus of previous research, which has adapted field experiments to understand how these factors affect the conditions of tillage tool traction forces (Mouazen et al., 2002).

Koolen and Kuipers (1983) developed several techniques for determining the tensile forces of the tool based on observations and relationships between independent and dependent variables. However, these techniques depend on the field data and the means of the tensile force values and the independent variables were taken into consideration when developing an equation of the tensile requirement of the tools. Multiple regression equations indicated that the theoretical horizontal force of the soil varies linearly with moisture and not linearly with depth and apparent density. The latter is expressed by the following regression equation:

$$ET = -33{,}74\ w + 73{,}253\ D^2 + 2{,}5649\ \varphi^2\ (R^2 = 0{,}94)$$

Where ET is a tractor effort; d=density (kg/m3) and w = water content (%). The coefficient of determination of the tensile force (KN), D the working depth of the soil (m), high n of 0,94 indicates that the theoretical tensile force can be estimated according to w, and d, with a high degree of precision, for a given soil.

And theoretical = 21,360 w + 73,931 D^2 + 1,673 (R^2 = 0,95)

The comparison of equations (3) and (4) points out that the non-linearity of the variation in tensile force was a quadratic function of wet density, and a cubic function of dry density (Mouazen et al., 2002).

A comparison of the calculated tensile force and the measured ground tensile force showed a slight difference in the calculated values over the measured tensile force values.

This was attributed to the slight divergence in soil texture and length of the working part of the tool in the soil (Mouazen et al., 2002).

Type of Tools and Their Characteristics

Tillage tools are essentially based on three types: ploughs with ploughs carrying out ploughing with overturning, toothed tools such as chisel carrying out work without overturning and disk tools such as disk offset and disk plough.

Several researchers have been interested in reducing tillage and many comparisons have been made between tillage with these different types of equipment, particularly with regard to effects on soil properties and yield. Most of the research on mechanical traction in tillage has involved efforts on narrow-toothed chisels.

Important work was done by Payne (1956) and Payne and Tanner (1959). They described the shearing of the ground in front of the chisel's narrow teeth.

Soil deformation is considered to be produced by three different mechanisms (Aluko et al., 2000), shear fracture, tensile fracture and plasticity fracture.

Shear fracture occurs during compression of forces created by forward movement of the teeth. Sometimes the soil can break at a certain depth; while it is only in plastic deformation below that depth (Kostritsyn, 1956; Godwin et al., 1977). The differences in the demands of mechanical drive of the plough, chisel and disk offset to a large extent can be attributed to differences in working depth. Traction is generally highest for chisel and lowest for plough and disk offset. The difference is likely due to differences in tool geometry and ground failure mode (Arvidsson et al., 2004).

Grisso et al. (1994) examined the different work presented by researchers to measure the tensile and power requirements required for the most commonly used tools in tillage. They pointed out that the traction required to pull a tool is essentially a function of the width of the tool, the working depth and the speed of advance. They reported that the pull forces of the tools significantly believe with the speed of advance and the relationship varies linearly to quadratic if the other parameters are kept constant.

METHOD OF PLANTING VINES

Planting or replanting a vineyard is not done lightly. The new plantation is in place for decades and must be productive quickly enough to be profitable. However, it is necessary to take the time to know the soil and the subsoil of his plot to promote the planting and the good development of the vine afterwards

Taking into Account Cultural Precedent

Cultural precedent is one of the first criteria to be taken into account. When the new plantation succeeds a vineyard or orchard, it is important to observe the health of the old plantation. The main risks are indeed the rot and the short-tied. Whether it is one or the other of these diseases, the grubbing up must be treated, trying to recover as many roots as possible. These sources of contamination can infect young plants and cause the disease to reappear years after planting. In the case of rot, a wetland favors the development of this fungus: this is why drainage of these zones is recommended to curb its diffusion.

In the case of the short-knotted vine, a devitalisation of the old vine with a herbicide proves to be effective, especially when it is carried out early (just after the harvest) and when the time between the devitalisation and the grubbing is important. This devitalization deprives the nematode vector of the short-knotted virus of food. The IFV advises a rest of the plot from seven to ten years: during this rest, the winegrower will be able to set up a green fertilizer, to limit erosion, control weeds, enrich in organic matter and restructure in depth. The choice of green fertilizer must be made according to the nature of the soil and the objectives of the winemaker.

Know the Soil and Basement

Once the vine has been torn off, making several soil pits scattered in the plot can be decisive in the choice of tools for the preparation. Some winegrowers use resistivity maps to strategically position these pits. These make it possible to determine the thicknesses of each horizon: the superficial horizon rich in organic matter, the deeper horizon poorer and the mother rock. In the case of a later use of a plough to prepare the planting, the winemaker will be able to determine the working depth if he does not want to disturb the horizons and conserve the organic matter in the area.

The profile also makes it possible to highlight physical barriers preventing root sinking. Depending on the depth of these barriers, a work of under-solage will allow subsequently to break them and the vines to draw more in depth.

Soil pits also provide an opportunity to sample different horizons for analysis.

Facilitate Root Penetration

A good soil preparation starts with deep work. The plough is one of the tools for this work. Be careful, however, not to mix up horizons, warns Manuel Blondy of the Gironde Chamber of Agriculture. You have to keep the organic matter, so the microbial life on the surface. But if the organic horizon measures 50 centimeters, there is no problem to draw at this depth."

In addition, this tool traces roots that would have remained deep. A passage with a toothed tillage tool allows them to be raised to the surface. "We offer large rakes on the tractor to carry the vines and roots to the edge of the plot," says Cedric de Bourayne of Kirpy. However, the plough can be a plough sole and thus a physical barrier to root sinking. Ploughing in a relatively dry soil limits the smoothing and therefore the ploughing surface. "This applies to all soil work," says Cedric de Bourayne.

In the absence of a plough and/or to work deeper, the sub-sun ensures a cracking of the ground without disturbing the different horizons.

Amend Soils

From the soil analyses, the winegrower will be able to select the appropriate rootstocks and grape varieties. It can also determine the amendments and quantities to be made per hectare to rebalance the pH of the soil and possibly improve microbial life by bringing organic manure. Contrary to what is sometimes observed on the ground, these inputs are to be carried out once the ploughing is carried out, to avoid too deep burying the organic manure.

Grinding the Stones

In stony plots, replanting offers the opportunity to refine the soil using a stone crusher, up to a depth of 25 to 40 centimeters. At first, the living organisms of the soil took the hit, «but very quickly, life resumed its course in a layer of soft soil more important than before», assures Cedric de Bourayne. "A precaution to take in stony limestone soils, paying attention to the fact that grinding does not increase the level of active limestone too much and raises chlorosis problems," says Manuel Blondy.

Refining the Surface Soil

When exposed, the plot is more suitable for drainage operations. In the Mediterranean regions, replanting provides an opportunity for underground micro-irrigation.

It remains to till the soil on the first 15 or 20 centimeters, to encourage the recovery of the plant. This work can be done by any surface tillage tool, toothed or disc, powered (milling cutters or harrows) or not powered.

On loamy soils, however, care should be taken not to crumble the soil too much to avoid generating a crust of beat at the first rain: animated tools are rather prohibited in this case.

The establishment of a vineyard is a decisive step in the success of a farm, it engages the winemaker for about thirty years. The administrative procedures, the technical choices, the financial investment require anticipating a plantation of at least 3 years. The reasoning of a plantation starts from the definition of the production objective in relation to the situation, the future of the holding, and the regulatory constraints.

PLANTING VINES ON TEENAGERS

Definition of Teen Planting Technique

The Teen is a surface deliberately raised above ground level, curved, held or not held at the base by planks, optimized by a living or spreading permanent vegetal cover (Figure 1)

Figure 1. Planting vines culture on teenagers

The Different Planting Steps

The aim of soil preparation is to ensure the optimal rooting of young plants, in particular by allowing a good circulation of air and water in the exploitable horizons. The ideal is to make a complete diagnosis beforehand, following several complementary approaches:

- General situation of the plot: altitude, slope, concavity, exposure, heterogeneity zones, etc.
- In the case of replanting: take into account the observation of previous vines (fragmentary heterogeneity, vigor, sensitivity to drought, deficiencies, asphyxiation zones, rootstock behavior, etc.).
- Interpretation of soil pits: examination of different horizons (texture, moisture, structure, compactions, etc.)
- Perform physico-chemical analyses on representative horizons (soil and subsoil).
- Tillage should start with a subsurface to create cracks inside the soil and avoid the formation of clods. This sub-layer is applied by a robust sub-layer equipped with a decomposing roller to prevent the removal of clods from the ground (Figure 2).

Figure 2. Substract tool

Construction of teenagers is done after the operation of undermining by a specific tool (figure 3) called tool of construction of teenagers.

Figure 3. Teen building tool

EFFECT OF PLANTING VINES ON TEENAGERS

Effect on Soil Properties

- It increases the depth of arable land and improves water and air circulation and therefore promotes plant rooting
- By exposing one side to the morning sun and the other side to the afternoon heat (for a north/south teen orientation), the hill heats up faster than a classic culture board
- When the ground is raised, it rises faster after a downpour, so sowing or planting can be done right away
- Increases soil fertility by storing organic matter digested by earthworms as humus over the entire height of the mound.
- Improves growing conditions by:
- More soil height for roots,
- Effective drainage under wet conditions,
- Optimum solar capture (spring warming),
- Increased gas exchange with the atmosphere,

Effect on Yield

The technique of planting on teenagers showed a gain in vine yield.

About 16% more yield compared to the control technique (conventional) is recorded in the experimental plots.

Effect on Energy Parameters

The technique of planting on teenagers is less consuming in traction energy by offering an energy gain of 25%. The hourly volume measured during tillage by this new technique recorded an energy gain of 12% in comparison with the conventional technique.

CONCLUSION

The impact of the multitude of mechanized operations for the installation and operation of various operations for the cultivation of vines may be the cause of soil compaction problems. Such problems have effects on production.

Indeed, this phenomenon can be remedied by new tillage techniques (planting on teenagers). This showed good results in terms of working time, energy savings and improvement of the yield of the vines.

REFERENCES

Aluko, O. B., & Seig, D. A. (2000). An experimental investigation of the characteristics of and conditions for brittle fracture in two-dimensional soil cutting. *Soil & Tillage Research*, *57*(3), 143–157. doi:10.1016/S0167-1987(00)00156-2

Arvidsson, J., Keller, T., & Gustafsson, K. (2004). Specific draught for mouldboard plough, chisel plough and disc harrow at different water contents. *Soil & Tillage Research*, *79*(2), 221–231. doi:10.1016/j.still.2004.07.010

Asselin, C. (2010). *Choisir le bon pneu pour diminuer la compaction du sol.* La référence en nouvelles technologies agricoles au Québec.

Chehaibi, S., Hannachi, C., Pieters, J. G., & Verschoore, R. A. (2008). Impacts de la vitesse d'avancement du tracteur sur la structure du sol et le rendement d'une culture de pomme de terre. *Trpiculture*, *26*(3), 195–199.

Dwyer, M. J., Evernden, D. W., & Mcallister, M. (1976). Handbook of Agricultural Tyre Performance (2nd ed.). NIAE.

Elaoud, A., & Chehaibi, S. (2011). Soil compaction due to tractor traffic. *Journal of Failure Analysis and Prevention*, *11*(5), 539–545. doi:10.100711668-011-9479-3

Godwin, R. J., & Spoor, G. (1977). Soil failure with narrow tines. *Journal of Agricultural Engineering Research*, *22*(3), 213–228. doi:10.1016/0021-8634(77)90044-0

Grisso, R. D., Yasin, M., & Kocher, M. F. (1994). Tillage implement forces operating in silty clay loam. ASAE paper n° 94-1532, ASAE, St. *Joseph, Michigan, USA, 1994*, 17.

Grogan, J. D., Moris, A., Searcy, S. W., & Stout, B. A. (1987). Microcomputer based tractor performance monitoring and optimisation system. *Journal of Agricultural Engineering Research, 38*(4), 227–243. doi:10.1016/0021-8634(87)90091-6

Koolen, A. J., & Kuipers, H. (1983). *Agricultural Soil Mechanics*. Springer. doi:10.1007/978-3-642-69010-5

Kostritsyn, A. K. (1956). *Cutting of a Cohesive Soil Medium with Knives and Cones*. National Institute of Agricultural Engineering.

Lebert, M., Burger, N., & Horn, R. (1989). Effects of dynamic and static loading on compaction of structured soils. In *Mechanics and related processes in structured agricultural soils* (pp. 73–80). Springer. doi:10.1007/978-94-009-2421-5_7

Lefebvre, M. P. (2010). *Spatialisation de modèles de fonctionnement hydromécanique des sols appliquée à la prévision des risques de tassement à l'échelle de la France* [Thèse de doctorat]. Université d'Orléans.

Michel, J. A., Fornstrom, K. J., & Borrelli, J. (1985). Energy requirements of two tillage systems for irrigated sugarbeets, dry beans and corn. *Transactions of the ASAE. American Society of Agricultural Engineers, 28*(6), 1731–1735. doi:10.13031/2013.32508

Mouazen, A. M., Ramon, H., & De Baerdemaeker, J. (2002). SW-Soil and water: Effects of bulk density and moisture content on selected mechanical properties of sandy loam soil. *Biosystems Engineering, 83*(2), 217–224. doi:10.1006/bioe.2002.0103

Payne, P. J. C. (1956). The relationship between the mechanical properties of soils and the performance of simple cultivation implements. *Journal of Agricultural Engineering Research, 1*, 23–50.

Payne, P. J. C., & Tanner, D. W. (1959). The relationship between rake angle and the performance of simple cultivation implements. *Journal of Agricultural Engineering Research, 4*, 312–325.

Perdok, U. D., & Van de Werken, G. (1983, March). Power and Labour requirements in soil tillage. *Soil & Tillage Research, V3*(1), 3–25. doi:10.1016/0167-1987(83)90013-2

Pouya, M. B., Bonzi, M., Gnankambary, Z., Traoré, K., Ouédraogo, J. S., Somé, A. N., & Sédogo, M. P. (2013). Pratiques actuelles de gestion de la fertilité des sols et leurs effets sur la production du cotonnier et sur le sol dans les exploitations cotonnières du Centre et de l'Ouest du Burkina Faso. *Cahiers Agricultures, 22*(4), 282–292.

Schrock, M., Matheson, D., Blumanhourst, M., & Thompson, J. (1986). A device for aiding gear selection in agricultural tractor. *Transactions of the ASAE. American Society of Agricultural Engineers, 29*(5), 1232–1236. doi:10.13031/2013.30301

Smith, L. A. (1993). Energy requirements for selected crop production implements. *Soil & Tillage Research, 25*(4), 281–299. doi:10.1016/0167-1987(93)90028-N

Vitlox, O., Loyen, S. (2002). *Conséquences de la mécanisation sur le compactage du sol et l'infiltration d'eau, actes de la journée d'étude: érosion hydrique et coulées de boue en Wallonie*. Academic Press.

Weill, A. (2009). *Les profils de sol agronomiques, un outil de diagnostic de l'état des sols. Centre de références en agriculture et agroalimentaire du Québec*. CRAAQ.

Chapter 10
Emblica officinalis Orchard-Based Agroforestry for Augmented Productivity

Kumud Dubey
https://orcid.org/0000-0001-6367-4617
Forestry Research Centre for Eco-Rehabilitation, Prayagraj, India

K. P. Dubey
Forest Department Uttar Pradesh, India

ABSTRACT

Emblica officinalis (Indian Gooseberry or Aonla) is a horticultural commercial crop. Among various agroforestry, a number of fruit orchard-based agroforestry systems are practiced all over parts of the country. These systems not only provide fruits but also fulfill the necessity of timber, food, fuel, and fodder. Aonla's orchard-based agri-horticultural system has immense potential for the betterment of people's livelihoods. Due to its thin canopy and leaf shedding nature in the winter season, orchards of this fruit tree species are well suited for agro-forestry. Aonla orchard-based agroforestry was discussed in the present study. Agricultural crop (wheat and paddy) grain and straw production data were recorded under Aonla orchard-based agroforestry. Though the agricultural crop production was reduced, Aonla fruits gave an additional income to the farmers. Overall, a good economic return is obtained from Aonla orchard-based agroforestry. Utilization of Aonla orchard interspace with the traditional cropping system was profitable in the Gangetic region of Uttar Pradesh, India.

INTRODUCTION

Acceleration in human and livestock population necessitated acquisition of more and more land under cultivation to meet the ever increasing demand for food, fodder, vegetables, fuel wood, timber, medicines etc. This increasing demands for food and wood mainly has resulted in over exploitation of forests. This has made planners, foresters and environmentalists to consider alternate system of land use for serving

DOI: 10.4018/978-1-6684-2423-0.ch010

society (Khosla and Khurana 1987). Further, demographic pressure has forced man to seek unconventional methods of agriculture to utilize land to the maximum extent. The practice of agroforestry can meet all the multiple demands of growing population (Mutanal *et. al.*,2009). Therefore in the quest of optimizing productivity, the multi-tier system came into existence. The origin of agro-forestry practices, i.e. growing trees with food crops and grasses, is believed to have been during Vedic era (Ancient period, 1000 B.C.), however the agro-forestry as a science is introduced only recently. Agro-forestry is the science of designing and developing integrated, self-sustainable, land-management systems that involves the introduction and retention of woody components such as trees, shrubs, bamboos, canes and palms along with agricultural crops including pastures or animals, simultaneously or sequentially on the same unit of land. Agroforestry systems also provide significant sustainable development benefits such as food security and secure land tenure in developing countries, increasing farm income, restoring and maintaining above-ground and below ground biodiversity, maintaining watershed hydrology, and soil conservation. Agroforestry also mitigates the demand for wood and reduces pressure on natural forests (Pandey, 2002, Dubey, 2010). Moreover, Trees have a special role in the ethos of the people of India and have a long historical tradition of tree-growing on farms and around homes (Pandey, 2007; Dubey, 2010). Tree has the dominant role to play in all agro-forestry systems for sustainable agriculture and environmental protection. Tree planting reduces salinity, improves soil fertility, controls and prevents erosion, controls water logging, reduces catchment eutrophication and possible checks acidification, increases local bio-diversity and reduces the green-house effect. Unfortunately, agroforestry systems and practices are little known in terms of functioning and dynamics and little progress has been made in the study and promotion of agroforestry and other tree-based systems. Being a traditional land use system, agroforestry is able to satisfy a large diversity of socio-economic needs in a sustainable way and in many different agro-ecological conditions (Depommier, 2003). Agroforestry is also being practiced in fruit orchard systems. Horticultural crops cover 13 per cent cropped area of the country (Rai *et.al.*,2017). Ecological agricultural works like agro-forestry are being encouraged to overcome various problems of agriculture, land, food, fodder and fuel production. Among various agroforestry, number of fruit orchard based agroforestry systems are practiced in all over parts of the country. These systems not only provide fruits but also fulfill the necessity of timber, food, fuel and fodder. In addition, this fruit orchard based agroforestry also supports livelihood, offers nutritional security, mitigates climate change related problems, ameliorates soil health and provide economic stability in local populace. Fruit-based agroforestry has also been proven as an important tool for crop diversification (Kumar and Chaturvedi, 2017; Kumar *et. al.*, 2019). Several drought hardy fruit trees like *Capparis decidua*, *Salvadora oleoides*, *Cordia dichotoma*, *Cordia gharaf*, *Ziziphus nummularia*, *Z. mauritiana* are suitable in low rainfall receiving area. In addition to provide fruits, these fruit plants produce fodder for livestock. Several other fruits such as *Emblica officinalis*, *Punica granatum*, *Aegle marmelos*, *Phoenix dactylifera*, *Mangifera indicum* and *Tamarindus indica* based agroforestry may be opted in the irrigated area (Kumar *et. al.*, 2019).

Agro forestry system of farming has come up in big way throughout India but eastern part of Utter Pradesh is lagging in this regard. Extension of agro-forestry in eastern part of Utter Pradesh (U.P.) is the need of the day because this region has only about 4.0% of area under tree cover against the figure of 33% recommended by National Forest Policy. Increase in agroforestry practices is only option to increase forest cover in U.P. Agroforestry system of farming has good scope also in this region. Fruit orchards like *Psidium guajava* (Guava), *Mangifera indica* (Mango), *Ziziphus mauritiana* (Ber), *Moringa oleifera* (Sahajan), *Emblica officinalis* (Aonla) are commonly used for orchard agroforestry in U.P. Present study is focused on *E. officinalis* orchard based agroforestry.

Emblica Officinalis

E. officinalis is indigenous to Indian sub-continent. India ranks first in the world in area and production of this crop. Apart from India naturally growing trees are found in different parts of the world like Sri Lanka, Cuba, Puerto Rico, USA (Hawai & Florida), Iran, Iraq, Pakistan, China, Malaysia, Bhutan, Thailand, Vietnam, Philippines, Trinidad, Panama and Japan. Aonla (*Emblica officinalis*, Gaertn Syn., Phyllanthus emblica Linn) belong to the family Euphorbiaceae. Aonla is also known by its several vernacular names such as 'amla' or 'aura' in Hindi, 'dhatri' or 'amlaki' in Sanskrit, 'amla' or 'amlaki' in Bengali or 'Indian gooseberry' in English. Aonla is a common home yard tree throughout India. The natural distribution of wild aonla is found on the Himalayas, Chota Nagpur, Bihar, Orissa, West Bengal, North Circars, Deccan, Karnataka and in Western Ghats. Uttar Pradesh, Gujarat, Rajasthan, Tamil Nadu, Haryana, Maharashtra, Andhra Pradesh, Punjab, Karnataka and Himachal Pradesh are the major aonla growing states in India. Uttar Pradesh ranks first in its area and production. Aonla is more popular in Uttar Pradesh, India where it is largely cultivated as commercial orchards in Pratapgarh, Azamgarh, Varanasi, Faizabad, Sultanpur, Raibareli and Bareilly district (Rai *et. al.,*2016). It is a medium sized branched tree with height of 8-10 m with thin canopy. It is deciduous tree with defoliation of leaves. This quality makes it suitable for agroforestry. The fruit has been known for several medicinal values. It is used for making a variety of value added products which are used as popular neutralceuticals and food. It is considered as an effective remedy for cardiac disease. Due to high content of anti oxidents, it has a revitalizing effect. Aonla is accepted hair tonic in traditional recipe for enriching hair growth and hair pigmentation (Wali *et.al.,*2015). Fruit powder is also used for making synthetic vitamin C. Due to rich content of Vitamin C, particularly in this post COVID Era, this fruit has gained extreme popularity and importance. In India, India ranks first in production of aonla. It occupies an area of 103.55 thousand ha with a production of 1221.25 thousand metric tones (Pathak *et al.* 2003; Wali *et.al.,*2015; Singh *et. al.* 2019). Well drained fertile loamy soil is ideal for aonla cultivation although, it can be successfully grown in light as well as heavy soils except highly sandy soil. Recently, this species is recognized suitable for various kinds of wastelands like moisture stress, eroded, ravenous land with undulated topography. Its deep root system makes it helpful in providing vertical drainage and improving soil properties. Its plantation has found suitable for rehabilitation of rangelands, particularly through popularization of agri-silvi-horticulture land use system. It is also salt tolerant which make it an ideal crop for salt affected wastelands (Negi *et.al.,*2013; Wali *et.al.,*2015). Due to its hardy nature, high productivity, nutritive and therapeutic values, and its suitability for various kinds of value added products, Aonla has been popular as important commercial fruit crop (Pathak 2003). Some important cultivars of this species like Banarasi, Francis, Chakaiya are known to be cultivated in Uttar Pradesh. Trees usually begin to bear fruit when 5 to 6 years old and normally bear for about 50 years under well managed conditions (Orwa *et. al.*, 2009). Aonla fruits are ready to harvest during November-December. An aonla tree may bear 1-3 q./tree, giving 15-20 tonnes/ha with about 100 trees per ha with 100 trees per ha (Wali *et.al.,* 2015).

METHODOLOGY

Survey of farmer fields for established aonla orchard based agroforestry being practiced by the farmers in Uttar Pradesh was conducted in Prayagraj and Pratapgarh area of U.P. where this aonla based agroforestry practicing are prevalent. After survey, suitable aonla orchard based agroforestry plots were selected for the study. While selecting the study plots, two age groups of aonla orchards of 3 years and 6 years old trees were taken. Study plots were selected on willing farmers field to avoid any further hindrance in study. In selected agroforestry plots, aonla with wheat and paddy crop intercropping was studied. Productions of both agricultural crops (Straw production/m^2 and Grain production/ m^2) was observed. Production was recorded at a distance of 1m and 5m from the Aonla plant. Productivity of the crop was measured in 3 replications. Production of adjoining fields of pure agriculture crop was recorded as control. Soil samples at different distances from tree (1m and 5m) were also collected and analyzed. Triplicate soil samples were collected from selected agroforestry plots. Each soil samples were prepared by thoroughly mixing of soils from these triplicate soil samples. Soil samples from area were collected randomly. The sample were analyzed for pH, conductivity, moisture content, nitrogen, available phosphorous, potassium and organic carbon. Soil samples from adjoining fields with sole agricultural crop were recorded as control. Physical growth data of orchard tree species was also recorded.

RESULTS AND DISCUSSION

Selection of Aonla Orchard Agroforestry in U.P.

Aonla agroforestry was found preferred agroforestry by farmers in the region. Due to its thin canopy and leaf shedding nature in winter season, this fruit tree species is a favoured by the farmers in agroforestry. It is very important fruit tree species and gives good economic return. It can easily grow in dry conditions. It acts as soil improver also. The branches are lopped for green manure. They are said to remediate excessively alkaline soils.

Two plots with two age groups, 3 years and 6 years old aonla trees, were selected for the study. The selected plotes were located in Prayagraj District of U.P., India. Its physical details were as following Table 1:

Table 1. Physical parameters of selected Aonla orchard Agroforestry plots

Tree sp.	Age	Spacing	Average Height (m)	Average Girth (cm.)	Branch	Canopy (m)
Aonla	6	6mX6m	5.50	55	4 to 8	5m
Aonla	3	6mX6m	4.50	35	3 to 5	3m

Branching in case of 6 years old aonla tree was more than that of 3years old aonla tree. Average height, girth and canopy were also had similar trends (Table 1). It shows that aonla tree was still in growing phase and not attained full growth. Aonla tree attains full growth in 8 to 10 years.

Agricultural crop (wheat and paddy) grain and straw production data were recorded. It was observed that crop production was reduced with the age of the tree. Crop production was also affected with distance and production was significantly higher in case of 5m distance. It was observed that wheat crop production was increase with increase in distance from the tree line, however the effect was more pronounced in case of grain production in comparison to straw production. This effect was most probably due to shade effect. Statistically, wheat production data was analyzed and found significant. F value for wheat grain production is 29.43, wheat straw production is 71.74, paddy grain production is 49.638, paddy straw production is 57.32. The production data for wheat and paddy is depicted in Figure 1, respectively. Similar trend was also observed in case of paddy, but the reduction was more in straw production in comparison to grain production (Figure 1). Soil analysis data were depicted in figure 2. It was found that moisture content was more under agroforestry in comparison to control. It may be due to shade effect which protected the soil from direct sun light. It has no significant effect on pH and EC. Soil nutrient condition was improved under aonla agroforestry. Organic carbon, nitrogen, potassium and phosphorus were more under aonla agroforestry except of soil from 5 meter distance under 3 years aonla aged agroforestry. This may be due to low leaf litter from 3 years old aonla (Figure 2). It showed that the aonla tree ameliorates the soil quality.

Per aonla tree average 1.0 to 1.5 Q fruit production was reported from 6 years old tree and @ of 20 to 25/- Rs./Kg, about 2000/- to 2500/- of Indian Rs obtained from each tree. In spacing of 6m X 6m, about 275 trees may be planted, from which about 5.50-6.87 Lakh/year of Indian Rs. economic return may be obtained from aonla fruits in agroforestry, in addition to agricultural crop production. As far as agricultural crop production was concerned. It was observed that in case of wheat 10 to 15% reduction was observed under aonla orchard in comparison to sole wheat crop. More reduction was observed in case of paddy crop ant it was about 20 to 25%. Production of wheat under 6 years old aonla orchard was about 4.5 tonn/ha in comparision to pure wheat field which was about 6.0 tonnes/ha. Production of paddy under 6 years old aonla orchard was about 4.7 tonn/ha in comparision to pure wheat field which was about 8.2 tonnes/ha. But this reduction in agricultural crop production was compensated by higher economic gain obtained from aonla fruit. Even the production profit may be increased per year by adopting proper management practices.

Figure 1. (a) Production of wheat grain (b) Production of wheat straw (c) Production of Paddy grain (d) Production of Paddy straw in Aonla Agroforestry

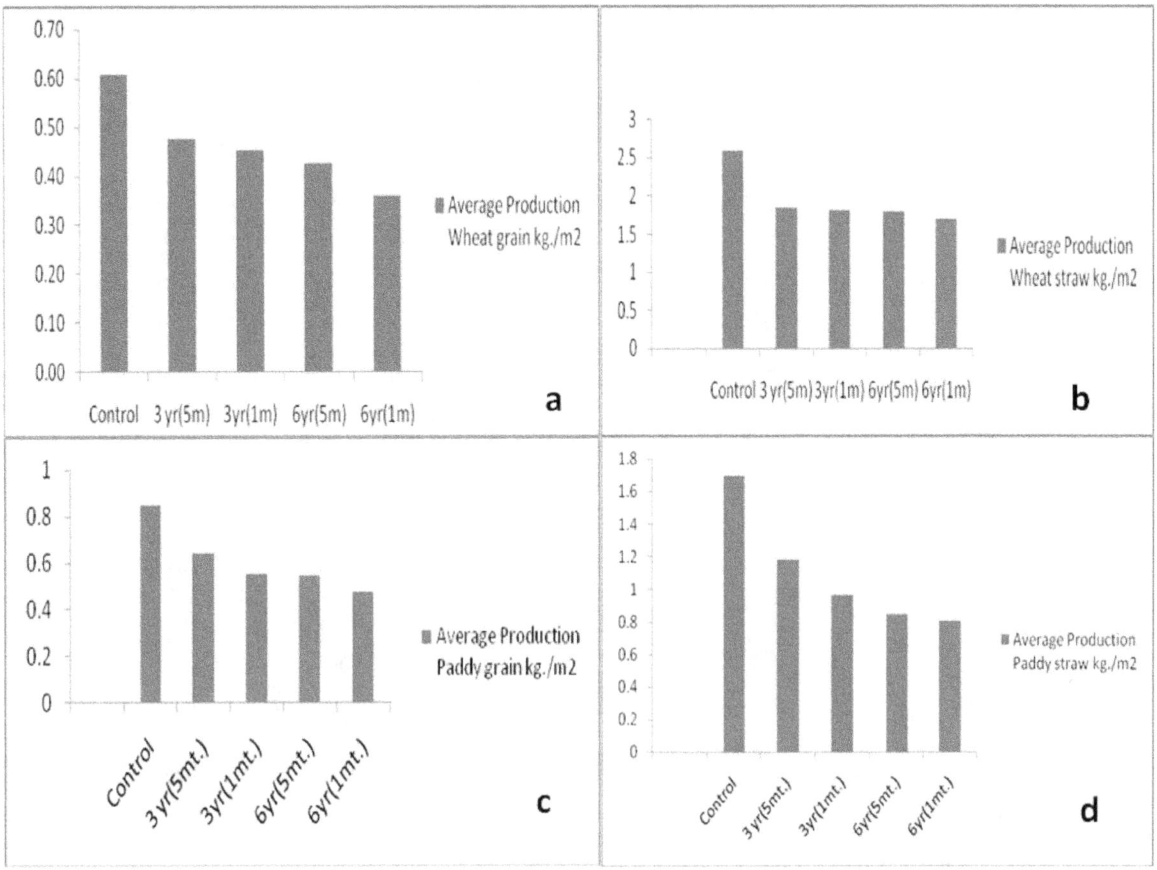

Figure 2. Physico-chemical characteristics of soil (a) pH, EC and moisture contents (b) Organic carbon (c) Nitrogen (d) Phosphorus and Potassium in Aonla Agroforestry system

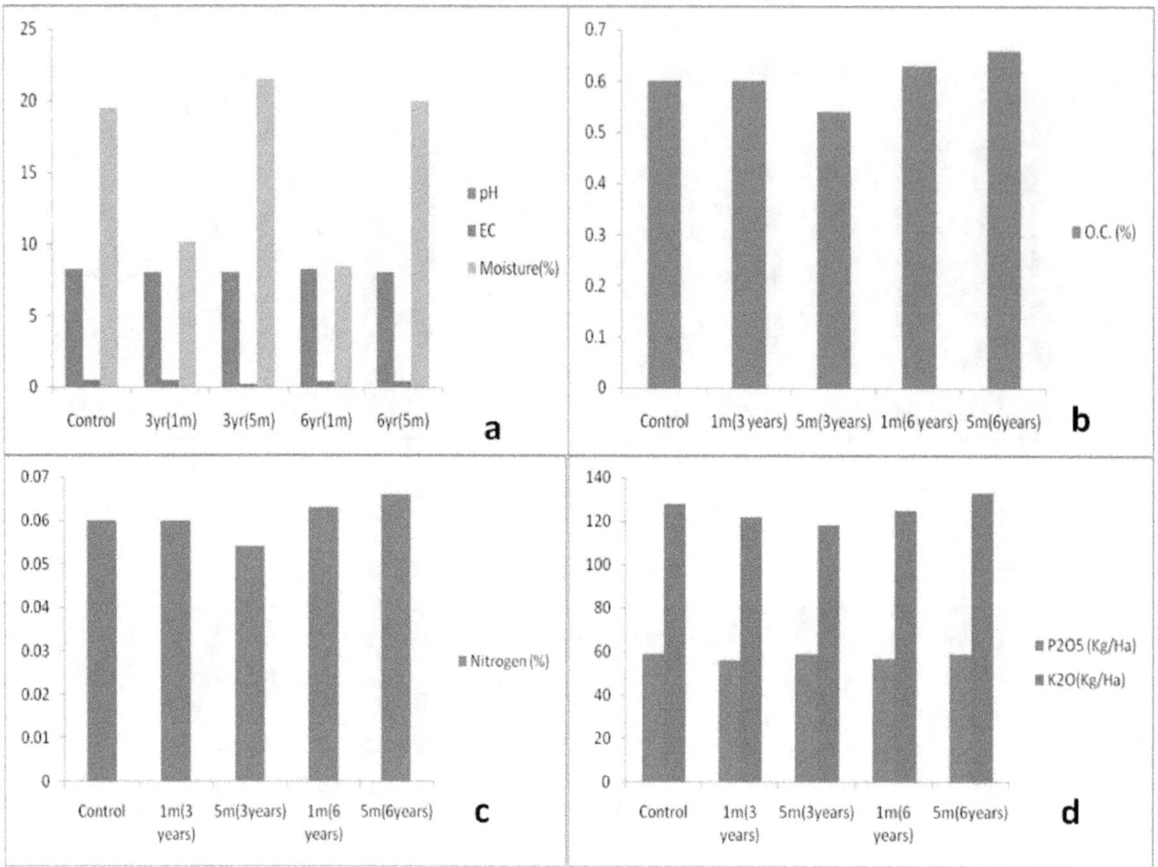

Aonla based agroforestry was most preferred agroforestry. Utilization of its interspace with traditional cropping system was profitable in Gangetic region of Uttar Pradesh. The fact was also supported by the report of Pant (2007). Contribution of agro forestry in rural economy was discussed by the author and stated that increasing population has put higher demands on the forest produce which brings the reserve forests under increased pressure and so agroforestry is the only viable alternative. Author gave socioeconomic status, existing and preferred agroforestry models and choice of species for trees in four villages of Allahabad District, Uttar Pradesh, India. Motivation is the only factor to increase the area under agro forestry in these villages. Among all agroforestry models, the maximum number of rural people preferred to adopt silvohorticultural system because of short term gains by horticulture prospects. The images of Aonla Orchard based Agroforestry with wheat and paddy are depicted in Figure 3and 4.

In the present study, crop production was reported less under agroforestry. Competition for water and nutrients is often reported. It is reported that there is an increase of growth and mineral nutrition of the grass stratum due to tree shade. Shade enhances the availability of nutrients in the soil, and changes microclimatic conditions favouring the biological activity of the soil. Proper management can be practiced to minimize competition for nutrients and moisture in agroforestry systems for better utilization of belowground resources. Crop production also depends on the site characteristics (Dubey, 2010).

In present study, it was observed that the agricultural crop production of both wheat and paddy under aonla orchard was negatively affected and production was reduced in comparison to sole crop. It was in contrary to the findings of Awasthi *et. al.,* (2009) who studied the performance of intercrops during establishment phase of aonla (*Emblica officinalis*) orchard and reported a higher grain and straw yield in mothbean-chickpea (497, 1250 kg/ha) and mothbean-fenugreek (465, 1161 kg/ha) crop sequence under aonla orchard. Amongst the winter (*rabi*) crops, grain yield of fenugreek, chickpea, mustard and cumin were higher by 28.05, 38.11, 19.96 and 36.50%, respectively, when grown in association with aonla compared to its sole crops. It might be due to aonla grop was in establishment phase and its growth was not enough to affect the crop production. In the present study,the crop production was increased with increase in the distance from the tree row plantation. The similar findings were also reported by Nekar *et. al.,* (2007).

Figure 3. (a and b) Aonla Orchard Agroforestry with Paddy (c and d) Aonla Orchard Agroforestry with wheat

Figure 4. Aonla Orchard Agroforestry after harvesting of Paddy and Aonla tree with fruits

CONCLUSION

Over utilization of natural resources is a major challenge for sustainable production and livelihood security. Agroforestry with components like timber trees, fruit trees, agricultural crops, grasses, livestock, etc. provides all kind of life supports. Agroforestry has great future role for ensuring environmental sustainability and meeting sociological/economic objectives for achieving sustainable development goals in farming system (Srinidhi et. al. 2007). Adoption of horticulture-based production systems can improve the socio-economic conditions of resource deficient farming community. Many underutilized fruit species play an important role in the social economy and livelihoods of tribal, small, marginal and landless farmers. Produce of trees provides additional income to these farmers and substantial livelihood support in addition to the nutritional security to the children and women (Malik et al., 2013). In present scenario of increasing population (human and animal), unemployment, malnutrition, shrinking land and deterioration of soil fertility, changing climate, degradation of environment and threatening food security, the fruit-based agroforestry systems offers great potential to fight against these problems. Notable improvement of total nitrogen and organic matter were obtained in the soil under agroforestry system. Therefore, fruit-based agroforestry system can be a promising alternative to provide higher economic benefits, boosting food and nutritional security, improving soil nutrient status and utilization of land (Das *et. al.,* 2020). In agroforestry most of the research works have been done on timber species like Teak, Poplar, Eucalyptus etc, but works have not been reported on orchard based agroforestry. In the present study it was found that Aonla may be used a suitable fruit species for orchard based agroforestry. Moreover due to its hardy and salt tolerant nature, it is suitable for reclamation of salt affected wasteland which covers about an area of 7.0 million hectares in India and other degraded and unproductive lands. It has been proven a very good source of income under rainfed conditions (Newaj and Rai, 2005). Therefore the establishment of Aonla orchard may address three aims simultaneously, Climate Resilient Agriculture, Degraded land Reclamation and poverty alleviation of rural people.

ACKNOWLEDGMENT

The authors are indebted to and duly acknowledge ICFRE, Dehradun for enprovisioning financial support for the study.

REFERENCES

Awasthi, O. P., Singh, I. S., & More, T. A. (2009). Performance of intercrops during establishment phase of aonla (*Emblica officinalis*) orchard. *Indian Journal of Agricultural Sciences*, *79*(8), 587–591.

Das, A. K., Rahman, M. A., Saha, S. R., Sarmin, N. S., Hoque, M. A., & Bhuiyan, F. (2020). Transforming Malta orchard into agroforestry system with different crops for improving productivity, profitability and land uses. *Ann. Bangladesh Agric., 24*(1), 113-125. www.doi.org/10.3329/aba.v24i1.51940

Depommier, D. (2003). The tree behind the forest: Ecological and economic importance of traditional agroforestry systems and multiple uses of trees in India. *Tropical Ecology*, *44*(1), 63–71.

Dubey, K. (2010). *Development of Agro-forestry models for Eastern Uttar Pradesh. Project Report*. ICFRE.

Khosla, P. K., & Khurana, D. K. (1987). *Agroforestry for Rural Need* (Vol. 1). Indian Soc. of Tree Sci.

Kumar, S., Prasad, R., Anil Kumar, A., & Dhyani, S. K. (2019). Integration of fruit trees in agroforestry for sustainability and profitability of farming systems in arid and semi-arid regions. *Indian J. of Agroforestry*, *21*(1), 95–99.

Malik, S. K., Bhandari, D. C., Kumar, S., & Dhariwal, O. P. (2013). Conservation of multipurpose tree species to ensure ecosystem sustainability and farmers livelihood in Indian arid zone. In S. Nautiyal, K. S. Rao, K. Herald, K. V. Raju, & S. Ruediger (Eds.), *Knowledge Systems of Societies for Adaptation and Mitigation of Impacts of Climate Change. Environmental Science and Engineering* (pp. 257–270). Springer. doi:10.1007/978-3-642-36143-2_16

Mutanal, S. M., Patil, S. J., Shahapurmath, G., & Maheswarappa, V. (2009). Performance of arable crops in a teak based agroforestry system. *Karnataka Journal of Agricultural Sciences*, *22*(4), 854–856.

Negi, R. S., Baghel, B. S., & Gautam, U. S. (2013). Effect of method of orchard establishment and propagation on growth and development of aonla (*Emblica officinalis* Gaertn.) plants in wastelands. *Journal of Applied Horticulture*, *15*(3), 227–231. doi:10.37855/jah.2013.v15i03.46

Nekar, M. M., Goroji, P. T., & Channakeshava, S. (2007). Growth and yield of greengram (Vigna radiata) near teak tree row. *Journal of Ecobiology*, *21*(3), 289–294.

Newaj, R., & Rai, P. (2005). Aonla-based agroforestry system: A source of higher income under rainfed conditions. *Indian Farming*, *55*(9), 24–27.

Orwa, C., Mutua, A., Kindt, R., Jamnadass, R., & Simons, A. (2009). *Agroforestree Database: A tree reference and selection guide version 4.0.* https://www.worldagroforestry.org/af/treedb/

Pandey, D. N. (2002). Carbon sequestration in agroforestry systems. *Climate Policy*, *2*(4), 367–377. doi:10.3763/cpol.2002.0240

Pandey, D. N. (2007). Multifunctional agroforestry systems in India. *Current Science, VOL.*, *92*(4), 455–463.

Pant, N. (2007). Contribution of agro forestry in rural economy - a case study. *Vegetos*, *20*(1), 89–93.

Pathak, R. K. (2003). *Status Report on Genetic Resources of Indian Gooseberry-Aonla (Emblica officinalis Gaertn.) in South and Southeast Asia. IPGRI Office for South Asia National Agriculture Science Centre (NASC)*. DPS Marg, Pusa Campus.

Rai, J., Singh, S. P., & Singh, A. K. (2016). Economics of marketing and processing of aonla in district Pratapgarh, Uttar Pradesh. *International Journal of Commerce and Business Management*, *9*(2), 209–213. doi:10.15740/HAS/IJCBM/9.2/209-213

Singh, A. K., Singh, S., Saroj, P. L., Mishra, D. S., Singh, P. P., & Singh, R. K. (2019). Aonla (Emblica officinalis) in India: A review of its improvement, production and diversified uses. *Indian Journal of Agricultural Sciences*, *89*(11), 1773–1781.

Srinidhi, H. V., Chauhan, S. K., & Sharma-S, C. (2007). SWOT analysis of Indian agroforestry. *Indian-Journal-of-Agroforestry*, *9*(1), 1–11.

Wali, V. K., Bakshi, P., Jasrotia, A., Bhushan, B., & Bakshi, M. (2015). *Aonla*. SKUAST-Jammu.

Chapter 11
Good Agricultural Practices for Date Palms (*Phoenix dactylifera L.*)

Shrief Fathy Sharabasy
https://orcid.org/0000-0002-6616-1793
Agriculture Research Center, Egypt

Hesham Sayed Ghazzawy
Date Palm Research Center of Excellence, King Faisal University, Saudi Arabia & Central Laboratory for Date Palm Research and Development, Agriculture Research Center, Giza, Egypt

ABSTRACT

Date palm (Phoenix dactylifera L.), commonly grown in the hot arid zones predominantly in the Middle East and North Africa, provides highly healthy fruits in addition to multiple uses for all parts of the tree. With the interest of many countries of the world in the direction towards planting palm trees, it was therefore significant to point out the importance of agricultural practices that are concerned with serving the palm crown, such as pollination, pruning, fruit thinning, bunch covering, and bagging fruits. The paramount importance of soil service and irrigation operations that will obtain the highest productivity of trees and achieve remunerative profits for the stakeholders and those interested in the palm sector from breeders and factory owners are shown. The latest technologies and modern programs that serve this essential agricultural sector were reviewed.

INTRODUCTION

The date palm (*Phoenix dactylifera* L.) is considered as one of the world's oldest cultivated fruit trees. The Arecaceae or Palmae family includes the genus Phoenix. It consists of 14 species, the most well-known of which is the widely cultivated date palm, *Phoenix dactylifera* (Figure 1a, WCSP, 2013). It has a value of great importance, which makes it one of the most critical major economic crops, as it has many uses that vary between food and industry, as the industrial processes based on this crop vary,

DOI: 10.4018/978-1-6684-2423-0.ch011

starting from the exploitation of the fruits in various manufacturing processes, whether in the form of date molasses or jam, and through the exploitation of the seeds are used to make coffee, as well as the date burger, dates of religion, date ketchup, in addition to pickled dates from the fruits of the baser, and we also use the palm stalk in the work of the ceiling, and cellulose is extracted from the leaves, and recently bioplastics can be produced from the biomass unit with it, there are several advantages for the different parts of date palm tree in industrial products, trunk can be used for furniture, leaves have particular importance in making shades and roofs, and the fibers of trunk and leaves used for manufacturing of mats, papers, bags, baskets, ropes, cords (El Hadrami & Al-Khayri, 2012). In addition to it. To the possibility of producing charcoal and activated charcoal for various uses in laboratory purposes, in food farms, and in medical purposes (Al-Mssallem et al., 2019; Ashraf & Hamidi-Esfahani, 2011; El Hadrami & Al-Khayri, 2012; Gantait et al., 2018; Kulkarni et al., 2010; Tang et al., 2013). It is also used for ornamental purposes and ornamental works by using papers to make bouquets and some religious occasions in some non-Islamic countries; among these species especially *P.canariensis*, *P.roebelenii*,

*Figure 1. Distribution of the Phoenix species. Based on; (Map by (*Carreño et al., 2007*;* Gros-Balthazard, 2013*;* Munier, 1973*;* Pintaud et al., 2010*)*

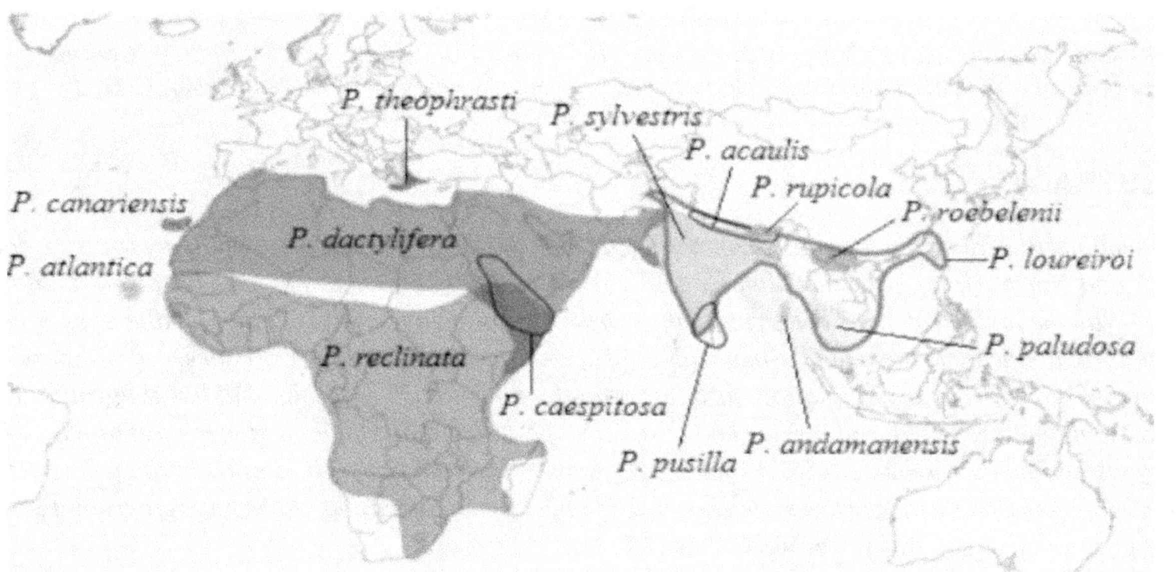

P. canariensisor and *P. theophrasti*.

Botany and Taxonomy

Phoenix dactylifera L. is a date palm, as it is known throughout the world. All palm trees grow in tropical and subtropical climates. Date palm is a perennial tree that is diploid (2n = 36). Date palm trees are classified as Angiosperms, monocotyledonous, Arecaceae in the plant Kingdome, including 200 genera. The date palm genus is known as the Phoenix genus, and dactylifera is the date palm species among more than 1500 species that are related. The majority of the Phoenix genus's relatives are considered ornamental trees (Chao & Krueger, 2007; El Hadrami & Al-Khayri, 2012). The suggested native zones are Mesopotamia (Southern Iraq), Egypt, and Western India.

Since antiquity, date palm trees have been suggested to be native to the Mesopotamia zone (Southern Iraq), Egypt, and Western India (Gantait et al., 2018; Tengberg, 2012).

Date palm cultivation areas are primarily found in the Arabian Peninsula, the Middle East, and North Africa deserts. Date palm cultivation has spread to other parts of the world via germplasm exchange, including Southern Africa, the United States, Mexico, South America, and Australia (Al-Alawi et al., 2017; Jain, 2012; Johnson, 2011).

The botanical discretions of the date palm tree revealed unique characteristics, owing to its status as a monocotyledon plant. The trunk of the date palm tree is vertical and lignified, with no ramification (lacking cambium). It has a grass-like appearance and is quite adaptable to strong desert winds. It is brown, has a cylindric shape, and can grow to a height of up to 30 m when mature. The trunk of the date palm is covered with the bases of all the dried-up old leaf fronds (rachis base) and carries the upper crown, which ranges in length from 6 to 10 m and has a moderately dense 60 – 80 of green leaves about 3 – 6 m in length.

The leaves are 4 - 5 meters long, pinnate, and spiral upwards when they grow. The midrib (rachis) of a date palm frond contains approximately 150 stiff leaflets (2 cm wide and 30 cm long), modified leaflets identified as sharp spines, located at 10 –15 cm on each side of the midrib base. Every year, approximately twelve new leaves emerge. Every leaf has an auxiliary bud, which can be vegetative, floral, or intermediate. The date palm root system has been greatly improved to reach deeper water resources, up to 25 m from the main tree, at a depth greater than 6 m. Histologically, the root system is monocotyledon, fibrous in nature, lacks cambium, a primary root that emerges directly from its monocotyledonous; secondary roots appear on it, which in turn produces adventitious roots of comparable size. Because of the presence of pneumatics in date palm roots, they were thought to be respiratory organs.

The date palm is a dioecious plant, meaning that the (male) and (female) flowers are carried on separate trees. The spathe is a rigid fibrous cover that surrounds the inflorescence strands and protects petite male or female flowers during the early stages of flower development. Male spathes are shorter and broader than female spathes. Female flowers are typically yellowish-green in color, whereas male flowers are waxy-white in appearance. In general, both male and female flowers have three sepals and three petals. Date palm fruit develops from the female flowers. The fruits are ovoid to oblong in shape, with a smooth thin fruit skin (pericarp) and a fleshy mesocarp that contains only a single elongate seed and is covered by a fibrous parchment (endocarp)(Abahmane, 2011; Bekheet, 2013; El Hadrami & Al-Khayri, 2012; Gantait et al., 2018; Zaid & De Wet, 1999).

Importance and Utilization

The date palm tree has become one of the essential cultivated palms in the world, in addition to playing a significant role in the economy of many Middle Eastern and North African countries (Al-Alawi et al., 2017; El Hadrami & Al-Khayri, 2012; El Hadrami & El Hadrami, 2009; Jain, 2012).

The date palm is indeed known as the "tree of life" because of its importance in human settlement, prosperity, and food security in hot and arid regions of the world. The date palm tree has a wide range of uses, not only for its sugary fruits and high nutritional value but also for the entire tree's parts. Date fruits have a high nutritional value due to several essential minerals such as calcium, iron, magnesium, copper, potassium, phosphorus, silicon, sulfur, chlorine, and low sodium content. In addition, date fruit contains significant amounts of vitamins such as riboflavin, thiamine, folic acid, biotin, and ascorbic acid.

Carbohydrates are the primary component, accounting for 65 – 80 percent of the sugar content in date flesh, while fiber (2.5%), protein (2%), and fat (2%) are present in minor amounts (Al-Harrasi et al., 2014).

Moreover, the researchers evaluated the previously obtained results. Through the above, we find that the science of bioeconomic has recently been interested in increasing and activating the use of science and highlighting the role of modern industrial technologies in contributing to the production of many new and renewable biological resources and then reusing them in the production of many outputs, including food products and feed products used. In animal feed and others to produce vital energy, which increases the added value of this essential strategic crop. Researchers have stimulated finding new ways to exploit the waste of dates, which exceeds two million wastes annually at the global level, to find new industries in which these quantities are exploited, especially that they are environmentally friendly and do not affect human heal.

Among the outputs obtained is the production of amino acids from the products of microbial fermentation, the production of biofuels and enzymes (Chauhan, 2007; Abd-Alla, 2012 and Alsafadi, *et al.,* 2020).

There are also scientific revolutions represented in studying ways of re-exploiting palm tree residues from several angles, such as the extraction and production of cellulose materials from biomass and the production of bioplastics that are used in many industries such as environmentally friendly plastic bags from palm oil stems (Wellenreuther, 2020). Production of some materials that can be added to worn fabrics for reuse and the work of covering seats and leather jackets.

DATES NUTRITIONAL AND HEALTH BENEFITS

It is well known that the date palm is known as the "tree of life" because of its importance in human stability, prosperity, and food security in all hot and dry areas around the world. Furthermore, the food of backpackers in the desert indicates the high nutritional value of the fruits of dates and their health benefits.

Studies over the past years have shown that date fruits have a high nutritional value, as they contain many essential minerals needed to build cells, such as calcium, magnesium, silicon, copper, potassium, iron, phosphorous, sulfur, and low content of sodium and chlorine. It also contained large amounts of vitamins such as thiamine, biotin, ascorbic acid, folic acid, and riboflavin. By measuring the proportions of the components, it was found that the proportion of the carbohydrate component is the most abundant component, as it represents 65 - 80% of the sugar content in the flesh of the date fruits, while the proportions of the presence of fiber (2.5%), protein (2%) and fat (2%) were estimated present in small quantities, and fiber (Al-Hooti *et al.*, 1997; Myhara *et al.*, 1999 and Al-Harrasi *et al.* 2014). Thus, functional foods provide health benefits and essential nutrition (IFICF, 1998; Al-Farsi and Lee, 2008).

From a medical point of view, many epidemiological studies conducted by the medical sector have shown that there are many clear positive correlations between consumption rates of fruits and vegetables, and their association with lower death rates from heart disease and various common types of cancer, and the presence of many antioxidants and dietary fiber. ., and other vital compounds found in the eaten fruit reinforced the importance of eating the fruit. Prophetic in the Islamic religion the safety of that fact (Joseph *et al.*, 1999; Dillard and German, 2000; Prior and Cao, 2000; Wargovich, 2000; Al-Farsi and Lee, 2008). The date fruits are mainly composed of many substances such as carbohydrates, soluble sugars, and dietary fiber, with low proteins and fats. Recently, Dates also contained many important nutrient groups with wide effects, for example, antioxidants and immunostimulants, and there are many studies published on the possibility of some extracts that reduce the effect of myocardial infarction from some varieties of dates such as Khalas and Riziz cultivars (Baliga *et al.*, 2011).

The fruits of dates pass during their physiological growth stages through four stages of development and formation, divided according to physiological age, starting from the first chimeric stage, followed by the Khalal stage, then the wet and dates (Sawaya *et al.*, 1982&1983; Mustafa *et al.*, 1986; Siddiqui and Gupta, 1994; El-Zoghbi,1994; Al-Hooti *et al.*, 1997; Myhara *et al.*, 1999; Al-Shahib and Marshall, 2003; Al-Farsi and Lee, 2008). In the Kumari stage, we find a rapid increase in the size and weight of the fruits, and the moisture content can reach approximately 85 percent. Then the fruits begin to turn yellow or red depending on the variety by the end of this stage. The percentage of weight gain is slow during the khalal stage, and the conversion of sucrose into glucose and fructose begins, the moisture content decreases, and the tannins begin to precipitate and lose their viability and dissolution fruit palatable in the khalal stage. At the beginning of entering the wet stage, at that time, the ends of the fruit begin to ripen to turn brown, which is characterized by a decrease in the weight of the fruit due to the loss of moisture and reaches about 35%, and the transformation of sucrose to varying degrees, and some varieties may be sold as fresh fruit at this stage, and others are left to be allowed to ripen More on the palm or dry in the sun, they reach the stage of tamarind {(Barreveld, 1993)

Nutritional Composition

Dates are a storehouse of nutrients, and the date palm provides many nutritional, environmental, economic, and aesthetic benefits (Al-Farsi and Lee, 2008; Biglari, 2009; Hussain *et al.*, 2020).

Bioactive Compounds

Previous studies have shown that date palm fruits contain a wide variety of biologically active substances (Al-Mssallem *et al.*, 2020; Tang *et al.*, 2013; Al-Farsi and Lee, 2008), including secondary metabolites. The date fruits of *Phoenix. dactylifera* have been shown to contain biologically active components such as carotenoids, flavonoids, and phenolic acids (Al-Turki *et al.*, 2010; Echegaray *et al.*, 2020; Al-Farsi *et al.*, 2005; Hussain *et al.*, 2020). The most abundant bioactive constituent detected in the fruits of *Phoenix. dactylifera* was phenolic acids, which are believed to be the main contributor to antioxidant activity. During the different maturation stages, a significant difference in the number of bioactive components was detected in different cultivars of *Phoenix. dactylifera*. Carotenoids and phenolic acids are components that are of critical importance in avoiding oxygen stress in the human body (El Hadrami *et al.*, 2012). Health studies are currently showing the growing interest in the usefulness of bioactive materials through human cellular, animal, and clinical research. Recently, these fruit by-products of P. dactylifera have been used and littered to produce industrial ingredients such as biopolymers, antibiotics, and organic acids (Al-Mssallem *et al*, 2020; Chandrasekaran and Bahkali, 2013).

DATE PALM PROPAGATION

Seeds propagate the date palm as a sexual propagation method, an old method used in the past. However, there are many limitations to this method because of its apparent drawbacks. For example, the percentage of male plants is higher than the female plants, which may appear after a long period of seed propagation, and they are usually slow-growing. Therefore, asexual propagation methods were used, such as propagation with offshoots that grow next to the mothers from trees, which is a suitable method. However, it is faulted by the small number of offshoots obtained from trees. Moreover, in this regard, all research and studies have adopted the trend towards finding solutions that increase multiplying plants. The science of biotechnology has adopted this trend in multiplication, as tissue culture has shown a new value to increase the number of plants produced through the use of this modern technology, which is a method Good for obtaining large numbers of plants without being restricted to a specific growing season and in controlled conditions (Ghazzawy 2013)

Seed Propagation

One of the methods used sexually in the propagation of date palms is usually done from seeds, and it is called (sexual reproduction), which is an old method of propagation. The dates of seedlings show characters inherited from both parents. The resulting plant, but the determination of the sex appears only at the beginning of flowering at the age of 6 -7 years, although beneficial for breeding reasons, is not a suitable technique of date palm vegetative proliferation and should be discouraged for some reasons like Female seedlings frequently yield varied. Generally, inferior late-ripening fruits in comparison with the clonal palms developed. It is unusual for seedlings to produce Fruit of a good grade more than 10 percent of palms; Date palms are heterozygous so that the progeny will vary a great deal, and the desirable qualities of the parent palm may vary be lost. In other words, the propagation of the seedlings is not natural, and there are no two palms; Seedlings fluctuate significantly from one harvest to another in terms of production capacity, quality of fruits, and harvest times, making marketing challenging; It results in waste of time, space and money for the reasons mentioned above (Zaid 2002; Johnson 2011).

Offshoot Propagation

The following advantages are provided by offshoot propagations, often known as asexual or vegetative propagation: The process of asexual propagation (vegetative propagation) is the second method of propagation and it is carried out through the development of the offshoots from the axillary buds located at the base of the stem of the mother plant, and scattered around the trunk in all directions. The same tissue as the original mother plant. Fruiting processes usually start from 2 to 3 years. The lifespan of a palm is divided into two distinct developmental stages: vegetative, in which the buds that form the leaf veins grow into offshoots. From the time the axillary bud of the leaf differentiates into a branch, it takes up to three years (at 18-36 months) to reach the ideal division size and planting for three or four years (Hilgeman, 1954).

Offshoots are generally generated, at the earliest (20 to 30 years from the date of planting), according to the variety and to pre-fertilisation, irrigation and earthing, in a limited number (20 to 30 years from the date of planting) (Nixon and Carpenter, 1978). Although a palm produces 20 or 30 offshoots, only 3 or 4 offshooting offshoots may be planted within 1 year and must be kept for 1 or 2 years before planting in the nursery. The varieties Zahidi, Berim and Hayani generate high number of offshoots whereas the variants Mectoum and Barhee produce very less offshoots, Offshoots have their curved shape whereas the seedlings have a straight shape. Another method of distinguishing between them is that plantings have roots all around the base and no point connecting the palm, while an offshoot has no roots on the side where it is linked to the mother plant. In addition, on one side, a branch is always marked by its parent palm separation.

Tissue Culture Propagation

Over the past half century, scientific efforts from all over the world have combined to develop agricultural production by devising and developing new methods of reproduction that contribute to increasing the number of plants, and consequently increasing and increasing the quantities of produced dates, and producing seedlings resistant to pests and diseases. Through the science of biotechnology, the so-called tissue culture has appeared, which contributes to an increase in numbers in a short time without being restricted to a specific agricultural season. (Al-Khayri 2005, 2007; Zaid 2002 a) reviewed the areas of tissue propagation in palm trees have been before.

Propagation of healthy chosen female (disease and pestilence-free), bayoud-based, resistant or superior pollen men with beneficial metaxenic properties that can be replicated quickly and simply; Multiplication on a large scale; No seasonal influence on plants since it can be reproduced year-round in the laboratory under controlled circumstances; Genetically homogeneous plant production; Clons to be reproduced from existing elite cultivars or from F1 hybrids and palms originating exclusively from seed; Ensure a simple and quick plant material interchange, without the danger of spreading diseases and plagues, across various areas of one nation or between countries. Only very few herbotic species have had the success of growing monocotyledones in *in vitro*. Likewise, the majority of dicotyledons, were cultivated successfully, were herbaceous species. The capacity to regenerate plants utilizing tissue culture techniques has been suggested to be lesser compared to herbaceous plants in woody plants. In palms, limited success in induction and maintenance of excellent skalks was obtained till 20 years ago. Techniques of plant tissue culture were used to clone a variety of plants including commercially significant palms like, for example, coconut, oil and date palms (Al-Khayri 2005, 2007; Zaid 2002b).

PALM CROWN SERVICE OPERATIONS

The agricultural practices that are carried out on date palm trees are usually divided according to the time of the year, the productive age, the place of the operation, and accordingly they are divided according to the place of the procedure into practices above the soil surface, and practices under the soil. Accordingly, those practices above the soil surface have been divided into Several practices, including pruning, and it is usually performed on specific dates that will be dealt with in detail in terms of when it will be performed and the purpose of it. Also, the vaccination process is carried out. The methods and dates of its procedure vary according to the nature of the different varieties, and the date of the exit of the female puppies, and the process of bagging the fertilized taste also comes after the vaccination process In order to avoid losing the fruit, the process of slipping the fruit is also sometimes carried out, the methods of which vary according to the purpose of it, where it is in the way of removing certain fruits in order to raise the marketing value in some cases, and it may take place in other ways that will be reviewed later in the chapter, and finally a procedure Field harvesting practices and following the correct methods for that. The date of harvest varies according to several factors, including the nature of the variety, the region and the nature of its demography.

Pollination

Pollination is an important process in the date palm production chain because it influences the availability of raw materials, fruit development, quality, and yield. Commercial date palm production requires cross-pollination and external pollination to produce adequate fruiting because it is a dichotomous plant, Furthermore, the male parent employed has a significant influence on date palm characteristics, including the flavor and scent of the fruit.

Metaxenia is the direct influence of pollen on the physical and chemical features of fruit, affecting ripening time and color (El-Ghayaty, 1983; Abdelal *et al.*1983 and Bekheet & El-Sharabasy, 2015). As a result, choosing the right male parents to act as pollinizers is critical for increasing the quantity and quality of dates. When Egyptian date palm growers recognized the effect of pollen grains from particular male cultivars on the size of fruits and seeds (xenia) as well as the period of fruit ripening (metaxenia) of female cultivars, they realized the relevance of artificial pollination. Male trees are chosen and shared between farmers, even across large distances. It's worth noting that the pollen production of male palms varies substantially from one place to the next. Male flower clusters are clipped in the morning after the spathe splits open in Egypt to prevent pollen loss due to wind or bees; clusters are then dried in heated rooms before being separated from the spadix. Delaying pollination after female spathe splitting lowered the percentage of fruit sets, according to Shaheen (1986). He also stated that pollination within 3 or 4 days after the female spathe opening produced the most set fruit for most Egyptian date palm cultivars. Pollination is done by hand in Egypt, with pollen picked from males blooming at the same time as females in order to supply fresh pollen for fruit set. Male flower strands are clipped from

In addition, artificial insemination can be done using traditional procedures or mechanical equipment. Cutting strands of male flowers from a newly opened area and laying 5-10 of these strands lengthwise and upside-down is the most popular pollination procedure (Zaid and de Wet 2002). Furthermore, this pollination approach necessitates the availability of a high number of male spathes, which is not always the case, especially for cultivars that flower early in the season, Outside of the embryo and endosperm, pollen type has a direct effect on some fruit traits as reported by Ben Salah and Hellali (1998). In modern plantations, the male/female ratio is roughly 1/50. (2 percent). Additionally, traditional pollination is a time-consuming procedure with high labor costs.

Pollen grain-water suspension spray is an alternate technology that is predicted to save labor and costs in fruit tree cultivation. As a result, various attempts to build spray pollination methods for a variety of fruit kinds have been made. A 10 percent sucrose solution and a wetting agent were utilized in spray pollination of peaches.

Pruning

One of the most misunderstood components of palm culture is pruning. Pruning a plant or tree correctly can help the plant's development and health. Palm trees are low-maintenance trees if they are properly cared for. For some reason, some people feel they may hack at palms indiscriminately, including removing most or nearly all of the fronds (leaves) several times a year, without harming the tree. Palms aren't exempt from basic trimming practices. Palms, like any other plant or tree, will be harmed by improper trimming procedures. A common fallacy is that the more a palm is clipped, the faster it grows. This isn't correct. Many palm experts advise against over-pruning unless absolutely necessary. The food required for optimum growth is produced by all green fronds, resulting in a healthy palm. Food production is reduced as the green leaf area is reduced, putting the palm's health and growth in peril. The more green leaves there are, the better. The more growth a plant has, the more growth it will produce. Plant food is the sole actual source of nutrition for plants. anything which is produced by the plant What is sold in stores is not a plant, tree, or shrub. Even though the package reads "plant food," it's palm food. It is a type of fertilizer (nutrients) that is used in agriculture. Water and sunlight are used by plants to create a plant, tree, or shrub. cuisine made from palms (Pfalgraz, 2002). Palm tree (Phoenix dactylifera L.) leaf life is declining, with the least activity at 5-6 years old, and about 15–25 years old leaves should be manually clipped based on the leaf pruning line once a year (Chao and Krueger, 2007; Alikhani& Aghdam, 2021). The 5 and 6-year-old leaf lines correspond to leaves that are 5 and 6 years old, respectively, and the 5 and 6-year-old leaf pruning lines correspond to the pruning of leaves that are older than 5 and 6 years old. To produce with suitable quantity and quality, date leaves are clipped from a distance of 30–40 cm above the leaf sheath to preserve the leaf to bunch ratio. Pollination, changing the leaf to bunch ratio within 6–8 weeks following pollination, and after harvest as the priority time are all times when leaf pruning can be done (Al-Khayri, 2005; Alikhani-Koupaei et al., 2020). The intensity and precision of leaf removal can be used to assess the involvement of varied leafy numbers in providing photosynthetic materials, their transfer into the plant, and flower induction and abscission rates. Flower induction occurs from October to November, and its intensity is influenced by leaf starch reserves and the leaf-to-bunch ratio from July to October (Nixon, 1957; Omar et al., 2013; Sidhu, 2012).

Fruit Thinning

Thinning is a significant managerial activity in date palm to improve fruit size, weight, and quality while reducing the likelihood of bunchbreaking and alternate bearing. Date palm trees were thinned using a variety of techniques, including bunch thinning, bunch strand thinning, and individual fruit removal. The removal of individual fruits and strands in combination significantly increased fruit quality (Dinar., et al 2002; Marashi, & Mousavi, 2006 and Abdel-Hamid, 2000). Flower or fruit thinning is an important cultural activity in the date palm production chain since it impacts fruit development, quality, yield, and tree annual bearing. The thinning process is a natural one. Fruit thinning treatments may reduce total yield, soluble tannins percent, crude fibers percent, and total acidity percent while increasing fruit weight, size, and dimensions, pulp weight percent, total soluble solids percent, and total and sugar contents (Khalifa et al., 1984 and Awad, 2006). When compared to un-thinning, thinning by eliminating 10-30% of bunches increased bunch weight, advanced ripening, and produced the finest fruit quality (Ghazzawy, 2013). and Mostafa & El Akkad, 2011).

Bunch Covering (Bagging)

Bagging date palm spathes throughout blooming and fruiting phases boosted fruit set and yield, as well as expedited ripening and fruit quality. Such The highest fruit weight, flesh weight, total soluble solids, and total sugars % were found in the treatment.and the lowest tannin content (El-Salhy, 2000; Rabah and Kassam, 2003 and Moustafa, 2007). Furthermore, they stated that date palm bunch covers had various advantages in terms of protecting fruits from extreme temperatures. Humidity and rain, as well as damage from insects and viruses, are all threats.

Date palm bunch covering has various advantages and is often used to protect fruits from excessive humidity, rain, birds, and insects in date palm grown areas. During the flowering and fruit setting seasons, bagging spathes of date palm cultivars boosted fruit set and yield, as well as expedited ripening and improved fruit quality. The maximum fruit weight, flesh weight, total soluble solids, total sugars percentage, and lowest tannins percentage were all found in this treatment (El-Salhy, 2000, Rabah and Kassam, 2003. Moustafa, 2007 and Mostafa et al., 20014). Canvas, gauze, or palm fibers, as well as polyethylene bagging with removal at a later growth stage, are all good options.

PALM LAND SERVICE OPERATIONS

Irrigation

According to all research, date palms require a lot of water in order to produce good-quality fruit with a high yield. In Egypt's date palm orchards, a variety of irrigation technologies are used. The effectiveness of these systems is determined by the availability and quality of water, the age of the palm trees, and the soil composition. Furrow and basin irrigation are the contemporary irrigation systems in use. Date palm farms in Egypt have traditionally been irrigated by splitting them into separate basins of 10 m by 10 m. These basins give farmers enough water management to ensure uniform irrigation, even in fields that aren't perfectly level. This irrigation system is usually used in old areas, especially in the south. In Egypt, drip irrigation has recently been deployed on date palm fields. This technology is widely used in Nobaria's freshly reclaimed territory as well as the vacation areas of south Sinai and the Red Sea. The main source of irrigation water in oasis, on the other hand, is naturally flowing springs or water pumped from wells. This form of irrigation is carried out using a customized system of side channels that was created to deal with a lack of water supply. Allam *et al.*, (1973). conducted a variety of field trials in Egypt to assess the water requirements of date palms. They found that dates cultivated in Lower Egypt consume an average of 10,280 m3/ha, while dates grown in Upper Egypt consume 14,880 m3/ha. In addition, increasing the frequency of irrigation improved the size, weight, moisture content, and total soluble solids of the fruits. Irrigation should be done 12 times each year, according to the final recommendation.

Fertilization

Adopting a suitable date palm fertilization program, which includes enough rates, appropriate sources, and efficient application methods and timing, is a key strategy for increasing fruit yield and quality. The amount of fertilizer required by a single tree varies depending on the soil type, depth, date cultivar, and age. Fertilizer trials on various Egyptian date palm varieties have been conducted in this regard in order to increase productivity and improve fruit quality (Attalla et al. 1988; Soliman and Osman 2003; Soliman and Shaaban 2006). In terms of potassium fertilization, it has been found that applying potassium to several Egyptian date palm cultivars promotes growth, production, and fruit quality. El-Hammady et al. (1991) discovered that adding 2 kg potassium sulfate/palm annually resulted in the maximum yields and fruit quality of Siwy cv. dates.

HARVESTING

Fresh dates are normally washed before being divided into different groups based on texture, moisture content, and appearance. Overripe fruits, as well as damaged or undersized mature fruits, are removed. Then, depending on the target market, dates are pitted before to packing. Postharvest treatments, according to research, play a significant influence in improving fruit quality during cold storage and during the marketing period. The fruits should be stored at a temperature of 0 to 4 °C and at a relative humidity of 85–90% for a month or more, depending on the variety (Al-Redhaiman 2005). A recent study was conducted on the Zaghloul, Samany, Amhat, and Siwy cvs. to see how effective it is to utilize hot water and sodium carbonate instead of chemicals to reduce postharvest damage during cold storage (Hafez et al. 2012).

Weather factors, consumer desire, cultivar, processing technology, and picking method all influence the stage of ripeness at which date palm fruit is gathered. Furthermore, because not all female flowers are generated at the same time, the stages of maturity of the dates for the various bunches are staggered. When the bulk of dates are ready, khalal, rutab, and tamar dates are picked as entire bunches in Egypt. Gathering might last 3 to 4 weeks in the same cultivar. In most situations, workers must scale the date palm using a variety of ways in order to reach the fruit brunches. The climber is normally barefoot and uses a rope tied to the wall. In both treated and untreated fruits, cold storage treatment inhibits pathogenic molds for up to 40 days. In all of the date palm cultivars studied, hot water and sodium carbonate treatments are more successful at reducing weight loss (percent) and fruit degradation (percent) than untreated fruits. In Amhat and Siwy cvs, these treatments also prevent fruit degradation for up to 20 days with cold storage. Otherwise, Nazam El-Din and Abd El-Hameed, 2001, conducted research on the storage of Siwy cv. reported that treating dates with sulfate dioxide resulted in the best color, and that dates packaged in plastic containers had an unsatisfactory hue. Plastic, cardboard boxes, and timber crates are among the materials used. Dates are typically packed in open-top cartons weighing 1, 3, or 5 kg (fresh date), 5 or 10 kg (semidry dates), and 10 or 25 kg (large dates) (dry dates). Despite the importance of conventional date palm farming and its extensive cultural areas, field postharvest losses are considerable, and techniques for monitoring product quality and the utilization of date products and by-products need to be improved as a different types of packaging.

CONCLUSION

The time Palm trees are crucial in agriculture, and they play a key role in the national reclamation program. The date palm and its by-products are used on a daily basis, in addition to their nutritious and calorific qualities. For many Egyptians, date palm is now their primary source of income and the foundation of their economy, as the business supports one million families. From Alexandria in the North to Aswan in the South, and from the Red Sea in the east to the New Valley and the Oasis in the West, the date palm tree is grown all over Egypt. Although Egypt has several date palm cultivars, only about 20 are commercially available. Complementary propagation, conservation, genetic improvement, and disease and pest control strategies for the most economically valuable cultivars should be implemented, employing both conventional and sophisticated technologies. In addition, contemporary date palm harvesting, postharvest, and processing equipment should be incorporated into Egypt's date palm business. Farmers and extension personnel should be educated on date palm cultivation and processing through workshops, seminars, and trips in this regard.

REFERENCES

Abahmane, L. (2011). Date palm micropropagation via organogenesis. In *Date palm biotechnology* (pp. 69–90). Springer.

Abd-Alla, M. H., & El-Enany, A. W. E. (2012). Production of acetone-butanol-ethanol from spoilage date palm (Phoenix dactylifera L.) fruits by mixed culture of Clostridium acetobutylicum and Bacillus subtilis. *Biomass and Bioenergy*, *42*, 172–178. doi:10.1016/j.biombioe.2012.03.006

Abdel-Hamid, N. (2000). Effect of time, rate and patterns of thinning, leaf/bunch ratio and male type on" Zaghloul" date yield and quality. *Arab Universities Journal of Agricultural Sciences*, *8*(1), 305–317.

Abdelal, A. F., Mahmoud, H. M., & El-Agamy, S. Z. (1983). The effect of pollen source on fruit characteristics of Zaghloul dates (Phoenix dactylifera, L.). *Assuit J Agr Sci*, *14*, 347–355.

Ahmad, N., Rab, A., & Ahmad, N. (2016). Light-induced biochemical variations in secondary metabolite production and antioxidant activity in callus cultures of Stevia rebaudiana (Bert). *Journal of Photochemistry and Photobiology. B, Biology*, *154*, 51–56. doi:10.1016/j.jphotobiol.2015.11.015 PMID:26688290

Al-Alawi, R. A., Al-Mashiqri, J. H., Al-Nadabi, J. S., Al-Shihi, B. I., & Baqi, Y. (2017). Date palm tree (Phoenix dactylifera L.): Natural products and therapeutic options. *Frontiers in Plant Science*, *8*(845), 1–12. doi:10.3389/fpls.2017.00845 PMID:28588600

Al-Alawi, R. A., Al-Mashiqri, J. H., Al-Nadabi, J. S., Al-Shihi, B. I., & Baqi, Y. (2017). Date palm tree (Phoenix dactylifera L.): Natural products and therapeutic options. *Frontiers in Plant Science*, *8*, 845.

Al-Farsi, M., Alasalvar, C., Morris, A., Baron, M., & Shahidi, F. (2005). Comparison of antioxidant activity, anthocyanins, carotenoids, and phenolics of three native fresh and sun-dried date (Phoenix dactylifera L.) varieties grown in Oman. *Journal of Agricultural and Food Chemistry*, *53*(19), 7592–7599. doi:10.1021/jf050579q PMID:16159191

Al-Farsi, M. A., & Lee, C. Y. (2008). Optimization of phenolics and dietary fibre extraction from date seeds. *Food Chemistry*, *108*(3), 977–985. doi:10.1016/j.foodchem.2007.12.009 PMID:26065761

Al-Harrasi, A., Rehman, N. U., Hussain, J., Khan, A. L., Al-Rawahi, A., Gilani, S. A., Al-Broumi, M., & Ali, L. (2014). Nutritional assessment and antioxidant analysis of 22 date palm (Phoenix dactylifera L.) varieties growing in Sultanate of Oman. *Asian Pacific Journal of Tropical Medicine*, *7*, S591–S598. doi:10.1016/S1995-7645(14)60294-7 PMID:25312188

Al-Harrasi, A., Rehman, N. U., Hussain, J., Khan, A. L., Al-Rawahi, A., Gilani, S. A., ... Ali, L. (2014). Nutritional assessment and antioxidant analysis of 22 date palm (Phoenix dactylifera) varieties growing in Sultanate of Oman. *Asian Pacific Journal of Tropical Medicine*, *7*, S591–S598.

Al-Hooti, S., Sidhu, J. S., & Qabazard, H. (1997). Physicochemical characteristics of five date fruit cultivars grown in the United Arab Emirates. *Plant Foods for Human Nutrition (Dordrecht, Netherlands)*, *50*(2), 101–113. doi:10.1007/BF02436030 PMID:9201745

Al-Khayri, J. M. (2005). Date palm Phoenix dactylifera L. In S. M. Jain & P. K. Gupta (Eds.), *Protocols of somatic embryogenesis in woody plants* (pp. 309–319). Springer, Netherlands. doi:10.1007/1-4020-2985-3_25

Al-Khayri, J. M. (2007). Date palm Phoenix dactylifera L. micropropagation. In S. M. Jain & H. Häggman (Eds.), *Protocols for micropropagation of woody trees and fruits* (pp. 509–526). Springer, Netherlands. doi:10.1007/978-1-4020-6352-7_46

Al-Mssallem, M. Q., Alqurashi, R. M., & Al-Khayri, J. M. (2020). Bioactive compounds of date palm (Phoenix dactylifera L.). *Bioactive Compounds in Underutilized Fruits and Nuts*, 91-105.

Al-Mssallem, M. Q., Elmulthum, N. A., & Elzaki, R. M. (2019). Nutritional security of date palm fruit: An empirical analysis for Al-Ahsa region in Saudi Arabia. *Scientific Journal of King Faisal University*, *20*, 47–54.

Al-Redhaiman, K. N. (2005). Modified atmosphere extends storage period and maintains quality of (Barhi) date fruits. *Acta Horticulturae*, (682), 979–986. doi:10.17660/ActaHortic.2005.682.127

Al-Shahib, W., & Marshall, R. J. (2003). The fruit of the date palm: Its possible use as the best food for the future? *International Journal of Food Sciences and Nutrition*, *54*(4), 247–259. doi:10.1080/09637480120091982 PMID:12850886

Al-Turki, S., Shahba, M. A., & Stushnoff, C. (2010). Diversity of antioxidant properties and phenolic content of date palm (Phoenix dactylifera L.) fruits as affected by cultivar and location. *Journal of Food Agriculture and Environment*, *8*(1), 253–260.

Ali-Dinar, H. M., Alkhateeb, A. A., Al-Abdulhadi, I., Alkhateeb, A., Abugulia, K. A., & Abdulla, G. R. (2002). Bunch thinning improves yield and fruit quality of date palm (Phoenix dactylifera L.). Egypt. *J. Applied Sci*, *17*(11), 228–238.

Alikhani-Koupaei, M., & Aghdam, M. S. (2021). Defining date palm leaf pruning line in bearing status by tracking physiological markers and expression of senescence-related genes. *Plant Physiology and Biochemistry*, *167*, 550–560. doi:10.1016/j.plaphy.2021.08.035 PMID:34454314

Alikhani-Koupaei, M., Aghdam, M. S., & Faghih, S. (2020). Physiological aspects of date palm loading and alternate bearing under regulated deficit irrigation compared to cutting back of bunch. *Agricultural Water Management*, *232*, 106035. doi:10.1016/j.agwat.2020.106035

Allam, H. M., Hussein, F., & Abdalla, K. M. (1973). Seasonal requirements of water by Sakkoti dates grown at Asswan. *Moshtohor Fac Res Bull*, *39*, 1–10.

Ashraf, Z., & Hamidi-Esfahani, Z. (2011). Date and date processing: A review. *Food Reviews International*, *27*(2), 101–133.

Attalla, A. M., Haggag, M. N., Attia, M. M., & Ibrahim, A. M. F. (1988). Fruit and leaf mineral content of four date palm varieties grown in Alexandria. *J Agr Res Tanta Univ*, *14*, 319–331.

Awad, M. A. (2006, May). Accelerating ripening of date palm fruit (Phoenix dactylifera L.) cv." Helaly" by some pre and post-harvest treatments. In *International Conference on Date Palm Production & Processing Technology* (*Vol. 13*). Academic Press.

Awad, M. A. (2010). Pollination of date palm (Phoenix dactylifera L.) cv. Khenazy by pollen grain-water suspension spray. *Journal of Food Agriculture and Environment*, *8*, 313–317.

Baliga, M. S., Baliga, B. R. V., Kandathil, S. M., Bhat, H. P., & Vayalil, P. K. (2011). A review of the chemistry and pharmacology of the date fruits (Phoenix dactylifera L.). *Food Research International*, *44*(7), 1812–1822. doi:10.1016/j.foodres.2010.07.004

Bekheet, S. (2013). Direct organogenesis of date palm (Phoenix dactylifera L.) for propagation of true-to-type plants. *Scientia Agrícola*, *4*(3), 85–92.

Bekheet, S. A., & El-Sharabasy, S. F. (2015). Date palm status and perspective in Egypt. In *Date palm genetic resources and utilization* (pp. 75–123). Springer. doi:10.1007/978-94-017-9694-1_3

Ben Salah, M., & Hellali, R. (1998). Metaxenic effects of nine pollinators on three palm date varieties (Phoenix dactylifera, L.) growing in Tunisia coastal oasis. In *First international conference on date palms*. United Arab Emirates Univ.

Biglari, F. (2009). *Assessment of antioxidant potential of date (phoenix dactylifera) fruits from iran, effect of cold storage and addition to minced chicken meat* (Unpublished Master thesis). University Sains Malaysia.

Carreño, E., Inocencio, C., Alcaraz, F., Ríos, S., Palazón, J., Vázquez, L., . . . Laguna, E. (2007). Morphological systematics of date-palm diversity (Phoenix, Arecaceae) in Western Europe and some preliminary molecular results. *V International Symposium on the Taxonomy of Cultivated Plants, 799.*

Chandrasekaran, M., & Bahkali, A. H. (2013). Valorization of date palm (Phoenix dactylifera) fruit processing by-products and wastes using bioprocess technology–Review. *Saudi Journal of Biological Sciences*, *20*(2), 105–120. doi:10.1016/j.sjbs.2012.12.004 PMID:23961227

Chandrasekaran, M., & Bahkali, A. H. (2013). Valorization of date palm (Phoenix dactylifera) fruit processing by-products and wastes using bioprocess technology–Review. *Saudi Journal of Biological Sciences*, *20*(2), 105–120. doi:10.1016/j.sjbs.2012.12.004 PMID:23961227

Chao, C. T., & Krueger, R. R. (2007). the date palm (Phoenix dactylifera L.): Overview of biology, uses, and cultivation. *HortScience*, *42*(5), 1077–1082. doi:10.21273/HORTSCI.42.5.1077

Chao, C. T., & Krueger, R. R. (2007). The date palm (Phoenix dactylifera L.): Overview of biology, uses, and cultivation. *HortScience*, *42*(5), 1077–1082.

Dillard, C. J., & German, J. B. (2000). Phytochemicals: Nutraceuticals and human health. *Journal of the Science of Food and Agriculture*, *80*(12), 1744–1756. doi:10.1002/1097-0010(20000915)80:12<1744::AID-JSFA725>3.0.CO;2-W

Echegaray, N., Pateiro, M., Gullon, B., Amarowicz, R., Misihairabgwi, J. M., & Lorenzo, J. M. (2020). Phoenix dactylifera products in human health–A review. *Trends in Food Science & Technology*, *105*, 238–250. doi:10.1016/j.tifs.2020.09.017

El-Ghayaty, S. H. (1983). Effect of different pollinators on fruit setting and some fruit properties of the Siwi and Ahmadi dates. In *Proceedings 1st symposium on date palm.* King Faisal Univ.

El Hadrami, A., & Al-Khayri, J. M. (2012). Socioeconomic and traditional importance of date palm. *Emirates Journal of Food and Agriculture*, *24*(5), 371–385.

El Hadrami, A., & Al-Khayri, J. M. (2012). Socioeconomic and traditional importance of date palm. *Emirates Journal of Food and Agriculture*, *24*(5), 371.

El Hadrami, I., & El Hadrami, A. (2009). Breeding date palm. In *Breeding plantation tree crops: tropical species* (pp. 191–216). Springer.

El-Hammady, A. M., Khalifa, A. S., & Montasser, A. S. (1991). Effect of potash fertilization on Sewy date palms. II. Effect on yield and fruit quality. *Egyptian Journal of Horticulture*, *18*, 199–210.

El-Salhy, A. M. (2000). Effect of bagging the spathes of Zaghloul date productivity under Assiut conditions. *Assiut J. Agric. Sci.*, *31*(3), 123–134.

El-Zoghbi, M. (1994). Biochemical changes in some tropical fruits during ripening. *Food Chemistry*, *49*(1), 33–37. doi:10.1016/0308-8146(94)90229-1

Gantait, S., El-Dawayati, M. M., Panigrahi, J., Labrooy, C., & Verma, S. K. (2018). The retrospect and prospect of the applications of biotechnology in Phoenix dactylifera L. *Applied Microbiology and Biotechnology*, *102*(19), 8229–8259.

Ghazzawy, H. S. (2013). Effects of some applications with growth regulators to improve fruit physical, chemical characteristics and storage ability of Barhee date palm cultivar. *Int. Res. J. Plant Sci*, *4*(7), 208–213.

Gros-Balthazard, M. (2013). Hybridization in the genus Phoenix: A review. *Emirates Journal of Food and Agriculture*, 831–842.

Hafez, O. M., Saleh, M. A., & Naguib, M. M. (2012). Quality improvement and storability of some date palm cultivars by safe postharvest treatments. *Australian Journal of Basic and Applied Sciences*, *6*, 542–550.

Hilgeman, R. H. (1954). *The differentiation, development and anatomy of the axillary bud, inflorescence and offshoot in the date palm.* Academic Press.

Hussain, M. I., Farooq, M., & Syed, Q. A. (2020). Nutritional and biological characteristics of the date palm fruit (Phoenix dactylifera L.)–A review. *Food Bioscience*, *34*, 100509. doi:10.1016/j.fbio.2019.100509

IFICF. (1998). International Food Information Council Foundation. Backgrounder: functional foods. In Food Insight Media Guide. IFICF.

Jain, S. M. (2012). Date palm biotechnology: Current status and prospective-an overview. *Emirates Journal of Food and Agriculture*, 386–399.

Johnson, D. (2011). Introduction: date palm biotechnology from theory to practice. In Date palm biotechnology (pp. 1-11). Springer.

Joseph, J. A., Shukitt-Hale, B., Denisova, N. A., Bielinski, D., Martin, A., McEwen, J. J., & Bickford, P. C. (1999). Reversals of age-related declines in neuronal signal transduction, cognitive, and motor behavioral deficits with blueberry, spinach, or strawberry dietary supplementation. *The Journal of Neuroscience: The Official Journal of the Society for Neuroscience*, *19*(18), 8114–8121. doi:10.1523/JNEUROSCI.19-18-08114.1999 PMID:10479711

Khalifa, A. S., Hamdy, Z. H., El-Masry, H. M., Tadros, M. R., & Said, G. (1984). Effect of some growth regulators on thinning" Amhat" date palm. *Agricultural Research Review*, *62*(3A), 255–266.

Kulkarni, S., Vijayanand, P., & Shubha, L. (2010). Effect of processing of dates into date juice concentrate and appraisal of its quality characteristics. *Journal of Food Science and Technology*, *47*(2), 157–161.

Marashi, S., & Mousavi, A. (2006, February). Effects of different methods and degrees of fruit thinning on yield and fruit characteristics of Barhee date cultivar. In *III International Date Palm Conference 736* (pp. 187-192). Academic Press.

Mostafa, R. A. A., & El Akkad, M. M. (2011). Effect of fruit thinning rate on yield and fruit quality of Zaghloul and Haiany date palms. *Australian Journal of Basic and Applied Sciences*, *5*(12), 3233–3239.

Mostafa, R. A. A., El-Salhy, A. M., El-Banna, A. A., & Diab, Y. M. (2014). Effect of bunch bagging on yield and fruit quality of Seewy date palm under New Valley conditions (Egypt). Middle East. *Journal of Agricultural Research (Lahore)*, *3*, 517–521.

Moustafa, A. A. 2007. Effect of bagging period of spathe after pollination on fruit set, yield and fruit quality of "Seewy" dates under Fayoum Governorate conditions. *Proceeding of the Fourth Symposium on the Date Palm in Saudi Arabia, Al-Hassa*, 123.

Munier, P. (1973). The date palm. *Techniques Agricoles et Productions Tropicales,* (24).

Mustafa, A. B., Harper, D. B., & Johnston, D. E. (1986). Biochemical changes during ripening of some Sudanese date varieties. *Journal of the Science of Food and Agriculture*, *37*(1), 43–53. doi:10.1002/jsfa.2740370107

Myhara, R. M., Karkalas, J., & Taylor, M. S. (1999). The composition of maturing Omani dates. *Journal of the Science of Food and Agriculture*, *79*(11), 1345–1350. doi:10.1002/(SICI)1097-0010(199908)79:11<1345::AID-JSFA366>3.0.CO;2-V

Nazam El-Din, A. M. M., & Abd El-Hameed, A. K. E. (2001) Study on the storage of Egyptian Siwi date variety (semi-dry date). *The second international conference on date palm.*

Nixon, R. W. (1957). *Effect of age and number of leaves on fruit production of the date palm.* Academic Press.

Nixon, R. W., & Carpenter, J. B. (1978). *Growing dates in the United States (No. 207)*. Department of Agriculture, Science and Education Administration.

Omar, A. E. D. K., Soliman, S. S., & Ahmed, M. A. (2013). Impact of leaf/bunch ratio and time of application on yield and fruit quality of Barhi date palm trees (Phoenix dactylifera L.) under Saudi Arabian conditions. *Journal of Testing and Evaluation*, *41*(5), 813–817. doi:10.1520/JTE20120340

Pfalgraz, K.E. (2002). Loss of a Legacy, Fusarium oxysporum in Ornamental Phoenix canariensis. *Principes: Journal of The International Palm Society, 46*(3).

Pintaud, J.-C., Zehdi, S., Couvreur, T., Barrow, S., Henderson, S., Aberlenc-Bertossi, F., . . . Billotte, N. (2010). Species delimitation in the genus Phoenix (Arecaceae) based on SSR markers, with emphasis on the identity of the date palm (Phoenix dactylifera L.). *Diversity, phylogeny, and evolution in the monocotyledons*, 267-286.

Prior, R. L., & Cao, G. (2000). Antioxidant phytochemicals in fruits and vegetables: Diet and health implications. *J. Hort. Sci., 35*(4), 588–592. doi:10.21273/HORTSCI.35.4.588

Rabeh, M. R. M., & Kassem, H. A. (2003). The effect of bagging the spathes after pollination on yield and quality of Zaghloul and Samany dates. *Zagazig J. Agric. Res., 21*(3B), 935–944.

Sawaya, W. N., Khalil, J. K., Safi, W. N., & Al-Shalhat, A. (1983). Physical and chemical characterization of three Saudi date cultivars at various stages of development. *Canadian Institute of Food Science and Technology Journal, 16*(2), 87–92. doi:10.1016/S0315-5463(83)72065-1

Sawaya, W. N., Khatchadourian, H. A., Khalil, J. K., Safi, W. M., & Al-Shalhat, A. (1982). Growth and compositional changes during the various developmental stages of some Saudi Arabian date cultivars. *Journal of Food Science, 47*(5), 1489–1492. doi:10.1111/j.1365-2621.1982.tb04967.x

Shaheen, M. R. (1986). Pistil receptivity in three cultivar of date palm (Phoenix dactylifera L.). In *Proc 1st Hort Sci, conference*. Tanta Univ.

Siddiqui, S. A., Subharwal, M., Gupta, S., & Balkrishan. (1994). Finite range survival model. *Microelectronics and Reliability, 34*(8), 1377–1380. doi:10.1016/0026-2714(94)90154-6

Sidhu, J. S. (2012). Production and Processing of Date Fruits. Handbook of Fruits and Fruit Processing, 629-651. doi:10.1002/9781118352533.ch34

Soliman, S. S., & Osman, S. M. (2003). Effect of nitrogen and potassium fertilization on yield, fruit quality and some nutrients content of Samany date palm. *Ann Agr Sci Ain Shams Univ Cairo, 48*, 283–296.

Soliman, S. S., & Shaaban, S. H. A. (2006). Macronutrient changes of Samany date palm leaflets and fruits as affected by nitrogen and potassium application. *Egypt J Appl Sci, 21*, 641–660.

Tang, Z. X., Shi, L. E., & Aleid, S. M. (2013). Date fruit: Chemical composition, nutritional and medicinal values, products. *Journal of the Science of Food and Agriculture, 93*(10), 2351–2361.

Tengberg, M. (2012). Beginnings and early history of date palm garden cultivation in the Middle East. *Journal of Arid Environments, 86*, 139–147.

Wargovich, M. J. (2000). Anticancer properties of fruits and vegetables. *J. Hort. Sci., 35*, 573–575.

Zaid, A., & De Wet, P. (1999). *Climatic requirements of date palm. Date palm cultivation*. Academic Press.

Zaid, A., & de Wet, P. F. (2002). Pollination and bunch management. In A. Zaid (Ed.), *Date palm cultivation* (pp. 145–175). FAO.

Zaid, A., & de Wet, P. F. (2002a). Botanical and systematic description of the date palm. In A. Zaid (Ed.), *Date palm cultivation. Food and agriculture organization plant production and protection* (pp. 1–25). Food and Agriculture Organization of the United Nations.

Zaid, A., & de Wet, P. F. (2002b) Climatic requirements of date palm. In Date palm cultivation. Food and Agriculture Organization Plant Production and Protection. Food and Agriculture Organization of the United Nations.

Chapter 12
In Vivo Date Palm

Rasmia Darwesh
Central Laboratory for Research and Development of Date Palm, Agriculture Research Center, Egypt

ABSTRACT

Date palm (Phoenix dactylifera L.) plants demand reclamation of new areas or renewal of the oldest trees. Different methods to produce these plants as seeds that gave only 50% of seedlings will be female. The dates from seedling plants are often smaller and of poorer quality, so the propagation of date palm plantlets will occur with different methods. In vitro culture produced true-to-type huge numbers of plantlets under control conditions that led to many differences in morphology, physiology, and anatomy characteristics. The acclimatization was done to modify these abnormalities, and the plantlets after acclimatization need to improve growth by different treatments. Offshoots are the other method for propagation. These offshoots that grow from the base of the tree need to improve new rooting at the trunk base by growth regulators substances. This work shed light on the different methods of date palm propagation, the crucial role of the acclimatization greenhouse stage, and improved plantlets planted in the open field.

INTRODUCTION

Date palm (*Phoenix dactylifera* L.) is a dioecious propagated by seeds which are not used as the commercial method for reasons;

The production plants are not true to type mother plants

Only 50% of seedlings will be female

The dates from seedling plants are often smaller and of poorer quality. So, these methods have done not in the research aspect (Khaled et al. 2017)

Date palm tissue culture technique is the considerable tool can be produce the huge numbers of plants to farmers in order to increasing dates which have high nutritional value, in vitro technique become used for large scale due to

- Annihilation of genetic variation by sexual propagation
- Production of large number of genetically true –to type plantlets

DOI: 10.4018/978-1-6684-2423-0.ch012

- Disease-free plants especially Bayoud (lethal disease)
- important for scientific purpose (Junaid and Khan, 2009; Bhattacharjee 2006, Fki 2011, Rajmohan 2011 and Zaid et al. 2011)

Several factors can be affecting successfully in vitro technique i.e.
Type of explant
Nutrients and growth regulators
Culture conditions as temperature, light intensity and light duration (Muthukrishnan et al. 2013).
These culture conditions contribute to different morphological, physiological and anatomical characters

IN VITRO AND EX VITRO PLANTLETS DIFFERENCES

Morphological Effects

In vitro shoots and plantlets are tiny compared to the open field plants, this tiny shape refers to a high concentration of cytokinins specially in the shooting stage. Saturated water, plants *in vitro* have increasing water contents and high concentration of cytokinins, this makes reducing dry matter accumulation per area, all of the organs more fragile. Plantlets from *in vitro* have shortest shoots and roots and lowest numbers of leaves and roots than *ex vitro,* plant height, number of leaves, total dry mass and leaf mass area ratio of *Nicotiana tabacum* (Pospisilova *et al.*1999), Curled dark leaves of date palm derived *in vitro* condition (El- Bahr *et al.* 2003). Leaf length and shoot number and shoot and root dry weights were increased after ex vitro than in vitro of sea oats (*Uniola paniculata* L.) (Aracama *et al.* 2008). Stem diameter, shoots numbers of *Alpinia purpurata* were increased in ex vitro compared to in vitro (Medina *et al.* 2007), ex vitro survived-plantlets when acclimatized had varied survivability from 73-93% with 82% in average and increasing in leaf and root number and length and height of shoots of *Ruscus hypophyllum* L. (Winarto and Setyawati 2014), after 45 days from acclimatization stage in the greenhouse the roots numbers, mean root length and shoot height of Teak (*Tectona grandis*) were increased (Yasodha *et al.* 2005), numbers of leaves and axillary shoots, fresh and dry weights of leaves and shoots of *Malus hupehensis* were increased under glasshouse (Ur-Rahman*et al.* 2007), in ex vitro of *Pterocarpus santalinus* L. the plant height, number of leaves/plant, fresh and dry weight of leaves and shoot and root/plant were increased (Rajeswari and Paliwal 2008), 12 weeks of acclimatization is the best treatment for increasing roots length which penetrated out from the pots through the holes at the bottom of the pots or grew in circle on the bottom of the pots, and also highest shoot height, leaf number an shoot diameter of oil palm plantlets *Elaeis guineensis* Jacq in *ex vitro* condition was by dipping the basal end of the plantlets in 2 -8 mM IAA,IBA and NAA solution (Sumaryono and Riyadi 2011), Pineapple (*Ananas Comosus*.L.Meer. cv.Del Mont) plantlets grown in culture medium containing only peat moss was significantly increased in plant length, number of leaves /plant, fresh weights of vegetative growth and roots (AL-Taha 2013).

Physiology Effects

Shoots *in vitro* conditions were exposed to
High sugars concentration which is a source of carbohydrates, constant light, constant temperature

Upon this fact the metabolism process and carbon dioxide fixation is depressed, low photosynthetic pigments, pigments synthesis may be impaired under constant and low light intensity also concentrations of starch is very low, when un change in light, temperature and water the evapotranspiration is very low or lacking, CO2 uptake is relatively low, Therefore, the growth of plantlets *in vitro* is often greater under photoautotrophic conditions than under heterotrophic conditions, provided that the *in vitro* environment is properly controlled for promoting photosynthesis (Kozai et al. 2005), leaves formed during the acclimatization period may still have a lower photosynthetic capacity than leaves of greenhouse-grown plants (Carvalho *et al.* 2001), leaf yellowing is attributed to photo inhibition, nutrient deficiency, premature senescence (Sicher 2008), implying that higher CO2 enrichment (3000 mmol mol21) could have a positive effect on the photoautotrophic growth of orchid plantlets grown at 10 000 mmol mol21 CO2 under high PPFD showed decreased photosynthetic capacity and total Rubisco activity tended to decline (Sharma *et al.* 1999, Norikane and Tanaka 2010 and Norikane *et al.* 2013), plantlets of *Swertia Chirata* derived in vitro are heterotrophic in their mode of nutrition must to be acclimatized to can be culture in the open field (Pant *et al.* 2010) recently, high concentration CO2 increased photosynthetic parameters, highest photosynthetic fluorescence, and the relative expression of photosynthesis related genes were significantly up-regulated during acclimation and transplantation, exogenous CO2 laid to a good acclimation and transplantation by promoting the photosynthesis of plantlets and adjusting the source-sink relationship of leaf and stem photosynthetic products (Qi et al. 2021)

Anatomical Effects

- Reducing mechanical support tissues as fewer collenchyma and sclerenchyma cells

Thin cell wall, environment without a lot of air movement lead to inhibit cell wall deposition and sclerenchyma, collenchyma formation

- Absent of cuticular wax and epicuticular refers to low light relatively and no changes in light
- Thinner epidermal layer or irregular epidermal cells
- Limited palisade layer and sometimes strangely shaped palisade cells, palisade development is affected by different light levels and reduced as a consequence of relatively low light intensity
- Stomata had slow response times or impaired function (many researcher make anatomy for
- -Stomata of in vitro leaves are found closed which refer to unchanged in the light, loosely organized spongy mesophyll, also refers to absorption of water when plants absorption they evaporate
- Roots in vitro were irregular, epidermal and subepidermal layers could not regular also compacted and darkly stained were found, cortical cells are smaller in size and tightly packed with smaller intercellular spaces between cells, most cortical zone were disrupted, xylem vessels were not bedded, pericycle and endoderms were darkly stained.

Plantlets in vitro

Shoots and roots showed different characteristics compared to plants in the open field, these problems could be summarized as follows: the shoots were very thin, very long internodes and a few shoots were vitrified.

Table 1. Comparison between in vitro and ex vitro

Organ	*In vitro*	*ex vitro*
Shoots	tiny size, fragile	thickness
1-leaves	Small, succulent, brittle and fragile	Normal size and shape
2-Epidermis	Deformed thin cell walls irregular shaped	normal cell wall
3- Cuticle	thin and discontinuous	thick and continuous
4-stomata	irregular guard cells, thin cell walls, larger stomatal aperture	normal
5- mesophyll	spongy parenchyma and irregular, loosely organized spongy mesophyll	regular structure
6- palisade	fewer palisade strangely shaped palisade cells	natural numbers for species
7- wax layer	reduction or absent of cuticular wax , Poor epicuticular formation	thick cuticular and normal structure
8-Chloroplasts	abnormal, impaired functional chloroplasts	Lowest chlorophyll content
9-roots	thin, limited and limited function	normal

in this respect, low efficient rooting and low rate of survival upon transfer to soil, plantlets were very thin, higher and have non – functional roots and they could survive and eventually died in the greenhouse. Thus, it was needed to acclimatize the plants *in vitro*, where they receive a special treatment before they can be transferred to greenhouse (Hassanen and Khalil 2013). Deposition of protective epicuticular wax on the surface of the leaves is the most important factor responsible for excessive loss of water, leading to poor transplantation success (Hazarika 2003). Poorly formed of vascular bundles of shoots and roots, the restricted water uptake from the root into the shoot, leaves were developed in culture deteriorated rapidly after transplanting to ex vitro conditions, and new leaves were formed in the second week of acclimatization, less number and size of stomata was observe each stoma was surrounded by four epidermal cells differing in shape and size from the other epidermal cells, roots of date palm offshoots showed epidermal and sclerenchyma cell thickness also cortical region showed thickness and greatest maximum vascular region thickness (Fatima *et al.* 2010; Trejgell and Tretyn 2011 and Abd- El Baky 2012 on date palm cvs),), in vitro microshoots of *Castanea sativa* had wide open, spherical stomata and higher stomatal density than nursery plants and they had almost no epicuticular wax, high stomatal conductance and high transpiration rate, all of these anatomic and functional leaf characteristics are causes of the low survival rates of plantlets after ex vitro transfer (Saez et al. 2012) recently, *Camellia oleifera in vitro* plantlets showed highest chlorophyll content, the thickest palisade and spongy tissues, and consequently, the thickest leaves under combined red and blue (4:1) light (Chaoyin et al. 2020). To overcome these problems and fulfil the demand for planting material, it is necessary to use the acclimatization stage

ACCLIMATIZATION OF PLANTLETS

In vitro propagated plantlets of date palm facing problem in the acclimatization stage as high mortality concerning failure under this stage. In vitro condition as relatively low light intensity, constant temperature 20-28 C^0, high humidity, water continuously found, this in vitro condition lead to deficiency of plants morphology, physiology and anatomy, these plantlets when transferred to ex vitro become stressful by different condition compared in vitro, these plantlets must be acclimatize in the greenhouse for modified these morphology, physiology and anatomy effects in order to increasing survive percentage of plantlets. Healthy date palm plantlets 3-4 leaves, 10-15 cm in length and 2-3 roots were transferred into ex vitro for acclimatized in the greenhouse, these plantlets washed carefully with tap water and treated with fungicide, plantlet cultured in peat moss + perlite 2:1 and kept in the plastic covered inside 100% humidity and 28 C^0 till new leave or root was grown, the plastic covers were opening daily for 5 – 10 minutes to renewal air in the plastic covers (Darwesh, Rasmia,S.S. 2015)

Pre-Acclimatization Stage

Many treatments was done in the pre- acclimatization in vitro to prepare plants to withstand conditions which lead to increasing rates of survival percentage, therefore, GA_3 at 0.15, 0.25 and 0.5 m/l and 4 mg/l Pbz with different strength of MS in the pre acclimatization of date palm increased, shoot and root lengths, numbers of leaves and roots also stem thickness, and indole content moreover highly rising of survival percent in the greenhouse (Table 2,3 and Fig 1 Pic. 1) as found by (Darwesh et al. 2011),

Table 2. Effect of GA_3 (mg/l), ammonium nitrate (mg/l) and pbz (4mg/l) on shoot and root length/plantlet of date palm plantlets

A	Shoot length/plantlet					Root length/plantlet				
GA_3 / NH_4NO_3 B	Con	0.15	0.25	0.5	Mean	Con	0.15	0.25	0.5	Mean
1650(con)M1	11.5	12.1	12.7	13.7	12.5	2.3	2.6	3.6	4.1	3.2
825 M2	12.1	12.6	12.9	14.1	12.9	3.2	3.3	3.4	4.2	3.5
412.5 M3	11.1	11.6	11.6	12.7	11.8	3.1	3.1	3.1	3.6	3.2
206.25 M4	11.0	10.6	11.4	12.5	11.4	2.4	2.7	2.7	3.4	2.8
Mean	11.4	11.7	12.2	13.3		2.8	2.9	3.2	3.8	
L.s.d.			A= 0.4	B= 0.4	AB= 1.5			A= 0.02	B= 0.4	AB= 0.05

Table 3. Effect of GA_3 (mg/g), ammonium nitrate (mg/l) and pbz (mg/l) on number of leaves and roots/ plantlet of date palm plantlets

A GA_3 NH_4NO_3 B	Number of leaves/plantlet					Number of roots/plantlet				
	Con	0.15	0.25	0.5	Mean	Con	0.15	0.25	0.5	Mean
1650(con)	2.3	2.2	2.8	2.3	2.4	2.2	2.8	2.5	2.5	2.5
825	2.8	3.2	3.3	3.0	3.0	2.5	3.2	3.2	2.9	2.9
412.5	2.2	3.4	3.6	3.3	3.1	3.3	3.3	4.0	3.4	3.4
206.25	2.1	3.4	3.7	3.4	3.2	3.3	4.0	4.5	3.4	3.8
Mean	2.4	3.1	3.4	3.0		2.8	3.3	3.6	3.2	
L.s.d.	A= 0.4 B= 0.1 AB= 0.4					A= 0.4 B=0.2 AB= 0.9				

Figure 1. Effect of ammonium nitrate (mg/l) and GA3 (mg/l) on growth of date palm plantlets (from Darwesh et al. 2011)
0= con 1= M 2+ 0.15 GA3 2= M2 + 0.25 GA3 3= M2 + 0.5 GA3
4= M3+ 0.15 GA3 5= M3 + 0.25 GA3 6= M3 + 0.5 GA3
7= M4 + 0.15 GA3 8= M4 + 0.25 GA3 9= M4 + 0.5 GA3

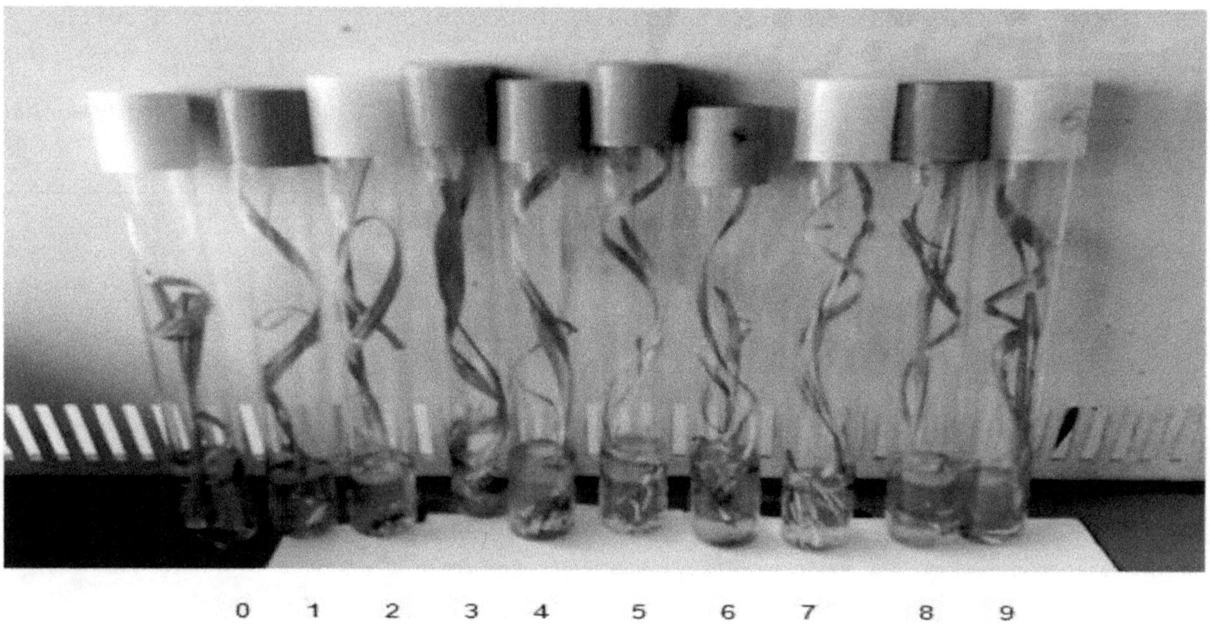

Figure 2. Ammonium nitrate and GA3 affecting plantlets acclimatization % (Darwesh et al. 2011)

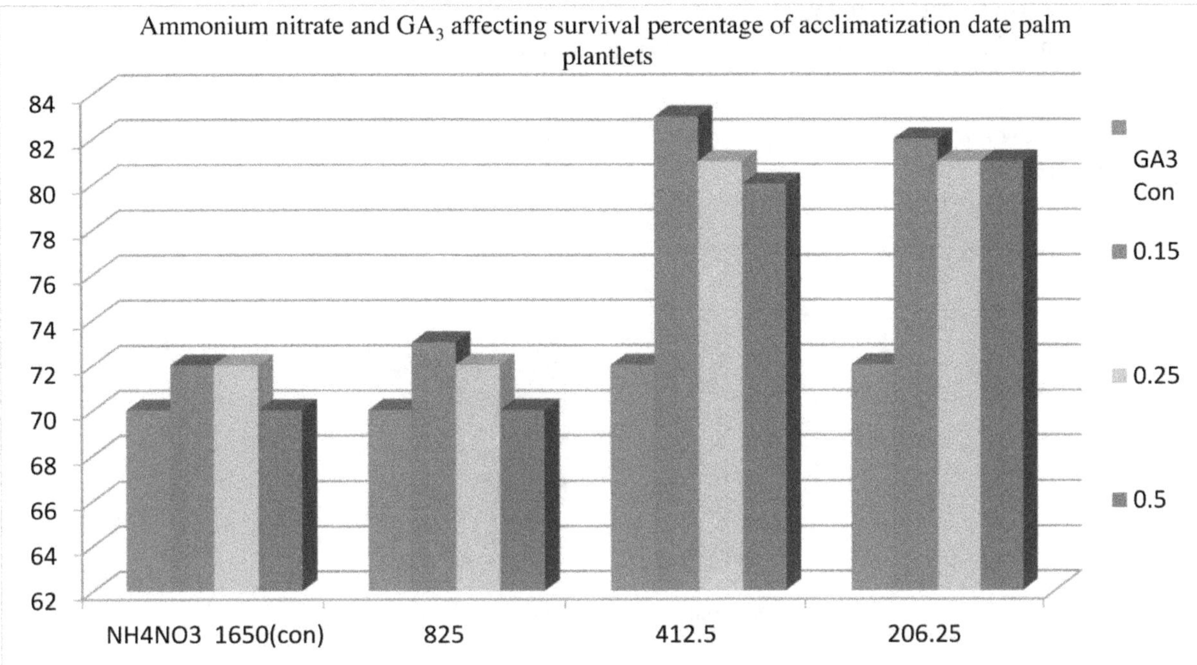

The pre-acclimatization of date palm cv Barhee 950 mg/l KNO_3 and 45 g/l sucrose have left 82.8% survival in the greenhouse (Fig 2) as found by **(Ibrahim et al. 2012)**,, 2000 and 3000 lux and peatmoss 25% + vermiculite 25% + sand 50% increased survival percentage of date palm cv Barhee.

Figure 3. Potassium nitrate and sucrose affecting ex vitro plant height

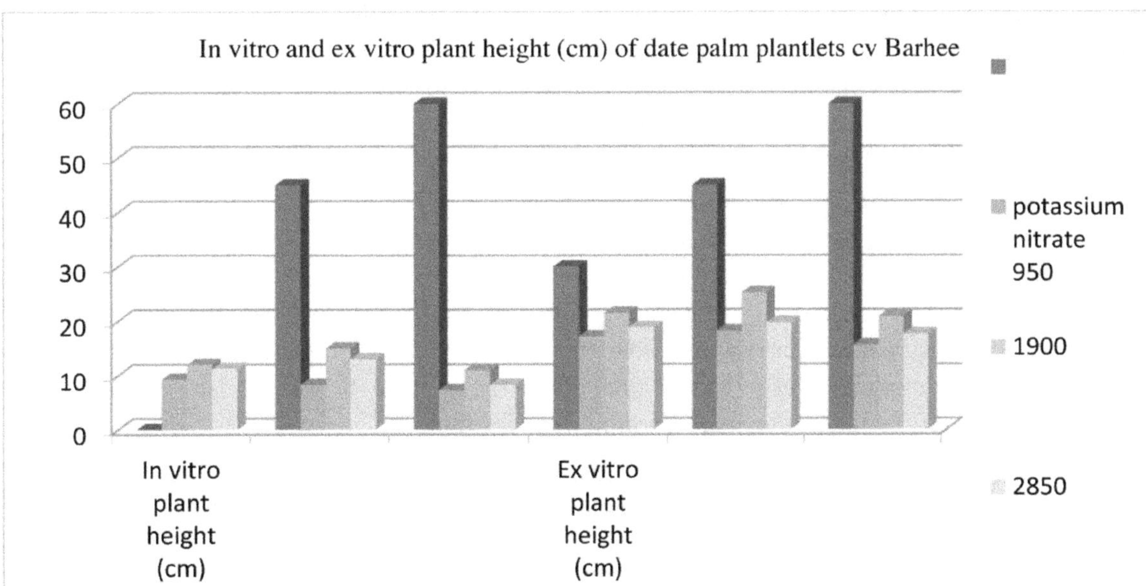

On the other hand different types of sugars were influenced the survival of acclimatization plantlets in the greenhouse as date palm syrup, sucrose and glucose (Fig 3), in the greenhouse and under plastic covers plant height, leaves numbers and survival percentage% (**Darwesh** et al. 2021)

Figure 4. The evaluation units of different treatments types of sugars regarding acclimatization stage measurements

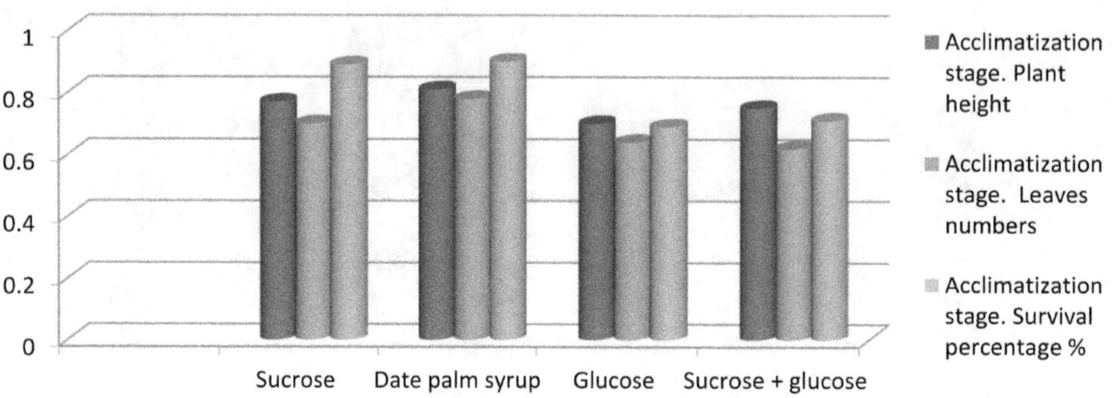

Growth retardants increased the survival percentage in acclimatization stage, in this concern, (Ezz et al. 2008) stated that, 1.0 or 0.5 mg/l ABA recorded higher value of roots numbers of date palm cv. Malakaby, Ancymidol at .05 mg/l increased root length, 2.0 mg/l Ancymidol increased significantly the plantlet thickness. ABA at 0.5 mg/l and paclobtrazol at 1.0 mg/l increased significantly the content of phenols and indols. Rooted plants have been successfully acclimatized with a success rate 93%

Many other treatments were used as pre- acclimatization stage for increasing survival percentage of greenhouse acclimatization of plantlets ; date palm cv. Karama plantlets treated with PEG in media at 4.0g/L or 8.0g/L which was effective to enhance root hardening, 72.72 percent of plantlets and increasing root thickness and length (Mahdia and Abd- Alla 2010), PEG at 30% led to increase the percentage of acclimatization (100%) of date palm; while treatment with PEG at 25% led to increase the percentage of acclimatization 80% (Al-Meer *et al.* 2008) .

Increasing relative humidity around the newly transferred plantlets of date palm by micro-tunnel covered by transparent plastic was used. The plantlets were sprayed with a fungicide every 2-3 days to prevent crown and leaf rot. Under such conditions, plantlets having 2 to 3 leaves, a well formed and closed crown and 3 to 4 roots showed a high survival rate of about 80-90% (Abahmane 2013) date palm *Phoenix dactylifera* L. plantlets were transferred to the liquid medium (1/4 MS) as the pre acclimatization stage for three subcultures (3 weeks for each one). Then these plantlets were transferred to the green-house for acclimatization under tunnels (90% humidity) for three month, with 82% survival (Darwesh and Mohamed 2009 on cv Bartoumoda, Othmani *et al.* 2009 on

date palm c. Deglet Nour). Acclimatizing, hardening-off, or conditioning plantlets from the *in vitro* to the ambient environment can be a challenge that may result in death or damage to a large percentage of micropropagated plants (Preece 2010). In the greenhouse; in particular, differences in light, relative humidity, nutrients and other growth promoters,the gaseous composition and the medium substrate

(Seelye et al., 2003 and Chandra *et al.,* 2010). Different culture soil media was used as, date palm cv Kheneizi plantlets were hardened and acclimatized and planted in pots, containing 1:1:1 peat, sand and dehydrated cow manure, which resulted in over 60% ex vitro plant survival (Kurup et al. 2014). date palm cv. Maktoom (plantlets) were transplanted in peat moss and perlite (2:1) and placed in plastic tunnels in a greenhouse with 85% survival percentage (Khierallah and Bader 2007). Date palm plantlets showed highest survival efficiency 83% after 8 weeks of ex vitro transplantation and great thickness (Khan and Bi Bi 2012), successful adaptation of vitro plantlets of date palm cv. Zaghlool when planted in pots contained equal volumes of peat moss and vermiculite under high humidity conditions (Bekheet 2013).

Enhancing plantlets characteristics

Inorganic fertilizers

Improving acclimatized date palm plantlets to be necessary for increasing demand numbers of trees in the new areas by many treatments of chemical fertilizers NPK, increase in N and K contents of pinnae in Zaghloul date palm due to increasing of potassium fertilizer rate, while decrease Ca and Mg contents (Kassem et al. 1997), potassium sulfate (48% K2O) at 1, 2, and 3 kg/palm increase the number of new growing leaves increased the pinnae contents of N, P, K, Fe, Mn, Cu and Zn of Zaghloul date palm *Phoenix dactylifera,* L (Harhash and Abdel-Nasser 2007), Potassium sulfate 3 kg/tree (48% K_2O) and potassium nitrate increased number of leaves of date palm cv. Kabkab (Abdi and Hedayat 2010), increment in the number of date palm leaves rate and also increment the value contents of chlorophyll and carbohydrate and protein value in leaves under N.P.K at ratio (2:1:2) and (2:1:1) with depth of 30cm (Ebtihaj 2012), foliar fertilization as commercial product Biofert (8:9:9) with substrate fertilization K2O at 100 g and P2O5 at 1.8 kg significantly greater plant height of Lady palm or raffia palm (*Rhapis excelsa* (Thunberg) Henry ex. Rehder) and stem diameter (Luz et al. 2008), Ammonium sulfate, ammonium nitrate and urea at 350 to 1050 g/tree increased N,P, and K leaves contents of date palm (Saleh 2009 and Kassem 2012), Increasing growth of date palm plantlets cv. Barhee by irrigated with NPK and Hogland solutions (Muhsen et al. 2013), nitrogen formula as (T3) potassium nitrate > (T2) ammonium sulfate > (T4) urea were recommended to highly increasing growth (Table 4) of date palm plantlets cv. Baromouda in the green house (Zayed et al. 2014)

Table 4. Effect of different sources of nitrogen fertilizer on chlorophyll a, b and carotenoides mg/g f.w.,

Vegetative growth / treatments	Chlorophyll a mg/g f.w		Chlorophyll b mg/g f.w		Carotenoides mg/g f.w		Indoles mg/g f.w.		Proline mg/g d.w.	
	first season	Second season	first season	Second season	first season	Second season	first season	Second season	first season	Second season
T1	0.54	0.56	0.17	0.18	0.40	0.41	6.2	6.3	0.3	0.4
T2	0.85	0.87	0.28	0.30	0.49	0.53	19.4	20.4	2.2	2.6
T3	0.88	0.91	0.45	0.46	0.51	0.55	22.5	23.3	3.2	3.6
T4	0.61	0.65	0.28	0.30	0.45	0.47	17.4	18.1	2.9	3.1
L.S.D.	= 0.02	= 0.03	=0.03	= 0.04	= 0.01	= 0.01	= 1.2	=1.4	= 0.1	= 0.1

From (Zayed et al. 2014)

indoles mg/g f.w. and proline mg/g d.w. of date palm plantlets (*Phoenix dactylifera* L) in the green house in the 1st and 2nd seasons

The plantlets of date palm (*Phoenix dactylifera* cv. Bartomouda) at acclimatization stage in the greenhouse. The different treatments of NPK fertilizer (33:6:43) at three levels 1.5,2.5 and 3.5 g/l in the irrigation water one time per week, and two levels 1.0 and 1.5 cm/l of foliar spraying of isogreen (N 8%, P 5%, K 7%,Fe 150 ppm, Zn 100 ppm, Mn 200 ppm, B 5% and amino acids 5%) one time/3weeks, biofertilizer as potasiomag (*Bacillus circulans*) mixing with soil increased plant height, leaves number /plantlet, root length, number of roots/plantlet, and fresh and dry weight of leaves, in addition nitrogen, potassium, phosphorus, also Fe, Mn, Zn and B leaves content (Darwesh et al. 2016), Leave length (cm), numbers of leaflets/leaf, leaflet length (cm), leaflet width(cm). total chlorophyll of date palm trees c.v Barhee were enhancing with different treatment of Soil application at 3 kg/palm, Foliar application at 3% K2SO4 and Injection into the trunk at 3% K2SO4 (Elsayd et al. 2018), vegetative growth of date palm cv Sewy was increased by 18.1%, leaf NPK content by 26% under soil injection (500,750 and 1000 ml/palm tree /month) and trunk injection (100,200 and 300ml/palm tree/month) from NPK 2:1:2 (Zaen El – Daen 2019)

ORGANIC AND BIO-FERTILIZERS

Organic fertilizers as compost, farm yard manure (FYM), slurry, peat, greenmanure, wormcastings, urine, dried blood, bonemeal, fishmeal, and feathermeal (Haynes and Naidu 1998)

Organic Residue

sheep manure (1.83% N) as 38 gm N/tree and urea (46% N) and 200 gm S/tree as agriculture sulphure (95% S) induced rising of peach (*Prunus persica* CV. Dixired) available amount of N, P, K and S in the soil, and its concentration in the leaves, leaves chlorophyll content, carbohydrate concentration in the leaves, leaves area, trees height, main stem diameter, branches length (Al- Aa'reji 2010), goat manure+ sandy soil (1:2, v: v) under 100 ml EM produced the highest offshoots survival ratio, rooting length, diameter and numbers and leave length and numbers of date palm cv. Hayani (Sheren 2014)

Rooting, growth characters and mineral content of Zaghloul date palm offshoots in Olive solid waste + sandy soil 1:2, v/v (El- Kosary et al. 2008), Date Palm Leaves Compost (DPLC) relative to the peat moss (Ali 2008), date palm leaves as a compost individually or in combined with PGPR viz. *Bacillus megaterium, B. cereus* and *Pseudomonas fluorescens* increased of nodulations, growth and yield parameters and controlling root rot disease in faba bean (Abdel-Monaim et al. 2018), height, stem diameter and shoot biomass, root volume and biomass of *Pinus ponderosa* Douglas ex Lawson & C. Lawson increased with Sphagnum peat – sawdust (PS) medium compared to 1:1 v:v, Sphagnum peat – vermiculite (PV) or a 7:3 v:v (Dumroese et al. 2011), height and leaves numbers/flower of Banana were highest with NPK + micronutrients and also with composted dairy cow manure CDM+ 10% and 30% date palm straw DPS or girth with NPK + micronutrients and un-composted dairy cow manure FDM (Al-Busaidi 2013)

Biofertilizers

Biofertilizers are the formulations of living microorganisms, which are able to fix atmospheric nitrogen in the available form to plants, either by living freely in the soil or being associated symbiotically with plants (Chandrasekar et al. 2005), they are living cells of microorganism which convert nutritionally important elements from unavailable to available from through biological processes (Vessey, et al, 2003), in addition they are low cost also renewable sources of plant nutrients that supplement chemical fertilizers and have a great importance role for they are components of integrated nutrient management, effective and renewable source of energy for plants and to help in reducing the use of chemical fertilizers for sustainable agriculture (Rana et al,2013),

Bacteria

Azotobacter spp and *Rhizobium japonicum* showed excellent growth in both the morphological as well asbiochemical parameters (Amit and Satish 2013), biological nitrogen fixation is carried out by both symbiotic and free living bacteria and blue green algae. Azotobacter, a free-living diazotroph has also been reported to produce beneficial effects on crop yields through a variety of mechanisms including biosynthesis of biologically active substances, stimulation of rhizospheric microbes, modification of nutrient uptake and ultimately boosting biological nitrogen fixation (Somers et al. 2004), Rhizosphere bacteria (*Azospirillum brasilense, Bacillus megaterium* and *Klebsiella pneumoniae*) at 10, 20 and 30 cm/l with irrigation water with 2.5 g/l NPK increased significantly (Table 5) plant height, number of leaves/plantlet, root length, number of roots/plantlet, indole mg/g (f.w.) and chlorophyll (Hala et al. 2011)

Table 5. Effect of different bacterial supernatant on root length (cm) and roots number/plant of date palm at the 1st and 2nd seasons

Bacterial strains	Root length (cm)							
	First season				**Second season**			
	10	**20**	**30**	**mean**	**10**	**20**	**30**	**mean**
Control	6.1	6.4	6.8	6.4	6.6	6.8	7.1	6.8
Azospirillium brasilense	11.5	13.0	13.3	12.6	12.1	13.6	14.1	13.3
Bacillus megaterium	11.7	13.0	13.5	12.8	12.3	13.5	14.1	13.3
Klebsiella pneumonia	13.1	14.1	15.6	14.3	13.5	14.6	16.2	14.8
Mean	10.6	11.6	12.3		11.1	12.1	12.9	
L.S. D	A= 0.3 B= 0.3 AB=0.5				A= 0.3 B=0.3 AB=0.6			
	Roots number/plant							
	First season				Second season			
Control	3.7	3.8	4.0	3.8	4.1	4.2	4.4	4.3
Azospirillium brasilense	3.8	6.0	6.5	5.4	4.2	6.5	6.9	5.9
Bacillus megaterium	4.8	6.4	7.1	6.1	5.1	6.6	7.5	6.4
Klebsiella pneumonia	5.3	7.0	7.9	6.7	5.6	7.3	8.3	7.1
Mean	4.4	5.8	6.4		4.8	6.1	6.8	
L.S. D	A=0.4 B=0.5 AB=0.8				A=0.4 B=0.5 AB=0.8			

From (Hala et al. 2011)

Yeast

Yeast contains important substances as Protein 47%, carbohydrates 33%, nucleic acid 8%, lipids 4%, different minerals 8% such as Na,Fe,Mg,K,P,S,Zn,Mn,Cu,Si,Cr,Ni,Va,and,Li, as well as vitamins including thiamin, riboflavin, pyridoxine in addition to hormones and other growth regulating substances, biotin,B12,folic acid and tryptophan (Nagodawithana 1991).

(Rasmia Darwesh 2010) found that yeast "*Saccharomyces cervisais*" (at 20 and 40 ml/l) and NPK fertilizers (14:6:43), N as 33% NH_4NO_3, Pas P_2O_5 and Kas K_2O increased significantly (Table 6) plant height, leaves number, root length, root number, fresh and dry weights of leaves, in appending N, P and K and indole of *Phoenix dactylifera* cv. Medjol plantlets

Table 6. Effect of different concentrations of yeast (cm/l) and complete fertilizers NPK (g/l) on plant height (cm) and leaves number /plantlet of Phoenix dactylifera L. cv. Medjol

Yeast A / NPK B	Plant height cm.								Leaves number/plantlet							
	First season				Second season				First season				Second season			
	con	20	40	mean	con	20	40	Mean	con	20	40	mean	con	20	40	Mean
Con	9.7	16.2	18.4	14.8	14.2	19.6	23.6	19.1	2.3	3.0	3.3	2.9	3.3	3.7	4.1	3.7
2.5	14.5	27.8	33.2	25.2	16.2	31.1	36.1	27.8	3.0	3.7	4.7	3.8	3.3	4.5	5.3	4.4
3.0	19.4	38.2	49.4	35.7	22.8	48.7	61.5	44.3	4.0	5.9	6.1	5.3	4.4	6.5	7.0	6.0
3.5	22.8	39.2	49.6	37.2	24.2	49.8	62.5	45.5	4.2	6.1	6.3	5.5	4.6	6.7	7.3	6.2
Mean	16.2	30.1	37.7		19.4	37.3	45.9		3.4	4.7	5.1		3.9	5.4	6.0	
l.s.d	A= 1.2 B= 1.4 AB= 2.4				A=1.1 AB=2.3			B=1.3	A=0.5 AB=1.1			B=0.6	A=0.4 AB=0.8			B=0.5

A=yeast B= NPK AB= yeast x NPK A=yeast B= NPK AB= yeast x NPK
A=yeast B= NPK AB= yeast x NPK A=yeast B= NPK AB= yeast x NPK

Effective microorganisms EM application increased total bacteria, total actinomyces and total fungi which produced of indole acetic acid and gibberellins leads to improvement growth of root system that reflected on enhanced the uptake of nutrients (Higa 1991), EM effective microorganisms at 1.0 cm3/

palm and 90 ml/palm enhancing growth and leaf mineral content of date palm cvs as of "Bartamuda, "Sewy", "Zaghloul" and "Hayany" (Osman et al. 2011, El-Khawaga 2013, Amro et al. 2014), increased growth parameters (shoot length, number of leaves/shoot and leaf area) of 'Canino' apricot with different treatments of humic acid as Actosol which contain (2.9% humic acid + 10, 10, 10% NPK) as soil + foliar application (Fathy et al. 2010), 50 to 75% inorganic N with 50 to 200 ml humic acid and EM significantly can be increase the area of pinnae and leaf, N, P, K and Mg of leaves Zaghloul date palm (Ahmed et al 2014).

ARBUSCULAR MYCORRHIZAL

Many investigators exhibited that Arbuscular mycorrhizal increased the absorbtion of nutrient elements through its hypha and transport them to the plant, increased the height mint root by inoculation mycorrhyza (VAM) fungus *Glomus fasiculatum* (Gupta et al. 2002), mycorrhiza with 5-20 mg/kg per active ingredient increased fresh and dry weight of shoots date palm, especially on nutrient poor soil, had significant effect on the absorption of phosphorus, zinc, and copper (Esmaeilifar A 2013)), however the AMF have the different strains which also affected plant growth as Plant height, leaf length and numbers, stem diameter and fresh and dry weights of leaves of Phoenix canariensis with inoculation of AMF (*G. mosseae, G. deserticola* and *G. intraradices*) under fertilized or unfertilized plants (Morte and Honrubia 2002) and *Glomus* (*G. aggregatum, G. intraradices, G. verriculosum, G. mosseae, G. fasciculatum*) to data palm cvs cultivars: Tijib and Nakhla Hamra can be enhancing shoot and root growth (Diatta et al. 2014), aerial fresh weight of date palm number of leaves, the plant length, and the stem diameter were higher with inoculated by AMF and *Trichoderma harzianum* (Sghir et al. 2014), (Abo- Rekab et al. 2010) on date palm cv Bartomouda they found that, arbuscular mycorrhizal fungi (Glomus spore) with 1.5, 2.0 and 2.5 g/l, increased plant height (cm), number of leaves/ plantlet, length of root (cm), and number of roots/plantlet, also increasing N,P,K, Fe, Zn, Mn and Cu

Amino Acids

5- aminolevulinic increased date palm cvs Khalas and Sukkariat- Yanbo growth as plant height, leaves numbers and chlorophyll (Awad MA 2008), *Olea europaea* L growth parameters were significant increased with 5- aminolevulinic acid from 200-250 ppm (Al-Qurashi et al. 2011 and Aml et al. 2011), (Darwesh et al. 2014) showed that different spraying of amino acids had significant positive effect on the vegetative growth (Table 7) and chemical composition of acclimatized plantlets cv. Bartomouda

Table 7. Effect of 5- aminolevulinic acid and different amino acids on plant height (cm), leaves numbers and leaves stem diameter at 1st and 2nd seasons

Treatments		Plant height (cm)			Leaves numbers/plant			Stem diameter (cm)		
		1st season	2nd season	Mean	1st season	2nd season	Mean	1st season	2nd season	Mean
Control		30.3	62.0	46.2	5.3	8.0	6.7	5.8	8.3	7.1
ALA mg/l	100	32.3	78.7	55.5	5.7	9.3	7.5	7.3	8.4	7.9
	200	65.3	86.0	75.7	8.0	10.7	9.4	10.2	11.4	10.8
	300	75.3	94.7	85.0	8.7	11.7	10.2	11.4	12.8	12.1
	Mean			72.1			9.0			10.3
glutamine mg/l	100	32.0	78.3	55.2	5.7	9.3	7.5	8.0	8.7	8.4
	200	64.0	85.7	74.9	8.0	10.3	9.2	10.8	11.5	11.2
	300	75.0	94.3	84.7	8.7	11.7	10.2	12.6	13.2	12.9
	Mean			71.6			9.0			10.8
asparagine mg/l	100	31..6	77.0	54.3	6.0	9.0	7.5	7.9	8.4	8.2
	200	61.0	84.0	72.5	6.7	9.6	8.2	10.8	11.2	11.0
	300	72.3	90.0	81.2	7.3	10.7	9.0	12.0	13.5	12.8
	Mean			69.3			8.2			10.7
arginine mg/l	100	31.0	76.7	53.9	6.0	9.0	7.5	7.4	8.4	7.9
	200	59.3	83.0	71.2	6.3	10.0	8.2	10.1	11.3	10.7
	300	72.0	89.0	80.5	7.3	10.3	8.8	11.4	12.3	11.9
	Mean			68.5			8.2			10.2
proline mg/l	100	32.0	78.3	55.2	5.7	9.3	7.5	7.4	8.4	7.9
	200	64.7	87.0	75.9	7.3	10.7	9.0	10.4	11.5	11.0
	300	76.0	95.0	85.5	8.3	11.7	10.0	11.3	12.3	11.8
	Mean			72.2			8.8			10.2
tyrosine mg/l	100	30.7	64.3	47.5	5.3	8.0	6.7	5.9	8.3	7.1
	200	59.0	76.3	67.7	5.3	9.6	7.5	9.3	10.3	9.8
	300	61.3	89.0	75.2	6.0	10.0	8.0	10.0	10.7	10.4
	Mean			63.5			7.4			9.1
tryptophan mg/l	100	32.0	78.3	55.2	6.0	9.3	7.7	7.3	8.4	7.9
	200	65.0	86.0	75.5	8.0	10.3	9.2	10.2	11.3	10.8
	300	75.7	94.3	85.0	9.0	11.7	10.4	11.7	12.7	12.2
	Mean			71.9			9.1			10.3
L.s.d.		A= 2.9 B= 2.2 AB= 5.8			A= 0.6 B= 0.5 AB= 1.1			A= 0.6 B= 0.5 AB= 1.2		

A= amino acids B= levels AB= A x B (from Darwesh Rasmia et al. 2014)
(Darwesh Rasmia et al. 2014)

ROOTING OFFSHOOTS

Date palm off-shoots are the axillary buds on the trunk base of palm, offshoots are true to type to the parent palm, date palm tree produced limited numbers of off-shoots (20-30) during juvenile stage (10 – 15) years, meanwhile these off-shoots have the lowest transplanting survival and needs to improve new rooting at the trunk base. The root system of palms are adventitious and composed of numerous, small- to medium-sized, non woody roots, the primary roots are less constant diameter and arise independently at near base of the stem called the root initiation zone (RIZ), these root system characteristics were ability to store water and carbohydrates palms are relatively easy to transplant, (Tomlinson 1990), date palm cv Hillawi, viz., small (1-4 kg), medium (5-11 kg) and large (12-20 kg) were treated with 0, 500, 1000 and 2000 mg/l of IBA using quick-dip method, the supremacy of large sized rooted ones over medium and small sizes (Nasir 1996), *Phoenix canariensis* canary Island root regeneration of juvenile plants was found to be abundant in transplanted date palm, (Hodel *et al.* 2003), IBA and NAA at 1000,2000 and 3000 mg/l improved root length and numbers, root hairs and root thickness of date palm cv Hillawi (Al-Najm 2009 and Afzal *et al.* 2011 on date palm cv Hillawi), more than 96% of offshoots rooted if treated with 10 or 15 mg/l IBA solutions, while only 50% rooting was obtained with 2 mg/l IBA treated and control offshoots. These rooted offshoots produced an average of 12 visible roots per offshoot. Furthermore, 90% of offshoots of less than 2 kg rooted well if treated with 10 or 15 mg/l IBA solutions (Zirari and Ichir 2010 on date palm cv. Najda offshoots), Two types of date palm (*Phoenix dactylifera* L.) offshoots occur on a date palm tree: the lower and older ones, and the upper and younger ones. It is believed that low offshoots are more active physiologically than high ones; they probably grow faster (the number of leaves produced increases with age). In fact, the high offshoots have less carbohydrate than low offshoots, resulting in low roots production and consequently low survival rate, these offshoot cannot be planted into the field directly after removal from the mother plant, one to two years in a nursery is essential in order to ensure an optimum survival rate and to avoid uneven development of the plantation, (Zaid and De wet 2012). 500ml of auxins mixture (500ppm IBA + 500ppm NAA) and 2.5g/l licorice root extract rooting percentage, growth characteristics of roots and vegetative growth characteristics (El-Dengawy et al. 2017), also localized IBA powder improved offshoots rooting (Bitar et al.2019). Offshoots weights (4, 8 and 12 kg) treated with IBA and Pbz, 72.2% of offshoots successfully rooted if treated with 4000 ppm IBA furthermore 77.7 and 86.1% successfully rooted with 2000 and 4000 ppm/l IBA + 0.4 mg/l Pbz, while 16.7% rooted offshoots with control treatment, the treatment with the IBA at 4000 ppm/l led to highest length of leaves and roots, numbers of new leaves, roots/ offshoots, weight of offshoots and offshoot diameter, to regard the effect of Pbz at 0.4 mg/l on these parameters, the treatments 1000, 2000 and 4000 ppm IBA + 0.4 Pbz significant increased all growth parameters, significant higher was to be at 4000 ppm IBA + 0.4 mg/l Pbz. The offshoots at 12 kg induced highest significant (Table 8) successfully rooting 83.3% if treated with 4000 ppm IBA, while 91.7% at 4000 ppm IBA + 0.4 mg/l Pbz, (Rasmaia S. Darwesh et al. 2013)

Table 8. Effect of IBA alone and in combination with 0.4 pbz treatments on survival percentage % and leaf length (cm) of date palm off-shoots cultivar Zaghloul

Offshoots measure / Treat IBA + Pbz	Survival percentage %				Leaf length (cm)			
	4 kg	8 kg	12 kg	Mean	4 kg	8 kg	12 kg	Mean
Con	0.0	0.0	50.0	16.7%	0.0	0.0	150.0	50.0
1000 ppm IBA	0.0	33.3%	66.7%	33.3%	0.0	75.0	120.0	65.0
2000 ppm IBA	33.3%	58.3%	75.0%	55.5%	100.0	120.0	152.0	124.0
4000 ppm IBA	58.3	75.0%	83.3%	72.2%	110.0	130.0	250.0	163.3
Mean	22.9%	41.7%	66.7%		52.5	81.3	168.0	
1000 ppm IBA + 0.4 Pbz	33.3%	66.7%	83.3%	61.1%	80.0	120.0	135.0	111.7
2000 ppm IBA + 0.4 Pbz	66.7%	83.3%	83.3%	77.7%	120.0	125.0	150.0	131.7
4000 ppm IBA + 0.4 Pbz	75.0%	91.7%	91.7%	86.1%	115.0	140.0	238.0	164.3
Mean	43.8	60.4%	77.1%		78.8	96.3	168.3	
L.s.d.					A= 1.2	B= 0.8	AB= 2.1	

REFERENCES

Abahmane, L. (2013). Recent achievements in date palm (*Phoenix dactylifera* L.) micropropagation from inflorescence tissues. *Emirates Journal of Food and Agriculture*, *25*(11), 863–874. doi:10.9755/ejfa.v25i11.16659

Abd-El Hamed, Rasmia, & Zayed. (2017). Evaluation Physical and Chemical Characteristics of Some Seedlings Date Palm Fruits (Maghal) in the North Delta Egypt. *International Journal of Advances in Agricultural Science and Technology*, *4*(7), 13–32.

Abd-ElBaky, M. A. (2012). Using morphological and Anatomical features as taxonomical evidences to differentiate between some soft and semidry Egyptian cultivars of date palm. *Journal of Horticultural Science & Ornamental Plants*, *4*(2), 195–200.

Abd-ElHamied (2014). Effect of Some Agro-Management Systems on Growth and Production of Date Palm Off-Shoots under North Sinai Conditions. *IOSR Journal of Agriculture and Veterinary Science*, *7*(11), 41.

Abdel-Monaim, M. F., Khalil, M. S. M., & Sahar, H. A. (2018). Application of date palm leaves compost (DPLC) and plant growth promoting rhizobacteria(PGPR) for controlling faba bean root rot disease in New Valley, Egypt. *Journal of Phytopathology and Pest Management*, *5*(1), 88–100.

Abdi, G. H., & Hedayat, M. (2010). Yield and fruit physiochemical characteristics of 'Kabkab' date palm as affected by methods of potassium fertilization. *The Free Library Science and Technology*. www.google.com

Abo-Rekab, Darwesh, & Hassan. (2010). Effect of arbuscular mycorrhizal fungi, NPK complete fertilizers on growth and concentration nutrients of acclimatized date palm plantlets. *Mesopotomia J. of Agric*, *38*(1), 1–11.

Afzal, M., Khan, M. A., Pervez, M. A., & Ahmed, R. (2011). Root induction in the aerial offshoots of date palm (*Phoenix dactylifera* L.) cultivar Hillawi. *Pakistan Journal of Agricultural Sciences*, *48*(1), 11–17.

Ahmed, F. A., Hamdy, I., Ibrahim, I. M., & Kamel, M. K. h. (2014). Reducing Inorganic N Partially in Zaghloul Date Palm Orchards by Using Humic Acid and Effective Microorganisms. *World Rural Observ*, *6*(2), 102–110.

Al-Aa'reji. (2010). Effect of organic fertilizer, urea and sulphur on vegetative growth and concentration of some nutrient of young peach trees cv. Dixired. *J.Tikrit Univ.Agric.Sci*, *10*(2), 76–86.

Al-Busaidi, K. T. S. (2013). Effects of organic and inorganic fertilizers addition on growth and yield of banana (*Musa* AAA cv. Malindi) on a saline and non-saline soil in Oman. *Journal of Horticulture and Forestry*, *5*(9), 146–155.

Al-Meer, Ahmed, Al-Najim, & Yaseen. (2008). Effect of treatment by polyethelene glyol on acclimatization of date palm (*Phoenix dactylifera* L.) cv. Barhee propagated in vitro. *El-Basra Journal for Date Palm Research,* *7*(1), 1-9.

Al-Najm, A.R.A. (2009). Effect of Indole butyric acid and Indole- 3-acetic acid on root initiation data palm (*Phoenix dactylifera*)c.v Hillawi. *Basrah Journal for Date Palm Research*, *8*(1), 13–20.

Al-Qurashi, A. D., & Awad, M. A. (2011). 5-aminolevulinic acid increases tree yield and improves fruit quality of (Rabia) and (Sukkariat- Yanbo) date palm cultivars under hot arid climate. *Scientia Horticulturae*, *129*, 441–448.

Al-Taha, H.A. (2013). Effect of Different Treatments on Regeneration, Growth and acclimatization of Ananas Plantlets Produced By Plant Tissue Culture 2-Application Effect of Culture Media and Urea Fertilizer on Growth and Acclimatization of In vitro Propagated Pineapple (*Ananas Comosus* L. Meer. cv.Del Mont) Plantlet. *Tekret Journal*, *13*(1), 230–23.

Ali, Y. S. S. (2008). Use of Date Palm Leaves Compost as A Substitution to Peatmoss. *American Journal of Plant Physiology*, *3*(4), 131–136.

Aml, R. M. Y., Hala, S. E., & Saleh, M. M. S. (2011). Olive seedlings growth as affected by humic, amino acids, macro and trace elements applications. *Agriculture and Biology Journal of North America*, *2*(7), 1101–1107.

Amro, El- Sayed, & El Gammal. (2014). Effect of Effective Microorganisms (EM) and Potassium Sulphate on Productivity and Fruit Quality of "Hayany" Date Palm Grown Under Salinity Stress IOSR. *Journal of Agriculture and Veterinary Science (IOSR-JAVS), 7*(6), 90-99.

Aracama, C. V., Michael, E. K., Sandra, W. B., & Nancy, P. L. (2008). Comparative growth, morphology, and anatomy of easy- and difficult-to-acclimatize Sea Oats (*Uniola paniculata*) Genotypes During *in vitro* Culture and Ex Vitro acclimatization. *Amer Journal Soc. Hort. Sci., 133*(6), 830–843. doi:10.21273/JASHS.133.6.830

Awad, M. A. (2008). Promotive effects of a 5- aminolevulinic acid- based fertilizer on growth of tissue culture- derived date palm plants (*Phoenix dactylifera* L.) during acclimatization. *Scientia Horticulturae, 118*, 48–52.

Bekheet, S. (2013). Direct Organogenesis of Date Palm (*Phoenix dactylifera* L.) for Propagation of True-to-Type Plants. *ScientiaAgriculturae, 4*(3), 85–92.

Bhattacharjee, S. K. (2006): *Advances in Ornamental Horticulture.* Pointer.

Bitar, Abu-Qaoud, & Al-Said. (2019). Studies on Date Palm Propagation by Offshoots. *Palestinian Journal of Technology & Applied Sciences,* (2), 62-68.

Carvalho, L. C., Osorio, M. L., Chaves, M. M., & Amancio, S. (2001). Chlorophyll inflorescences as an indicator of photosynthetic functioning of in vitro grapevine and chestnut plantlets under ex vitro acclimatization. *Plant Cell, Tissue and Organ Culture, 67*(3), 271–280. doi:10.1023/A:1012722112406

Chandra, S., Bandopadhyay, R., Kumar, V., & Chandra, R. (2010). Acclimatization of tissue cultured plants: From laboratory to land. *Biotechnology Letters, 32*(9), 1199–1205. doi:10.100710529-010-0290-0 PMID:20455074

Chandrasekar, B. R., Ambrose, G., & Jayabalan, N. (2005). Influence of biofertilizers and nitrogen source level on the growth and yield of *Echinochloa frumentacea* (Roxb.). *Link. J. Agric. Technol., 1*, 223–234.

Darwesh. (2015). Morphology, physiology and anatomy in vitro affected acclimatization ex vitro date palm plantlets: A Review. International Journal of Chemical. *Environmental & Biological Sciences, 3*(2), 175–182.

Darwesh, El-Banna, & Zayed. (2016). Enhancing growth of date palm plantlets by different fertilizer treatments in the acclimatization greenhouse. *International Journal of Scientific Research & Engineering Trends*, 2(1), 42–49.

Darwesh, R. S. S., & Mohamed, H. f. (2009). The adverse effect of GA3 (Gibberellic acid) on salinity of of *Phoenix dactylifera* L. Plantlets in vitro rooting stage. *4th conference on Recent Technology in Agriculture*, 1-8.

Darwesh, R. S. S., Zaid, Z. E., & Sidky, R. A. (2011). Effect of Ammonium Nitrate and GA_3 on Growth and Development of Date Palm Plantlets in Vitro and Acclimatization Stage. *Research Journal of Agriculture and Biological Sciences, 7*(1), 17–22.

Darwesh Rasmia, S. S., Abeer Abd-El Kareim, H. E., & Mona, H. M. (2014). Effect of Foliar Spraying With 5- Aminolevulinic Acid and Different Types Amino Acids on Growth of Date Palm of Plantlets afte Acclimatization in the Green House. *International Journal of Plant and Soil Science*, *3*(10), 1317–1332.

Darwesh Rasmia, S. S., Eman Zayed, M., & Hala Farag, M. A. (2021). Influence of different carbohydrate source on improving in vitro rooting, growth and ex vitro survival of date palm Plantlet. *Plant Cell Biotechnology and Molecular Biology*, *22*(29&30), 109–122.

Diatta, I. L. D., Kane, A., Agbangba, C. E., Sagna, M., Diouf, D., Aberlenc-Bertossi, F., Duval, Y., Borgel, A., & Sane, D. (2014). Inoculation with arbuscular mycorrhizal fungi improves seedlings growth of two sahelian date palm cultivars (*Phoenix dactylifera* L., cv. Nakhla hamra and cv. Tijib) under salinity stresses. *Advances in Bioscience and Biotechnology*, *5*, 64–72.

Dumroese, R. D., Deborah, S., Dumroese, P., & Brown, R. E. (2011). Allometry, nitrogen status, and carbon stable isotope composition of Pinus ponderosa seedling in two growing media with contrasting nursery irrigation regimes. *Canadian Journal of Forest Research*, *41*, 1091–1101.

Al-Temimi. (2012). Effect the addition of equivalent ratio of mechanical firtilizarers on growth of date palms (*Phoenix dectylifera* L.) Barhi cultivar. *Basra Science Journal*, *38*(4), 60–73.

El-Bahr, M. K., Ali, Z. A., & Saker, M. M. (2003). A comparative anatomical study of date palm vitro plants. *Arab Journal of Biotechnology*, *7*(2), 219–228.

El-Daen. (2019). Effect of fertilization by injection of soil and trunk with NPK on productivity and fruits quality of Sewy date palm. *J. Agric. Res, & Dev., 33*(1).

El-Dengawy, E. F. A., Wanas, A. L. E., & Mervat, H. M. (2017). Improvement of the rooting efficiency and vegetative growth in date palm offshoots by Licorice root extract and auxins mixture applications. J. Plant Production, Mansoura Univ., 8(7), 789 – 796.

El-Khawaga, A. S. (2013). Effect of anti-salinity agents on growth and fruiting of different date palm cultivars. *Asian Journal of Crop Science*, *5*(1), 65–80.

El-Kosary, Bakr, Hussien, & Sheren. (2008). Effect of the irrigation system and soil components on the growth of date palm offshoots in the nursery. *Egyptian Journal of Applied Sciences, 23*(8), 602-620.

Elsayd, El-Merghany, & El-Dean. (2018). Influence of Potassium Fertilization on Barhee Date Palms Growth, Yield and Fruit Quality Under Heat Stress Conditions. *J. Plant Production*. *Mansoura Univ.*, *9*(1), 73–80.

Esmaeilifar, A. (2013). Influence Of Irrigation, Phosphorus And Mycorhiza On Date Palm. *Advances in Environmental Biology*, *7*(1), 123–130.

Ezz Gadalla, G., Rasmia Darwesh, S., Rehab Sidky, A., & Zeinab Zaid, E. (2008). Growth and Developing of date palm plantlet for pre- acclimatization stage: The effect of growth retardants and vessel types. *Egyptian Journal of Biotechnology*, *29*(6), 173–189.

Fathy, M. A., Gabr, M. A., & El Shall, S. A. (2010). Effect of Humic Acid Treatments on 'Canino' Apricot Growth, Yield and Fruit Quality. *New York Science Journal*, *3*(12), 109–115.

Fatima, G., Khan, I. A., Jaskani, M. J., & Ul-Ain Rasool, Q. (2010). studies on different cultivars of date palm (Phoenix dactylifera L.) and their comparative root anatomy. *Science International (Lahore)*, *24*(2), 177–180.

Fki, L., Masmoudi, R., Kriaa, W., Mahjoub, A., Sghaier, B., Mzid, R., Mliki, A., Rival, A., & Drira, N. (2011). Date Palm Micropropagation via Somatic Embryogenesis. In Date Palm Biotechnology. Springer.

Growth Promotion of Date Palm Plantlets. (n.d.). *Ex vitro* by Inoculation of Rhizosphere Bacteria. *ARAB PALM Conference*.

Gupta, M. L., Prasad, A., Ram, M., & Kumar, S. (2002). Effect of the vesicular-arbuscular mycorrhizal (VAM) fungus *Glomus fasiculatum* on the essential oil yield related characters and nutrient acquisition in the crops of different cultivars of menthol mint (*Mentha arvensis*) under field conditions. *Bioresource Technology*, *81*, 77–79.

Harhash, M. M., & Abdel-Nasser, G. (2007). Impact of potassium fertilization and bunch thining on Zaghloul date palm. In *The Fourth Symposium on Date Palm in Saudi Arabia*. King Faisal University.

Hassanen, S. A., & Khalil, R. M. A. (2013). Biotechnological Studies for Improving of Stevia (*Stevia rebaudiana* Bertoni) in vitro Plantlets. *Middle East Journal of Scientific Research*, *14*(1), 93–106.

Haynes, R. J., & Naidu, R. (1998). Influence of lime, fertilizer and manure applications on soil organic matter content and soil physical conditions: A review. *Nutrient Cycling in Agroecosystems*, *51*(2), 123–137. doi:10.1023/A:1009738307837

Hazarika, B. N. (2003). Acclimatization of tissue cultured plants. *Current Science*, *85*, 1704–1712.

He, C., Zeng, Y., Fu, Y., Wu, J., & Liang, Q. (2020). Light quality affects the proliferation of in vitro cultured plantlets of Camellia oleifera Huajin. *PeerJ*, *8*, 10016. doi:10.7717/peerj.10016 PMID:33083122

Higa, T. (1991). Effective microorganisms: A biotechnology for mankind. *Proc First International Conference on Kyusei Nature Farming*, 8–14.

Hodel, D. R., Pittenger, D. R., & Downer, A. J. (2003). Effects of leaf removal and tie up on juvenile, transplanted of canary Island date palms (*Phoenix dactylifera* L.) and queen palms (*Syagrus ramanzoffiana*). *Palms*, *47*(4), 177–184.

Ibrahim, A. E., Mona, H. M., & Darwesh, R. S. S. (2012). Pre- acclimatization stage of date palm (*Phoenix dactylifera*) plantlets cv. Barhee as affected by potassium nitrate and sucrose. *Journal of Biological Chemistry & Environmental Sciences*, *7*(3), 1–20.

Junaid, A., & Khan, S. A. (2009). In vitro micropropagation of 'khalas' date palm (*Phoenix dactylifera* l.) an important fruit plant. *Journal of Fruit and Ornamental Plant Research*, *17*(1), 15–27.

Kassem, H. A. (2012). The response of date palm to calcareous soil fertilization. *Journal of Soil Science and Plant Nutrition*, *12*(1), 45–58. doi:10.4067/S0718-95162012000100005

Kassem, H. A., El-Sabrout, M. B., & Attia, M. M. (1997). Effect of nitrogen and potassium fertilization on yield, fruit quality and leaf mineral content in some Egyptian soft varieties. *Alexandria Journal of Agricultural Research*, *42*(1), 137–157.

Khan, S., & Bi Bi, T. (2012). Direct shoot regeneration system for date palm (*Phoenix dactylifera* L.) cv. Dhakki as a means of micropropagation. *Pakistan Journal of Botany*, *44*(6), 1965–1971.

Khierallah, H. S. M., & Bader, S. M. (2007). Micropropagation of Date Palm (*Phoenix dactylifera* L.) var. Maktoom through Direct Organogenesis. Proc. IIIrd IC on Date Palm. *Acta Horticulturae*, (736), 213–224. doi:10.17660/ActaHortic.2007.736.19

Kozai, T., Afreen, F., & Zobayed, S. M. A. (2005). *Photoautotrophic (sugar-free medium) micropropagation as a new micropropagation and transplant production system*. Springer. doi:10.1007/1-4020-3126-2

Kurup, S. S., Aly, M. A. M., Lekshmi, G., & Tawfik, N. H. (2014). Rapid in vitro regeneration of date palm (*Phoenix dactylifera* L.) cv. Kheneizi using tender leaf explant. *Emirates Journal of Food and Agriculture*, *26*(6), 539–544. doi:10.9755/ejfa.v26i6.18051

Luz, Paiva, & Tavares. (2008). Effect of Foliar and Substrate Fertilization on Lady Palm Seedling Growth and Development. *Journal of Plant Nutrition*, *31*, 1311–1318.

Mahdia, F. G., & Abd-Alla, M. M. (2010). Micro Propagation of *Phoenix dactylifera* L. var karama. *New York Science*, *3*(12), 64–69.

Medina, O., Anaya, L. A., Aguilar, A. C., Llaven, A. O., Talavera, T. A., Dendooven, L., Miceli, F. G., & Figueroa, M. S. (2007). *Ex vitro* Survival and Early Growth of *Alpinia purpurata* Plantlets Inoculated with Azotobacter and Azospirillum. *Pakistan Journal of Biological Sciences*, *10*(19), 3454–3457. doi:10.3923/pjbs.2007.3454.3457 PMID:19090169

Morte, A., & Honrubia, M. (2002). Growth response of Phoenix canariensis to inoculation with Arbuscular mycorrhizal fungi. *Morte & Honrubia*, *46*(2), 76- 80.

Muhsen, K. A., Ibrahim, A., & Jasim, A. M. (2013). Effect of NPK and Hoagland solution on accumulation of date palm plantlets Phoenix dactylifera L. cv. Barhee production in vitro. *Journal of Basrah Researches (Sciences)*, *39*(4B), 32–45.

Muthukrishnan, Benjamin, & Sathishkannan, Kumar, & Rao. (2013). *In vitro* propagation of genus Ceropegia and Retrosynthesis of cerpegin- A Review. *International Journal of Pharmaceutical Sciences and Research*, *22*(2), 315–330.

Nagodawithana, W.T. (1991). *Yeast technology. Universal foods cooperation.* Van Nostrand.

Nalawde & Bhalerao. (2013). Comparative account of Effect of Biofertilizers on the growth and biochemicalparameters of *Vigna mungo* (L.Hepper). *International Journal of Advanced Research in Biological Sciences*, *2*(5), 62–66.

Nasir, M. J. (1996). *Effect of IBA on rooted and un-rooted offshoots of date palm (Phoenix dactylifera L.) cv. Hillawii* [M.Sc. Thesis]. Dept. Hort., Univ. Agric., Faisalabad, Pakistan.

Norikane, Teixeira da Silva, & Tanaka. (2013). Growth of in vitro Oncidesa plantlets cultured under cold cathodefluorescent lamps with super-elevated CO_2 enrichment. *Journal of Plant Sciences*, *22*, 1–9.

Norikane, A., Takamura, T., Morokuma, M., & Tanaka, M. (2010). In vitro growth and single-leaf photosynthetic response of Cymbidium plantlets to super-elevated CO2 under cold cathode fluorescent lamps. *Plant Cell Reports*, *29*(3), 273–283. doi:10.100700299-010-0820-1 PMID:20094885

Osman, Moustafa, El-Galil, & Ahmed. (2011). Effect of yeast and Effective Microorganisms (Em1) application on the yield and fruit characteristics of Bartamuda date palm under Aswan conditions. *Assiut J. of Agric. Sci.*, *42*, 332-349.

Othmani, A., Bayoudh, C., Drira, N., Marrakchi, M., & Trifi, M. (2009). Regeneration and molecular analysis of date palm (Phoenix dactylifera L.) plantlets using RAPD marker. *African Journal of Biotechnology*, *8*, 813–820.

Pant, M., Bisht, P., & Gusain, M. P. (2010). *In vitro* propagation through axillary bud culture of *Swertia chirata* Buch- Ham ex Wall: An endangered medicinal herb. *International Journal of Integrative Biology*, *10*(1), 48–53.

Pospíšilová, J., Synková, H., Haisel, D., Čatský, J., Wilhelmová, N., & Šrámek, F. (1999). Effect of elevated CO_2 concentration on acclimation of tobacco plantlets to ex vitro conditions. -. *Journal of Experimental Botany*, *50*(330), 119–126. doi:10.1093/jxb/50.330.119

Preece, J. E. (2010). *Acclimatization of Plantlets from In vitro to the Ambient Environment. Encyclopedia of Industrial Biotechnology: Bioprocess*. Bioseparation, and Cell Technology. doi:10.1002/9780470054581.eib593

Qi, Ying, Xin, & Li-N., Shi-Jin, Wen, Juan, Zong, & Ma. (2021). Effects of CO_2 on transplantation of grape plantlets cultured in vitro by promoting photosynthesis. *Scientia Horticulturae*, *287*(20), 110286.

Rajeswari, V., & Paliwal, K. (2008). In vitro plant regeneration of red sanders (Pterocarpus santalinus L.f) from cotyledonary nodes. *Indian Journal of Biotechnology*, *7*, 541–546.

Rajmohan, K. (2011). Date palm tissue culture: A pathway to rural development. In S. M. Jain, J. M. Al-Khayri, & D. V. Johnson (Eds.), *Date palm biotechnology*. Springer. doi:10.1007/978-94-007-1318-5_3

Rana, R. R., & Kapoor, P. (2013). Biofertilizersand Their Role in Agriculture. *Popular Kheti*, *1*(1), 56–61.

Rasmaia Darwesh, S., Essam Madbolly, A., & Ezz Gadalla, G. (2013). Impact of Indole Butyric Acid and Paclobutrazol on Rooting of Date Palm (Phoenix dactylifera L.) Off-Shoots Cultivar Zaghloul. *Journal of Horticultural Science & Ornamental Plants*, *5*(3), 145–150.

Rasmia Darwesh, S. S. (2010). *Phoenix dactylifera* cv. Medjol plantlets as affected by yeast extract and NPK fertilizers. *The Six International Conference Of Sustainable Agricultural Development*, 115-130.

Saez, P. L., Bravo, L. A., Saez, K. L., Sánchez-Olate, M., Latsague, M. I., & Ríos, D. G. (2012). Photosynthetic and leaf anatomical characteristics of Castanea sativa: A comparison between in vitro and nursery plants. *Biologia Plantarum*, *56*(1), 15–24. doi:10.100710535-012-0010-9

Saleh, J. (2009). Yield and chemical composition of 'Piarom' Date-Palm *Phoenix dactylifera* L. as affected by nitrogen and phosphorus Levels. *International Journal of Plant Production*, *3*(3), 57–64.

Seelye, J. F., Burge, G. K., & Morgan, E. R. (2003). Acclimatizing tissue culture plants: Reducing the shock. *Proc. Int. Plant Prop. Soc.*, *53*, 85–90.

Sghir, F., Chliyeh, M., Touati, J., Mouria, B., Ouazzani, A., Touhami, A., Filali-Maltouf, C., El Modafar Moukhli, A., Benkirane, R., & Douira, A. (2014). Effect of a dual inoculation with endomycorrhizae and *Trichoderma harzianum* on the growth of date palm seedlings. *Int. J. Pure App. Biosci.*, *2*(6), 12–26.

Sharma, M., Sood, A., Nagar, P. K., Prakash, O., & Ahuja, P. S. (1999). Direct rooting and hardening of tea microshoots in the field. *Plant Cell, Tissue and Organ Culture*, *58*(2), 111–118. doi:10.1023/A:1006374526540

Sicher, R. C. (2008). Effects of CO_2 enrichment on soluble amino acids and organic acids in barley primary leaves as a function of age, photoperiod and chlorosis. *Plant Science*, *174*(6), 576–582. doi:10.1016/j.plantsci.2008.03.001

Somers, E., Vanderleyden, J., & Srinivasan, M. (2004). Rhizosphere bacterial signaling: A love parade beneath our feet. *Critical Reviews in Microbiology*, *30*, 205–240.

Sumaryono & Riyadi, I. (2011). *Ex vitro* rooting of oil palm (*Elaeis guineensis* Jacq.) plantlets derived from tissue culture. *International Journal of Agricultural Sciences*, *12*(2), 57–62.

Tomlinson, P. B. (1990). *The structural biology of palms*. Oxford University Press., www.google.com

Trejgell, A., & Tretyn, A. (2011). Shoot multiplication and in vitro rooting of *Carlina onopordifolia* Basser. *Acta Biologica Cracoviensia. Series; Botanica*, *53*(2), 68–72. doi:10.2478/v10182-011-0026-z

Ur-Rahman, A. H., James, D. J., Caligari, P. D. S., & Wetten, A. (2007). Difference in competence for in vitro proliferation and ex vitro growth of genetically identical mature and juvenile clones of apomictic Malus species. *Pakistan Journal of Botany*, *39*(4), 1197–1206.

Vessey, J. K. (2003). Plant growth promoting rhizobacteria asbiofertilizers. *Plant and Soil*, *255*, 571–586.

Winarto, B., & Setyawati, A. S. (2014). Young shoot nodes derived organogenesis in vitro mass propagation of Ruscus hypophyllum L. *South Western Journal of Horticulture, Biology and Environment*, *5*(2), 63–82.

Yasodha, R., Sumathi, R., & Gurmurthi, K. (2005). Improved micropropagation methods for Teak. *Journal of Tropical Forest Science*, *17*(1), 63–75.

Zaid, A., El-Korchi, B., & Visser, H. J. (2011). Commercial date palm tissue culture procedures and facility establishment. In S. M. Jain, J. M. Al-Khayri, & D. V. Johnson (Eds.), *Date palm biotechnology*. Springer. doi:10.1007/978-94-007-1318-5_8

Zaid, A., & De wet, P.F. (2012). *Date Palm Propagation: Offshoot propagation*. www.agrihunt.com, www.google.com

Zayed, E. M. M., Rasmia, S. S. D., El-Din, A. F. M. Z., & Farrag, H. M. A. (2014). Impact of different sources of nitrogen fertilizers on performance growth of date palm (*Phoenix dactylifera* L. cv. Bartomouda). *Arab Universities Journal of Agricultural Sciences*, *22*(2), 371–379. doi:10.21608/ajs.2014.14742

Zirari, A., & Ichir, L. L. (2010). *Effect of exogenous Indole butyric acid (IBA) on rooting and leaf growth of small date palm offshoots (Phoenix dactylifera L.) derived from adult vitro plants of Najda cultivar.* International Society for Horticulture Science. www. Google .com

Chapter 13
Biological Control of Some Fruit Plants Under Organic Agriculture Systems:
Grapes

Mohamed Farouk Attya Ahmed
Central Laboratory of Organic Agriculture (CLOA), Agricultural Research Center, Egypt

ABSTRACT

The entire world has realized the danger of chemical pesticides because of their harmful impact on environment, biodiversity, and human health as a result of the accumulation of pesticides in the food and water chain and the outbreak of epidemic diseases such as cancer, kidney failure, and liver epidemic, which entails the cost of the state many funds to control these diseases. We must start changing the trend of the world by using biological control methods as alternatives to pesticides in order to eat healthy and safe food, raise soil fertility, increase biodiversity, and maintain the period of postharvest in the long term in addition to the booming economic aspect of the individual and society. The international organic agriculture law is concerned with controlling any diseases on plants and is prohibited from using any pesticides during the production, harvesting, processing, handling, and trading of organic crops.

BACKGROUND

Recently, human realized that using many chemical pesticides might have injury on the environment and human health because it highly toxic substances in agricultural, which led to great disturbance in the biological balance. This disturbance led to the appearance of new pests, caused reduction in number of natural enemies and increased the accumulated toxic chemicals in human food chain. In 2009, the new law Council Regulation of organic agriculture farming was issued in European Community (EC) No. 834/2007 in combination with 889/2008 on organic production and labeling of organic products and repealing regulation European Economic Community (EEC) No. 2092/91 came into force (Mikkelsen and Schluter, 2009). New group of food products appeared in the markets under different names, like

DOI: 10.4018/978-1-6684-2423-0.ch013

organic biodynamic and ecological food. All these names showed that no synthetic chemicals were added during production or processing.

The present work was designed to reduce using toxic chemicals in agriculture process to produce food of high quality in sufficient quantity, enhance biodiversity system, maintain and increase long-term fertility of soils. In addition, find out the most suitable non-chemical methods to protect the economic plants against soil-borne or foliar fungal diseases (Ahmed, 2013).

Defines Organic Farming: as an agricultural production system which avoids the use of mineral fertilizers and synthetic pesticides, which come into direct contact with the plant or soil.

Definition of Plant Pathology

The science of patient study plant and call it a scientific term Phytopathology derived from the ancient Greek name, which consists of three words and their meaning is Phyton meaning plant, Pathos meaning of the word disease and Logos meaning science. In additional to the scientific study of diseases in plants caused by pathogens (infectious organisms) and environmental conditions (physiological factors).

Definition of Plant Disease

Any deviation from the natural growth of the plant or any imbalance in the physiological functions of the plant, leading to the emergence of abnormal manifestations of the various members of the plant.

Factors which Occurring Plant Disease" The Principles of Plant Pathology: The Disease Triangle and Influence of the Environment"

Figure 1.

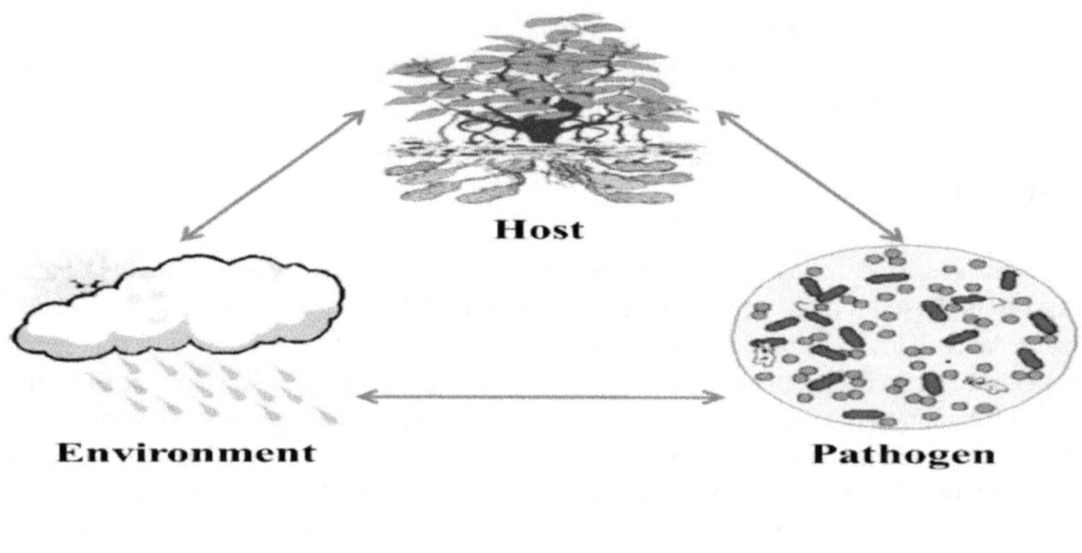

Definition of Biological control

It is defined as the use of micro or macro-organisms except plant or human to lose numbers or decrease the effect of another pathogen.

Figure 2.

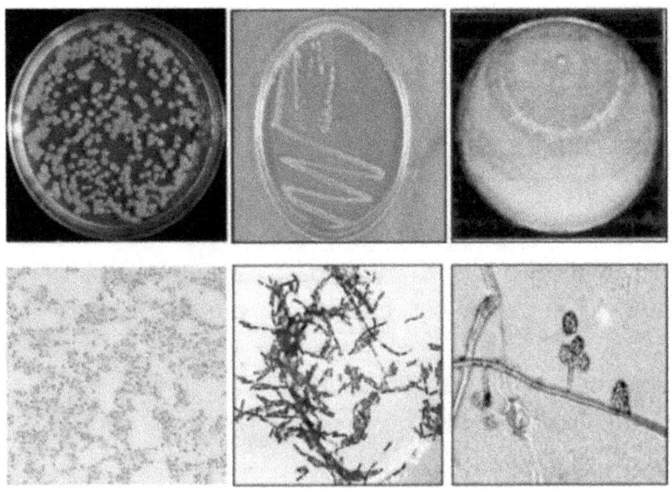

Serratia spp. بكتريا *Bacillus subtilis* بكتريا *Trichoderma* spp. فطر

The importance of using biological control and its relationship to sustainable development:

- The consumer's desire to obtain safe food that does not contain pesticide residues or chemicals, after the harmful effects of pesticides on human and animal health have been proven.
- Reducing environmental degradation: The current agricultural practices using pesticides and chemicals have led to the pollution of soil, water, and then the food chain (food and water).
- Preserving biological diversity.
- Maintain human health by avoiding the consumption of food and water contaminated with pesticides.
- Avoid depletion and pollution of natural resources, which have a wide working mechanism.
- It has a wide temperature range and acid medium.
- The speed of their reproduction and the cheap prices of their products.
- Non-toxic to human. Expensive, labor intensive and host specific.
- Not a water contaminant concern.
- Once colonized may last for years.
- Host specific.
- PH range are wide.

While Chemical pesticides are:

- Cost-effective, easy to apply and broad spectrum.
- Implicated in ecological, environmental, and human health problems.
- Require yearly treatments.
- Broad spectrum.
- Toxic to both beneficial and pathogenic species.

Definition of Biological Control

It is defined as the use of micro or macro-organisms except plant or human to lose numbers or decrease the effect of another pathogen.

Mechanisms of Biological Control of Plant Pathogens

- **Antibiosis** – inhibition of one organism by another as a result of diffusion of an antibiotic.
- Antibiotic production common in soil-dwelling bacteria and fungi.
- **Nutrient competition** – competition between microorganisms for carbon, nitrogen, O_2, iron, and other nutrients.
- Most common way organisms limit growth of others.
- **Destructive mycoparasitism** – the parasitism of one fungus by another.
    - Direct contact.
    - Cell wall degrading enzymes.
    - Some produce antibiotics.

Figure 3.

The nature of antagonist-pathogen interaction (morphological studies):

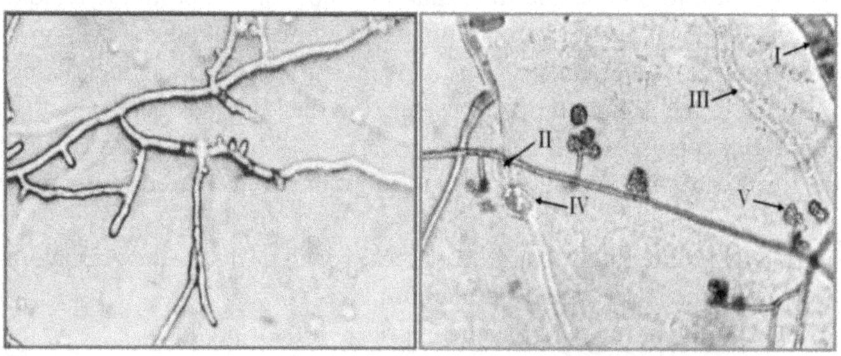

Fig. (12-a): Normal mycelium of *S. rolfsii* with clear cell wall. (X = 400)

Fig. (13-a): Malformation in *S. rolfsii* mycelium " X = 400"

The effect of *T. harzianum* on *S. rolfsii* due to:

I. Malformed hyphae show change in color and increase in wide with no clear cell walls.
II. Penetrating hyphae of *T. harzianum* Penetrate mycelium causing clear change of color at conduct point.
III. Lysised and decayed mycelium four days after subjection.
IV. Node formed on *S. rolfsii* mycelium of *T. harzianum*.
V. Conidiophore with conidia developed from infected *S. rolfsii* mycelium.

REQUIREMENTS OF SUCCESSFUL BIOCONTROL

1. **Highly effective biocontrol strain must be obtained or produced.**
   a. Be able to compete and persist.
   b. Be able to colonize and proliferate" Propagation ".
   c. Be non-pathogenic to host plant and environment.
2. **Inexpensive production and formulation of agent must be developed.**
   a. Production must result in biomass with excellent shelf live.
   b. To be successful as agricultural agent must be: -
      i. Inexpensive.
      ii. Able to produce in large quantities.
      iii. Maintain viability.
3. **Delivery and application must permit full expression of the agent.**
   a. Must ensure agents will grow and achieve their purpose.

Figure 4.

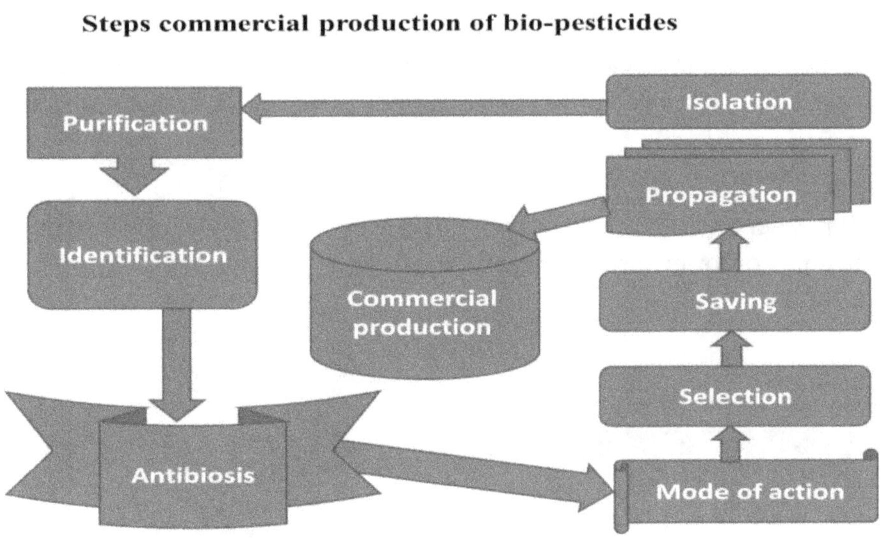

GRAPES

I took this fruit as examples for all fruit trees, its follow the same scheme of biological control for root rot diseases, soil-borne diseases or foliar diseases.

Fungal Diseases

Powdery Mildew Disease

The Causal Pathogen: Uncinula Necator

Powdery mildew is considered one of the most dangerous diseases that affect grapes in Egypt. This disease was recorded in Egypt in 1919 in an orchard in the Alexandria region, and then gradually spread. Along with the downy mildew, it is the most dangerous disease of grapes, but powdery mildew is more dangerous in destroying the fruits. While the danger of downy mildew is intensified in the northern delta, we find that powdery mildew spreads to the south and intensifies in some areas of Upper Egypt. And whiteness begins to appear from late April until the end of the season.

Most grape varieties are susceptible to severe infection with this disease, due to the delay in their ripening until the time when the air temperature and humidity are suitable for infection. As for the cultivars of the Sprior, the Earl of the Sprior and the Banati (Thomson Siddles), their fruits are spared from infection because they ripen before the appropriate conditions for infection are available, as they are among the early ripening cultivars.

Symptoms

Symptoms of infection with this disease appear on all parts of the plant above the surface of the earth (leaves - tender branches, flowers and fruits) in the various stages of its formation.

Symptoms on the Leaves

The appearance of powdery mildew on the leaves. The leaves appear on the leaves with minute grayish-white spots on the upper or lower surface or both surfaces.

Figure 5.

Together, but they are more visible on the upper surface. These spots extend under the appropriate conditions during hot, dry weather and with the progression of the infection, the color of the affected tissues turns into brown as a result of tissue death until it permeates the entire surface of the leaf and its fall.

The appearance of powdery mildew infection on ripe fruits, which is also susceptible to infection. If the flower clusters are infected, they wither and are unable to hold the fruits. But if the fruits start forming, their growth stopped, and they are covered with a grayish-white layer. But if the fruits are infected while they are in an advanced stage, they grow irregularly, dry out and take an unnatural color, and often crack and do not ripen, and when the infection intensifies, the affected areas emit a smell similar to the smell of rotten fish.

"Zaffa" as a result of the decomposition of the mycelium protein. Conditions suitable for injury to occur providing a high humidity of 80% and a temperature of 25° C.

Methods of Controlling

The most important methods of controlling in the light of integrated control:

First: Agricultural Control:

1. Use an appropriate breeding system that allows air and sun to enter the bushes and avoids shading.
2. Avoiding the increase in vegetative growth by controlling nitrogen fertilization and used organic composting and bio fertilizers.
3. Taking care of potassium fertilization has a significant role in reducing the incidence of disease.
4. Removing the basal leaves on the rootstock of shrubs, as they have a very important role in reducing infection.
5. Forecasting, which is based on linking meteorological information to the soil, which is useful in determining the dates of pesticide sprays, especially preventive ones.

Second: Biological Control:

- Bio-resistance must be carried out preventively before infection occurs, using the biological compound (*Ampelomyces quisqualis*) and sold commercially under the name (AQ10) at a concentration of 5 g / 100 liters of water, after the buds open and the arrival of modern vegetative growths of 10-15 cm in length at a rate of one spray every 15 day after that. The number of sprays depends on the variety grown.
- The recommended biofungicide which consists of (*Trichoderma* spp. 30×10^6) which used at the rate of 1 lit./50 Lit. water/fed., and repeat the spray after 2 weeks again, (Ahmed, 2018).

Third: Resistance by using alternatives to pesticides:

The resistance in this case must be done before the infection occurs, using sodium bicarbonate at a rate of 500 g / 100 liters of water at a sprinkle rate every 15 days, starting from the arrival of the vegetative growths to 10-15 cm in length until the flow of juice in the kernels.

Fourth: Chemical resistance:

Chemical resistance must be carried out preventively before infection occurs by using one of the following fungal antiseptics such as micronized sulfur, karathin - sorrel 80 - thiophyte, where the following program is followed:

1. The spraying begins with the use of micronized sulfur at a concentration of 250 g/100 liters of water sprayed on the wood when the buds swell until the germs lurking in the scales of the buds are eliminated. Micron sulfur is a useful nutritional element for the plant which helps to resist the static mites in the buds.
2. The spraying is repeated with micronized sulfur or any of the previous compounds, after the buds open and the arrival of modern vegetative growths of 10-15 cm in length at the rate of a spray every 15 days after that. Spraying with sulfur stops if the temperature rises above 29 degrees Celsius and is replaced by liquid certain at a rate of 60 cm^3/100 liters of water or powder at a rate of 100 g/100 liters of water. The number of sprays depends on the planted variety.

Downy Mildew Disease

The causal pathogen: *Plasmopara viticola*

This disease affects different types and varieties of grapes and is found in most of its cultivation areas, especially with humid weather conditions in Lower Egypt. The disease begins in June, especially in coastal areas, and continues to increase until November.

The economic importance of the disease:

The loss resulting from the disease is seasonal and its percentage does not exceed 10%. Most of the loss results from the effect of the disease on the vegetative parts and not from the direct rot of the fruits. If the infestation on the leaves is light, the damage is small. But if the infestation is severe, most of the leaves die and therefore the amount of nutrients decreases. Which is stored in the plant, and this causes a general weakness for it, especially if the infection is repeated annually and the sugar percentage is less than usual in the fruits taken from infected plants. The emergence of the disease begins under the Egyptian weather conditions as of mid-June, so it is not likely that the fruits will be infected because their maturity has been completed or is about to ripen.

Symptoms

1. **On the papers:**

It begins with the appearance of translucent pale-yellow spots with an oily appearance on its upper surface. In the case of severe infection, these spots extend and connect and permeate the entire surface of the leaf.

Figure 6.

A large part of the leaf may die between the main veins, offset on the lower surface by a white downy growth, which is the microbial carriers of the fungus, and it is difficult to distinguish in grape varieties with woolly lower surfaces gray. The disease also affects the petioles of the leaves and leads to their fall if they are severe.

2. **On the chicks and the tendrils:**

The infection leads to the shortening of the chicks and their increase in thickness than the normal branches, and the branch is covered with what is on it with the downy growth of the fungus. The infection leads to deformation and death of the branches.

Conditions Favorable to the Spread of the Disease

The spread of this disease is affected by temperature and air humidity. The moderately humid atmosphere in a continuous condition causes severe infection. While the dry weather stops the spread of the disease and the disease intensifies in areas where there is a lot of rain, because the occurrence of infection depends on the presence of water necessary for the spread and germination of the gift germs, even if the thick dew is sufficient to form the water membrane necessary for the spread of sporangia.

Methods of Controlling

The most important methods of resistance in the light of integrated control.
First: Agricultural resistance:

1. Pruning and appropriate breeding methods have a very important role in disease resistance.

2. Removing the basal leaves on the crowns of shrubs, as the presence of these leaves helps infection due to their proximity to the surface of the soil.
3. It is recommended not to plant any temporary crops under the bushes so as not to lead to an increase in moisture around the bushes, which increases the infection.
4. In the case of cultivating grapes on cubes, it is preferable that the cubes be as high as possible from the surface of the soil to reduce the chance of germs reaching the basal leaves by spraying rainwater or irrigation water.
5. Removing the weeds under the bushes, as their presence helps to increase the humidity around the bushes, which increases the infection.
6. Reducing the increase in vegetative growth by controlling the nitrogen fertilization by using composting and biofertilizers, as the increase in vegetative growth leads to an increase in the percentage of moisture and thus an increase in infection.
7. Taking care of potassium fertilization by using composting and biofertilizers, as it works to strengthen the cell walls, which hinders the penetration of the fungus that causes the cell walls of the plant (leaves - flowers - fruits). Also, potassium fertilization works to increase the contract rates, improve the properties of the fruits and increase the sugar percentage in the fruits.
8. Cultivation of resistant varieties, especially in areas where the disease is prevalent. It is known that European grape varieties in general are more resistant than American varieties. It was also found that the American Concord grape variety, strawberry grape, is resistant to downy mildew disease.
9. Pruning and culling the affected branches, collecting the fallen infected leaves and branches and burning them.

Second: Biological Control:

- Bio-resistance must be carried out preventively before infection occurs, using the biological compound of (*Trichoderma* spp. 30×10^6), with the rate of 1 Lit/50 Lit water/twice/every 2 weeks, which obtained from CLOA, (Ahmed, 2018).

Third: Chemical control:

Starting from mid-June, grape bushes protection begins by following the following program:
Spray the bushes with one of the following fungicides:

- Copper oxychloride at a rate of 300 g / 100 liters of water.

The bushes should be sprayed 3 sprays during the season, and if a fourth spray is needed, between two sprays and the other.

In areas where powdery mildew infestations appear and downy mildew is likely to appear, the recommended pesticides for the two diseases are added - and spraying is carried out starting from mid-June, taking care not to mix wettable pesticides with emulsified pesticides.

Grape Rot Diseases

1. **Aspergillus rot of grapes:**

Causal Pathogen Name: *Aspergillus flavus*

Symptoms

This fungus causes soft rot to the fruits of different grape varieties at relatively high temperatures from 25° C - 35° C. Contamination usually begins in the field, but mold usually spreads during storage, due to crowding of fruits and contact with healthy and infected ones. Ripe fruits are more susceptible to infection than immature fruits, as the latter have high acidity, which is not suitable for the growth of the causative fungus.

Figure 7.

2. **Rhizopus rot of grapes:**

The Causal Pathogen Name: *Rhizopus stolonifer*

It infects and damages the fruits of grapes at temperatures from 24 ° C - 30 ° C and causes similar symptoms to the fungus Aspergillus.

Figure 8.

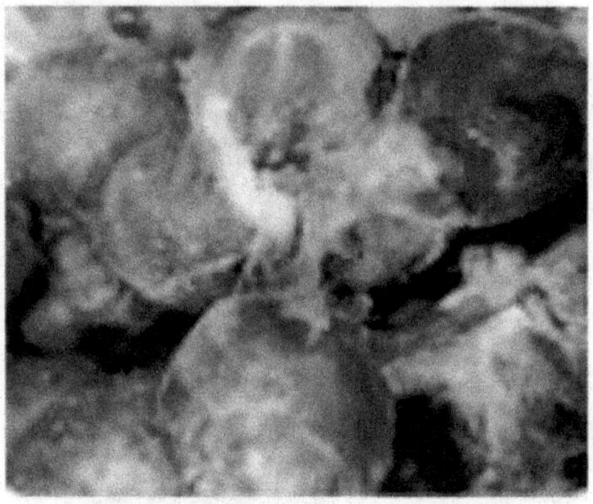

3. **Alternaria rot of grapes:**

The Causal Pathogen Name: *Alternaria alternata*
The disease appears in the field on the fruits and is in the form of dark, dark spots on the fruits.

Figure 9.

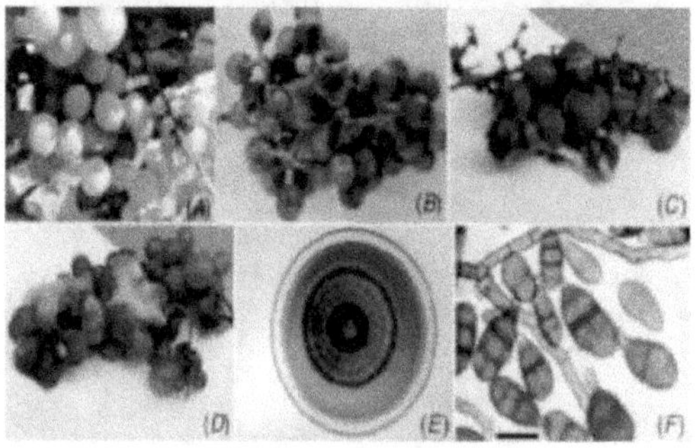

4. **Botrytis rot of grapes:**

The causal pathogen name: *Botrytis cinerea*
Symptoms:
The fungus attacks fruits, especially those stored at relatively low temperatures of 5°C - 25° and causes them to rot. Mushrooms grow on pregnant germs and spores. Likewise, severely infected fruits crack and watery secretions come out.

Figure 10.

Immature fruits are affected by a few, while ripe fruits are affected severely. This may be due to the fact that the immature fruits have a relatively high acidity and pH ranges from 2.4 to 2.6, and this does not suit the growth of the fungus that causes the disease. The appearance of fruit rot disease on the girl grape variety. The appearance of an advanced infection of the fruit rot disease on the girl grape variety. The appearance of fruit rot disease in an advanced stage. The appearance of the fruit rot disease on the film variety Sidles Clarification of the beginning of the infection stage of fruit rot disease

Methods of Controlling

The most important methods of resistance in the light of integrated control.
First: Agricultural control:

1. Thinning the leaves opposite the fruit clusters after the end of flowering or during the contract, taking care avoiding the sting of the sun, by thinning the leaves from the eastern side of the clusters only.
2. Removing the basal leaves around the stems of shrubs (*i.e.* the leaves on the rootstock) have a very important role in resisting the disease.
3. In the case of using cubes education, the cubes should be made high to increase ventilation and exposure to the sun.
4. Avoiding the increase in vegetative growth by controlling the use of nitrogen fertilization.
5. Attention to potassium fertilization.
6. Twisting plays a very important role in resistance, as it helps to increase ventilation and the entry of sunlight into the clusters.
7. Pruning and breeding methods have a very important role in the resistance to this disease.
8. Developing and modernizing the means of packing and collecting fruits and means of transportation and storage to reduce mechanical damage to avoid the spread of pathogens of post-harvest diseases.
9. Taking care of the resistance of insects and diseases that affect fruits in the field, as they pave the way for mold infection.

Second: Biological Control:

- To prevent infection with fruit rot, the grape vines are sprayed on the farm with mold biopesticides *i.e.* (*Trichoderma* spp. 30×10^6), with the rate of 1 Lit/500 Lit water/twice/every 2 weeks, which obtained from CLOA, (Ahmed, 2018), spray three or four times, depending on the severity of the infection and the length of storage period for the clusters, so that the spraying takes place on the following dates:
- The first spray: at the end of the flowering period or the beginning of the contract and varies according to the varieties.
- The second spray: before the beads touch the cluster.
- Third spray: At the beginning of maturity

Disease of Limbs Death of Grape

The Causal Pathogen Name: *Xyella fastidiosa*

This disease is a fungal disease caused by many different fungi and is considered one of the most dangerous diseases of grapes now in Egypt, as it affects the grapes and thus affects the crop and leads to a decrease and a decrease in its quality.

Figure 11.

Symptoms

Symptoms appear on the affected branches and branches from the previous season, which did not end with their death, with the appearance of small leaves, stunted, yellowish, wrinkled, with irregular edges. The flower clusters on the affected branches are small, incomplete, and may not mature, and the clusters remain attached to the branch until winter. The appearance of limb death disease on the leaves Appearance of death of the limbs.

Factors that lead to injury:

- High water level in the ancient lands.
- Not placing the dots in their proper places for the bushes, which causes an increase in humidity around the bushes or a dryness around the bushes as well.
- Potassium deficiency.
- Not to paint the places of wounds immediately after pruning with a fungal antiseptic.

Methods of Controlling

The method of resistance with this disease is followed in the light of integrated control as follows:

- The symptoms of the emergence of the disease in the spring must be observed, and their locations should be determined on the vine, and a distinctive mark should be placed on the vines.
- Removal of infected branches, 10-15 cm long, below the end of the affected part, any of the healthy growths, to ensure that the new growths on the branches are free of infection.
- In the event of a severe infection that reaches the entire stem up to the surface of the soil, the vines must be removed, cleansed, and replanted again.
- Avoid large pruning wounds as much as possible and pruning should be avoided during or before wet or rainy weather
- Immediately after pruning and removing the affected branches, the fungal disinfectant is sprayed with copper oxychloride at a rate of 300 g / 100 liters of water. Then the places of large wounds are painted with Bordeaux paste to ensure that the fungus spores do not enter the places of the wounds.

Dead Arm Disease in Grapes

The Economic Importance of the Disease

It was first seen in Egypt in 1995 in the Menoufia and Beheira region. It is noted that in the climatic conditions suitable for the disease, the infection causes a significant decrease in the amount of the crop.

Figure 12.

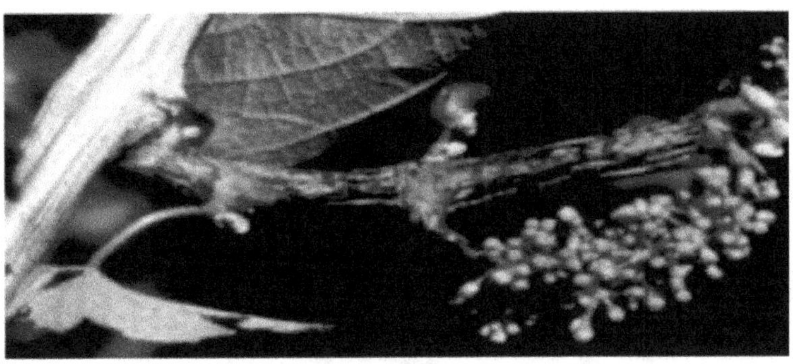

The Causal Pathogen Name: *Phomopsis viticola*

Symptoms

On twigs and branches affected in the previous season, which did not end with their death, in June and July of the current season, stunted, yellowish and wrinkled leaves with flimsy edges may appear on the leaves, and angular spots may appear on the leaves. As for the infection on the branches, it consists of ulcers that expand in two directions, up and down, and these ulcers have a dark middle, and these ulcers often unite with each other, so large rectangular brown areas appear on the branches, and these spots kill as the bud's advance. Also, the causative fungus goes deep into the tissues and damages the tissues that conduct the juice, causing dry rot to the wood of the branch, which leads to its death, hence the name of the disease (dead arm).

The fruits may be infected and have symptoms similar to the symptoms of black mold. The infected kernels darken in color, wither and mummify.

How the Injury Occurs

Infection occurs through fungus spores found inside dead branches, ulcerated twig tissues, and fallen leaf parts in the winter period. Late spring rain helps in spreading germs, so the germs move from the place of their production in perennial ulcers to modern tissues, revealing in branches or leaves.

Resistance

- Removing the infected parts first, so that the place of the cut is about 5-10 cm away from the nearest injury, then it is burned in a place far from the farm.
- Sterilization of tools used in pruning after each operation so that they are not a means of transmitting disease to healthy plants.
- Spraying with copper oxychloride at a rate of 300 g / 100 liters of water after pruning and removing the affected parts directly to disinfect the places of wounds.

Sooty Mold Disease in Grape

This disease arises from the presence of the black, pylori-like spores resulting from infection with a group of saprophytes.

Figure 13.

Disadvantages of This Disease

- This black powder prevents the sun and air from the grape vine, which hinders gaseous exchange and carbohydrate metabolism.
- These fungi parasitize on insect secretions such as mealybugs and scale insects, especially at high humidity as a result of the crowding of vegetative growths.
- Appearance of sooty mold disease on the stems of grape bushes Appearance of sooty mold disease on grape leaves.

Resistance in the Light of Integrated Control-

First: Agricultural resistance:

- The pruning must be done in a proper manner, in a way that suits the method of breeding and the type of grapes grown.
- Peeling the loose bark.
- Attention to potassium fertilization.
- Moderation in nitrogen fertilization and not excessive in order to avoid increased vegetative growth and thus reduce the increase in moisture.
- Paying attention to summer pruning and pruning.

Second: Biological Control:

- To prevent infection with sooty rot, the grape vines are sprayed on the farm with mold biopesticides *i.e.* (*Trichoderma* spp. 30×10^6), with the rate of 1 Lit/50 Lit water/twice/every 2 weeks, (Ahmed, 2018).

Third: Chemical resistance:

- Grape bushes are sprayed with copper oxychloride at a rate of 300 g/100 liters of water, in addition to summer mineral oil at a rate of 1.5 liters per 100 liters of water.
- Applying Bordeaux paste to the affected grape bushes after peeling the loose bark.

Root Rot Disease of Grape

The disease of root rot of seedlings and cuttings in nurseries and permanent land is one of the important diseases that have exacerbated in recent years, as this disease caused a heavy loss of cultivated grape bushes, especially in the newly reclaimed lands, where the environmental conditions helped the spread of this disease. In additional, it threatens the expansion of grape cultivation in these areas. It is caused by a group of fungi, including many soil fungi.

Figure 14.

Circumstances that Contributed to the Disease

1. Environmental conditions (salt soil and water used for irrigation) weakened grape bushes and prepared them for infection, as pathogenic fungi could easily penetrate them, especially in the initial period of transferring seedlings to permanent land.
2. Transporting silt and placing organic matter in contact with the roots of seedlings when planting plays a major role in attacking shrubs with various pathogens.
3. The high level of ground water plays an important role in the infection of this disease, as well as the presence of an error in placing the drops next to the trunk causes an increase in the percentage of moisture, which helps to contract this disease.
4. Nematode infection plays a major role in root rot infection, so nematodes must be treated with root rot. Symptoms of root rot on cuttings and seedlings.

Symptoms of Root Rot on Fruit Trees

1. **Symptoms of root rot on cuttings and seedlings in the nursery.**

The eyes do not go out on the cuttings and the parts buried in the soil do not rot.

Ease of cutting seedlings to decompose the root system. The decomposition of the tissues of the roots of seedlings and their fragmentation. The internal vascular bundles are discolored in different colors depending on the type of fungus that causes the disease. These colors and pigments are due to the toxic secretions secreted by the pathogens.

2. **Symptoms of root rot on grape bushes.**

Symptoms vary according to the condition of the infection:

In the case of a light injury:

1. A gradual decrease in the vigor of growth occurs for shrubs.
2. Leaves remain small in size and few and yellow in color and fall prematurely.

In case of severe injury:

1. The branches die starting from the ends.
2. Yellowing and wilting and falling of leaves.
3. The death of bushes in the case of complete wilting.
4. The internal vascular bundles are discolored in different colors according to the type of fungus that causes the disease. These colors and pigments are due to the toxic secretions secreted by the pathogens.

Resistance

Resistance in the Light of Integrated Control

First: Agricultural resistance:

1. Uprooting dead seedlings and purifying the hollows by adding quicklime.
2. Improving drainage and regulating irrigation periods, as well as not touching the irrigation water to the area of the tree trunk. It is considered one of the most important operations of the resistance.
3. Potassium fertilization has an important role in disease resistance.

Second: Biological Control:

- To prevent infection with root rot or damping off diseases, the grape vines are treated with biopesticides *i.e.* (*Bacillus subtilis* 30×10^6), with the rate of 2 Lit/fed./twice/every 2 weeks, which obtained from CLOA, (Ahmed, 2005,2013 & 2018).

Nematode Diseases of Grapes

They are often found in the soil and attack the roots of plants, and grapes are infected with many types of nematodes, which affects the efficiency of the roots and, accordingly, the vegetative total and the yield.

Figure 15.

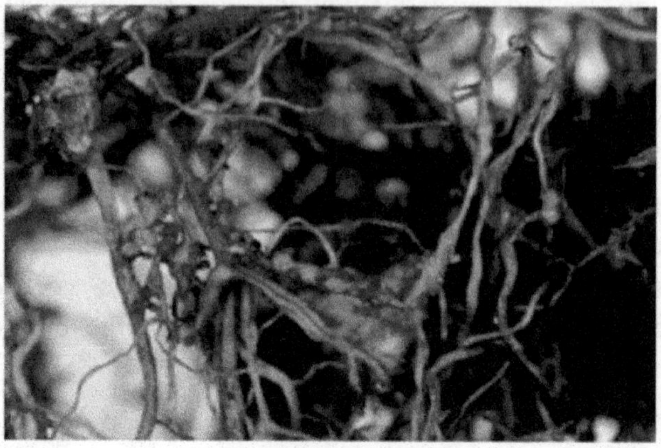

Among the most important types of nematodes spread in vineyards in Egypt are:

1. Root-knot nematodes:

It is widely spread in sandy and light lands, and its presence is little or no in clay and heavy lands in general, so it spreads in newly reclaimed lands.

2. Citrus nematodes:

This type infects grape trees and is abundant in medium and heavy lands and to a lesser extent in light lands, so its presence decreases in reclaimed lands.

3. Ulceration Nematodes:

It includes a group of species that are characterized by being mobile parasites, meaning that they penetrate the roots to feed on them and then leave them to the soil and this movement continues, causing ulcers to the roots and thus the growth and spread of other pathogens, especially root rot.

Precautions to be Followed to Prevent Nematode Contamination before Planting

1. Taking care the use of seedlings free of infection, or seedlings grafted on resistant roots, if possible, such as:
2. Not to establish the farm on land that was previously planted with vegetable crops or crops susceptible to nematode infection, except after plowing and solarization for a long period, with planting it with a grass crop to reduce the number of nematodes present.
3. Ensuring that the windbreak seedlings are free from nematode contamination before planting them.
4. Get rid of the existing weeds and burn them away from the farm site, as they are considered among the important hosts of nematodes that transmit infection with grape seedlings.

5. Not to transfer soil from the valley or any other source to the farm unless it is confirmed by laboratory analysis that it is free of nematodes.
6. Conducting a laboratory analysis of samples of municipal or organic fertilizer before adding it to ensure that it is free from nematode contamination.
7. Ensure by laboratory analysis that the farm to be established is free of contamination by conducting a laboratory analysis of the soil.

As for the existing farms, it is advisable to carry out integrated control operations to reduce the risk of nematodes, if they are present, by using the **following methods:**
First: Agricultural resistance:

1. Reducing the aggravation of the presence of the sources of pollution, such as not adding contaminated organic fertilizers or contaminated soil, removing weeds, and digging a tunnel with a depth of one meter separating the windbreaks, especially the kazurina, and the first line of the grape lines to prevent overlapping the roots.
2. Disinfecting agricultural machinery after its work has been transferred from one sector to another, to ensure that nematode contamination does not spread.
3. Some crops that repellent or kill nematodes such as garlic or some ornamental or medicinal plants can be planted next to the grape lines and then turned in the soil and minced these plants can be added directly during the month of March in the hollow around the plants.
4. Paying attention to the rates of fertilizing plants so that they can be strengthened in the face of infection, and it is possible to pay attention to potassium fertilizers, and vital organic fertilizers such as sepal and pigeon wither, as they have a role in reducing the risk of nematodes.
5. Paying attention to irrigation rates and ensuring that plants are not thirsty, and that the irrigation water, especially in the case of surface irrigation, is free of nematode contamination.
6. Some compounds that contain ascorbic acid, such as "Ascopene", can be used to spray plants according to the recommended rate, as it turns out that ascorbic acid is one of the compounds that stimulate resistance in plants. Planting garlic under grape bushes to resist nematodes

Second: Biological Control:

- To prevent infection with nematodes diseases, the grape vines are treated with biopesticides *i.e.* (*Serratia* spp. 30×10^6), with the rate of 2.5 Lit/fed./twice/every 2 weeks, which obtained from CLOA.

REFERENCES

Ahmed, M. F. A. (2005). *Effect of adding some biocontrol agents on non-target microorganisms in root diseases infecting soybean and broad bean plants* [M.Sc. Thesis]. Faculty of Agriculture Moshtohor, Benha Univ.

Ahmed, M. F. A. (2013). *Studies on non-chemical methods to control some soil borne fungal diseases of bean plants Phaseolus vulgaris L.* [Ph.D. Thesis]. Faculty of Agriculture, Cairo Univ., Giza, Egypt.

Ahmed, M. F. A. (2018). Evaluation of some biocontrol agents to control Thompson seedless grapevine powdery mildew disease. Ahmed Egyptian Journal of Biological Pest Control. *Ahmed Egyptian Journal of Biological Pest Control, 28*(93), 694–700.

Ahmed, M. F. A. (2018). Management of date palm root rot diseases by using some biological control agents under organic farming system. *Novel Research in Microbiology Journal, 2*(2), 37-47.

Mikkelsen, C., & Schluter, M. (2009). *The new EU Regulation for Organic Food and Farming, (EC) No. 834/2007*. International Federation of Organic Agriculture Movements IFOAM EU Group.

Chapter 14
Biointensive Integrated Pest Management (BIPM) Approaches in Orchards

Kalirajan Murugasridevi
Amrita School of Agricultural Sciences, Amrita Vishwa Vidyapeetham, India

Ravichandran Tamilselvan
Pandit Jawaharlal Nehru College of Agriculture and Research Institute, India

Rajendran Dhanapal
https://orcid.org/0000-0003-0031-8927
Adhiparasakthi Horticultural College, Tamil Nadu Agricultural University, India

Jayaraman Aravind
Amrita School of Agricultural Sciences, Amrita Vishwa Vidyapeetham, India

Sengodan Sekar
RVS Agricultural College, Tamil Nadu Agricultural University, India

ABSTRACT

This chapter discusses the biointensive integrated pest management strategies to be followed to combat the pest incidence in fruit orchards. The hostile habit of insects resulted in the elevated pesticide treatments and affected the destruction of the agroecosystem, which will be indicated by extremely resistant insect species with elimination of entomophages. This chapter highlights the idea of biointensive management tactics, that is, preliminary data collection through surveillance, accurate diagnosis, sampling and field scouting, and pest forewarning threshold level assessment. The major pests found in fruit crops such as mango, citrus, grapes, litchi, guava, apple, pear, and peach along with the BIPM techniques to mitigate the pest have also been debated. In addition, various management strategies, that is, cultural, mechanical, physical, bio-rational, biological, and significance of resistant cultivars in BIPM have been discussed in this chapter. The chapter concludes with a summary of approaches for implementing biointensive pest management programmes and its forthcoming areas.

DOI: 10.4018/978-1-6684-2423-0.ch014

INTRODUCTION

Insect pests are a persistent challenge to the cultivation of horticulture crops. They are to blame for the majority of yield losses caused by direct feeding injury and/or disease vectored thrust by viruses, fungi and bacteria. New pests, including the mango stone weevil, have revealed their severity in mango. Previously minor pests such as shoot gall maker and Anar butterfly, *Deudorix isocrates* are increasingly becoming a frequent concern in aonla, whereas the severity of fruit flies has worsened.

In order to control these pests, farmers experimented both cultural and mechanical techniques. Gradually, these tactics became a component of their approaches to manage pests. On the other hand, the utilization of predator ants, *Oecophylla smaragdina* F., is the first known example of biological control to suppress foliage feeding insects which was developed by the Chinese citrus growers in citrus Orchards. They employed bamboo stems to make it easier for ants to move between citrus trees (DeBach, 1964). Until approximately 2500 years ago, Chemical control was not used. However, from the late 1800 through the 1940, management practises began to shift, with the application of soaps, resins, oils and plant-derived products. Pesticides *viz.,* Dichloro Diphenyl Trichloroethane (DDT) (insecticide), ferbam (fungicide) and 2, 4-D (herbicide) were developed in early 20th century, heralded the start of the chemical era (Arneson and Losey, 1997). They became more popular in the 1940 to 1950 as a way to reduce crop pests and increase agricultural production. The negative consequences of these detrimental pesticides on flora and fauna, social hygiene and habitat were finally discovered in the year 1962 by Rachel Carson published the book *Silent Spring*.

Overuse of synthetic agrochemicals in plant protection strategy across the globe has created environmental disruption, contamination of water bodies, insect resurgence, pesticide resistance, and lethal and sublethal impacts upon non-target species, encompassing mankind (Mullen, 1995). Such adverse outcomes have sparked great global awareness regarding pesticide safety and frequent use. Simultaneously, rising population places ever-increasing demands on 'ecosystem services,' *viz.,* the supply of clean air, freshwater and animal domain for a plantation terrain. Reduced reliance on synthetic pesticides in support of environmental enhancements is a desirable approach for farmers in light of this anticipated future. In addition to habitat loss, pesticides also lead to a decline in biodiversity of flora and fauna in an agroecosystem (Simon *et al.,* 2011). Reduced pesticide use is therefore vital for the adoption of sustainable farming practices, particularly in orchard ecosystems that heavily rely on pesticides. Moreover, orchards are one of the most heavily sprayed agroecosystems in temperate climates to reduce insect and disease infestation and yield fruits that are free of visual defects to meet international commercial quality standards. Recently, consumer perceptions of fruits as fresh quality and healthy food have shifted as a result of knowledge on toxic residues of pesticides on fruits, leading to the development on zero residue strategies in various countries (Suárez-Jacobo *et al.*, 2017; Mebdoua, 2018; Tari *et al.*, 2020;). In a shifting regulatory context, meeting public demand for ecofriendly techniques and healthier fruits, as well as maintaining pests and diseases under economic threshold levels to retain cultivator's revenue is a huge challenge.

To increase crop yields from available fields, crop protection must be effective from sowing to harvesting. The significant challenge confronting the scientist is to accomplish a management without threatening the ecosystem or depleting the natural resources. This can be accomplished in a horticulture habitat by implementing a Biointensive Integrated Pest Management (BIPM) approach that is environmentally benign. BIPM portrays public knowledge about ecosystem quality and food protection by incorporating economic and environmental issues into agroecosystem infrastructure and decision-making. The advantages of executing BIPM *viz.,* minimized chemical input expenses, decreased environmental consequences on-farm and off-farm and more resilient and comprehensive pest control. An ecology-based IPM has the ability to reduce fuels, machineries and chemical inputs, which are all energy concentrated and becoming extremely expensive in respect of monetary and ecological effect. Such reduced costs will profit both the cultivator and community.

BIPM is stated as 'A systematic framework to manage the pest based on a knowledge of pest ecology'. This starts with procedures that precisely evaluate the origin as well as extent of pest issues and subsequently uses wide variety of prophylactic measures and biocontrol agents to maintain pest populations under control. Minimal-hazardous pesticides can be deployed as a last choice when previous approaches have failed to be effective, and with caution to limit dangers (Simon *et al.*, 2011). The main purpose of BIPM is to furnish recommendations and opportunities for sustainable pest and beneficial organism management in an ecological context. A BIPM technique can be useful in a variety of cropping systems owing to its flexibility and environmental compatibility. BIPM would almost certainly reduce chemical consumption and costs even more. Biorational agents and strategies will be vital in advancing the IPM tactics in order to tackle the societal challenges the world faces (Ishaaya *et al.,* 2005).

This chapter will discuss the urge for safer, environmentally sound pest management methods, as well as innovative ways for reducing pest resistance. The biointensive approaches for insect (and mite) pests' management *viz.,* proactive and reactive tactics will be discussed in this chapter. The management tactics include, (1) Biological control: conservation, augmentation and release of entomophages, (2) Cultural control: crop sanitation, crop rotations, cover cropping, timing planting, crop residue management, intercropping, mechanical weed control, (3) Synthetic agrochemicals replaced by Biopesticides *viz.,* microbial pesticides derived from fungi, viruses and bacteria, biochemical pesticides such as insect growth regulators, plant volatiles, pheromones and hormones which alter pest behavior and reproduction and botanical pesticides (4) Resistant Cultivars: varieties produced through traditional or conventional, modern biotechnological as well as transgenic methods. The successful transmission of BIPM technology and its application by professionals remain crucial in yielding healthy and residue free fruits with improving fruit orchard's productivity.

REQUIREMENTS OF BIOINTENSIVE INTEGRATED PEST MANAGEMENT APPROACH

Gathering of Preliminary Data by Survey and Surveillance

The preliminary data is utterly mandatory to fully comprehend the original depiction or condition of awareness by growers about the BIPM. The purpose of the standard study is to determine growers perceptions on pest and its management tactics and risk management strategy, as well as basic socio-economic indicators and other parameters. On the other hand, automatic detection traps have also been established to monitor fruit flies in fruit farms.

Precise Recognition of Important Pests in Fruit Orchard

In a fruit tree habitat, precise diagnosis of insect pests is crucial for pest control. It's critical to understand all phases of a pest, including eggs, larvae, location of pupation and the morphology of the adult, as well as the sort of fruit damage. An inaccurate analysis may result in redundant treatments and monetary losses. Many of the insects observed in the orchard are casual visitants, whilst few are productive (beneficial entomophages) and serve as biological pest controllers. Anyway the infestation may be induced by the external surrounding or by the pest, recognition or prognosis is the act of diagnosing the damage from their symptoms. The diagnostic process entails examining the whole plant as well as its individual components, thoroughly interpreting the findings, and seeking to determine why injuries or damage have happened. In addition, before making a pest management move, type of insects found in the field, its style of eating and their destructive behaviors should all be evaluated.

Pest Sampling

Insect pest population assessments are a key activity in ecology and the major cornerstone of integrated pest management (Pedigo, 2001). Cochran (1977) provides a thorough examination of the strategies for determining the optimal size sample unit. However, due to continuous changing in insect population density, a precise measurement of sample unit should not be placed too high on the priority list. The total number of samples necessary is determined by the precision required. For orchard trees, sampling methodologies should be standardized and obtained from various corners, entire peripheries, diagonally, zig-zag diagonally, circular and star pattern. Because of the large array of habitat present in the fruit tree ecosystem, sampling in this situation is a more complex matter than sampling in agroecosystem (Hare, 1994). Smith *et al.* (1997) provided a clear description of insect pest sampling methodologies in mango and citrus trees. Sampling and surveillance strategies are very well documented for certain pests causing direct damage *viz.,* fruit flies and borers and for few pests causing indirect damage namely, banana weevil, mites, leaf miner and defoliators (Smith *et al.*, 1997). In the apple orchards, sampling methodologies for pests, entomophages and beneficial organisms have been recorded by Beers *et al.* (2020).

In India, a systematic sampling technique for tracking mango hopper populations has been devised. The association amongst bud development and mite populations could be employed to predict action levels in the field to measure mite densities. Boavida *et al.* (1992) investigated the spatial distribution of a polyphagous mealybug, *Rastrococcus invadens* (Williams) that destructs leaves, flowers, fruits and devised binomial sampling techniques for assessing population numbers. In Brazil's semi-arid regions, the sample procedure and action levels over significant mango pests were also designed (Haji *et al.*, 2004).

Pest Forecasting in Orchards

Disease and pest epidemics are aided by favorable meteorological conditions, which allow them to multiply indefinitely. The amount and grade of food produced by the plant species and the pest species connected with them is extremely affected by the weather and climatic parameters. The role of different weather elements on pests and pathogens varies depending on the area and crop. Temperatures of 35°C and 23°C, respectively, in combination with relative humidity of 50 to 80 per cent and vapour pressure of 20–24 mm of Hg, were shown to be favorable to mango hopper mating (Pandey *et al.*, 2003) and, as a result, useful in enhancing population accumulation in the following months. Generally, most of the orchard cultivators use a chronological schedule or phenological timeline to assess insect occurrences and time of pest management in their plantations. Because pest growth does not always occur on the exact dates on every year and hence the calendar technique is highly incorrect. The phenological method, on the other hand, whereas more precise, wrongly implies that pest numbers spike matched the phase of pest growth each year. Degree-day assemblage are a much accurate and precise approach of predicting insect pest proliferation and occurrence in fruit orchards. Usually, most of the fruit growers employ a chronological or phenological calendar to predict insect occurrences and pest treatment times in their plantations. Whereas, the calendar technique is highly incorrect, in which pest growth does not always happen on an exact chronological days every year. Ayont, the phenological technique is precise to a long extent, but, wrongly implies that population of pests spike at the same phase of host growth each year. A degree-day (DD) principle is stated to be an index of insect pest growth as development of several season-arising pests are supported by temperature. There are a range of criterion and levels for degree-day accumulation for insect establishment, just as there are for plants. On the account of related variable weather parameters, the presence of pest or pathogen can be forecasted with excellent precision both temporally and spatially (Huda and Luck, 2008).

Automatic meteorological centers have been utilized to track the microenvironment of fruit orchards at different span of time and based on the need. On a weekly or biweekly basis, data from meteorological stations can indeed be saved immediately in a CD or a desktop computer. Degree days and disease predictions are calculated using a computer software application. Several agencies have developed software tools that use meteorological data from a specific field to run disease models such as fire blight, powdery mildew and apple scab. Meteorological information at the micro scale will aid not just in the early detection of pests and pathogens, as well as in determining the best time to apply pesticides, because the accumulation and retainment of spray and dust droplets of insecticides on canopy are reliant upon the forecast status in the fruit orchard on the hour of treatment. Occurrence of the tea mosquito bug, *Helopeltis antonii* (Sign) may be predicted in cashew in a month ahead of time based on the weather, which can be used to make IPM decisions (Prasada Rao and Beevi, 2008). Morgan *et al.* (1995) constructed a forecasting tool named 'PESTMAN" to monitor the pests of apple and pear in England. Samietz *et al.* (2007) designed a forecasting tool called 'SOPRA' for fixing the surveillance and control and management methods pertaining to pests of fruit orchards in Switzerland.

Field Surveillance and Monitoring of Pest Population

Field monitoring is essential in combating the changing patterns of numerous insect pests and enables producers to make judgments about insect pest control actions (Barnes, 1990). Insect pest injury necessitates survey and surveillance as part of the monitoring strategy. To track the early stages of pest complex establishment in endemic regions, survey pathways must be developed, and state extension workers must focus their attention at the village level to encourage farmers to conduct field reconnaissance. Farmers must be organized for field scouting to monitor the incidence of insect pests and diseases at periodic intervals. The crop protection measures are being employed only when insect pests and diseases exceed the ETL, depending upon the field scouting. Also, the farmers must do it once a week to determine out their economic threshold level (ETL). In order to deliver high-quality fruits, the sampling density and awareness of Economic threshold levels are important (Pena, 2004). Hence, to plan, assess and properly implement IPM strategies, field observation of pests and natural enemies are required. Based on pest population, execute roving surveys every 10 km at intervals of 7 to 10 days can be done. Maintain records of key pests and pathogens on trees and alternate hosts in fruit orchards. At each location, count 20 trees arbitrarily and take 5 representations from each tree. On three leaves of these plants' young shoots, observe all the pest population. Keep a documentation of biocontrol specie's population ability. Agro ecosystem analysis (AESA) is a method that extension workers and farmers can use to examine field conditions including pests, beneficial insects, and condition of the plant health, soil, impact of climate variables and their interrelationships in order to develop healthy crops. Such a thorough examination of the reality on the ground will aid in making suitable management decisions (Sathyapal *et al.,* 2014). Pheromone trap surveillance can be done by installing 4 to 5 traps for each pest demarcated by more than 75 feet around the surrounding of the concerned area and the pheromone traps can be fixed to a pole of height equals to middle of the canopy. After 2 to 3 weeks, lures have to changes and each week, surveillance data should be maintained as the moths per trap for a week and observed throughout the year. In addition, the moths collected in the traps should be eliminated and killed after documentation (Sathyapal *et al.,* 2014).

Assessment of Threshold Level

Hoyt and Burts proposed an economic threshold for temperate fruits (1974). There is a scarcity of data about the ecological repercussions of equatorial fruit arthropods. The scant data supplied is only a speculation. This strand combined with insufficient sampling methodologies, indicates a small effort has gone into defining economic threshold values. As a consequence, pesticides are still routinely used as preventative measures. When spraying is required as a management approach, take into account all monitoring data as well as other parameters namely stage of the tree, fruit, insect, variety of the fruit, weather factors and so on. The economic injury level is described as an intensity of an insect pest that, provided with these conditions, may lead to crop monetary loss that overcomes the amount of the management approach (EIL). EIL have only been experimentally proven for a few pests, thus researchers have developed thresholds for numerous pests and fruit types centered on past encountering, amount of abhorrence to discerned danger of crop loss, and other factors. The action threshold is the level at which appropriate activity must be performed to restrict irreversible agro-ecosystem damage. If the loss does not warrant the cost of management, a modest degree of insect pest infestation may be overlooked. The economic, ecological, and artistic aspects, as well as the social significance of the crops farmed, are the most important variables to consider when making management decisions. The establishment of pest resistance and recurrence, as well as the ascent of minor insect pests to destructive pest level, have hampered chemical control of mango fruit pests (Cunningham, 1984). Sampling, mortality and economic thresholds in agricultural system was laid as groundwork for IPM by Flint and van den Bosch (2012). In North America, Whalon and Croft (1984) ascertained the ETL for the pests of apple and found that it varies among the nation owing to sampling systems variations, status of the pest, market specifications, control features, annual control period, orchard background, and human factors, among other things.

Proper Record Keeping

In simple terms, keeping a record is a method of learning from previous experiences in a systematic manner. Fruit growers can now use a range of software applications to keep a record of and obtain information on their farm. The farm's ready estimator device is the monitoring process, which runs concurrently with record keeping. Records should include data on cultural practices namely proper irrigation timing, appropriate fertilizer enactment, intercultural operations and so on) and its impact over target and non-target organisms, in addition to the knowledge about timing and occurrence of the pest. The impact of abiotic elements, particularly climatic conditions and biotic elements such as insect pests and its entomophages can be documented for forecasting the insect pest. Systematic data collection in the field takes time, but it is crucial for efficiency in reviewing and interpreting the outcome. If field measures are problematic and a rating scale namely, trace, low, medium, high, extremely high can be used instead, assigning numerical values from 1 to 5 is appropriate. After field sampling, the data is organized in summary form, allowing for detection, analysis, and evaluation of individual orchard circumstances. The data collected during the monitoring phase is used to assist plan for future pest management approach.

Decision Support Systems in Orchards

A Decision Support System (DSS) is a collection of tools, such as computer programmes, simulations, and optimization algorithm (*i.e.*, farmers', inspectors', and managers' experience and expertise), that work together to aid decision-making. DSS's have primarily been developed to manage insects of perennial plants and agricultural crops in insect control (Pontikakos *et al.*,2012). DSS tools are also commonly used in pest management across large areas. Modeling climatic suitability and periodical action of invasive insect pest can aid stakeholders such as growers, natural resource administrators, and surveillance groups in detecting and preventing their occurrence, slowing their distribution and managing enduring populations in a much sustainable and also cost-effective manner. Pest monitoring is used in fruit orchards to determine the pest status by collecting data on the pest's incidence, abundance, and spread (Preti *et al.*, 2021). In apple orchards, An automatic trap developed by combining data gathering and data transfer systems (Pomotrap®, now Carpo® by Isagro S.p.A., Milan, Italy) was employed to detect the codling moth, *Cydia pomonella* L. (Lepidoptera: Tortricidae) (Guarnieri *et al.* 2011). In vineyards, Unlu *et al.* (2019) presented a comparable prototype to watch the European grapevine moth, *Lobesia botrana* Denis and Schifermüller (Lepidoptera: Tortricidae). Shaked *et al.* (2018) described G sticky delta traps, "Jackson trap" was outfitted with a camera to track the Mediterranean fruit fly, *Ceratitis capitata* (Diptera: Tephritidae). Liao *et al.* (2012) developed an automatic forewarning device to monitor massive *Bactrocera dorsalis* (Hendel) epidemics in orchards. Doitsidis *et al.* (2017) developed Bucket traps which employs camera fixed McPhail trap to track olive fruit fly, *Bactrocera oleae* Gmelin's. The DSS employed in different orchards for managing various pest species have been furnished below in Table 1.

Table 1. DSS utilized in orchards to control insect pests

S. No.	Type of DSS	Fruit crop	Targeted pest	Reference
1.	Location aware system for spray control	Olive	Fruit fly, *Bactrocera oleae*	Pontikakos *et al.* (2012)
2.	CitrusVol – Pesticide volume adjustment	Citrus	California red scale, *Aonidiella aurantia* Maskell	Fonte *et al.* (2021)
3.	Spatial Decision Support System	Citrus	Mediterranean fruit fly, *Ceratitis capitata*	Cohen *et al.* (2008)
4.	SOPRA – Sustainable Orchard Pest Management with the Swiss Forecasting System	Apple, Plum, Pear and Cherry	Aphid, *Dysaphis plantaginea*, apple sawfly, *Hoplocampa testudinea*, Codling moth, *Cydia pomonella*, smaller fruit tortrix, *Grapholita lobarzewskii*, European cherry fly *Rhagoletis cerasi*, red spider mite *Panonychus ulmi*, apple blossom Weevil, *Anthonomus pomorum*, pear psylla *Cacopsylla pyri*, and plum fruit tortrix *Grapholita funebrana*	Samietz *et al.* (2015)
5.	Pheromone based Decision Support Tool	Apple	Brown marmorated stink bug, *Halyomorpha halys* (Stal),	Short *et al.* (2017)
6.	Washington State University (WSU) -Decision Aid System (DAS) Environmental data, weather forecasts and management recommendations	Apple, Cherry and Peach	Apple maggot, *Rhagoletis pomonella* Walsh, Sanjose scale, *Quadraspidiotus perniciosus* (Comstock), Codling moth, *C. pomonella* (L.), Western cherry fruit fly *Rhagoletis indifferens* Curran and Peach twig borer *Anarsia lineatella* Zeller	Jones *et al.* (2010)
7.	An automatic infield monitoring system	Orange, Guava, Pear, Mango and Peach	*B. dorsalis*	Jiang *et al.* (2013)
8.	Knowledge based DSS	Apple	Woolly apple aphid *Eriosoma lanigerum*	Bangels *et al.* (2021)
9.	Web Decision Support System (SSD Manzano)	Apple	Pests of Apple	Mondino and Andujar (2019)
10.	Decision Support System (DSS)	Olive	Olive fruit fly *Bactrocera oleae* (Rossi)	Miranda *et al.* (2019)
11.	DSS based on a semi-automatic pest monitoring	Peach, Apple and Fig	Mediterranean Fruit Fly, *Ceratitis capitata* (Wiedermann)	Sciarretta *et al.* (2019)
12.	Spatial DSS	Cherry	Cherry fruit fly, *Rhagoletis cerasi*	Ioannou *et al.* (2019)
13.	DSS based on population monitoring and infestation	Melons	Ethiopian fruit fly (EFF), *Dacus ciliates*	Nestel *et al.* (2019)
14.	Decision support systems (DSSs) for pesticide dose adjustment	Vineyards	Pests of Grapes	Román *et al.* (2022)

MAJOR PESTS ATTACKING FRUIT CROPS IN ORCHARD

Mango

Mango (*Mangifera indica* L.) (Family: Anacardiaceae) is a tropical and subtropical fruit that originated in the Indo-Burma region and has since been naturalized and adaptable around the world. The major insect pests infesting mango fruits in orchard are furnished below in Table 2.

Table 2. Major insect pests attacking mango

S. No.	Insect pest	Scientific name	Family & Order	Affected part	Reference
1.	Leafhoppers	*Idioscopus clypealis* (Lethierry), *I. nagpurensis* (Pruthi), *I. nitidulus* (Walker), and *Amritodus atkinsoni* (Lethierry)	Cicadellidae and Hemiptera	Inflorescence and Tender Shoots	(Veeresh 1989; Peña *et al.*, 2002)
2.	Thrips	*Scirtothrips dorsalis* (Hood), *Anaphothrips sudanensis* (Trybom), *Aeolothrips collaris* (Priesner), *Selenothrips rubrocinctus* (Giard), *Caliothrips indicus* (Bagnall), *Rhipiphorothrips cruentatus* (Hood), (Schmutz), *Ramaswamiahiella subnudula* (Karny) and *Frankliniella occidentalis* (Pergande),	Thripidae and Thysanoptera	Foliage and buds	(Butani 1979; Wysoki *et al.* 1993)
3.	Mango Leaf Webbers	*Orthaga euadrusalis* Walker	Pyralidae and Lepidoptera	Foliage	Butani (1979)
4.	Mealybug	*Drosicha mangiferae* (Green), and *Rastrococcus iceryoides* (Green)	Pseudococcidae and Hemiptera	Panicles and shoots	Tandon and Verghese (1985)
5.	Mango midge,	*Erosomya indica* Grover	Cecidomyiidae and Diptera	Flowers	(Ahmed *et al.*, 2005)
6.	Mango leaf gall midge	*Procontarinia matteiana* (Kieffer & Cecconi)	Cecidomyiidae and Diptera	Leaf	Patel and Kumar (2020)
7.	Scales	*Aspidiotus destructor* Signoret	Diaspididae and Hemiptera	Leaf	(Srivastava, 1997)
8.	Shoot Gall Psyllid	*Apsylla cistellata* Buckton	Psyllidae and Hemiptera	Shoots and leaves	(Tandon and Verghese, 1985)
9.	Mango Shoot Borers	*Chlumetia transversa* Walker, *Anarsia melanoplecta* Meyrick,	Noctuidae and Lepidoptera	Shoot and stem	Reddy *et al.* (2015)
10.	Mango stemborers	*Batocera rufomaculata* (De Geer), *B. roylei* (Hope) and *B. rubus* (Linnaeus),	Cerambycidae and Coleoptera	Stem	(Fletcher 1914; Butani, 1979
11.	Fruitflies	*Bactrocera dorsalis* (Hendel) and *B. zonata* (Bezzi),	Tephritidae and Diptera	Fruits	(Verghese *et al.* 2011)
12.	Stone weevil	*Sternochetus mangiferae* (F.)	Curcurlionidae and Coleoptera	Fruits	Reddy *et al.* (2018)
13.	Mango Pulp Weevil,	*S. frigidus* (F.)	Curcurlionidae and Coleoptera	Fruits	Reddy *et al.* (2018)
14.	Mango fruit borer	*Deanolis albizonalis* (Hampson),	Pyralidae and Lepidoptera	Fruits	Reddy *et al.* (2018)

Citrus

Citrus is cultivated in a variety of soils and environments, and it is afflicted by a diverse range of pests, leading to significant economic losses. On diverse Citrus spp. have been observed with over 300 species of insects pests and mites. The major pests attacking citrus have been enlisted in Table 3.

Table 3. Major insect pests attacking citrus

S. No.	Insect pest	Scientific name	Family & Order	Affected part	Reference
1.	Citrus leafminer	*Phyllocnistis citrella* Stainton	Gracillaridae and Lepidoptera	Leaves	Debbarma and Hath (2021)
2.	Psyllids	*Diaphorina citri* Kuwayama	Psyllidae and Hemiptera	Leaves, tender shoots and flowers	Debbarma and Hath (2021)
3.	Mealy bugs	*Planococcus citri* Risso, *P. pacificus* Cox and *Icerya purchasii* Maskell	Pseudococcidae and Hemiptera	Leaves and shoots	Debbarma and Hath (2021)
4.	Citrus aphid	*Toxoptera aurantii* (Boyer de Fonscolombe)	Aphididae and Hemiptera	Leaves, tender shoots	Debbarma and Hath (2021)
5.	Lemon butterfly	*Papilio demoleus* L.,	Papilionidae and Lepidoptera	Leaves	Debbarma and Hath (2021)
6.	Whitefly	*Dialeurodes citri* (Ashmead)	Aleyrodidae and Hemiptera	Leaves	Debbarma and Hath (2021)
7.	Blackfly	*Aleurocanthus woghlumi* Ashby	Aleyrodidae and Hemiptera	Leaves	Debbarma and Hath (2021)
8.	Fruit sucking moth	*Othreis fullonica, O. materna, Achoea janata*	Noctuidae and Lepidoptera	Fruits	Satyagopal *et al.* (2014)
9.	Bark borer	*Indarbela quadrinotata*	Metarbelidae and Lepidoptera	Bark	Debbarma and Hath (2021)
10.	Trunk borer	*Anoplophora versteegi* Ritsema	Cerambycidae and Coleoptera	Trunk	Nath and Sikha (2019)

Guava

Out of the 80 insect pest species found on guava trees, few of them are considered pests that cause substantial damage to the trees on a regular basis. The major pests attacking guava have been enlisted in Table 4.

Table 4. Major insect pests attacking guava

S. No.	Insect pest	Scientific name	Family & Order	Affected part	Reference
1.	Bark-eating caterpillars	*Indarbela* spp.	Metarbelidae and Lepidoptera	Bark	Haseeb (2007)
2.	Fruit flies	*Bactrocera* spp.	Tephriditae and Diptera	Fruit	(Kapoor, 2000)
3.	Mealybugs	*Maconellicoccus hirsutus, Ferrisia virgata, Nipaecoccus viridis, Planococcus citri* and *P. lilacinus*	Pseudococcidae and Hemiptera	Leaves	(Butani, 1979)
4.	Scales	*Chloropulvinaria psidii*	Coccidae and Hemiptera	Leaves, shoots and twigs	Haseeb (2007)
5.	Tea mosquito bug	*Helopeltis antonii* Signoret	Miridae and Hemiptera	Fruits	Haseeb (2007)
9.	Spiralling Whitefly	*Aleurodicus dispersus* Russel	Aleyrodidae and Hemiptera	Leaves	(Charati *et al.*, 2003)
10.	Guava fruit borer	*Deudorix isocrates* (Fabricius)	Lycaenidae and Lepidoptera	Fruits	(Gupta and Arora, 2001)
11.	Fruit borer	*Dichocrocis punctiferalis* (Guenee)	Crambidae and Lepidoptera	Fruit	(Shasank *et al.* 2015)
12.	Stem Borer	*Aristobia testudo* Voet	Cerambycidae and Coleoptera	Stem	(Firake *et al.* 2013)
13.	Trunk Borer	*B. rufomaculata*	Cerambycidae and Coleoptera	Trunk	Haseeb (2007)
14.	Thrips	*Selenothrips rubrocinctus* Giard and *Rhipiphorothrips cruentatus* Hood	Thripidae and Thysanoptera	Leaves	(Butani, 1979)

Pome Fruits

Pome fruits, are infested by a various pests that are endemic in nature. The massive extension of the region under pome fruits has concluded in the introduction of several indigenous major and minor insect pests. Around 70 pests have been identified worldwide damaging pome fruits. The major pests attacking pome fruits have been enlisted in Table 5.

Table 5. Major insect pests attacking apple and pear

S. No.	Insect pest	Scientific name	Family & Order	Affected part	Reference
1.	San Jose scale	*Quadraspidiotus pernicious* (Comstock),	Diaspididae and Hemiptera	Leaves, twigs fruits and Bark	(Sherwani *et al.*, 2016)
2.	Apple woolly aphids	*Eriosoma lanigerum* (Hausmann)	Aphididae and Hemiptera	Trunk, stems, branches, leaves petioles, twigs, and fruit stalks	(Misra, 1920)
3.	The codling moth	*Cydia pomonella* (Linnaeus)	Torticidae and Lepidoptera	Fruits	(Sherwani *et al.*, 2016)
4.	The Apple Root Borer	*Dorysthenes hugelii* (Redtenbacher)	Cerambycidae and Coleoptera	Root	(Sherwani *et al.*, 2016)
5.	Apple Stem Borer,	*Apriona cinerea* (Cheverlot)	Cerambycidae and Coleoptera	Shoot, Stem and trunk	(Sherwani *et al.*, 2016)
6.	The Bark Beetle,	*Scolytus nitidus* (Schedl)	Scolytidae and Coleoptera	Bark	(Sherwani *et al.*, 2016)
7.	Stem Borer	*Aeolesthes sarta* (Solsky)	Cerambycidae and Coleoptera	Bark	(Sherwani *et al.*, 2016)
8.	The tent caterpillar	*Malacosoma indica* Walker	Lasiocampidae and Lepidoptera	Leaves	(Sherwani *et al.*, 2016)
9.	Indian gypsy moth	*Lymantria obfuscata* (Walker),	Lymantriidae and Lepidoptera	Leaves	(Sherwani *et al.*, 2016)

Grapevine

Grapevine (*Vitis vinifera* L.) is an economically important fruit crop grown in temperate conditions which actually got acclimated to both tropical and subtropical conditions. It is grown commonly in many countries *viz.*, France, America, Australia, Italy, Chile, Africa, India, Algeria and so on. So far, 132 insect pests have been documented to infest grapevine across the globe. The major pests attacking grapevine have been enlisted in Table 6.

Table 6. Major insect pests attacking grapevine

S. No.	Insect pest	Scientific name	Family & Order	Affected part	Reference
1.	Grape phylloxera	*Daktulosphaira vitifoliae*	Phylloxeridae and Hemiptera	Roots and leaves	(Tello *et al.*, 2019)
2.	Thrips	*Scirtothrips dorsalis* Hood *Rhipiphorothrips cruentatus* Hood, *Retithrips syriacus* Mayet *Thrips hawaiiensis* Morgan, *and Thrips palmi* Karny	Thripidae and Thysanoptera	Leaves and berries	Mani *et al.* (2008)
3.	Leafhoppers	*Arboridia viniferata* Sohi and Sandhu, *Typhalocyba sp., Empoasca (Chlorita) lybica* (Bergevin and Zanon), *Empoasca minor* Pruthi, *Flata ferrugata* Fab., *Unna intracta* Walker *and Amrasca biguttula biguttula* (Ishida)	Cicadellidae and Hemiptera	Leaves	Mani *et al.* (2008)
4.	Pink Hibiscus Mealybug	*Maconellicoccus hirsutus*	Pseudococcidae and Hemiptera	Leaves and shoots	Kulkarni (2020)
5.	Citrus Mealybug	*Planococcus citri*	Pseudococcidae and Hemiptera	Leaves and shoots	Mani and Kulkarni (2007)
6.	Striped Mealybug	*Ferrisia virgata*	Pseudococcidae and Hemiptera	Leaves and shoots	(Mani, 1986)
7.	Long-Tailed Mealybug	*Pseudococcus longispinus*	Pseudococcidae and Hemiptera	Leaves and shoots	Kulkarni (2020)
8.	Spherical Mealybug	*Nipaecoccus viridis*	Pseudococcidae and Hemiptera	Leaves and shoots	(Mani, 1986)
9.	Soft scales	*Pulvinaria maxima* Green	Coccidae and Hemiptera	Leaves and shoots	Kulkarni (2020)
10.	Hard scales	*Hemiberlesia lataniae* (Signoret)	Diaspididae and Hemiptera	leaves, shoots, petioles and bunches	Kulkarni (2020)
11.	Aphids	*Aphis gossypii* Glover	Aphididae and Hemiptera	shoots and bunches	Kulkarni (2020)
12.	The orange spiny whitefly	*Aleurocanthus spiniferus* (Quaintance)	Aleyrodidae and Hemiptera	Leaves	(Mohanasundaram, 1974)
13.	Spiralling Whitefly	*Aleurodicus dispersus* Russell	Aleyrodidae and Hemiptera	Leaves	Kulkarni (2020)
14.	Flea beetle	*Scelodonta strigicollis* Westwood	Chrysomelidae and Coleoptera	Roots and Leaves	Kulkarni (2018)
15.	Shot-Hole Borer	*Xylosandrus crassiusculus* (Motsch.)	Scolytidae and Coleoptera	Trunk	Kulkarni (2018)
16.	Stem Girdler	*Sthenias grisator* F.	Cerambycidae and Coleoptera	Trunk	Kulkarni (2018)
17.	Bark eating Caterpillar	*Indarbela* spp.,	Metarbelidae and Lepidoptera	Bark	(Satyanarayana, 1981)

Techniques of BIPM

Cultural Control

Clean culture and ploughing reduces weeds and other alternate hosts of various pests, making conditions unsuitable for insect pest development. They have been shown to be effective in preventing pests from concealing the hibernating stages of pests deep in the soil or subjecting those to adverse climatic factors. Sucking pests including mealy bugs and Scales on guava, mango and citrus, grapes shot hole borer and thrips, apple San Jose scale, citrus fruit sucking moths, and peach leaf curl aphid have been found to be managed promising (Kulkarni, 2020). Plant density influences the microclimate vegetation in which pest communities thrive, such as aphids on peaches, which establish a foliage advantageous for aphid proliferation in a dim humid environment and also for scales, mealy bugs and mango hoppers as well as for citrus white flies and black flies. Pruning is employed to ensure high quality and maximum yield of fruit while also preserving vigour, shape and size of the tree. Furthermore, good orchard practices such as weed control, pruning and cleanliness aids in lowering the pest incidence and also boost the beneficial species in fruit orchards (Hoyt and Burts, 1974). Legumes and alfa-alfa as cover crop at crucial junctures is effective for improving diversity of habitat for the sustenance of entomophages which established potential efficacy against aphid and mite pests in many ecosystems. Nevertheless, Hansen and Amstrong (1990) discovered as cleanliness in fruit orchard does not decrease mango stone weevil damage in Hawaii. Fruit trees that have been left unchecked and ignored are a main cause of the fruit fly infestation. Disposal of alternate host plants of insect pests is another useful cultural practice for pest reduction. When a vineyard is covered with vinyl film, the intensity of thrips is reduced in contrast to a control plot (Kulkarni,2020). Bagging of fruit is the most effective ways of keeping fruit flies away from mangoes. In mango orchards, Ndiaye *et al.* (2008) evaluated hygienic strategies viz., weeding and eliminating rotten fruit materials as a component of an IPM program which actually limited *B. invadens* in Senegal.

Mechanical and Physical Control

Pest disposal generally entails the expulsion as well as eradication of phytophagous pests, primarily through the use of manual labour. It is very successful and efficient on a small scale, especially for kitchen garden maintainers. The gathering and elimination of invertebrate stages is one of the dominant and potential method for controlling the spread of various pests. Fruits of citrus and pomegranate are bagged with butter sheet or cloth material instantly following fruit development to avoid infestation from *Deudorix Isocrates* F., and *Othreis* spp., respectively. To thwart specific phases of pests from ascending to canopy from the soil surface, an alkathene sheet or greasy banding have to be wrapped surrounding the tree bark (Shankar, 2017). This technique is widely employed to check mealy bug infestations on mango orchard tree trunks Disposal of various pests by utilizing physical components such as heat, radiation, sound, cold, and so on also shown promise when using artificial or natural sources. Debarking and scraping the vine barks with a sturdy fabric, then applying it with a mix of copper oxychloride + chlorpyrifos may help reduce population of mealybugs in grapes. Debarking done to eliminate mealybugs have decreased fruiting season incidence by 40 per cent. Application of insecticides in absence of debarking is pointless. The application of 'TackTrap,' a sticky substance which contains 76 per cent polyisobutylene, over the shoot region on peduncle for 5 cm lowers affliction of mealybug to 50 per cent.

Host Plant Resistance (HPR) in BIPM

In spite of significant progress in agricultural crop's resistance to pests, such facets have only received slight recognition in orchard crops. Excluding the historical case study of Winter Majetin, a resistant apple variety against *Eriosoma lanigerum* (Hausmann) and wild American grapes against *Phylloxera vitifolia* (Fitch) little progress has been so far made in orchard fruit crops, whereas both these instances still remain iconic in orchard pest management. Resistant or tolerant fruit cultivars of various fruit crops such as mango, citrus, guava, peach, apple, plum, grapes, banana, sapota, ber and date palm against certain insect pests have been found, eventhough, little effort has been taken to inculcate the tolerant and resistant biotypes in molecular breeding (Sharma and Singh, 2006). The majority of mango cultivars are vulnerable to fruit fly infestation, whereas its damage is comparatively low (Jothi *et al.*, 1994) in Langra, Dashehari, and Mumbai Green varieties. Verheji (1991) documented that in the Philippines, *Mangifera altissima* appears to be unaffected by leafhoppers and seed borers.

Table 7. Resistant or tolerant stock of various fruit crops against various pests of fruit crops

Fruit crop	Insect/ mites pest	Resistant or tolerant root stock
Citrus	Whitefly, *Dialeurodes citri* Ashmed	Nova mandarin, Redblush, Fallglo Star ruby and Marsh seedless
	Psyllid, *Diaphorina citri* Kuwayama	Nagpur mandarin, Rubidoux, Cleopatra and Gal-gal
	Leaf-miner, *Phyllonistis citrella* Stainton	Eureka lemon, Jatti Khatti, Carrizo, Coorg citron, Citrumelo and Savage
	Lemon caterpillar, *Papiliodemoleus* Linnaeus	Citron and Marsh seedless
	Aphid, *Toxoptera citricida* Boyer de Fonscombe, *Aphis gossypii* Glover	Washington Naval and Nagpur mandarin
	Scales, *Coccus hesperidium* L., *Aonidiella aurantii* Maskell,	Jenru tenga, Washington Naval, Hazara
	Mites, *Panonychus citri* McGreger and *Eutetranychus orientalis* Klein,	Coorg, Wilking, Blood red, Chakotra, Kinnow, Malta, Mosambi and Kagzi lime
Guava	Fruit fly, *Bactrocera zonatus* Saunders and *B. dorsalis* Hendel	Behat Coconut, Red flesh and Smooth Green hybrid
	Shoot borer, *Indrabela tetraonis* Moore	Red flesh
	Tea mosquito bug	Bangalore Round, Bapatala, AC 10
Mango	Leaf Hoppers, *Amritodes atkinsoni* Lthiery	Chinnarasam, Pulhora, Baneshan Bangalora and Annanas,
	Gall insect, *Procontarinia matteriana* Kieff & Cecec	Maharaja of Mysore, Delicious, Gulabkhas, Salem, Banglora, Annanas, KO7, KO11 and Vellakachi Anain,
	Fruit fly, *Bactrocera dorsalis* Hendel	Manjurad, Monteiro and Toranjo

Novel Approach

Resistance of major pests to Insecticides may remain a significant thrust for the development of new insecticides. With this view, limited range Insect Growth Regulator (IGR) formulations and safe pesticide encapsulations are used in pest management and its resistance management programmes, especially considering codling moth (Blomefield, 1997). The use of small quantity of bait sprayed over fruit flies eliminates requirement for full-cover sprays, that leading to a greater detrimental effect on beneficial organisms (Barnes, 1999). The fact that these new products act over biological development of insects, such as ecdysis, is a significant advantage. In contrast to the wide spectrum insecticides, many possess higher specificity to specific species and found safer to entomophages. Currently, novel insecticides that are in use such as those that target lepidopteran pests, sap feeders, dipterans, leafminers and also includes insect growth regulators, all of which manage a broad range of insects. One disadvantage of these type of insecticides includes, limited range of action, which controls a restricted insect pests, farmers required to apply supplementary insecticides to manage secondary pest with impoverished biocontrol agents, raising the treatment numbers per hectare and also the amount of pest management.

Trapping Tools and Pheromone Lures

Disruption of mating using pheromones are potentially significant and cost-effective in controlling *C. pomonella* and *B. dorsalis* of fruit crops (Barnes and Blomefield, 1997). Whereas hitherto, sex attractants and nutrition have only been established to combat fruit flies. Some of the traps are species specific, such as those used for weevils and lepidopteran moths (Gold *et al.*, 2003). Few of the pests necessitate the placement of different types of trapping devices, such as pheromones, yellow pans and yellow and blue sticky traps, to track the preliminary insect pest escalation. Incandescent lights installed in fruit orchards deter nocturnal fruit sucking moths occurrence (Shimoda *et al.*, 2013). Thus this idea wants to be familiarized among farmers, government agencies should take steps for pest observation using specific pheromone trapping devices, based on the below mentioned details.

Pheromone Based Devices

The insect pests attacking horticultural crops especially orchards can be successfully eliminated or managed using the pheromone-based tools (Welter *et al.* 2005). Nowadays pheromone dispensers are widely accessible and they sprays pheromones from a microencapsulation, reservoir from which it can be applied by hand and porous layer to modulate discharge, hollow fibers and wires and twist-tie strands (Onufrieva *et al.* 2018).

Research reports showed several amount of sex pheromones of females especially for the lepidopteran moths and those pheromones are mainly utilized for the processing of disrupting the mating for nearly 20 pest species across the globe (Carde, 2007; Miller and Gut, 2015). In addition, mating disruption pheromones have also been established for lepidopteran storage pests *viz.,* Indian meal moth *Plodia interpunctella* (Hübner), Almond moth, *Cadra cautella* (Walker) and Mediterranean flour moth, *Ephestia kuehniella* Zeller (Lepidoptera: Pyralidae) Wijayaratne and Burks (2020). Furthermore, pheromone dispenser utilization have potentially managed various horticultural pests namely *Tuta absoluta* (Meyrick) (Lepidoptera: Gelechiidae) and gypsy moth *Lymantria dispar* (L.) (Lepidoptera: Erebidae) (Thorpe *et al.*, 2006; Cocco *et al.*, 2020). Apart from this, few formulations have efficaciously evolved to manage some of the orchard pests *viz.,* apple pests such as *C. pomonella* and *Grapholita molesta* (Busck) (Angeli *et al.*, 2007; Stelinski *et al.*, 2007; Kovanci *et al.*, 2015; McGhee *et al.*, 2016 and Louis *et al.*, 2001). Installation of these kind of pheromone lures and traps in vineyards have reduced the population intensity of *L. botrana* (Ioriatti *et al.*, 2011).

Yellow Pan/Sticky Traps

Establish yellow pan or sticky traps for detecting whiteflies at a rate of ten yellow pans or sticky trap for an acre. Blank yellow tins available locally that have been smeared with grease or castor oil or vasline on the outside can be followed.

Light Trap

For black flies tracking, yellow colored traps with reflection light at a wavelength of 550 nm implemented for two hours in the evening. Florescent lights strategically installed in fruit orchards also restricts the movement of nocturnally active pests particularly fruit sucking moth, *Othreis* spp., (Whitehead and Rust, 1972).

Efficiency of Biocontrol Agents in BIPM

Pest management by means of biological control is not only a crucial element for long-term agricultural sustainability, but it is also environmentally friendly and safe for non target organisms (Murugasridevi *et al.*, 2021a). Macrobial agents including natural enemies *viz.,* parasitoids and predators. Microbial agents namely entomopathogenic fungus, entomopathogenic bacteria, entomopathogenic viruses and entomopathogenic nematodes are established and artificially proliferated in this technique, rather than relying on nature to overcome the target pests. Biological control is a time-consuming procedure which needs a long span to attain the target level of pest management. In contrast to synthetic insecticides, biocontrol agents are not commercially obtainable in many countries, despite more than 55 years of research in this field, excluding few parasitoids such as *Trichogramma* spp., *Leptomastix dactylopii, Bracon brevicornis* and *Parasierola nephantidis,* and few of the predators such as *Crytolaemus montrouzieri* Pharos and *Chilocorus nigritus,* (Dhanapal *et al.*, 2019). The protection of biocontrol agents has recently gained a lot of attention. There are four methods of biocontrol: natural, conservation, augmentation, and classical biological control (Bentley and O'Neil, 1997). For resource-poor farmers' fundamental food crops, importation is an alternative and cost effective rather than the chemical control. Although there are some technical issues with augmenting, it is often an ecologically safe and virulent choice to chemicals, as well as a source of jobs (Murugasridevi *et al.*, 2021 b).

Conservation technique of classical biological control helps the farmers to conserve the native species, despite the fact saving the labor, cost effective and lowers the detrimental effect of chemical insecticides over the environment (Murugasridevi *et al.*, 2019). Achievements in biocontrol encourages the farmers/researchers to attempt the conservation of entomophages. Those practices includes conservation, habitat manipulation, augmentation and genetic modification of natural enemies. There were several predators, parasitoids and pathogens have been identified in orchard crops and which seems suitable for the managing of sucking pests, defoliators and borers.

Generally plant species supplies plenty amount of pollen, honey dew, extra floral nectar, flower nectar and seeds as a great nutritional source for the adults of both parasitoids and predators (Wackers, 2005; Araj and Wratten, 2015), and also renders a proper inhabiting environment for some of the alternate hosts. Wackers (2005) reported the potential of nectar on survival of natural enemies when they don't have access to the host and it also increases fitness when the hosts are present. Earlier studies have also evidenced the screening and testing of plants for enhancing the conservation of natural enemies in orchard crops. Gontijo *et al.* (2013) documented the potential of brassicaceous plant, sweet alyssum, *Lobularia maritima* (Roux) which elevated the intensity of predators and decreased the infestation caused by *D. plantaginea* in apple orchards. Cahenzli *et al.* (2019) evidenced that strip cropping of dicotyledonous species and grasses positively enhanced the reduction of aphid infestation in apple orchards of Europe. In addition, the existence of flowers actually elevated the population of predatory anthocorid bugs and helped in managing hemipteran pear psyllid, *Cacopsylla pyri* L. (Psyllidae) (Fitzgerald and Solomon, 2004; Winkler *et al.* 2007). Wan *et al.* (2014a, 2014b) confirmed that cover cropping of fabaceous plant, *Trifolium repens* L. strengthened the generalistic predator population especially in canopies of peach orchards and lowered the infestation of aphids and torticid moth, *Grapholita molesta* (Busck) in China.

It has been challenging to fully exploit the biocontrol possibilities in orchards which are sprayed with broad-spectrum insecticides on a regular basis and/or have strict quality standards. Minor pests won't induce direct fruit loss are the ideal targets for biological management in tree fruits which includes leaf miners, mites and aphids. Pest communities that sustain on the fruit are often too large for specific biological control agents to reproduce (Bale *et al.*, 2008). In natural ecosystems, insect and mite pest numbers are limited by natural enemies and environmental conditions. Natural control typically fails when entomophages are destroyed by human activity in an agroecosystem, or during which the pests are transferred to new locations in absence of their natural enemies, resulting in pest epidemics.

Although biocontrol is frequently believed as a biopesticide in which a only one species of beneficial arthropod is delivered or protected, the best outputs are usually obtained when many natural enemies contribute in lowering the pest communities at various time span of the season and at various developmental phases (Niu *et al.*, 2014). While insecticide resistance in the some of the beneficial arthropods such as black ladybeetle predator *Stethorus punctum* and different predatory mite species is often considerably sluggish to establish. The great majority of biocontrol potential could not be achieved without acquiring resistance to pesticides. Some of the potential biocontrol agents of various orchard pests are enlisted in Table 8.

Table 8. Important biocontrol agents for control of orchard pests

Predators	
Rodolia cardinalis	Cottony cushion scale and Citrus mealybug
Platymeris laevicollis	Coconut Rhinoceros beetle
Cryptolaemus montrouzieri	Mealy bug- *Ferrisia virgata, Planococcus citri, Maconellicoccus hirsutus,* Grapevine soft scale
Mallada boninensis	Citrus blackfly, Mango hoppers, Grapes thrips
Crytolaemus nigritus	Citrus red scale
Scymnus coccivora, Cryptolaemus montrouzieri and *Anagyrus dactylopii*	Grapes mealybugs
Chilocorus nigrita	Grapevine hard scale
Parasitoids	
Aphelinus mali	Apple woolly aphid
Encarsia perniciosi	San Jose Scale
Trichogramma cacoecia and *T. embryophagum*	Apple codling moth
Tetrastichus radiates	Citrus psyllid
Leptomastrix dactylopi and *Anagyrus dactylopii*	Grapevine Mealy bug
Pathogens	
Lecanicillium lecanii	Mango hoppers and grapes mealybugs
Lecanicillium lecanii (Zimm.) or *Beauveria bassiana* and *Metarhizium anisopliae*	Grapes thrips
Beauveria bassiana	Grapes leafhopper
Entomopathogenic Nematode *Heterorhabditis indicus*	Grape flea beetle

CONCLUSION

There is a great requirement to establish attention and enable the growers to be perceive the new techniques by involving them through participation, training and demonstration. Nurturing different horticultural crops and maintaining nursery can be taught through trainings and demos. Thus, strengthened commercial system of fruits have to be installed by exploitation of modern crop production and protection tactics. The research as well as development of biological control agents and biorational products for management of pests in orchards in some agricultural institutes and private organizations led to noteworthy reduction in an usage of synthetic pesticides. The exporting of fruits to other countries are greatly impairing by the usage of chemical pesticides. The BIPM's has been placed as a best tactic for the control of various orchard pests especially in the farm where the fruit crops were growing for export purpose. Finally, as governments and producers throughout the world become more aware of the economic and environmental benefits of utilizing biocontrol and biorational products, the demand for BIPM is growing. As a result, pesticide use is expected to drop in the next decades, lessening the biochemical risks they represent to our health. However, how well BIPM methods are sustained as a viable discipline will be determined by our collective reactions to these problems.

REFERENCES

Ahmed, W., Nawaz, M. A., Saleem, B. A., & Asim, M. (2005). Incidence of mango midge and its control in different mango growing countries of the world. In *International Conference on Mango and date Palm: Culture and Export* (pp. 98-101). Academic Press.

Angeli, G., Anfora, G., Baldessari, M., Germinara, G. S., Rama, F., De Cristofaro, A., & Ioriatti, C. (2007). Mating disruption of codling moth *Cydia pomonella* with high densities of Ecodian sex pheromone dispensers. *Journal of Applied Entomology*, *131*(5), 311–318. doi:10.1111/j.1439-0418.2007.01172.x

Araj, S. E., & Wratten, S. D. (2015). Comparing existing weeds and commonly used insectary plants as floral resources for a parasitoid. *Biological Control*, *81*, 15–20. doi:10.1016/j.biocontrol.2014.11.003

Arneson, P. A., & Losey, J. E. (1997). *Integrated pest mangement-lecture notes.* Cornell University.

Bale, J. S., Van Lenteren, J. C., & Bigler, F. (2008). Biological control and sustainable food production. *Philosophical Transactions of the Royal Society of London. Series B, Biological Sciences*, *363*(1492), 761–776. doi:10.1098/rstb.2007.2182 PMID:17827110

Bangels, E., Alhmedi, A., Akkermans, W., Bylemans, D., & Belien, T. (2021). Towards a Knowledge-Based Decision Support System for Integrated Control of Woolly Apple Aphid, Eriosoma lanigerum, with Maximal Biological Suppression by the Parasitoid *Aphelinus mali*. *Insects*, *12*(6), 479. doi:10.3390/insects12060479 PMID:34063971

Barnes, B. N. (1990). *FFTRI monitoring manual for orchard research institute.* Fruit and Fruit Technology Research Institute.

Barnes, B. N. (1999). *Fruit fly baiting on deciduous fruit. Doing the right things right.* Deciduous Fruit Grower.

Barnes, B. N., & Blomefield, T. L. (1997). Goading growers towards mating disruption: the South African experience with Grapholita molesta and Cydia pomonella (Lepidoptera, Tortricidae). *Goading growers towards mating disruption: the South African experience with Grapholita molesta and Cydia pomonella (Lepidoptera, Tortricidae)*, *20*(1), 45-56.

Beers, E. H., Hull, L. A., & Jones, V. P. (2020). Sampling pest and beneficial arthropods of apple. In *Handbook of Sampling Methods for Arthropods in Agriculture* (pp. 383–416). CRC Press. doi:10.1201/9781003067900-18

Bentley, J. W., & O'Neil, R. J. (1997). On the ethics of biological control of insect pests. *Agriculture and Human Values*, *14*(3), 283–289. doi:10.1023/A:1007477300339

Blomefield, T. L. (1997). Managing resistance of codling moth, Cydia pomonella (L.) in South African pome fruit orchards. Combating Resistance, IARC-Rothamsted, Harpenden, 14-16.

Boavida, C., Neuenschwander, P., & Schulthess, F. (1992). Spatial distribution of *Rastrococcus invadens* Williams (Hom., Pseudococcidae) in mango trees. *Journal of Applied Entomology*, *114*(1-5), 381–391. doi:10.1111/j.1439-0418.1992.tb01141.x

Butani, D. K., & Butani, D. C. (1979). Insects and fruits (No. 634 B8). Delhi, India: Periodical Export Book Agency.

Cahenzli, F., Sigsgaard, L., Daniel, C., Herz, A., Jamar, L., Kelderer, M., Jacobsen, S. K., Kruczyńska, D., Matray, S., Porcel, M., Sekrecka, M., Świergiel, W., Tasin, M., Telfser, J., & Pfiffner, L. (2019). Perennial flower strips for pest control in organic apple orchards-A pan-European study. *Agriculture, Ecosystems & Environment*, *278*, 43–53. doi:10.1016/j.agee.2019.03.011

Cardé, R. T. (2007). *Using pheromones to disrupt mating of moth pests. In Perspectives in ecological theory and integrated pest management.* Cambridge University Press.

Charati, S. N., Pokharkar, D. S., & Ghorpade, S. A. (2003). Abundance of spiralling whitefly, a newly introduced pest in Maharashtra State. *Journal of Maharashtra Agricultural Universities (India).*

Cocco, A., Deliperi, S., & Delrio, G. (2013). Control of Tuta absoluta (Meyrick)(Lepidoptera: Gelechiidae) in greenhouse tomato crops using the mating disruption technique. *Journal of Applied Entomology*, *137*(1-2), 16–28. doi:10.1111/j.1439-0418.2012.01735.x

Cochran, W. G. (2007). *Sampling techniques*. John Wiley & Sons.

Cohen, Y., Cohen, A., Hetzroni, A., Alchanatis, V., Broday, D., Gazit, Y., & Timar, D. (2008). Spatial decision support system for Medfly control in citrus. *Computers and Electronics in Agriculture, 62*(2), 107-117.

Cunningham, I. C. (1984, November). Mango insect pests. In *Proceedings of the First Australian Mango Research Workshop* (pp. 211-224). Academic Press.

DeBach, P. (1964). *Biological control of insect pests and weeds.* Academic Press.

Debbarma, T., & Hath, T. K. (2021). Pest Complex of Lemon CV. Assam Lemon (*Citrus limon* L. Burm) in Terai Region of West Bengal, India. *International Journal of Current Microbiology and Applied Sciences*, *10*(04), 736–744. doi:10.20546/ijcmas.2021.1004.075

Dhanapal, R., Kumar, D. V. S. R., Lakshmipathy, R., Rani, C. S., & Kumar, R. M. (2019). Pathogenicity testing of indigenous isolates of entomopathogenic fungus, *Lecanicillium lecanii* against tobacco caterpillar, *Spodoptera litura. Journal of Experimental Zoology India*, *22*(2), 753–756.

Doitsidis, L., Fouskitakis, G. N., Varikou, K. N., Rigakis, I. I., Chatzichristofis, S. A., Papafilippaki, A. K., & Birouraki, A. E. (2017). Remote monitoring of the Bactrocera oleae (Gmelin)(Diptera: Tephritidae) population using an automated McPhail trap. *Computers and Electronics in Agriculture*, *137*, 69–78. doi:10.1016/j.compag.2017.03.014

Firake, D. M., Behere, G. T., Deshmukh, N. A., Firake, P. D., & Thakur, N. A. (2013). Recent scenario of insect-pests of guava in North East India and their eco-friendly management. *Indian Journal of Hill Farming*, *26*(1), 55–57.

Fitzgerald, J. D., & Solomon, M. G. (2004). Can flowering plants enhance numbers of beneficial arthropods in UK apple and pear orchards? *Biocontrol Science and Technology*, *14*(3), 291–300. doi:10.1080/09583150410001665178

Fletcher, T. B. (1914). *Some South Indian Insects and Other Animals of Importance: Considered Especially from an Economic Point of View/by T. Bainbrigge Fletcher*. Bishen Singh Mahendra Pal Singh.

Flint, M. L., & Van den Bosch, R. (2012). *Introduction to integrated pest management*. Springer Science & Business Media.

Fonte, A., Garcerá, C., Tena, A., & Chueca, P. (2021). Volume Rate Adjustment for Pesticide Applications against Aonidiella aurantii in Citrus: Validation of CitrusVol in the Growers' Practice. *Agronomy (Basel)*, *11*(7), 1350. doi:10.3390/agronomy11071350

Gold, C. S., Nankinga, C., Niere, B., & Godonou, I. (2003). *IPM of banana weevil in Africa with emphasis on microbial control. Biological control in IPM systems in Africa*. CABI Publishing.

Gontijo, L. M., Beers, E. H., & Snyder, W. E. (2013). Flowers promote aphid suppression in apple orchards. *Biological Control*, *66*(1), 8–15. doi:10.1016/j.biocontrol.2013.03.007

Guarnieri, A., Maini, S., Molari, G., & Rondelli, V. (2011). Automatic trap for moth detection in integrated pest management. *Bulletin of Insectology*, *64*(2), 247–251.

Gupta, R. K., & Arora, R. K. (2001). Lepidopteran fruit borers on guava in Jammu. *Insect Environment*, *7*(2), 83–84.

Haji, F. N. P., Barbosa, F. R., Lopes, P. R. C., Moreira, A. N., de Alencar, J. A., & Ferreira, R. C. F. (2002, September). Monitoring mango pests within an integrated production program in Brazil. In *VII International Mango Symposium 645* (pp. 163-165). Academic Press.

Hansen, J. D., & Armstrong, J. W. (1990). The failure of field sanitation to reduce infestation by the mango weevil, *Cryptorhynchus mangiferae* (F.) (Coleoptera: Curculionidae). *International Journal of Pest Management*, *36*(4), 359–361.

Hare, J. D. (2020). Sampling arthropod pests in citrus. In *Handbook of sampling methods for arthropods in agriculture* (pp. 417–431). CRC Press. doi:10.1201/9781003067900-19

Haseeb, M. (2005, December). Current status of insect pest problems in guava. In *International Guava Symposium 735* (pp. 453-467). Academic Press.

Hoyt, S. C., & Burts, E. C. (1974). Integrated control of fruit pests. *Annual Review of Entomology*, *19*(1), 231–252. doi:10.1146/annurev.en.19.010174.001311

Huda, A. K. S., & Luck, J. (2008, February). Early warning of pest/diseases for selected crops using climate information and crop simulation modeling approach under climate change scenarios. In *Proceeding of International Symposium on Agrometeorology and Food Security* (pp. 18-21). Academic Press.

Ioannou, C. S., Papanastasiou, S. A., Zarpas, K. D., Miranda, M. A., Sciarretta, A., Nestel, D., & Papadopoulos, N. T. (2019). Development and Field Testing of a Spatial Decision Support System to Control Populations of the European Cherry Fruit Fly, *Rhagoletis cerasi*, in Commercial Orchards. *Agronomy (Basel)*, *9*(10), 568. doi:10.3390/agronomy9100568

Ioriatti, C., Anfora, G., Tasin, M., De Cristofaro, A., Witzgall, P., & Lucchi, A. (2011). Chemical ecology and management of *Lobesia botrana* (Lepidoptera: Tortricidae). *Journal of Economic Entomology*, *104*(4), 1125–1137. doi:10.1603/EC10443 PMID:21882674

Ishaaya, I., Kontsedalov, S., & Horowitz, A. R. (2005). Biorational insecticides: Mechanism and cross-resistance. *Archives of Insect Biochemistry and Physiology: Published in Collaboration with the Entomological Society of America*, *58*(4), 192–199. doi:10.1002/arch.20042 PMID:15756702

Jiang, J. A., Lin, T. S., Yang, E. C., Tseng, C. L., Chen, C. P., Yen, C. W., Zheng, X.-Y., Liu, C.-Y., Liu, R.-H., Chen, Y.-F., Chang, W.-Y., & Chuang, C. L. (2013). Application of a web-based remote agro-ecological monitoring system for observing spatial distribution and dynamics of *Bactrocera dorsalis* in fruit orchards. *Precision Agriculture*, *14*(3), 323–342. doi:10.100711119-012-9298-x

Jones, V. P., Brunner, J. F., Grove, G. G., Petit, B., Tangren, G. V., & Jones, W. E. (2010). A web-based decision support system to enhance IPM programs in Washington tree fruit. *Pest Management Science: Formerly. Pesticide Science*, *66*(6), 587–595. doi:10.1002/ps.1913 PMID:20127866

Jothi, D. B., Tandon, P. L., & Verghese, A. (1994). Hot water immersion as a quarantine treatment for Indian mangoes infested with the oriental fruitfly, Bactrocera dorsalis (Hendel)(Diptera: Tephritidae). *FAO Plant Protection Bulletin*, *42*(3), 158–159.

Kapoor, V. C. (2000). Fruit flies (Diptera: Tephritidae): status, biology and management strategies. *IPM System in Agriculture, 7.*

Kovanci, O. B. (2015). Co-application of microencapsulated pear ester and codlemone for mating disruption of Cydia pomonella. *Journal of Pest Science*, *88*(2), 311–319. doi:10.100710340-014-0619-x

Kulkarni, N. S. (2018). Pests of Grapes. In *Pests and Their Management* (pp. 517–557). Springer. doi:10.1007/978-981-10-8687-8_16

Kulkarni, N. S. (2020). Sucking Pests of Grapes. In *Sucking Pests of Crops* (pp. 425–450). Springer. doi:10.1007/978-981-15-6149-8_14

Liao, M. S., Chuang, C. L., Lin, T. S., Chen, C. P., Zheng, X. Y., Chen, P. T., Liao, K.-C., & Jiang, J. A. (2012). Development of an autonomous early warning system for *Bactrocera dorsalis* (Hendel) outbreaks in remote fruit orchards. *Computers and Electronics in Agriculture*, *88*, 1–12. doi:10.1016/j.compag.2012.06.008

Louis, F., & Schirra, K. J. (2001). Mating disruption of *Lobesia botrana* (Lepidoptera: Tortricidae) in vineyards with very high population densities. *IOBC/WPRS Bulletin*, *24*, 75–79.

Mani, M. (1986). *Distribution, Bioecology And Management Of The Grape Mealybug, Maconellicoccus hirsutus (Green) With Special Reference To Its Natural Enemies* (Doctoral Dissertation). University Of Agricultural Sciences, Bangalore.

Mani, M., & Kulkarni, N. S. (2007). Citrus mealybug *Planococcus citri* (Risso) Homoptera; Pseudococcidae)-a major pest of grapes in India. *Entomon*, *32*, 235–236.

Mani, M., Kulkarni, N. S., Banerjee, K., & Adsule, P. G. (2008). *Pest management in grapes*. Extension.

McGhee, P. S., Miller, J. R., Thomson, D. R., & Gut, L. J. (2016). Optimizing aerosol dispensers for mating disruption of codling moth, Cydia pomonella L. *Journal of Chemical Ecology*, *42*(7), 612–616. doi:10.100710886-016-0724-9 PMID:27369280

Mebdoua, S. (2018). *Pesticide residues in fruits and vegetables.* Bioactive molecules in food. Reference series in phytochemistry. Springer.

Miller, J. R., & Gut, L. J. (2015). Mating disruption for the 21st century: Matching technology with mechanism. *Environmental Entomology*, *44*(3), 427–453. doi:10.1093/ee/nvv052 PMID:26313949

Miranda, M. Á., Barceló, C., Valdés, F., Feliu, J. F., Nestel, D., Papadopoulos, N., Sciarretta, A., Ruiz, M., & Alorda, B. (2019). Developing and Implementation of Decision Support System (DSS) for the Control of Olive Fruit Fly, *Bactrocera Oleae,* in Mediterranean Olive Orchards. *Agronomy (Basel)*, *9*(10), 620. doi:10.3390/agronomy9100620

Misra, C. S. (1920). Index to Indian fruit pests. Report. *Proc. 3rd Ent. Management*, 564-595.

Mohanasundaram, M. (1974). *Occurrence of Kerria lacca (Kerr.) and Aleurocanthus spiniferus (Quain.) on grape vine in Tamil Nadu*. Academic Press.

Mondino, P., & González-Andújar, J. L. (2019). Evaluation of a decision support system for crop protection in apple orchards. *Computers in Industry*, *107*, 99–103. doi:10.1016/j.compind.2019.02.005

Morgan, D., & Solomon, M. G. (1995). PEST-MAN: A forecasting system for apple and pear pests 1. *Bulletin OEPP*, *23*(4), 601–605. doi:10.1111/j.1365-2338.1993.tb00556.x

Mullen, J. D. (1995). *Estimating environmental and human health benefits of reducing pesticide use through integrated pest management programs* (Doctoral dissertation). Virginia Tech.

Murugasridevi, K., Jeyarani, S., & Kumar, S. M. (2021, December 4). Incidence of Groundnut Leafminer (GLM), *Aproaerema modicella* (Deventer) (Lepidoptera: Gelechiidae) and its Parasitic Fauna on Alternate Leguminous Hosts in Tamil Nadu, India. *Legume Research.* Advance online publication. doi:10.18805/LR-4672

Murugasridevi, K., Jeyarani, S., & Mohankumar, S. (2019). *Occurrence of groundnut Leafminer (GLM), Aproaerema modicella Deventer (Lepidoptera: Gelechiidae) and its parasitoid fauna in various groundnut growing areas of Tamil Nadu*. Academic Press.

Murugasridevi, K., Jeyarani, S., Nelson, S. J., Kumar, S. M., & Nakkeeran, S. (2021 a). Assessment of Diversity Indices and DNA Barcoding of Parasitic Fauna Associated with Groundnut Leafminer (GLM), *Aproaerema modicella* Deventer (Lepidoptera: Gelechiidae). *Legume Research*. Advance online publication. doi:10.18805/LR-4579

Nath, R., & Sikha, D. (2019). Insect pests of citrus and their management. *Int J Pl Pr*, *12*(2), 188–196. doi:10.15740/HAS/IJPP/12.2/188-196

Ndiaye, M., Dieng, E. O., & Delhove, G. (2008). Population dynamics and on-farm fruit fly integrated pest management in mango orchards in the natural area of Niayes in Senegal. *Pest Management in Horticultural Ecosystems*, *14*(1), 1–8.

Nestel, D., Cohen, Y., Shaked, B., Alchanatis, V., Nemny-Lavy, E., Miranda, M. A., Sciarretta, A., & Papadopoulos, N. T. (2019). An Integrated Decision Support System for environmentally-friendly management of the Ethiopian fruit fly in greenhouse crops. *Agronomy (Basel)*, *9*(8), 459. doi:10.3390/agronomy9080459

Niu, J. Z., Hull-Sanders, H., Zhang, Y. X., Lin, J. Z., Dou, W., & Wang, J. J. (2014). Biological control of arthropod pests in citrus orchards in China. *Biological Control*, *68*, 15–22. doi:10.1016/j.biocontrol.2013.06.005

Onufrieva, K. S., Hickman, A. D., Leonard, D. S., & Tobin, P. C. (2019). Relationship between efficacy of mating disruption and gypsy moth density. *International Journal of Pest Management*, *65*(1), 44–52. doi:10.1080/09670874.2018.1455116

Pandey, V., Patel, M. G., Chaudhari, G. B., Patel, J. R., Bhatt, B. K., Vadodaria, R. P., & Shekh, A. M. (2003). Influence of weather parameters on the population dynamics of mango hopper. *Journal of Agrometeorology*, *5*(1), 51–59. doi:10.54386/jam.v5i1.586

Patel, A. T., & Kumar, S. (2020). Chemical control of mango leaf gall midge *Procontarinia matteiana. Ann. Entomol, 38*(1-2), 21-26.

Pedigo, L. P. (1996). Entomology and pest management (2nd ed.). Prentice-Hall Inc.

Peña, J. E. (2002, September). Integrated pest management and monitoring techniques for mango pests. In *VII International Mango Symposium 645* (pp. 151-161). Academic Press.

Peña, J. E., Sharp, J. L., & Wysoki, M. (2002). *Tropical Fruit Pests and Pollinators Biology, Economic Importance, Natural Enemies and Control*. CABI. doi:10.1079/9780851994345.0000

Pontikakos, C. M., Tsiligiridis, T. A., Yialouris, C. P., & Kontodimas, D. C. (2012). Pest management control of olive fruit fly (Bactrocera oleae) based on a location-aware agro-environmental system. *Computers and Electronics in Agriculture*, *87*, 39–50. doi:10.1016/j.compag.2012.05.001

Prasada Rao, G. S. L. H. V., & Beevi, S. P. (2008). Forewarning tea mosquito bug *Helopeltis antonii* Sign.(Miridae: Hemiptera) in cashew. In *Proceeding of the International Symposium on Agrometeorology and food security* (pp. 18-21). Academic Press.

Preti, M., Verheggen, F., & Angeli, S. (2021). Insect pest monitoring with camera-equipped traps: Strengths and limitations. *Journal of Pest Science*, *94*(2), 1–15. doi:10.100710340-020-01309-4

Reddy, P. V. R., Gundappa, B., & Chakravarthy, A. K. (2018). Pests of mango. In *Pests and their management* (pp. 415–440). Springer. doi:10.1007/978-981-10-8687-8_12

Reddy, P. V. R., Varun Rajan, V., Thangam, D., & Chakravarthy, A. K. (2015, April). Stem borers in mango: species diversity and damage patterns. In *4th congress on Insect Science, Punjab Agricultural University, Ludhiana* (pp. 16-17). Academic Press.

Román, C., Peris, M., Esteve, J., Tejerina, M., Cambray, J., Vilardell, P., & Planas, S. (2022). Pesticide dose adjustment in fruit and grapevine orchards by DOSA3D: Fundamentals of the system and on-farm validation. *The Science of the Total Environment*, *808*, 152158. doi:10.1016/j.scitotenv.2021.152158 PMID:34871680

Samietz, J., Graf, B., Höhn, H., Schaub, L., & Höpli, H. U. (2007). Phenology modelling of major insect pests in fruit orchards from biological basics to decision support: The forecasting tool SOPRA. *Bulletin OEPP. EPPO Bulletin. European and Mediterranean Plant Protection Organisation*, *37*(2), 255–260. doi:10.1111/j.1365-2338.2007.01121.x

Samietz, J., Hoehn, H., Razavi, E., Schaub, L., & Graf, B. (2015). Decision support for sustainable orchard pest management with the Swiss forecasting system SOPRA. *Acta Horticulturae*, (1099), 383–390. doi:10.17660/ActaHortic.2015.1099.44

Satyagopal, K., Sushil, S. N., Jeyakumar, P., Shankar, G., Sharma, O. P., Sain, S. K., ... Latha, S. (2014). *AESA based IPM package for banana*. Academic Press.

Satyanarayana, G. (1981). Problems of grape production around Hyderabad. Andhra PRADESH GRAPE Growers Association.

Sciarretta, A., Tabilio, M. R., Amore, A., Colacci, M., Miranda, M. Á., Nestel, D., Papadopoulos, N. T., & Trematerra, P. (2019). Defining and evaluating a decision support System (DSS) for the precise pest management of the Mediterranean Fruit Fly, *Ceratitis capitata*, at the farm level. *Agronomy (Basel)*, *9*(10), 608. doi:10.3390/agronomy9100608

Shaked, B., Amore, A., Ioannou, C., Valdés, F., Alorda, B., Papanastasiou, S., Goldshtein, E., Shenderey, C., Leza, M., Pontikakos, C., Perdikis, D., Tsiligiridis, T., Tabilio, M. R., Sciarretta, A., Barceló, C., Athanassiou, C., Miranda, M. A., Alchanatis, V., Papadopoulos, N., & Nestel, D. (2018). Electronic traps for detection and population monitoring of adult fruit flies (Diptera: Tephritidae). *Journal of Applied Entomology*, *142*(1-2), 43–51. doi:10.1111/jen.12422

Shankar, U. (2017). Integrated pest management in horticultural crops. *Technological Innovations in Integrated Pest Management Biorational and Ecological Perspective*, 307.

Sharma, D. R., & Singh, S. (2006). Management of insect pests of temperate fruits. *Proceedings of advance training course on Emerging Trends in Economic Entomology,* 21-13.

Shashank, P. R., Doddabasappa, B., Kammar, V., Chakravarthy, A. K., & Honda, H. (2015). Molecular characterization and management of shoot and fruit borer *Conogethes punctiferalis* Guenee (Crambidae: Lepidoptera) populations infesting cardamom, castor and other hosts. In *New horizons in insect science: towards sustainable pest management* (pp. 207–227). Springer. doi:10.1007/978-81-322-2089-3_20

Sherwani, A., Mukhtar, M., & Wani, A. A. (2016). *Insect pests of apple and their management. Insect pest management of fruit crops*. Biotech Books.

Shimoda, M., & Honda, K. I. (2013). Insect reactions to light and its applications to pest management. *Applied Entomology and Zoology*, *48*(4), 413–421. doi:10.100713355-013-0219-x

Short, B. D., Khrimian, A., & Leskey, T. C. (2017). Pheromone-based decision support tools for management of Halyomorpha halys in apple orchards: Development of a trap-based treatment threshold. *Journal of Pest Science*, *90*(4), 1191–1204. doi:10.100710340-016-0812-1

Simon, S., Bouvier, J. C., Debras, J. F., & Sauphanor, B. (2011). Biodiversity and pest management in orchard systems. *Sustainable Agriculture*, *2*, 693–709. doi:10.1007/978-94-007-0394-0_30

Smith, D., Beattie, G. A., & Broadley, R. (1997). *Citrus pests and their natural enemies: integrated pest management in Australia*. Academic Press.

Srivastava, R. P. (1997). Mango insect pest management. International Book Distributing Co.

Stelinski, L. L., Miller, J. R., Ledebuhr, R., Siegert, P., & Gut, L. J. (2007). Season-long mating disruption of *Grapholita molesta* (Lepidoptera: Tortricidae) by one machine application of pheromone in wax drops (SPLAT-OFM). *Journal of Pest Science*, *80*(2), 109–117. doi:10.100710340-007-0162-0

Suárez-Jacobo, A., Alcantar-Rosales, V. M., Alonso-Segura, D., Heras-Ramírez, M., Elizarragaz-De La Rosa, D., Lugo-Melchor, O., & Gaspar-Ramirez, O. (2017). Pesticide residues in orange fruit from citrus orchards in Nuevo Leon State, Mexico. *Food Additives & Contaminants: Part B*, *10*(3), 192–199. doi:10.1080/19393210.2017.1315743 PMID:28374639

Tandon, P. L., & Verghese, A. (1985). *World list of insect, mite and other pests of mango*. Academic Press.

Tello, J., Mammerler, R., Čajić, M., & Forneck, A. (2019). Major outbreaks in the nineteenth century shaped grape phylloxera contemporary genetic structure in Europe. *Scientific Reports*, *9*(1), 1–11. doi:10.103841598-019-54122-0 PMID:31772235

Thorpe, K. W. (2006). A review of the use of mating disruption to manage gypsy moth, *Lymantria dispar* (L.). Academic Press.

Ünlü, L., Akdemir, B., Ögür, E., & Şahin, İ. (2019). Remote monitoring of European Grapevine Moth, Lobesia botrana (Lepidoptera: Tortricidae) population using camera-based pheromone traps in vineyards. *Turkish Journal of Agriculture-Food Science and Technology*, *7*(4), 652–657. doi:10.24925/turjaf.v7i4.652-657.2382

Veeresh, G. K. (1985, May). Pest problems in mango—world situation. In *II International Symposium on Mango 231* (pp. 551-565). Academic Press.

Verghese, A., & Devi Thangam, S. (2011). Mango hoppers and their management. *Extension folder*, (71-11), 31-11.

Verheij, E. W. M., & Coronel, R. F. (1991). Edible fruits and nuts. Plant resources in South- *East Asia*. Academic Press.

Wackers, F. L. (2005). Suitability of (extra-) floral nectar, pollen, and honeydew as insect food sources. *Plantprovided Food for Carnivorous Insects*, 17-74.

Wan, N. F., Ji, X. Y., Gu, X. J., Jiang, J. X., Wu, J. H., & Li, B. (2014). Ecological engineering of ground cover vegetation promotes biocontrol services in peach orchards. *Ecological Engineering*, *64*, 62–65. doi:10.1016/j.ecoleng.2013.12.033

Wan, N. F., Ji, X. Y., & Jiang, J. X. (2014). Testing the enemies hypothesis in peach orchards in two different geographic areas in eastern China: The role of ground cover vegetation. *PLoS One*, *9*(6), e99850. doi:10.1371/journal.pone.0099850 PMID:24963719

Welter, S., Pickel, C., Millar, J., Cave, F., Van Steenwyk, R., & Dunley, J. (2005). Pheromone mating disruption offers selective management options for key pests. *California Agriculture*, *59*(1), 16–22. doi:10.3733/ca.v059n01p16

Whalon, M. E., & Croft, B. A. (1984). Apple IPM implementation in North America. *Annual Review of Entomology*, *29*(1), 435–470. doi:10.1146/annurev.en.29.010184.002251

Whitehead, V. B., & Rust, D. J. (1972). Control of the fruit-piercing moth Serrodes parfifa (Fabr.)(Lepidoptera: Noctuidae). *Phytophylactica*, *4*(1), 9–12.

Wijayaratne, L. K., & Burks, C. S. (2020). Persistence of mating suppression of the Indian meal moth Plodia interpunctella in the presence and absence of commercial mating disruption dispensers. *Insects*, *11*(10), 701. doi:10.3390/insects11100701 PMID:33066462

Winkler, K., Helsen, H. H. M., & Wackers, F. (2007). Kader 2. Functionele biodiversiteit in boomgaarden. *Entomologische Berichten, 67*(6), 236-237.

Wysoki, M., Ben-Dov, Y., Swirski, E., & Izhar, Y. (1992, July). The arthropod pests of mango in Israel. In *IV International Mango Symposium 341* (pp. 452-466). Academic Press.

Chapter 15
Insect Pest Control in Orchards

Mohamed Abdel-Raheem
National Research Center, Egypt

ABSTRACT

The species of pests in our orchards have natural predators and parasites. These natural enemies of pests play an important role in regulating pest populations. One way of reducing pest damage, therefore, is to create an orchard habitat that encourages biodiversity and these natural enemies. Many pesticides will kill these beneficial species when we are trying to control the pests, and the first line of defense is removed the next time the pest species attacks fruit trees. Agriculture is more sustainable and resilient to change when it mimics biological systems. Codling moth is a common pest of apple fruit. It feeds on the fruit for many weeks. Many pests invest in orchard crops such as apple maggot, gypsy moth, peachtree borer, pecan weevil, psyllid, sawfly, and scale insect.

INTRODUCTION

The apple maggot fly, *Rhagoletis pomonella* (Walsh) is a quarantine pest of apples in the Pacific Northwest of the U.S. The species evolved on hawthorns (Crataegus spp.) in eastern North America (Bush 1966; Bush and Smith 1998) and Mexico (Rull et al. 2006) and moved from hawthorn onto apple about 150 years ago in the eastern U.S.A. (Walsh 1867), differentiating over time into hawthorn and apple host races (Feder et al., 1988; Bush and Smith 1998).

Gypsy moth is a major pest of hardwood trees, control is necessary to prevent damage. Pecan weevil (*Curculio caryae*) is a serious late season pest of pecan and hickory trees, the pecan weevil chew holes through the shuck and shell of nuts. The preferred hosts for Gypsy moth are oaks, apple, sweet gum, speckled alder, basswood, birch, popular and willow. However, Gypsy moth larvae are polyphagous and they will feed on any plant when the population is sufficiently high (USDA, 1995). A list of over 600 species and their susceptibility to Gypsy moth is reproduced from the draft environmental impact statement on Gypsy moth management in the United States (USDA, 1995).

Psyllids feed on most fruit trees. Both adults and nymphs feed by piercing the leaf surface and extracting cell sap. This causes foliage to turn yellow, curl and eventually die. Honeydew secreted by the psyllids encourages the growth of dark sooty molds. Sawflies (*Caliroa cerasi*), known as cherry or

DOI: 10.4018/978-1-6684-2423-0.ch015

pear slugs. On citrus, numerous species of thrips has been recorded but few of them are considered as serious pests in different parts of the world (Lewis, 1997 and Childers & Nakahara S., 2006). Among these, *Pezothrips kellyanus* has emerged as an economic key pest of citrus fruits in New Zealand (Blank & Gill, 1997), southern Australia, and some areas of the Mediterranean Basin such as Italy in Eastern Sicily (Mound & Jackman, 1998.) and Spain (Navarro, et al., 2011), during the last decade. This species damage appeared as scabby, silvery scars on the rind that lower the fruit market grade (Mound & Jackman, 1998 and Navarro, et al., 2008).

APPLE MAGGOT (RHAGOLETIS POMONELLA)

Apple maggots attack plum, apricot, pear, cherry and hawthorn. If trees are neglected, 100% of the crop can be wormy rendering the fruit unfit to eat and suitable only for livestock feed.

Adult flies emerge in late spring and begin to lay eggs just under the apple skin. The eggs hatch, and the larvae begin to tunnel through the fruit.

Control

1. Maggots leave the fruit many days after it has fallen from the tree. Control can be achieved by picking up and discarding the dropped apples.
2. **Red Sphere Traps** reduce damage and work well to capture and reduce number of eggs laying adults. Hang spheres high in the brightest areas of the tree, 6-7 feet from the ground. Set out one trap for every 150 apples (2 traps per dwarf tree). Over a period of 30 years, traps on apple, crabapple, and hawthorn trees in Kittitas and Yakima Counties were positive for *R. pomonella*, implying that human enacted management measures and/or natural factors have prevented fly populations from reaching levels seen in western Washington (Tracewski et al., 1987; Yee et al., 2005; Yee and Goughnour 2008, Yee et al.,2014).

The traps intercept and kill a large proportion of the females on the perimeter of the plot and allow few to penetrate into the interior of the block. The net result is too few flies in the interior of the plot to cause injuries of economic significance (Noubar and Gaétan, **2001**)

3. **Beneficial nematodes** are hunt, penetrate and destroy the pupal stage of this pest. Apply in the early spring or fall around the base of trees, out to the drip line. One application will continue working for 18 months. EPNs such as *Steinernema riobrave*, *S. carpocapsae*, and *S. feltiae* have significant potential to suppress *R. pomonella* pupal populations. A possible approach would be to apply the nematodes under the tree canopy in spring when the soil temperature is conducive, before pupae emerge into adults in the summer. *Rhagoletis pomonella* normally pupate in the fall and remain inside the soil until late June or July, thus providing sufficient time for EPNs to infect, (Muhammad, et al., 2020).
4. **Surround WP** — made from kaolin clay — will suppress a broad range of insects and has shown over 90% control of apple pests. It also has a positive effect on fungal diseases like fire blight, sooty blotch and flyspeck.

5. **Botanical insecticides** these natural pesticides have fewer harmful side effects than synthetic chemicals and break down more quickly in the environment.

CODLING MOTH (CYDIA POMONELLA)

The codling moth is the major pest infecting the apple, both in Croatia and abroad. Besides apple this pest attacks pear, walnut, quince and some stone fruits causing economic losses in fruit production (Ciglar, 1998). The pest was originally present in Eurasia, but during the last two centuries it dispersed around the world with the spread of the cultivation of apples and pears (Franck et al., 2007).

The pest overwinters as a full grown larva within a thick, silken cocoon that can be found under loose scales of bark and in soil or debris around the tree base. The larvae pupate inside their cocoons in early spring, usually in March when spring temperatures exceed 10°C. Depending on temperature pupae development takes 7-30 days. For the development of adults the sum of Day Degrees (DD) of 100 is required (Wildbolz, 1965 cit Ciglar, 1998) that is usually at the end of April. The appearance of moths is associated with climatic conditions (temperature and humidity) as well as with fruit development period. The flight of the first generation of CM is much more intense from the middle or the end of May and lasts until the end of June. Moths are most active at sunset and before sunrise. In this period they mate. Female deposits 30 to 70 eggs singly on fruit or nearby leaves. After the eggs hatch, young larvae seek out and bore into a fruit. They feed on the interior of the fruit and reach full growth in four weeks. After completing their development they leave the fruit and drop from the trees to search out sites for pupation. The sum of Day Degrees of 610 is required for the development of a generation from the egg stage until the appearance of moths (Wildbolz, 1965 cit Ciglar, 1998). The second generation appears after ten days and its flight lasts from mid-July to mid-August. In this period they lay eggs again and so the development cycle is repeated. In Croatia there are two generations of this pest per year (Kovačević, 1952; Alford, 1984; Ciglar, 1998; Maceljski, 2002).

This moth has a widespread distribution, being found on six continents. Adaptive behavior such as diapause and multiple generations per breeding season have allowed this moth to persist even during years of bad climatic conditions, (Jackson1982).

The codling moth caterpillars bore into a fruit within 24 hours of hatching from their eggs, usually traveling between 1.5 m to 3 m in search of a fruit. Because they are susceptible to predation, drying up, or being washed away between the period of hatching and boring into a fruit, the caterpillars are prompt in finding a fruit to feed on (Reis, et al., 2004). Although apples are their dominant food source, they are polyphagous, feeding on a wide variety of fruits from pear, walnut, apricot, peaches, plums, cherries, and chestnuts. They are unable to survive by feeding on leaves of the fruit trees (Jackson1982).

They are photopositive, which means they move towards light. This is adaptive because fruits tend to be located at the ends of the branches where there is most sunlight. Therefore, by following light, the larvae are able to move closer to fruits.[9]

Once the caterpillar has located a fruit to feed on, it starts penetrating the epidermis of the fruit. As the caterpillar makes way into the fruit, scraps of the skin, pulp, and frass build up near the entrance of the hole. These pieces are glued together by silk threads released from the caterpillar to create a cap. This cap protects the caterpillar by blocking the entrance. It takes the caterpillar approximately 45 minutes to bore into the fruit and about 15 minutes to cap, (Reis, et al., 2004). The caterpillar bores through the fruit until it reaches the seminal chamber of the fruit. There, the caterpillar bites into the seeds and halts the growth of the fruit. The fruit ripens prematurely as a result. By doing so, the caterpillar gains beneficial resources, such as albumin and fat. Such feeding behavior lasts for 23 to 27 days and the caterpillar feeds on an average of one to two fruits during this time, (Jackson1982).

The number of yearly generations varies depending on the climatic conditions. The warmer climate is optimal for higher number of generations. In Denmark, only one generation was observed; in Palestine, 4 to 5 generations have been noted, (Jackson1982).

In most of Europe and North America there are usually 2 generations of moths in a given flight period. First generation moths emerge in July and are active through August. Eggs laid by the first generation moths are called second generation. Eggs hatch and the caterpillar undergoes pupation. The pupae go into diapause and "hibernate" over the winter. In April and May of the following year, these second generation pupae eclose and the second generation moths are active during the months of May and June. (Jackson1982).

The pupae are 10–12 mm long and can be as wide as 3 mm. The color changes as time passes, from the brown color of the caterpillar to light brown. The pupa's morphology varies with both sex and generation. Female pupae are generally longer and wider than the male pupae.

As soon as the moths emerge, they copulate, oviposit first generation eggs, and caterpillars that arise from the eggs bore into the fruits. The average life span of the moth is around 13 to 18 days. However, the longest living male observed lived for 38 days, and the oldest female lived for 37 days, (Jackson1982).

Control

Predators of the codling moth are mostly **birds**, accounting for nearly 80% of caterpillar killings. Woodpeckers are especially significant predators because they find caterpillars from hidden crevices beneath the bark and branches of host trees, (Jackson1982).

They are also highly preyed upon by arthropods from the following taxonomic groups Araneae (spiders), Opiliones (harvestman), Carabidae (groundbeetle), Cicindellidae, Formicidae (ant), Geocoridae (big-eyed bugs), Staphylinidae (rove beetle), and Coleoptera (other beetles), (Witzgall, et al., 1999).

Ants are among the most significant of the insect predators. They attack all the stages of the codling moth life cycle. The commonly known ant predators include *Solenopsis molesta, Lasius niger, Formica fusca, Formica pallidefulva schauffussi inserta, Aphaenogaster fulva aquia, Tetramorium caespitum, and Monomorium minimum*. *Solenopsis molesta* can kill 90% of caterpillars they attack, which are usually those moving between fruits or fifth instar caterpillars looking for a pupation site, (Jackson1982).

Thrips are also predators of various life stages of the codling moth. *Haplothrips faurei* feeds on eggs from all generations, while *Leptothrips mali* feeds on second generation eggs, (Jackson1982).

Parasitoids

Codling moth damage in walnuts in California has declined since the release of *M. ridibundus* in 1995, with parasitism of overwintering cocoons reaching 56% in some unsprayed orchards (Mills unpublished observations). The outcome of the project cannot be considered a dramatic success, as should be expected in the case of a direct pest (Gross 1991; Lloyd 1960), but as noted by Goldson et al. (1994), the value of parasitism and the contribution of partial biological control to the overall management of such notorious and intractable pests as the codling moth should not be underestimated.

Pathogens

Fungi

Beauveria bassiana is a parasite to the caterpillar and pupae of the codling moth. It has a killing rate of 13.1% in caterpillars. ***Hirsutella subulata*** is another entomophagous, or insect-eating, fungal parasite to the codling moth larvae. Unlike *B. bassiana,* this fungus type can grow even if the humidity is low, (Jackson1982).

Bacteria

Known bacteria that parasitize the codling moth are *Erwinia amylovora* and *Bacillus cereus. B. cereus* parasitizes the larvae of the codling moth, (Jackson1982).

Granulovirus

Baculoviruses are common viruses of lepidopteran insects, and divide into two genera: Nucleopolyhedrovirus and Granulovirus. Fast-killing granuloviruses usually kill the host during the same instar in which it was infected. *Cydia pomonella granulovirus* is a species of fast granulovirus that is fatally pathogenic to codling moths. Because *Cydia pomonella granulovirus* is a fast granulovirus, the codling moth larvae dies within the same instar as when infected. The complete genome of *Cydia pomonella granulovirus* has been sequenced and was found to have 123,500 bp, (Mathews, et al., 2004).

Three main *Cydia pomonella granulovirus* isolates have been identified: *Cydia pomonella granulovirus*-M, E, R. These can be categorized into four genome types: genomes A, B, C, and D. It is believed that genome C is ancestral to the other genomes. Genome C is also less pathogenic to codling moth neonates compared to other genome types, (Rezapanah, et al., 2008). Isolates from Iran have also been identified and were found to have same genome types as the other isolates, (Berling, et al., 2009).

As Bioinsecticide

Cydia pomonella granulovirus-M, which is a Mexican isolate strain of *Cydia pomonella granulovirus*, has been used as a bioinsecticide, (Ballard, et al., 2000). During the later stages of infection, the virus form clusters, which causes apoptosis, or programmed cell-death, of host cells and eventually host death. Death of the host occurs within 5–10 days, (Ansebo, et al., 2004).

Mechanical Control

Trunk banding consists of wrapping a corrugated cardboard strip around the tree trunk. Larvae making their way down the tree to pupate after exiting the infested fruits will use bands as pupation sites. Bands may then be removed and burned, (Jackson1982).

Mass-trapping consists of placing kairomonal lures on a high density of sticky traps in orchards. Both male and female moths are attracted to the lure and become stuck in the trap. An experiment conducted over 5 years showed a significant decrease in the number of apples damaged by codling moth, (Unruh, et al., 2000).

This trial demonstrates that autosterilisation could be a new method of codling moth control, but it does not bring prove that it's efficiency is higher than attract and kill technique. However, a simulation study on codling moth dynamics under different pheromone based control techniques suggests that autosterilisation should be about twice as effective as attract and kill (Potting at al., 2001).

Particle Films

Particle films are hydrophobic solutions used to spray crops and plants to prevent damage from pathogens and arthropod pests. A common type of particle film is composed primarily of kaolin clay and adjuvants. Kaolin is a non-abrasive, white material that is found commonly in paint, pharmaceuticals, and cosmetics. It is an aluminosilicate mineral, which means it is composed of aluminum, silicon, and oxygen. Kaolin clay particle films are used to slow down the activity of larvae and adults. On trees coated with particle films, the larvae displayed decreased walking speed, fruit scavenging activity, and penetration rate. Although hatching rate of the eggs did not differ between treated and untreated trees, female moths oviposited less on film-treated trees, (Joshi, et al., 2011).

Chemical Control

Synthetic Attractants

Successful development of synthetic fruit volatiles has led to increased control of codling moths. Codling moths can be managed and controlled with the use of synthetic apple volatiles, such as (Z)3-hexenol, (Z)3-hexenyl benzoate, (Z)3-hexenyl hexanoate, (±)-linalool and E, E-a-farnesene, and other synthetic attractants, such as pear ester ethyl (E,Z)-2,4-decadienoate and its corresponding aldehyde, E,E-2, 4-decadienal, (Hern, et al., 1999). Attractants are used as lures in codling moth traps, and are widely used in codling moth management programs in orchards, (Charmillot; et al., 2001).

Growth Inhibitors and Regulators

Insect growth inhibitors (IGIs) and insect growth regulators (IGRs) are used in insects, especially Lepidoptera, to prevent the synthesis of chitin during development. Chitin is one of the major components which constitute the exoskeleton of arthropods and cell walls of fungi. Without chitin, insects cannot develop properly. Ovicidal IGIs, such as diflubenzuron, hexaflumuron and teflubenzuron, have been shown to be effective against egg development. Fenoxycarb is an ovicidal IGR, while Tebufenozide is a larvicidal IGR. Flufenoxuron and Methoxyfenozide are an IGI and IGR respectively and are equally effective in preventing growth in eggs as in larvae, (Hull, et al., 2009). In addition to IGRs, reduced-risk pesticides are also used to control codling moth population in apple orchards, (Reyes, et al., 2007).

Biological Control

The eggs are susceptible to biological control by ***Trichogramma*** **wasps**. The wasps deposit their eggs into codling moth eggs, and the developing wasp larvae consume the moth embryo inside.

Another candidate for a biological control agent is the parasitoid wasp *Mastrus ridens,* also known as the *Mastrus ridibundus*, (Mills, 2005).

Predators and Parasitoids

Codling moth hosts a number of predators and parasitoids in Australia. These natural enemies occasionally cause high mortality of codling moth, but none have reduced its pest status in Australia (Waterhouse and Sands 2001). Gorse pod moth, *C. succedana* has been tested as a potential biological control for gorse, *Ulex europaeus*, in Australia. It is anticipated that an application for its release from quarantine will be submitted in 2012. *C. succedana* was included in host-specificity testing in New Zealand and was not found to be a suitable host for *M. ridens* (Pipfruit NZ Inc. 2011). *M. ridens* has no observed hyper parasitism potential (Hennessey et al. 1995), and adults are not likely to compete with native species for floral food resources (Pipfruit NZ Inc. 2011).

Genetic Control

During the past 40 years, a number of pest insects including CM, have been sterilized by irradiation or chemicals used in genetic control programs (Hoy, 2003). This approach to pest management has been called the sterile insect release method (SIRM) or the sterile insect technique (SIT). It involves the colonization and mass rearing of the target pest species, sterilization through the use of gamma radiation and releasing them into the field on a sustained basis and in sufficient numbers to achieve appropriate sterile to wild insect over flooding ratios. Here the sterile males find and mate with fertile females, transferring sterile sperm (Hoy, 2003). It results in no off spring, thereby causing a reduction in the natural pest population. Unlike non-selective insecticide-based control, SIT represents a biologically-based tool for pest control in view of the species specificity involved. As a result of its species specificity, SIT can be effectively used to replace insecticides for control of insect pests.

NATURAL EXTRACTS

Natural extracts as well as mineral oils should be considered as potential tools for the control of this widespread pest, as mentioned by different authors for other insect pests (CHEN et al., 1996a; KOUL et al., 2000; NASSEF, 1999; KUMAR & BHATT, 1999; RENIPRABHA et al, 1999; BAUTISTA et al., 1998; DWIVEDI & GARG, 1997; RAJAPAKSE & SENANAYAKE, 1997; GRANT & LANGEVIN, 1995; GE & WESTON, 1995, Ismail,et al., 2014, 2015, 2016, Mohamed Abdel-Raheem and Ahmed Abdel-Salam, 2016, Salem, et al., 2017, 2021, Mohamed Abdel-Raheem, et al., 2018, 2020 a & b).

- Gypsy Moth

Gypsy moth has a single generation per year, with female moths laying eggs in mass between July and August. After winter diapause, larvae of gypsy moth hatch out of eggs usually in April to May of the following year. Gypsy moth caterpillars damage over 600 plant species, (Barber, et al., 1993). The foliage of hardwood trees, particularly of oaks, is preferred by this insect. For instance, in the forests of Crimea gypsy moth damages almost all trees giving preference to pubescent oak Quercus pubescens Willd., and eastern hornbeam Carpinus orientalis Mill., in gardens, to apple trees Malus domestica Borkh., and pear trees Pyrus communis L. The larvae are voracious feeders, consuming a total of about 1m2 of foliage during their caterpillar stage. In the period of population outbreaks, which last 1-3 years, larvae may defoliate host trees completely and then switch to cereal crops and even vegetables.

Infested by gypsy moth, the trees tend to weaken, fail to grow and fructify. Due to the absence of leaves, the physiological functions of the affected trees, such as photosynthesis and transpiration, are disrupted. The continual disruption of the photosynthetic processes depletes energy supplies in the plant tissues. Long-term shortage of nutrition may lead to dieback of the damaged trees, (Belickaya, 2011). Also gypsy moth defoliation may make trees susceptible to attack from secondary or "opportunistic" organisms, such as borers and fungi. Thus, there is a constant demand for insecticides, both safe and effective, to control the propagation of gypsy moth.

Control

- ***Bacillus thuringiensis*** **var. kurstaki (Btk)**

Apply *Bacillus thuringiensis, var. kurstaki* or Monterey Garden Insect Spray (Spinosad*)* to the leaves of trees to kill gypsy moth caterpillars. For best results, sprays must be applied when caterpillars are young, less than one inch long. In instances where populations are high, two applications five days apart might be needed. Many different strains of Bt have been isolated throughout the world which are toxic to a variety of insect and other species. A lepidopteran-active strain, *Bacillus thuringiensis* kurstaki (Btk), was isolated in 1962 and is lethal to the GM larval stage (Reardon et al., 1994). It is the most widely used control agent for GM in the world and has been used in the eradication campaigns in the USA and Canada. It is also registered as a control agent in New Zealand, mainly for use against leaf rollers, and has been used in the eradication programme for tussock moth in Auckland.

- AzaMax contains azadirachtin, the key insecticidal ingredient found in neem oil. This concentrated spray disrupts growth and development of pest insects and has repellent and antifeedant properties. Best of all, it's non-toxic to honey bees and many other beneficial insects.
- Least-toxic botanical insecticides should be used as a last resort. Derived from plants which have insecticidal properties, these natural pesticides have fewer harmful side effects than synthetic chemicals and break down more quickly in the environment.
- Peach tree Borer

The peach tree borer (*Synanthedon exitiosa)* does damage to a variety of stone fruit trees. It attacks not the fruit, but the tree itself, burrowing into its trunk near or beneath ground level and devouring its way into the living cambium layer underneath.

Heavy infestations, recognizable from the circle of oozing frass around the trunk, can completely girdle trees, killing them if left untreated. Young trees are especially vulnerable. Older trees that survive attacks will show less vigor and inferior fruit quality. Adults begin emerging in late spring and early summer from larvae that have overwintered inside the tree an inch or two beneath the soil line. They continue to emerge throughout the summer and into the fall. The moths begin breeding almost as soon as they emerge. Females lay as many as 400 eggs on the trunk of the tree near the soil line or in the soil against the tree. The eggs take 10 days to hatch. Larvae immediately seek entrance into the tree's bark, often through cracks, chips and other damage. They tunnel through the bark into the vulnerable cambium layer beneath it and beyond, growing as they do. Some may pass through to adult stage in a single season. As cold weather sets in, larvae will reduce activity and overwinter inside the tree. They begin feeding again when temperatures warm.

In spring, the larvae migrate from their holes, pupating near the entrance to their burrow or in nearby soil, creating a gummy cocoon of silk threads and bits of wood. The cocooning and pupation stages before moths emerge takes as much as four weeks. Moths begin laying eggs within minutes of taking to the air.

Peach tree borers bring major harm to important fruit-crop trees, destroying the tree's vascular system through boring and girdling while inducing plant pathogens to invade the weakened tree.

Borer damage inhibits the conduction of water and nutrients up the trunk to the tree's branches, leaves and fruits. One or two borers will harm its growth and fruiting. Several burrowing into the same tree, fouling their tunnels with their waste, can kill a newly infested tree in a single season.

Control

- Most trees will survive attack from small numbers. But large numbers of larvae can completely girdle a tree, killing it. Once the larvae move into the bark, they are difficult to manage. They're most vulnerable at the surface before they chew their way into the tree's bark and cambium layer.
- Wild trees provide a year-to-year home for borers. If you have wild plum, cherry or other stone fruits in your woodlot, consider clearing them. Female adults are attracted to diseased, damaged and otherwise stressed trees. Removing and replacing older, stressed trees that harbor borers can help make control issues easier to deposit their eggs.
- Healthy, adequately watered trees are less likely to invite infestation. Borer treatment begins early in spring even before larvae become active. Probe small holes in trunks near the soil line, especially those with evidence of frass, with the point of a knife or stiff wire to crush larvae beneath the bark. In severe infestations, scoop soil out from around the crown of the tree where frass collects and use a sharp-pointed object to dig out the larvae, taking care not to harm the tree.
- Monitor trees for moth activity daily and keep journal records for future years. Use pheromone traps or Tangle-Trap® Insect Trap Coating to capture adults. Inspect traps daily.
- Begin spraying organic neem oil in the highest recommended concentrations around the crown of the tree and up the first 6- 12 inches of the trunk when adults are anticipated. Saturate both bark and soil. The oil will disrupt the moths' breeding cycle and discourage them from leaving eggs. It will also neutralize eggs that may already have been laid as well as penetrate the bark and inhibit development of larvae already in the tree. Spray twice a month throughout the breeding season which can last until September.
- Citrus extract sprays will repel adults and discourage egg laying. Begin spraying trunks and around the crown of the tree just ahead of moth hatch.
- Paint tree trunks and exposed roots with a paste of Surround WP, a powder made of kaolin clay. Coat base of tree, exposed roots and trunk up to 12 inches. Once dry, the coating deters adults and their egg laying.
- Applications of *Bacillus thuringiensis*, naturally occurring soil bacteria, will disrupt larvae and kill them depending on exposure. Spray *Bt* directly into borer holes after clearing out as much frass as possible.
- Spinosad, an OMRI listed pesticide, can also be sprayed on tree trunks as larvae hatch and directly into borer holes. Spraying can be repeated every five or six days up to two weeks before harvest.

- **Nematodes** attack eggs, larvae and pupae of numerous insect pests in soil. A spring application of nematodes suppressed 88% of orchard borer infestations. **Spring and fall applications were found to be 100% effective.**
- Parasitic wasps can help with lesser peach tree borer whose eggs can be found. They are not effective on the pupae of the common, greater peach tree borer because they're under the soil line out of the wasp's reach. But certain wasps will parasitize eggs found on bark and just-hatched larvae that have not yet worked their way into the tree.
- **Woodpeckers** and other **birds** will reduce numbers by grabbing larvae on and under the bark. Encourage them by providing suitable habitat and not spraying harmful pesticides.
- Cedar chips and bark spread around the base of stone fruit trees is said to repel egg-laying adult moths. In the south, spreading tobacco dust around the base of trees is a traditional method of discouraging pests
- Use of moth crystals to "gas" the larvae inside their burrows. It's suggested that crystals worked into the soil at the trees crown will emit vapor that penetrates into the tree. Moth crystals, like moth balls, are made from naphthalene, a suspected carcinogen which is also linked to liver failure and neurological damage in infants.
- **Pecan Weevil** (*Curculio caryae*)

Pecan weevil chews holes through the shuck and shell of nuts and is responsible for 2 types of damage.

The first type of damage occurs when adult weevils feeding on kernels prior to the shells hardening. Nuts punctured by weevils during this stage have a tobacco-juice-like stain around the feeding site. Damaged nuts often shrivel, turn black inside and drop prematurely.

The second type of damage occurs when females place eggs within the newly formed kernels. Larvae hatch and feed within the kernel causing further damage.

Control

Steinernematid and heterorhabditid nematodes have been infect over 200 insect species from several orders, including nine species in five genera of the Curculionidae (Poinar, 1979.). Third-stage infective juveniles (I J) contain cells of the mutualistic enterobacteria Xenorhabdus spp. in their intestines. The IJ penetrate the host's natural openings, enter the haemocoel, then release the bacteria, which multiply rapidly and kill the host by septicemia, generally within 48 hours. In field trials, the nematode Steinernema carpocapsae (Weiser) strain DD-136 (Poinar, 1990.) caused 59.6% mortality of pecan weevil larvae after 20 days (Tedders, et al., 1973.). Harp and Van Cleave, 1976, reported 20% parasitism of pecan weevil 4th larvae and pupae by Steinernema sp. in one group of laboratory specimens. Ring et al., 1988, screened two nematode species (*S. carpocapsae* strains A11 and Mexican and Heterorhabditis sp.) against pecan weevil under laboratory conditions and reported 80, 86, and 75% larval mortality, respectively, after 28-35 days.

- Sawfly

The sawfly adult emerges in May and June and lays eggs on larch leaves. The larvae feed actively from June to mid-July, then drop to the ground, burrow 2-10 cm into the soil around the tree, and become prepupae in late July. They overwinter as prepupae, and pupation occurs in April. About i month later they emerge as adults. A parasitic nematode, later identified as *Steinernema feltiae* (= Neoaplectana carpocapsae) UK strain (6), was reported in prepupae of *C. Iariciphila* (Billany and Brown, 1980.). Low-cost in vitro mass-rearing methods for this nematode has increased the feasibility of field experimentation on some important insect pests with modest success (Kaya, 1985, Poinar, 1986).

Control

Successful applications of steinernemarid nematodes have been made against certain insects in cryptic and soil habitats (Kaya, 1985, Poinar, 1986. but attempts to control foliage insects have been generally discouraging. The limiting factor cited most often is rapid desiccation of the nematodes. Although the nighttime relative humidity in our study was favorable for nematode survival (Kaya, 1985).

- Scale Insect

These sap-sucking pests attach themselves to the twigs, leaves, branches and fruits of host plants. Learn least-toxic methods of scale control here.

Control

Scale insects are the main prey of lady beetles, coccinellids have little impact on the control of *P. manihoti*, (Neuenschwander, 2001), showing that no global rule applies to the biological control of Coccoidea. Phylogenetic reconstructions also have a decisive role in pinpointing the native region of pests, as shown for several pest Ceroplastes spp. (Coccidae) of unknown origin (Qin and Gullan, 1998), Making such information available to managers can greatly speed up the diagnostic, control, and warning procedures. Then, applied research can assist the efficient control of scale insect pests. For instance, the study of sex pheromones has shown their potential role as kairomones for several scale insects predators and parasitoids. The negative effect of ants on parasitoids has been quantified for some scale insects, including *A. aurantii* (Martinez-Ferrer, et al., 2003), and should be considered for both ant-attended and non-antattended species. Several studies have shown that morphologically similar but genetically different species (cryptic or sibling taxa) occur in scale insects. (Malausa, et al. 2011), If undetected, this taxonomic diversity may interfere with the proper choice of a specific parasitoid and lead to control failure. The discovery of several haplotypes in such species complexes also raises questions about pathogen transmission in these different lineages. Manageable diagnostic tools must be developed to get around these biological limitations, which are probably unexpectedly diverse. For instance, the influence of the host-plant secondary compounds on parasitoid mortality must not be neglected, (Tena, 2012), especially since the polyphagy of most economically important scale insects amplifies the range of such compounds, and leads to negative interactions with biological control.

- Harrathi (2008) use of mineral oils against *P. ziziphi* instead of using methidathion since mineral oils are known to be harmless towards its natural enemies, such as the coccinellid predator *C. bipustulatus*, commonly found feeding on this armoured scale on citrus trees in Tunisia. The promising performance of mineral oils against *P. ziziphi* was also shown in Egypt, another South Mediterranean country like Tunisia. Coll and Abd-Rabou (1998) assessed the effect of two experimental spray oils on *P. ziziphi* and two associated primary parasitoid species in a citrus orchard located in northern Egypt. Triona oil was more effective than Shecrona oil and reduced populations of *P. ziziphi* by up to 99%. The same authors demonstrated that the two spray oils used did not significantly affect the activity of the parasitoid *Encarsia citrina* (Craw); moreover, Triona oil was slightly harmful to Habrolepis aspidioti Compere and Annecke (Hymenoptera: Encyrtidae), another primary parasitoid of *P. ziziphi*. Mangoud (2008) demonstrated that Misrona oil and buprofezin, showing a long-term insecticidal effect on *P. ziziphi*, significantly decreased densities of nymphs and adults of this insect on citrus in Egypt. Therefore, based on these results, the application of Triona (paraffin oil 81%) and/or Misrona (paraffin oil 95%) oils applied at a rate of 15 ml/l of water and buprofezin treatments could be tested before to be applied as an alternative option to other synthetic insecticides used for controlling *P. zizphi* in Tunisian citrus orchards.

REFERENCES

Alford, D. V. (1984). *A colour atlas offruit pests their recognition, biology and control*. Wolfe Publishing Ltd.

Ansebo, L., Coracini, M. D. A., Bengtsson, M., Liblikas, I., Ramírez, M., Borg-Karlson, A.-K., Tasin, M., & Witzgall, P. (2004, August 1). Antennal and behavioural response of codling moth Cydia pomonella to plant volatiles. *Journal ofApplied Entomology*, *128*(7), 488–493. doi:10.1111/j.1439-0418.2004.00878.x

Bautista, N., Morales, O., Carrillo, J., & Bravo, H. (1998). Mortalidad de Phyllocnistis citrella con un aceite mineral y nim. *Manejo Integrado de Plagas*, *50*, 29–33.

Ballard, J., Ellis, D. J., & Payne, C. C. (2000, October 01). Uptake of Granulovirus from the Surface of Apples and Leaves by First Instar Larvae of the Codling Moth Cydia pomonella L. (Lepidoptera: Olethreutidae). *Biocontrol Science and Technology*, *10*(5), 617–625. doi:10.1080/095831500750016415

Barber, K. N., & Kaupp, W. J. (1993). Specificity testing of the nuclear polyhedrosis virus of the gypsy moth, Lymantria dispar (L.) (Lepidoptera: Lymantriidae). Can Entomol, 125, 1055–1066.

Belickaya, M. N. (2011). Izmenenie radial'nogo prirosta duba pri defoliacii kron listogryzushchimi vreditelyami. In Bolezni i vrediteli v lesah Rossii: vek XXI. Materialy Vserossijskoj konferencii s mezhdunarodnym uchastiem i V ezhegodnyh chtenij pamyati O.A. Kataeva, IL SO RAN. Academic Press.

Berling, M., Blachere-Lopez, C., Soubabere, O., Lery, X., Bonhomme, A., Sauphanor, B., & Lopez-Ferber, M. (2009, February 15). Cydia pomonella granulovirus Genotypes Overcome Virus Resistance in the Codling Moth and Improve Virus Efficiency by Selection against Resistant Hosts". *Applied and Environmental Microbiology*, *75*(4), 925–930. doi:10.1128/AEM.01998-08 PMID:19114533

Billany, D. J., & Brown, R. M. (1980). The web1spinning larch sawfly Cephalcia lariciphiIa Wachtl. (Hy1menoptera: Pamphiliidae) a new pest of Larix in En1gland and Wales. *Forestry*, *53*(1), 71–80. doi:10.1093/forestry/53.1.71

Blank, R. H., & Gill, G. S. C. (1997). Thrips (Thysanoptera: Terebrantia) on flowers and fruit of Citrus in New Zeland. *New Zealand Journal of Crop and Horticultural Science*, *25*(4), 319–332. doi:10.1080/01140671.1997.9514023

Bush, G. L. (1966). The taxonomy, cytology, and evolution of the genus Rhagoletis in North America (Diptera, Tephritidae). *Bulletin of the Museum of Comparative Zoology*, *134*, 431–562.

Bush, G. L., & Smith, J. J. (1998). The genetics and ecology of sympatric speciation: A case study. *Researches on Population Ecology*, *40*(2), 175–187. doi:10.1007/BF02763403

Charmillot, Gourmelon, Fabre, & Pasquier. (2001). Ovicidal and larvicidal effectiveness of several insect growth inhibitors and regulators on the codling moth Cydia pomonella L. (Lep., Tortricidae). *Journal of Applied Entomology, 125*(3), 147–153.

Chen, C., Chang, S., Hou, R. F., & Cheng, L. (1996). Deterrent effect of the chinaberry extract on oviposition of the diamondback moth, Plutella xylostella (L.) (Lep., Yponomeutidae). *J. Appl. Ent.*, *120*(1-5), 165–169. doi:10.1111/j.1439-0418.1996.tb01585.x

Childers, C. C., & Nakahara, S. (2006). Thysanoptera(thrips) within citrus orchards in Florida: Species-distribution, relative and seasonal abundance within trees, and species on vines and ground cover plants. *Journal of Insect Science*, *6*(45), 3–7. doi:10.1673/031.006.4501 PMID:20233100

Ciglar, I. (1998). *Integrirana zaštita voćaka i vinove loze. Zrinski*. Čakovec.

Coll, M., & Abd-Rabou, Sh. (1998). Effect of oil emulsion sprays on parasitoids of the black parlatoria, Parlatoria ziziphi, in grapefruit. *BioControl*, *43*(1), 29–37. doi:10.1023/A:1009974330554

Conti, F. (2001). Monitoring of *Pezothrips kellyanus* on citrus in eastern Sicily. *Proceedings of the 7th International Symposium on Thysanoptera*.

Dwtvedi, S. & Garg, S. (1997). Screening of plant extracts for ovicidal effect on the rice moth, Corcyra cephalonica (Stainton). *Pest Management and Economic Zoology*, *5*(1), 53–55.

Feder, J. L., Chilcote, C. A., & Bush, G. L. (1988). Genetic differentiation between sympatric host races of Rhagoletis pomonella. *Nature*, *336*, 61–64. doi:10.1038/336061a0

Franck, P., Reyes, M., Olivares, J., & Sauphanor, B. (2007). Genetic archi1tecture in codling moth populations: Comparison between microsatellite and insecticide resistance markers. *Molecular Ecology*, *16*(17), 3554–3564. doi:10.1111/j.1365-294X.2007.03410.x PMID:17845430

Ge, X., & Weston, P. A. (1995). Ovipositional and feeding deterrent from chinese prickly ash against angoumois grain moth (Lepidoptera: Gelechiidae). *Journal of Economic Entomology*, *88*(6), 1771–1775. doi:10.1093/jee/88.6.1771

Goldson, S. L., Phillips, C. B., & Barlow, N. D. (1994). The value of parasitoids in biologi1cal control. *New Zealand Journal of Zoology*, *21*(1), 91–96. doi:10.1080/03014223.1994.9517979

Grant, G. G., & Langevin, D. (1995). Oviposition deterrence, stimulation, and effect on cluth size of Choristoneura (Lepidoptera: Tortricidae) species by extract fractions of host and nonhost foliage. *Environmental Entomology*, *24*(6), 1656–1663. doi:10.1093/ee/24.6.1656

Gross, P. (1991). Influence of target pest feeding niche on success rates in classical biological control. *Environmental Entomology*, *20*(5), 1217–1227. doi:10.1093/ee/20.5.1217

Harp, S. J., & Van Cleave, H. W. (1976). New records of natural enemies of the pecan weevil. *Southwestern Entomologist, 1*, 38.

Harrathi, A. (2008). *Le pou noir de l'oranger Parlatoria ziziphi Lucas (Hemiptera: Diaspididae): bioécologie et essai de lutte chimique* [Mémoire du Projet de Fin d'Etudes du Cycle Ingénieur]. Institut National Agronomique de Tunisie, Université de Carthage.

Hennessey, R., Mills, N., & Unruh, T. (1995). *Field release of an exotic parasitic wasp, Mastrus ridibundus (Hymenoptera: Ichneumonidae), for biological control of codling moth, Cydia pomonella (Lepidoptera: Tortricidae), in the United States*. Environmental Assessment.

Hern, A., & Dorn, S. (1999, July 1). Sexual dimorphism in the olfactory orientation of adult Cydia pomonella in response to α-farnesene. *Entomologia Experimentalis et Applicata, 92*(1), 63–72. doi:10.1046/j.1570-7458.1999.00525.x

Hoy, M. A. (2003). *Insect Molecular Genetics, An Introduction to Principles and Applications* (2nd ed.). Elsevier Science.

Hull, L. A., Joshi, N. K., & Zaman, F. U. (2009). Large plot reduced risk insecticide study for Lepidopteran pests infesting apples, 2008. *Arthropod Management Tests, 34*(1).

Ismail, I. A., Farag, N. A., Abdel-Rahman, R. S., Abdel-Raheem, M. A., & Radwan, H. M. (2014). Insecticidal activity of some plant extracts rich in coumarin against cowpea beetle, *Closobruchus maculatus* (Fab.) (Coleoptera: Bruchidae). *Egyptian Journal of Biological Pest Control, 24*(2), 465–469.

Ismail, I. A., Abdel-Rahaman, R. S., & Abdel-Raheem, M. A. (2015). Influence of some essential oils, chemical compounds and their mixtures against *Ceroplastes rusci* L. and *Asterolcanium pustolans* Cock on fig trees. *International Journal of Chemtech Research, 8*(9), 187–195.

Ismail, I. A., Abdel-Rahman, R. S., & Abdel-Raheem, M. A. (2016). Utilization of certain plant extracts and entomopathogenic fungi for controlling the black fig fly, *Lonchaea aristella* on fig trees. *International Journal of Chemtech Research, 9*(4), 35–42.

Jackson, D. M. (1982, May 15). Searching Behavior and Survival of 1st-Instar Codling Moths. *Annals of the Entomological Society of America, 75*(3), 284–289. doi:10.1093/aesa/75.3.284

Joshi, N. K., Hull, L. A., Rajotte, E. G., Krawczyk, G., & Bohnenblust, E. (2011, May 02). Evaluating sex-pheromone- and kairomone-based lures for attracting codling moth adults in mating disruption versus conventionally managed apple orchards in Pennsylvania. *Pest Management Science, 67*(10), 1332–1337. doi:10.1002/ps.2194 PMID:21538805

Kaya, H. K. (1985). Entomogenous nematodes for insect control in IPM systems. In M. A. Hoy & D. C. Herzog (Eds.), *Biological control in agricultural IPM systems* (pp. 282–302). Academic Press. doi:10.1016/B978-0-12-357030-7.50022-3

Koul, O., Jain, M., & Sharma, V. (2000). Growth inhibitory and antifeedant activity of extracts from Melia dubia to Spodoptera litura and Helicoverpa armígera larvae. *Indian Journal of Experimental Biology, 38*, 63–38. PMID:11233088

Kumar, S. & Bhatt, R. (1999). Field evaluation of plant leaf extracts, oil and neem products against mango hopper (Amritodus atikinsoni Lethierry) and thrips (Scirtothrips mangiferae Hood). *Allelopathy Journal*, *6*(2), 271–276.

Kovačević, Ž. (1952). *Primijenjena entomologija. II. Knjiga Poljoprivredni štetnici*. Sveučilište Zagreb.

Lewis, T. (1997). Flight and dispersal. In T. Lewis (Ed.), Thrips as crop pests (pp. 175–196). CAB International.

Lloyd, D. C. (1960). Significance of the type of host plant crop in successful biological control of insect pests. *Nature*, *187*(4735), 430–431. doi:10.1038/187430a0 PMID:14417718

Maceljski, M. (2002). *Poljoprivredna entomologija. II. Edition, Zrinski*. Čakovec.

Malausa, T., Fenis, A., Warot, S., Germain, J.-F., Ris, N., Prado, E., Botton, M., Vanlerberghe-Masutti, F., Sforza, R., Cruaud, C., Couloux, A., & Kreiter, P. (2011). DNA markers to disentangle complexes of cryptic taxa in mealybugs (Hemiptera: Pseudococcidae). *Journal of Applied Entomology*, *135*(1-2), 142–155. doi:10.1111/j.1439-0418.2009.01495.x

Mangoud, A. A. H. (2008). Insecticidal effect of some chemical and natural control agents against the black parlatoria scale, Parlatoria ziziphi (Lucas) and associated parasitoids on citrus trees. *Egyptian Journal of Agricultural Research*, *86*, 2157–2167.

Martinez-Ferrer, M. T., Grafton-Cardwell, E. E., & Shorey, H. H. (2003). Disruption of parasitism of the California red scale (Homoptera: Diaspididae) by three ant species (Hymenop1tera: Formicidae). *Biological Control*, *26*(3), 279–286. doi:10.1016/S1049-9644(02)00158-5

Mathews, C. R., Bottrell, D. G., & Brown, M. W. (2004). Habitat manipulation of the apple orchard floor to increase ground-dwelling predators and predation of Cydia pomonella(L.) (Lepidoptera: Tortricidae). *Biological Control*, *30*(2), 265–273. doi:10.1016/j.biocontrol.2003.11.006

Mills, N. (2005). Selecting effective parasitoids for biological control introductions: Codling moth as a case study. *Biological Control*, *34*(3), 274–282. doi:10.1016/j.biocontrol.2005.02.012

Abdel-Raheem & Abdel-Salam. (2016). *Citrus Pests Bionomics and Control in Egypt*. Lambert Academic Publishing.

Abdel-Raheem, M., Alghamdi, H., & Reyad, N. (2018). *Botanical Pesticide and Insect Pests*. Lambert Academic Publishing.

Abdel-Raheem, M., Dimetry, N. Z., & Amin, A. E.-R. (2020a). *Nano-Preparations from Botanical Products for Controlling Insect pests*. Lambert Academic Publishing.

Abdel-Raheem, M., Alghamdi, H. A., & Reyad, N. F. (2020b). Nano Essential oils against the red palm weevil, *Rhynchophorus ferrugineus* Olivier (Coleoptera: Curculionidae). *Entomological Research*, *50*(5), 215–220. doi:10.1111/1748-5967.12428

Mound, L. A., & Jackman, D. J. (1998). Thrips in the economy and ecology of Australia. *Proceedings of the Sixth Australian Applied Entomological Research Conference*, 472-478.

Usman, Gulzar, Wakil, Piñero, Leskey, Nixon, Oliveira-Hofman, Wu, & Shapiro-Ilan. (2020). Potential of entomopathogenic nematodes against the pupal stage of the apple maggot Rhagoletis pomonella (Walsh) (Diptera: Tephritidae). *Journal of Nematology, 52*.

Nassef, M. (1999). Juvenile hormone mimic and plant1derived oils as control agents against whitefly, Bemisia tabaci (Genn.), on cotton. *Egypt J. Agrie. Res.*, *77*(2), 691–699.

Navarro, C. C., Aguilar, A., & Marí, F. G. (2011). Population trend and fruit damage of *Pezothrips kellyanus* in Citrus orchards in Valencia (Spain). *Integrated Control in Citrus Fruit Crops IOBC Bulletin SROP*, *38*, 204–209.

Navarro, C.C., Aguilar, A. & Marí, F.G. (2008). Pezothrips Kellyanus, Thrips causante de daños en frutos de cítricos. *Levante agrícola: Revista internacional de cítricos, 299*, 298-303.

Tadic, M. (1957). *The Biology of the Codling Moth as the Basis for Its Control*. Univerzitet U Beogradu.

Neuenschwander, P. (2001). Biological control of the cassava mealybug in Africa: a review. *Biological Control, 21*, 214–229. doi:10.1006/bcon.2001.0937

Bostanian, N. J., & Racette, G. (2001). Attract and kill, an effective technique to manage apple maggot, Rhagoletis pomonella [Diptera: Tephritidae] in high density Quebec apple orchards. *Phytoprotection*, *82*(1), 25–34. doi:10.7202/706212ar

Naumann, I. D. (1991). Hymenoptera. In *The Insects of Australia* (2nd ed., Vol. 2, pp. 917–1000). Melbourne University Press.

Pipfruit N. Z. Inc. (2011) Application to import for release or to release from containment new organisms. Application number 0022269. EPA.

Poinar, G. O. Jr. (1979). *Nematodes for biologi1cal control of insects*. CRC Press.

Poinar, G. O., Jr. (1986). Entomophagous nematodes. In Biological plant and health protection. New York: Gustav Fi1scher Verlag.

Poinar, G. O. Jr. (1990). Biology and taxonomy of Steinernematidae and Heterorhabditidae. In R. Gaugler & H. K. Kaya (Eds.), *Entomopatho1genic nematodes in biological control* (pp. 23–61). CRC Press.

Potting, R. P. J., & Knight, A. L. (2001). Predicting the efficacy of modified modes of action of a pheromone-based attracticide: A bisexual attractant and autosterilisation. *IOBC/WPRS Bulletin*.

Qin, T. K., & Gullan, P. J. (1998). Systematics as a tool for pest man1agement: case studies using scale insects and mites. *Pest Management—Future Challenges, Vols 1 and 2, Pro1ceedings of the Sixth Australasian Applied Entomological Research Conference*, 479–488.

Reardon, R. C., & Podgwaite, J. D. (1994). Summary of efficacy evaluations using aerially applied Gypchek against gypsy moth in the USA. *Journal of Environmental Science and Health. Part B, Pesticides, Food Contaminants, and Agricultural Wastes*, *29*(4), 739–756. doi:10.1080/03601239409372902

Reis, D., & Vian, B. (2004). Helicoidal pattern in secondary cell walls and possible role of xylans in their construction. *Comptes Rendus Biologies*, *327*(9–10), 785–790. doi:10.1016/j.crvi.2004.04.008 PMID:15587069

Reniprabha, A., Muralidharan, S., & Subramanian, A. (1999). Impact of plant extracts on molting in the larvae of mosquito Culex sp. *Environment and Ecology*, *17*(4), 1022–1024.

Reyes, M., Franck, P., Charmillot, P.-J., Ioriatti, C., Olivares, J., Pasqualini, E., & Sauphanor, B. (2007, September 1). Diversity of insecticide resistance mechanisms and spectrum in European populations of the codling moth, Cydia pomonella. *Pest Management Science*, *63*(9), 890–902. doi:10.1002/ps.1421 PMID:17665366

Rezapanah, M., Shojai-Estabragh, S., Huber, J., & Jehle, J. A. (2008, December 1). Molecular and biological characterization of new isolates of Cydia pomonella granulovirus from Iran. *Journal of Pest Science*, *81*(4), 187–191. doi:10.100710340-008-0204-2

Ring, Schroeder, Nyczepir, Payne, & Snow. (1988). Nematodes screened against pecan weevil larvae, 1986. *Insectil cide and Acaricide Tests, 13*(184).

Rull, J., Aluja, M., Feder, J., & Berlocher, S. (2006). Distribution and host range of hawthorn1infesting Rhagoletis (Diptera: Tephritidae) in Mexico. *Annals of the Entomological Society of America*, *99*(4), 662–672. doi:10.1603/0013-8746(2006)99[662:DAHROH]2.0.CO;2

Salem, S. A., Abd-El Salam, A. M. E., & Ahmed, S. (2017). Evaluate and assess the use of some insecticides of plant origin against *Scritothrips citri* Moulton (*Thysanoptera, Thripidae*) in reducting distortions orange fruits for export. *Bioscience Research*, *14*(2), 354–361.

Salem, S. A., Abd El-Salam, A. M. E., & Abdel-Raheem, M. A. (2021). Field Evaluation of the Efficacy of *Moringa oleifera* in Controlling Two Main Pests, *Aphis gossypii* and *Bemisia tabaci* Infesting Tomato Plants, Academic. *Journal of Entomology*, *14*(1), 24–29.

Sandanayaka, M., Chhagen, A., Page-weir, N. E. M., & Charles, J. G. (2011). Colony optimisation of Mastrus ridens (Hymenoptera: Ichneumonidae), a potential biological control agent of codling moth in New Zealand. *New Zealand Plant Protection*, *64*, 227–234. doi:10.30843/nzpp.2011.64.5959

Tedders, W. L., Weaver, D. J., & Wehunt, E. J. (1973). Pecan weevil: Suppression of larvae with the fungi Metarrhizium anisopliae and Beauveria bassiana and the nematode Neoaplectana dutkyi. *Journal of Economic Entomology*, *66*(3), 723–725. doi:10.1093/jee/66.3.723

Tena, A., Beltra, A., & Soto, A. (2012). Novel defenses of Protop1ulvinaria pyriformis (Hemiptera: Coccidea) against its major parasitoid Metaphycus helvolus (Hymenoptera: Encyrtidae): implications for biological control of soft scales. *Biological Control*, *62*(1), 45–52. doi:10.1016/j.biocontrol.2012.03.005

Tracewski, K. T., Brunner, J. F., Hoyt, S. C., & Dewey, S. R. (1987). Occurrence of Rhagoletis pomonella (Walsh) in hawthorns, Crataegus, of the Pacific Northwest. *Melanderia*, *45*, 19–25.

Unruh, T. R., Knight, A. L., Upton, J., Glenn, D. M., & Puterka, G. J. (2000, June 01). Particle Films for Suppression of the Codling Moth (Lepidoptera: Tortricidae) in Apple and Pear Orchards. *Journal of Economic Entomology*, *93*(3), 737–743. doi:10.1603/0022-0493-93.3.737 PMID:10902324

USDA. (1995). *Gypsy moth mangement in the United States: a cooperaive approach. (Draft environmental impact statement)*. USDA Forest Service.

Walsh, B. D. (1867). The apple-worm and the apple maggot. *American Journal of Horticulture*, *2*, 338–343.

Waterhouse, D. F., & Sands, D. P. A. (2001). *Classical biological control of arthropods in Australia. ACIAR Monograph No. 77*. CSIRO Publishing.

Witzgall, P., Bäckman, A.-C., Svensson, M., Koch, U., Rama, F., El-Sayed, A., Brauchli, J., Arn, H., Bengtsson, M., & Löfqvist, J. (1999, June 1). Behavioral observations of codling moth, Cydia pomonella, in orchards permeated with synthetic pheromone. *BioControl*, *44*(2), 211–237. doi:10.1023/A:1009976600272

Yee, Klaus, Cha, Linn, Jr., Goughnour, & Feder. (2014). Abundance of apple maggot, Rhagoletis pomonella, across different areas in central Washington, with special reference to black-fruited hawthorns. *Journal of Insect Science, 12*.

Yee, W. L., Landolt, P. J., & Darnell, T. J. (2005). Attraction of apple maggot (Diptera: Tephritidae) and nontarget flies to traps baited with ammonium carbonate and fruit volatile lures in Washington and Oregon. *Journal of Agricultural and Urban Entomology*, *22*, 133–149.

Yee, W. L., & Goughnour, R. B. (2008). Host plant use by and new host records of apple maggot, western cherry fruit fly, and other Rhagoletis species (Diptera: Tephritidae) in western Washington state. *The Pan-Pacific Entomologist*, *84*(3), 179–193. doi:10.3956/2007-49.1

Chapter 16
Different Types of Diseases Infecting Orchid Plants:
The Most Important Diseases Infecting Orchids

Rehab Yassin Ghareeb
https://orcid.org/0000-0002-3428-0536
City of Scientific Research and Technological Applications, Alexandria, Egypt

Amira A. Ibrahim
https://orcid.org/0000-0002-5904-9736
City of Scientific Research and Technological Applications, Alexandria, Egypt

ABSTRACT

Many orchids species were threatened because of environmental changes and the appearance of microbial diseases. Reduction in orchard yield and productivity depends on biotic factors like bacteria, viruses and nematodes, and pests. The yield losses range from 20 to 40% in global orchard production. Diseases causing crop orchid losses were epidemics with long term and massive effects. Many species of microbes were associated with orchids that are pathogenically resulting in symptoms of many diseases like brown spot, root rot, soft rot, brown rot, wilt, and anthracnose. This chapter will discuss diseases associated with Orchidaceae species that resulted from pests and other microbes and different biological managements for these diseases.

INTRODUCTION

Orchidaceae is one of the most important therapeutic families of plants, includes more than 28,000 recognized plants with 763 genera. Some species of Orchidaceae are neglected due to Arctic parts and desert regions but are mainly rich in the wet humid regions of wide-reaching (Śliwiński et al., 2022; Sarsaiya et al., 2019; Lin et al., 2020; The Plant List, 2013). The Orchidaceae family is monocotyledonous contains a high percentage of vulnerable genera, with most holding threatened species. Orchidaceae

DOI: 10.4018/978-1-6684-2423-0.ch016

is occurred in nature habitat, mainly spread in East Asia, South-East Asia and Oceania and cultivated in different habitats from tropical to temperate zones (Evans and Hans, 2022; Pérez-Escobar et al., 2017; Chen et al., 2014). Orchids are very unique plants, which are very useful and beautiful. It is the best ornament, the best medicine, and also the best ingredient for making ice cream (´Sliwi´nski et al.,2022; . It is the King of Division Magnoliophyta (the flowering plants) in the Kingdom Plantae or botany. Other than as beautiful horticultural plants, the economic importance of orchids is rather minor. Some of the orchids have immense value in food sources, medicinal uses, superstition and horticultural uses. Orchids are considered as a significally economic crops. Numerous nations have started developing and trading orchids. Worldwide exchange have driven to the development of plants with the information of degree and structure of species hereditary differing quality is fundamental for the foundation of an effective preservation procedure since hereditary variables contribute to species termination chance through inbreeding depression (Guo et al., 2019; Hinsley et al., 2018). Productions of orchids are increased in large scale nowadays for industrial purposes. Due to vegetative propagation, the planting materials are the major source of inoculum, and pathogens are easily distributed, especially viruses. Major diseases of orchids and their control measures are listed below. There are different types of Orchid diseases differed in microorganism causing this disease as follow: fungal diseases, bacterial diseases, viral diseases and Nematode diseases.

FUNGAL DISEASES

Phytopathogenic fungi are the most infectious microorganisms for different Orchids. Fungal diseases considered the common diseases infected Orchids plants with many different symptoms as foliar blights, leaf spots, fungal rots, and flower blights. Most common fungal diseases in orchids were demonstrated in Figure (1).Fungi attack seeds; once fungi invade the Orchids' seeds germination process begin and protocorms were grown. Many fungal diseases with different symptoms were found in Orchids as follow:

Black Rot Disease

Orchid black rot disease found in many genera and species in humid environment with rain fall weather. Pathogenic species from Phytophtora and Pythium causing black root disease. Major pathogenic fungi are Phytophthora cactorum, and Phytophthora palmivora (Hine 1962; Burnett 1974; Uchida 1994; Uchida and Aragaki 1991a; Orlikowski and Szkuta 2006; Cating et al. 2013). Black rot is the common disease for Orchids in most regions all over the world distinctly in tropics and subtropics (Burnett 1974; Chen and Hsieh 1978; Hall 1989; Duff 1993; Uchida1994; Ilieva et al. 1998; Hill 2004; Orlikowski and Szkuta 2006; Tao et al. 2011).Different genera of orchids such as Aerides, Ascocenda, Brassavola, Dendrobium, Gongora, Maxillaria, Miltonia, Oncidium, Paphiopedilum, Phalaenopsis, Rhynchostylis, and Schomburgkia infected by black root disease caused by two major Phytophthora (Phytophthora palmivora & Phytophthora cactorum) (Alfieri et al.1994; Orlikowski and Szkuta 2006). Infection pathway of Phytophthora and Pythium pathogens begins in humid environment with free water rising zoospores that swim, invade the infected tissues. Zoospores were produced in the infected plat's lesions and fast spread in irrigated water and in uninfected tissues (Agrios 2005; Uchida 1994).

Black Rot Symptoms

In seedling and young plantlets, small black lesions were observed on roots and basal portion of bulbs through high humidity. These lesions caused damping off syndrome; the disease begins in leaves as brown lesion then converted to black (Figure 2). Phytophthora pathogens spread from rhizome to all portions of the infected plant causing death. Infected root appears firstly in brown rot then stunted and dry out. The infected leaves appeared as zebra like shape stripes (horizontal light and dark brown threads) (Ilieva et al. 1998).

This disease is a common one mostly infected Vanda (whole plant) and Cattleya (leaf shoot); this disease prefers high humidity and temperature conditions beside poor ventilation in tropical and subtropical regions (USA and Thailand). Main reasons for this disease are Phytophthora nicotianae infected Vanda and Phytophthora cactorum infected Cattleya causing rot for shots, leaves, rhizomes and flower buds (Daly et al. 2013; Kamjaipai 1983; McMillan et al. 2009). Phytophthora cactorum oomycete produces a sexual spore that survives under unfavorable conditions, also produces chlamydospores that spreads by lemon or pear shape motile oospore and penetrate plants through wounds causing infection under wet condition (Orlikowski and Szkuta 2006; Uchida 1994; Uchida and Aragaki 1991a, b).

Symptoms of Shoot Rot

Top rot or shoot rot disease includes death of new leaves turned to dark brown. The disease starts from the base of the stem and was spread up worded giving dark brown stem causing death of nice shoot side turning it into black. This disease spread along rhizome to the new shoot with the same symptoms. These all symptoms characteristic for only Venda and Cattleya plants.

Botrytis Spot

Fungal disease (Botrytis spot) is affecting different types of Orchids in cool, humid and poor air circulation conditions caused by Botrytis cinerea . Susceptible orchids for this fungus are Phalaenopsis and Cattleya and may be found in other genera of Orchids. Infection of this disease was reported in USA (Florida &Hawaii) and Thailand as tropical and temperate zones (Jones 2003; Kamjaipai 1983; Uchida and Aragaki 1991b). Air borne pathogenic fungi of Botrytis cinerea is found in nurseries. Conidia of this fungus were produced on grey branched chonidiophores producing sclerotia. Botrytis cinerea in dead tissues and dying plant material produced spores and dispersing it during cool ad distributed spores by wind, rain and running water that infected and invade healthy surrounding plants through 14 h (Agrios 2005; Sumbali 2005).

Symptoms

Most susceptible tissue for infection is old flowers of Orchids plant; started from brown necrotic or light spots. These spots are enlarged and increased with the progress of infection and surrounded by yellowish or pink margin depending on flower's colors as shown in Figure (3). Also fungal mycelia were visible in cool and high moisture and cold conditions.

1. Petal Blight Disease

This type of disease appear in cool and moist weather, it was characteristic as small crown circular spots on the petals and sometimes also infected sepals of Orchids. Petal blight disease infected the flowers of different types of Orchids for example: Oncidium, Vanda, Phalaenopsis, Cattleya and Dendrobium. Dendrobium plant is the famous one was infected by this disease in Thailand, USA and Australia (Burnett 1957, 1965; Daly et al. 2013; Kamjaipai 1983).Petal blight disease effect rapid falling in the bulb tissues extended to stem that resulted in soften leaf which damping off and separated from the healthy plant. To protect the other healthy parts; infected flowers must be cut off and destroyed. Responsible fungi for this disease are *Alternaria* and *Curvularia* species. Pathogenic fungi from Alternaria and *Curvularia* with dark hyphae and conidia. Spores of these fungi can easily transfer by wind infected a lot of Orchids (Sumbali 2005; Agrios 2005). Flower's petals infected with circular spots enlarged and transformed to necrotic lesions with moist and rainfall condition these lesion will be rot as appear in Figure (4).

Leaf Spots Disease

Leaf spot disease caused by different pathogenic fungi according the host and pathogenic agent disease symptoms differ. The most Orchids are susceptible to this disease infections are Oncidium, Phalaenopsis, Cymbidium, Dendrobium, Epidendrum Brassolaeliocattleya,Cattleya, Laelia, Laeliocattleya and Odontoglossum in different countries all over the world (Australia, Japan, Brazil, USA (Hawaii), the Netherlands, Thailand and Australia (Daly et al. 2013; Ichikawa and Aoki 2000; Jones 2003; Kamjaipai 1983; Silva and Pereira 2007; Silva et al. 2008; Uchida and Aragaki 1980; Uchida 1994).

Leaf spots usually caused various spots differ from yellow to black color on leaves with high moisture and low light weather conditions. These spot not kill the leaves but become present with long time resulted in leaves become unsightly.

Disease Symptoms

According to the pathogen and host symptoms differ as follow:

- Fusarium spp. pathogens

There are two types of pathogenic Fusarium species (F. subglutinans and F. proliferatum) infected Cymbidium cultivars. Fusarium subglutinans caused yellow leaf spot appeared firstly on the leaves as soaked patches then transformed into large lesions with sunken centers, finally these symptoms converted to reddish brown spots with yellowish borders in irregular shape. In case of F. proliferatum infected Cymbidium, symptoms appeared as small black spots in leaves at the first stage from infection then converted to enlarge lesions with black irregular angular spots (Figure 5).

- ***Phyllosticta* spp. pathogens**

These pathogens almost infected Orchids' leaf, flower or pseudopubl parts; infection started as very tiny yellow sunken spots in leaves, and then converted to large lesions and finally these lesions enlarged with circular or oval shape. During the stage of infection, lesions become brown in color and these infected leaves drops from plant (Figure 6).

- *Guignardia* **spp. pathogens**

These pathogenic caused leaf spot disease for Colletotrichum, Cercospora, Gloeosporium, Phyllostictina plants. The infection symptoms started as small dark purple with elongated lesions on the leaf surface. The lesions become elongated streaks with purple color parallel to leaf veins and with final infection stage become irregular in shape and larger. Leaf spot disease by Guignardia pathogens not kills the plant but it making leaves unsightly (Figure 7(A)).

- *Cercospora* **spp. pathogens**

Cercospora species infected a lot of types of Orchids as Dendrobium genus in tropical and subtropical zones with temperate conditions (Agrios, 1997; Burnett, 1974; Chupp, 1953; Dodge and Ricket, 1943). Leaf disease by these pathogens started as small spots separated in leaves and then enlarged led to leaf blight. The infected spots then enlarged with brown lightly sunken area and then converted to black lesions (Figure 7(B)) .Cercospora produced dark hyphae with multicellular conidia on short conidiophores (Agrios 2005; Sumbali 2005).

BACTERIAL DISEASES

Leaf Spot

Bacterial leaf spot disease caused by Acidovorax avena resulted in economic losses production in Orchid's yield and quality in many countries such as: USA (Florida), China, Italy, India, Poland, Taiwan, Korea and Australia (Borah et al. 2002; Cating and Palmeteer 2011; Ding et al. 2010; Hseu et al. 2012; Kim et al. 2015; Li et al. 2009; Lin et al. 2015; Pulawska et al. 2013; Scortichini et al. 2005; Stovold et al. 2001). Most Orchids infected by these pathogens are Phalaenopsis, Cypripedium, Oncidium, Cattleya, Vanda and Dendrobium.

- **Symptoms**

Disease started with tiny soft brown spots, then turned to black causing in appearance of cavities in leaf's parenchyma then the pathogens invade all plant resulted in the death. Pathogenic bacteria of Acidovorax avenae subsp. cattleyae is a gram negative with rod shaped moved by flagellum which spread fast by irrigation, rain and contaminated farmer's tool; this pathogen infected specific Phalaenopsis orchids due to its succulent leaves.

Soft Rot

This type of disease is a vigorous one causing big losses in exports and imports of Orchids in many countries reported in USA (Florida), Taiwan and China. The most susceptible orchids for infections are Miltoniaorchids, Vanda, Cattleman, Phalaenopsis, Tolumnia, Dendrobium and Oncidium (Cating et al. 2008; Cating et al. 2009b; Li et al. 2009; Cating et al. 2009b; Hseu et al. 2012; Keith and Sewake 2009; Zhou et al. 2012).

- **Symptoms**

The symptoms of this disease are similar to symptoms of other bacterial soft disease which started with soaked spot at infection region. For Oncidium host, initial infection started at the base of stem with spots then black lesions extended and enlarged to be elongated shape. At age stage from infection leaves become more watery soft and then it rot. Bacterial ooze can be observed easily by cutting the infected stem edge and examined it under microscope.

For Phalaenopsis host, infection started as sacked spots in leaves surrounded by yellow bright area, with increasing in temperature and humidity the spots enlarged fast and extended along the leaf with bright tan color which turned then to black. At final stage from infection, spots increased and enlarged and become darker with rotten odor.

For Miltonia, Cattleya, and Oncidium hosts, infection started with tiny soaked spots turned with lesions and then converted to enlarged dark black spots on the host's leaves. For Dendrobium cultivars hosts, the spots expanded to be large necrotic lesions led to leaf drying and death. For pseudobulbs and Vanda hosts, spots are watery soft and become brownish and finally rot.

Black Spot

This type of disease caused by Burkholderia gladioli causing corm black spot disease in many gladiolus. These infectious pathogens caused huge losses in cultivated orchids in tropical and subtropical regions in East Asian countries as China, Taiwan, Thailand and Indonesia, in addition to USA (Hawaii). Most popular susceptible Orchids are Miltonic, Dendrobium Cattleya, Odontioda, Oncidium and Phalaenopsis (Chuenchitt et al. 1983; Hseu et al. 2012; Joko et al. 2014; Kamjaipai 1983; Keith et al. 2005; Lee et al. 2013; Takahashi et al. 2004;Uchida 1995; You et al. 2016).

- **Symptoms**

Burkholderia gladioli is pathogenic a gram-negative bacteria with rod shape that may be straight or slightly curved causing this disease started from soaked spots and spread by water to become black large spot along the host's leaf.

For Phalaenopsis host, infection started initially with small dark green spots surrounded by soaked watery area then turned to brown spots. Sometimes infectious spots observed in circular shape and under condition from high temperature and humidity the spit increased and become rot.

For pseudobulbs hosts, the pathogenic bacteria can invade the stem and causing defoliation, with low humidity the leaves cracked and become dry. The infection increased rapidly in young plants led to death.

VIRAL DISEASES

Most infectious pathogens for Orchids are viruses causing losses in productions and yields. Orchids are infected by different types of viruses; the most common viruses are Cymbidium mosaic virus (CymMV) and Odontoglossum ringspot virus (ORSV), also known as Tobacco mosaic virus-orchid strain (TMV-O); and these two viruses are often found in mixed infections. Other orchid viruses of note include: Orchid fleck virus (OFV), Impatiens necrotic spot virus (INSV), Vanilla mosaic virus (VanMV), and Calanthe mild mosaic virus (CalMMV). Nemesia ring necrosis virus (NeRNV) has onlyrecently been described as a new virus. NeRNV is widespread,leading to foliar concentric necrotic ring spots and linepatterns as well as to sporadical flecking of flowers.

- **Symptoms**

Cymbidium mosaic virus (CymMV) caused Leaf necrosis, black streak disease with brown irregular shape necrosis in leaves. This necrosis increased to become streaked spots on both upper and lower leaf surface. At final stage these streaks become dry and damping off.

Tobacco mosaic virus (TMV) causing mild flower break disease in Cattleya started with spots and rings surrounded the center of infection then elongated area appeared with diamond shape led to flower malformation.

NEMATODE DISEASES

Foliar Nematodes

Many Orchids in Thailand, Singapore and Florida attacked by Aphelenchoides nematodes species ex.(A. besseyi, A. fragariae & A. ritzemabosi) (Esser et al. 1988; Kawate and Sewake 2014; Latha et al. 1999). Nematodes grow in wet and warm conditions and widespread rapidly causing noticeable loss of Orchid production reached to 90–95% (Kawate and Sewake 2014). Most susceptible orchids to be infected with nematodes are Oncidium, Paphiopedilum, Cymbidium, Dendrobium and Cattleya.

- **Symptoms**

First infection of Aphelenchoides spp. pathogens started from yellow green batches in leaves turned to brown lesions.(Uchida and Sipes 1998). They attack broad range of hosts from ornamental and cultivated types. At adult stage for host, mesophyll and epidermal tissue within leaves were destroyed by nematodes, and these tissues become collapse and brown. Females Aphelenchoides spp.layed a lot of eggs within green leaf tissue which extended to stems buds and all parts of plant. Adults and 4th stage juveniles can survive in association with living and desiccated plant tissues for long periods. Foliar nematodes can be disseminated by splashing water, infected plant propagative material, and debris (Kohl,2011).

CONCLUSION

Orchids are the most important plants in medicinal and industrial purposes; many orchids infected with various pathoges as nematodes, bacteria, fungi and viruses.Biological control for theses diseases were used in large scal to get rid off the yield losses. So, We recommend many laws in exports and imports if theses economic ornamental orchids.

REFERENCES

Agrios, G. N. (2005). *Plant pathology* (5th ed.). Elsevier Academic Press.

Agrios, G. N. (1997). *Plant pathology* (4th ed.). Academic Press.

Alfieri, S. A., Jr., Langdon, K. R., Kimbrough, J. W., El-Gholl, N. E., & Wehlburg, C. (1994). Diseases and Disorders of Plants in Florida. Florida Department of Agriculture & Consumer Services, Division of Plant Industry.

Borah, P. K., Kataky, M., Bhagabati, K. N., Pathak, J. J., & Bora, L. C. (2002). A new bacterial disease of orchid from India. *J Agri Sci Soc North-East India*, *15*, 1–4.

Brunt, A. A., Crabtree, K., Dallwitz, M. J., Gibbs, A. J., Watson, L., & Zurcher, E. J. (Eds.). (1997). *Plant viruses online: Descriptions and lists from the VIDE database*. http://biology.anu.edu.au/Groups/MES/vide/

Burnett, H. C. (1957). Orchid diseases. *Proceedings of the Annual Meeting of the Florida State Horticultural Society*, *70*, 380–385. Retrieved September 20, 2021, from http://fshs.org/proceedings-o/1957-vol-70/380385%20%28BURNETT%29.pdf

Burnett, H.C. (1965). *Orchid diseases*. Fla Dept Agric Consumer Serv Div Plant Ind.

Burnett, H.C. (1974). *Orchid diseases*. Fla. Dept. Agr. and Consumer Serv., Div. of Plant Industry.

Cating, R. A., Hong, J. C., Palmateer, A. J., Stiles, C. M., & Dickstein, E. R. (2008). First report of bacterial soft rot on Vanda orchids caused by Dickeya chrysanthemi (Erwinia chrysanthemi) in the United States. *Plant Disease*, *92*(6), 977–977. doi:10.1094/PDIS-92-6-0977A PMID:30769750

Cating, R. A., & Palmateer, A. J. (2011). *Bacterial soft rot of Oncidium orchids caused by a Dickeya sp. (Pectobacterium chrysanthemi) in Florida*. Academic Press.

Plant, Zhao, Cun, Chen, & Zhu. (2010). Detection of Acidovorax avenae subsp.cattleyae by real-time fluorescent PCR. Wuli Xuebao, 40, 235–241. https://www.cabdirect.org/abstracts/20103213302.html

Cating, R. A., Palmateer, A. J., & McMillan, R. T. Jr. (2009a). First report of Sclerotium rolfsii on Ascocentrum and Ascocenda orchids in Florida. *Plant Disease*, *93*(9), 963. doi:10.1094/PDIS-93-9-0963B PMID:30754542

Cating, R. A., Palmateer, A. J., McMillan, R. T., & Dickstein, E. R. (2009b). First report of a bacterial Soft rot on Tolumnia orchids caused by a Dickeya sp. in the United States. *Plant Disease*, *93*(12), 1354. doi:10.1094/PDIS-93-12-1354B PMID:30759519

Cating, R. A., Palmateer, A. J., Stiles, C. M., Rayside, P. A., & Daviso, D. A. (2013). *Black rot of orchids caused by Phytophthora palmivora and Phytophthora cactorum*. Plant Pathology Department, Florida Cooperative Extension Service, Institute of Food and Agricultural Sciences, University of Florida.

Chen, J. S., & Hsieh, S. P. Y. (1978). Phytophthora black rot of Phalaenopsis in Taiwan. *Plant Prot Bull*, *20*, 161–170.

Chuenchitt, S., Dhirabhava, W., Karnjanarat, S., Buangsuwon, D., & Uematsu, T. (1983). A new bacterial disease on orchids Dendrobium sp. caused by Pseudomonas gladioli. *Witthayasan Kasetsat Witthayasat*, *17*, 26–36.

Chupp, C. (1953). *A monograph of the genus Cercospora*. Cornell University Press.

Coleman, G. (2012). Transmission of cymbidium mosaic virus in oncidium orchids by periplaneta australasiae. Faculty of the Graduate School of the University of Maryland, College Park.

Daly, A., Condé, B., & Duff, J. (2013). Orchid disease in the Northern territory. Northern Territory Government.

Dodge, O. B., & Ricket, H. W. (1943). Orchidaceae (orchids). In Diseases and pests of ornamental plants. Jacques Cattell Press.

Duff, J. D. (1993). Orchid diseases of the Northern Territory. *Agnote Darwin*, *568*, 3.

Esser, R. P., O'Bannon, J. H., & Clark, R. A. (1988). Procedures to detect foliar nematodes from annual nursery or out of state inspections. Nematology Circular No. 160. Fla. Dept. of Consumer Service Division of Plant Industry.

Evans, A., & Jacquemyn, H. (2022). Range Size and Niche Breadth as Predictors of Climate-Induced HabitatChange in Epipactis (Orchidaceae). *Frontiers in Ecology and Evolution*, *10*, 894616. doi:10.3389/fevo.2022.894616

Engelmann, J., & Hamacher, J. (2008). Plant Virus Diseases: Ornamental Plants. In Encyclopedia of Virology (3rd ed.). Academic Press.

Guo, J. L., Cao, W. J., Li, Z. M., Zhang, Y.-H., & Volis, S. (2019). Conservation implications of population genetic structure in a threatened orchid *Cypripedium tibeticum*. *Plant Diversity*, *41*(1), 13–18. doi:10.1016/j.pld.2018.12.002 PMID:30931413

Hall, G. (1989). Unusual or interesting records of plant pathogenic oomycetes. *Plant Pathology*, *38*(4), 604–611. doi:10.1111/j.1365-3059.1989.tb01458.x

Han, K.-S., Park, J.-H., Back, C.-G., & Park, M.-J. (2015). First Report of *Fusarium subglutinans* Causing Leaf Spot Disease on Cymbidium Orchids in Korea. *Mycobiology*, *43*(3), 343–346. doi:10.5941/MYCO.2015.43.3.343 PMID:26539053

Hill, C. F. (2004). First report of Phytophthora multivesiculata on cymbidium orchids in New Zealand. *Australasian Plant Pathology*, *33*(4), 603–604. doi:10.1071/AP04070

Hine, R. B. (1962). Pathogenicity of Phytophthora palmivora in the Orchidaceae. *The Plant Disease Reporter*, *46*, 643–645.

Hinsley, A., de Boer, H., Fay, M., Gale, S., Gardiner, L., Gunasekara, R., Kumar, P., Masters, S., Metusala, D., Roberts, D., Veldman, S., Wong, S., & Phelps, J. (2018). A review of the trade in orchids and its implications for conservation. *Botanical Journal of the Linnean Society*, *186*(4), 435–455. doi:10.1093/botlinnean/box083

Hseu, S. H., Shentu, H., & Sung, C. J. (2012). Multiplex PCR-based rapid detection of bacterial diseases on Phalaenopsis orchid. *Plant Pathology Bulletin, 21*, 91–100.

Ichikawa, K., & Aoki, T. (2000). New leaf spot disease of Cymbidium species caused by *Fusarium subglutinans and Fusarium proliferatum. Journal of General Plant Pathology, 66*(3), 213–218. doi:10.1007/PL00012948

Ilieva, E., Manin't Veld, W. A., Veenbaas-Rijks, W., & Pieters, R. (1998). Phytophthora multivesiculata, a new species causing rot in Cymbidium. *European Journal of Plant Pathology, 104*(7), 677–684. doi:10.1023/A:1008628402399

Joko, T., Subandi, A., Kusumandari, N., Wibowo, A., & Priyatmojo, A. (2014). Activities of plant cell walldegrading enzymes by bacterial soft rot of orchid. *Archiv für Phytopathologie und Pflanzenschutz, 47*(10), 1239–1250. doi:10.1080/03235408.2013.838374

Jones, S. (2003). Orchids. *Bull Am Orchid Soc*. http://www.aos.org/Default.aspx? id=120

Kamjaipai, K. (1983). *Diseases and pest of orchids*. Bangkok Flower Center Co. Ltd.

Kawate, M., & Sewake, K. T. (2014). *Pest management strategic plan for potted orchid production in Hawaii. University of Hawaii at Manoa.*

Keith, L. M., & Sewake, K. T. (2009). Isolation and characterization of Erwinia spp. causing bacterial soft rot on orchids in Hawaii. *Pac Agr Nat Resour, 1*, 4–11.

Keith, L. M., Sewake, K. T., & Francis, T. Z. (2005). Isolation and characterization of Burkholderia gladioli from orchids in Hawaii. *Plant Disease, 89*(12), 1273–1278. doi:10.1094/PD-89-1273 PMID:30791304

Khamtham, J., & Akarapisan, A. (2019). *Acidovorax avenae* subsp. *cattleyae* causes bacterial brown spot disease on terrestrial orchid *Habenaria lindleyana* in Thailand. *Journal of Plant Pathology, 101*(1), 31–37. doi:10.100742161-018-0135-6

Kim, S. M., Lee, S., Moon, J. S., & Lee, B. J. (2015). Draft genome sequence of the plant quarantine bacterium Acidovorax cattleyae ATCC33619. Korean Soc Plant Pathol: A. *Bacteriol Bacterial Dis, A-521*(2), 108.

Kohl, L.M. (2011). Astronauts of the nematode world: an aerial view of foliar nematode biology, epidemiology, and host range. *APSnet Features*. doi:10.1094/APSnetFeature-2011-0111

Latha, S., Babu, D. V. N., Sathyanarayana, N., Reddy, O. R., & Sharma, R. (1999). Plant parasitic nematodes intercepted from imported ornamental plants. *Indian Phytopathology, 52*(3), 283–284.

Lee, Y. A., Chao, C. S., & Jung, C. H. (2013). Combination of a simple differential medium and tox A-specific PCR for isolation and identification of phytopathogenic Burkholderia gladioli. *European Journal of Plant Pathology, 136*(3), 523–533. doi:10.100710658-013-0184-9

Li, B., Qiu, W., Fang, Y., & Xie, G. L. (2009). Bacterial stem rot of Oncidium orchid caused by a Dickeya sp. (ex Pectobacterium chrysanthemi) in mainland China. *Plant Disease, 93*(5), 552–552. doi:10.1094/PDIS-93-5-0552B PMID:30764150

Lin, W., Wang, J., Xu, X., Wu, Y., Qiu, D., He, B., Sarsaiya, S., Ma, X., & Chen, J. (2020). Rapid propagation in vitro and accumulation of active substances of endangered *Dendrobium cariniferum* Rchb. f. *Bioengineered, 11*(1), 386–396. doi:10.1080/21655979.2020.1739406 PMID:32172675

Lin, Y. H., Lee, P. J., Shie, W. T., Chern, L. L., & Chao, Y. (2015). Pectobacterium chrysanthemi as the dominant causal agent of bacterial soft rot in Oncidium "Grower Ramsey". *European Journal of Plant Pathology, 142*(2), 331–343. doi:10.100710658-015-0618-7

Mcmillan, R. T., Palmateer, A. J., & Vendrame, W. A. (2008). Cercospora Leaf Spot Caused by *Cercospora dendrobii* on *Dendrobium antennatoum* Lindl. and Its Control. *Proceedings of the Annual Meeting of the Florida State Horticultural Society, 121*, 353–355.

McMillan, R. T. Jr, Palmateer, A., & Cating, R. A. (2009). Problems in controlling Phytophthora cactorum on catteleya orchids. *Proceedings of the Annual Meeting of the Florida State Horticultural Society, 122*, 426–428.

Meyers, M., & Hudelson. (2014). *Foliar nematodes University of Wisconsin Extension provides equal opportunities in employment and programming, including Title IX and ADA requirements*. https://pddc.wisc.edu

Moon, H., Parkb, H., Jeonga, A., Hanb, S.-W., & Parka, C.-J. (2016). Isolation and identification of Burkholderia gladioli on Cymbidium orchids in Korea. *Biotechnology, Biotechnological Equipment*. Advance online publication. doi:10.1080/13102818.2016.1268069

Orlikowski, L. B., & Szkuta, G. (2006). Phythophthora rot of some Orchidaceae – new disease in Poland. *Phytopathologia Polonica, 40*, 57–61.

Pérez-Escobar, O. A., Chomicki, G., Condamine, F. L., Karremans, A. P., Bogarín, D., Matzke, N. J., Silvestro, D., & Antonelli, A. (2017). Recent origin and rapid speciation of Neotropical orchids in the world's richest plant biodiversity hotspot. *The New Phytologist, 215*(2), 891–905. doi:10.1111/nph.14629 PMID:28631324

Pulawska, J., Mikicinski, A., & Orlikowski, L. B. (2013). Acidovorax cattleyae causal agent of bacterial brown spot of Phalaenopsis lueddemanniana in Poland. *Journal of Plant Pathology, 95*, 407–410.

Sarsaiya, S., Shi, J., & Chen, J. (2019). A comprehensive review on fungal endophytes and its dynamics on Orchidaceae plants: Current research, challenges, and future possibilities. *Bioengineered, 10*(1), 316–334. doi:10.1080/21655979.2019.1644854 PMID:31347943

Scortichini, M., D'Ascenzo, D., & Rossi, M. P. (2005). New record of Acidovorax avenae subsp. cattleyae on orchid in Italy. *Journal of Plant Pathology, 87*, 244.

Silva, M., & Pereira, O. L. (2007). First report of Guignardia endophyllicola leaf blight on Cymbidium (Orchidaceae) in Brazil. *Australasian Plant Disease Notes, Australasian Plant Pathology Society, 2*(1), 31–32. doi:10.1071/DN07015

Silva, M., Pereira, O. L., Braga, I. F., & Lelis, S. M. (2008). Leaf and pseudobulb diseases on Bifrenaria harrisoniae (Orchidaceae) caused by Phyllosticta capitalensis in Brazil. *Australasian Plant Disease Notes, Australasian Plant Pathology Society, 3*, 53–56.

Śliwiński, T., Kowalczyk, T., Sitarek, P., & Kolanowska, M. (2022). Orchidaceae-Derived Anticancer Agents: A Review. *Cancers (Basel), 14*(3), 754. doi:10.3390/cancers14030754 PMID:35159021

Stovold, G. E., Bradley, J., & Fahy, P. C. (2001). Acidovorax avenae subsp. cattleyae (Pseudomonas cattleyae) causing leafspot and death of Phalaenopsis orchids in New South Wales. *Australasian Plant Pathology, 30*(1), 73–74. doi:10.1071/AP00066

Sudarsono, S., Elina, J., Giyanto, & Sukma, D. (2018). Pathogen Causing Phalaenopsis Soft Rot Disease – 16S rDNA and Virulence Characterisation. *Plant Protect. Sci., 54*(1), 1–8.

Sumbali, G. (2005). *The fungi.* Alpha Science International.

Takahashi, Y., Takahashi, K., Watanabe, K., & Kawano, T. (2004). Bacterial black spot caused by Burkholderia andropogonis on Odontoglossum and intergeneric hybrid orchids. *Journal of General Plant Pathology, 70*(5), 284–287. doi:10.100710327-004-0127-6

Tao, Y. H., Ho, H. H., Wu, Y. X., & He, Y. Q. (2011). Phytophthora nicotianae causing Dendrobium blight in Yunnan Province, China. *International Journal of Plant Pathology, 2*(4), 177–186. doi:10.3923/ijpp.2011.177.186

The Plant List. (2013). *Version 1.1.* http://www.theplantlist.org/1.1/browse/A/Orchidaceae/

Uchida, J. (1995). Bacterial diseases of Dendrobium, Research extension series, vol 158. Institute of Tropical Agriculture and Human Resources. University of Hawaii.

Uchida, J. Y. (1994). Diseases of orchids in Hawaii. *Plant Disease, 78*(3), 220–224. doi:10.1094/PD-78-0220

Uchida, J. Y., & Aragaki, M. (1980). Nomenclature, pathogenicity, and conidial germination of *Phyllostictina pyriformis. Plant Disease, 64*(8), 786–788. doi:10.1094/PD-64-786

Uchida, J. Y., & Aragaki, M. (1991a). Phytophthora diseases of orchids in Hawaii. Research extension series, vol 129. College of Tropical Agriculture and Human Resources. University of Hawaii.

Uchida, J. Y., & Aragaki, M. (1991b). Fungal diseases of Dendrobium flowers. Research extension series, vol 133. College of Tropical Agriculture and Human Resources. University of Hawaii.

You, Lü, Zhong, Chen, Li, Liu, & Zhang. (2016). First report of bacterial brown spot in Phalaenopsis spp. caused by Burkholderia gladioli in China. *Plant Dis.*

Zhou, J. N., Lin, B. R., Shen, H. F., Pu, X. M., Chen, Z. N., & Feng, J. J. (2012). First report of a soft rot of Phalaenopsis aphrodita caused by Dickeya dieffenbachiae in China. *Plant Disease, 96*(5), 760–760. doi:10.1094/PDIS-11-11-0942 PMID:30727539

APPENDIX

Figure 1.

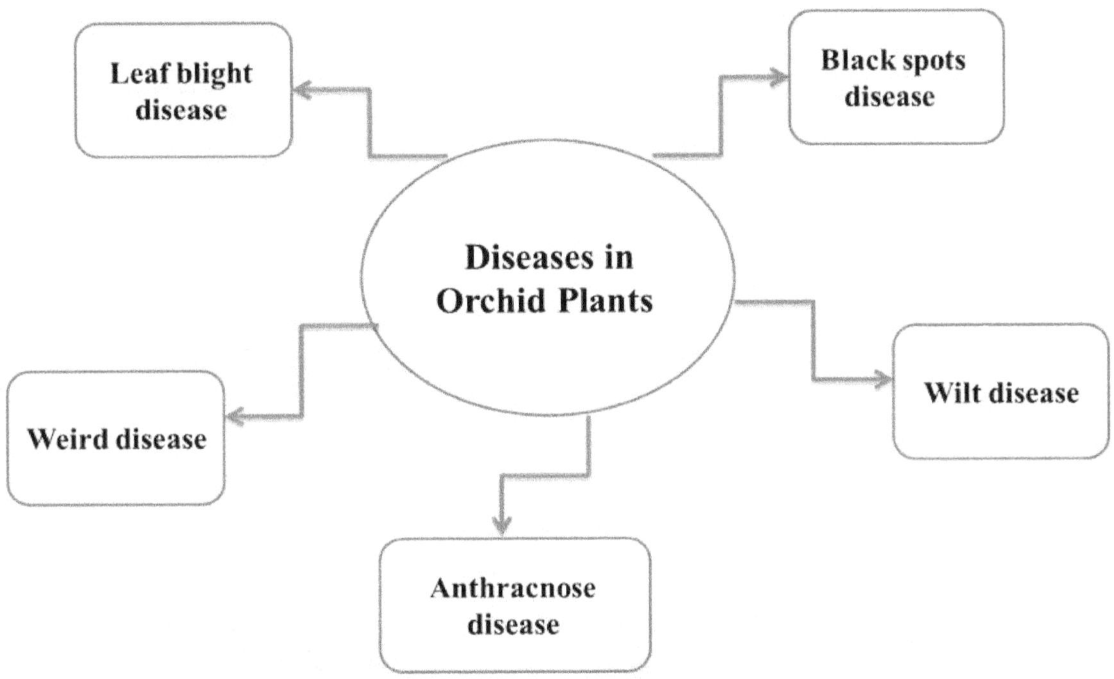

Figure 2.

Figure 3.

Figure 4.

Figure 5.

Figure 6.

Figure 7.

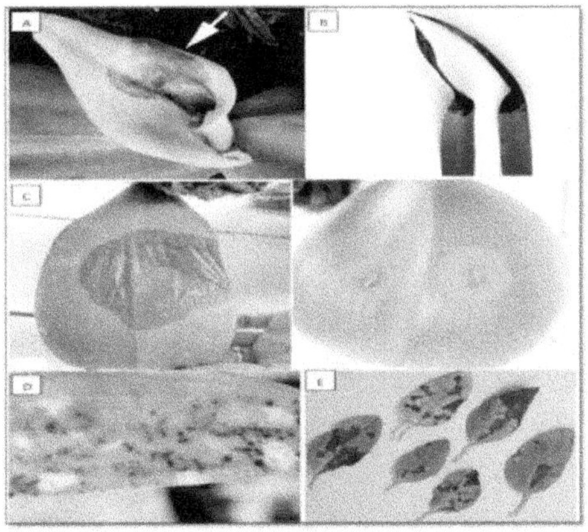

Compilation of References

Hu, Z., & Xia, X. (2004). Hydrogel nanoparticle dispersions with inverse thermoreversible gelation. *Advanced Materials, 16*(4), 305–309.

Abed, M., Haddad, A., Hassen, A., & Sultan, S. (2006). Preparation and evaluation of new hydrogels as new fertilizer delivery system. *Basrah J. Sci.(C), 24*(1), 103–114.

Arias, C., López-González, M. M. C., Fernandez-Garcia, M., Barrales-Rienda, J. M., & Madruga, E. L. (1993). Free-radical copolymerization of methyl acrylate with methyl methacrylate in benzene solution. *Polymer, 34*(8), 1786–1789.

Di Silvio, L., Gurav, N., Kayser, M. V., Braden, M., & Downes, S. (1994). Biodegradable microspheres: A new delivery system for growth hormone. *Biomaterials, 15*(11), 931–936.

Bolto, B. A. (1995). Soluble polymers in water purification. *Progress in Polymer Science, 20*(6), 987–1041.

Park, T. G., Cohen, S., & Langer, R. (1992). Poly (L-lactic acid)/pluronic blends: Characterization of phase separation behavior, degradation, and morphology and use as protein-releasing matrixes. *Macromolecules, 25*(1), 116–122.

Amiji, M. M. (1995). Permeability and blood compatibility properties of chitosan-poly (ethylene oxide) blend membranes for haemodialysis. *Biomaterials, 16*(8), 593–599.

Lee, J. M., Noh, G. Y., Kim, B. G., Yoo, Y., Choi, W. J., Kim, D. G., ... Kim, Y. S. (2019). Synthesis of poly (phenylene polysulfide) networks from elemental sulfur and p-diiodobenzene for stretchable, healable, and reprocessable infrared optical applications. *ACS Macro Letters, 8*(8), 912–916.

Cascone, M. G., Sim, B., & Sandra, D. (1995). Blends of synthetic and natural polymers as drug delivery systems for growth hormone. *Biomaterials, 16*(7), 569–574.

Adam, G. A., Cross, A., & Haward, R. N. (1975). The effect of thermal pretreatment on the mechanical properties of polycarbonate. *Journal of Materials Science, 10*(9), 1582–1590.

Wang, L., Cavaco-Paulo, A., Xu, B., & Martins, M. (2019). Polymeric hydrogel coating for modulating the shape of keratin fiber. *Frontiers in Chemistry, 7*, 749.

Yeo, J. K., Sperling, L. H., & Thomas, D. A. (1981). Poly (n-butyl acrylate)/polystyrene interpenetrating polymer networks and related materials. III. Effect of grafting level and molecular weight in semi-2 IPNs. *Journal of Applied Polymer Science, 26*(10), 3283–3294.

Fyvie, T. J., Frisch, H. L., Semlyen, J. A., Clarson, S. J., & Mark, J. E. (1987). Polymeric catenanes from crosslinked poly (2, 6-dimethyl-1, 4-phenylene oxide) and cyclic poly (dimethylsilozane). *Journal of Polymer Science. Part A, Polymer Chemistry, 25*(9), 2503–2509.

Eschbach, F. O., Huang, S. J., & Cameron, J. A. (1994). Hydrophilic-hydrophobic binary systems of poly (2-hydroxyethyl methacrylate) and polycaprolactone. Part II: Degradation. *Journal of Bioactive and Compatible Polymers*, *9*(2), 210–221.

Eschbach, F. O., & Huang, S. J. (1994). Hydrophilic-hydrophobic binary systems of poly (2-hydroxyethyl methacrylate) and polycaprolactone. Part I: Synthesis and characterization. *Journal of Bioactive and Compatible Polymers*, *9*(1), 29–54.

Rathna, G. V. N., Rao, D. M., & Chatterji, P. R. (1994). Water-induced plasticization of solution cross-linked hydrogel networks: Energetics and mechanism. *Macromolecules*, *27*(26), 7920–7922.

Kaur, H., & Chatterji, P. R. (1990). Interpenetrating hydrogel networks. 2. Swelling and mechanical properties of the (gelatin-polyacrylamide) interpenetrating networks. *Macromolecules*, *23*(22), 4868–4871.

Ma, J. T., Liu, L. R., Yang, X. J., & De Yao, K. (1995). Bending behavior of gelatin/poly (hydroxyethyl methacrylate) IPN hydrogel under electric stimulus. *Journal of Applied Polymer Science*, *56*(1), 73–77.

Sawhney, A. S., Pathak, C. P., & Hubbell, J. A. (1993). Bioerodible hydrogels based on photopolymerized poly (ethylene glycol)-co-poly (. alpha.-hydroxy acid) diacrylate macromers. *Macromolecules*, *26*(4), 581–587.

Sawhney, A. S., Pathak, C. P., van Rensburg, J. J., Dunn, R. C., & Hubbell, J. A. (1994). Optimization of photopolymerized bioerodible hydrogel properties for adhesion prevention. *Journal of Biomedical Materials Research*, *28*(7), 831–838.

Pérez-Luna, V. H. (2020). The Effect of Glutathione Incorporated as Chain Transfer Agent in Thermosensitive Hydrogels for Controlled Release of Therapeutic Proteins. In *Controlled Drug Delivery Systems* (pp. 53–75). CRC Press.

Duan, J., & Jiang, J. (2017). Structure and properties of hydrophobic aggregation hydrogel with chemical sensitive switch. *International Journal of Polymer Science*, 2017.

Zhao, H., Sterner, E. S., Coughlin, E. B., & Theato, P. (2012). o-Nitrobenzyl alcohol derivatives: Opportunities in polymer and materials science. *Macromolecules*, *45*(4), 1723–1736.

Hietala, S., Maunu, S. L., & Sundholm, F. (1999). Structure of styrene grafted poly (vinylidene fluoride) membranes investigated by solid-state NMR. *Macromolecules*, *32*(3), 788–791.

Matthew, I. R., Browne, R. M., Frame, J. W., & Millar, B. G. (1995). Subperiosteal behaviour of alginate and cellulose wound dressing materials. *Biomaterials*, *16*(4), 275–278.

Miller, J. S. (2000). Organometallic-and organic-based magnets: New chemistry and new materials for the new millennium. *Inorganic Chemistry, 39*(20), 4392-4408.

Svanberg, C., Adebahr, J., Ericson, H., Börjesson, L., Torell, L. M., & Scrosati, B. (1999). Diffusive and segmental dynamics in polymer gel electrolytes. *The Journal of Chemical Physics*, *111*(24), 11216–11221.

Djabourov, M., Maquet, J., Theveneau, H., Leblond, J., & Papon, P. (1985). Kinetics of gelation of aqueous gelatin solutions. *British Polymer Journal*, *17*(2), 169–174.

Bellamkonda, R., Ranieri, J. P., Bouche, N., & Aebischer, P. (1995). Hydrogel-based three-dimensional matrix for neural cells. *Journal of Biomedical Materials Research*, *29*(5), 663–671.

Lee, S. J., & Park, K. (1996, January). Glucose-sensitive phase-reversible hydrogels. In *ACS Symposium Series* (Vol. 627, pp. 11-16). Washington, DC: American Chemical Society.

Obaidat, A. A., & Park, K. (1996). Characterization of glucose dependent gel-sol phase transition of the polymeric glucose-concanavalin A hydrogel system. *Pharmaceutical Research*, *13*(7), 989–995.

Sen, N., Shaikh, T., Singh, K. K., Sirsam, R., & Shenoy, K. T. (2020). Synthesis of polyacrylamide (PAM) beads in microreactors. *Chemical Engineering and Processing-Process Intensification*, *157*, 108105.

Kim, J. J., & Park, K. (1998). Smart hydrogels for bioseparation. *Bioseparation*, *7*(4), 177–184.

Vashist, A., Vashist, A., Gupta, Y. K., & Ahmad, S. (2014). Recent advances in hydrogel based drug delivery systems for the human body. *Journal of Materials Chemistry. B, Materials for Biology and Medicine*, *2*(2), 147–166.

Brandl, F., Kastner, F., Gschwind, R. M., Blunk, T., Teßmar, J., & Göpferich, A. (2010). Hydrogel-based drug delivery systems: Comparison of drug diffusivity and release kinetics. *Journal of Controlled Release*, *142*(2), 221–228.

Ramanan, R. M. K., Chellamuthu, P., Tang, L., & Nguyen, K. T. (2006). Development of a temperature-sensitive composite hydrogel for drug delivery applications. *Biotechnology Progress*, *22*(1), 118–125.

Feil, H., Bae, Y. H., Feijen, J., & Kim, S. W. (1991). Molecular separation by thermosensitive hydrogel membranes. *Journal of Membrane Science*, *64*(3), 283–294.

Yoshida, R., Uchida, K., Kaneko, Y., Sakai, K., Kikuchi, A., Sakurai, Y., & Okano, T. (1995). Comb-type grafted hydrogels with rapid deswelling response to temperature changes. *Nature*, *374*(6519), 240–242.

Seida, Y., & Nakano, Y. (1995). Concept to control the phase behavior of stimuli-sensitive polymer gel. *Journal of Chemical Engineering of Japan*, *28*(4), 425–428.

Homma, M., Seida, Y., & Nakano, Y. (2000). Evaluation of optimum condition for designing high-performance electro-driven polymer hydrogel systems. *Journal of Applied Polymer Science*, *75*(1), 111–118.

Kaetsu, I., Uchida, K., Morita, Y., & Okubo, M. (1992). Synthesis of electro-responsive hydrogels by radiation polymerization of sodium acrylate. *International Journal of Radiation Applications and Instrumentation Part C Radiation Physics and Chemistry*, *40*(2), 157–160.

Hooper, H. H., Baker, J. P., Blanch, H. W., & Prausnitz, J. M. (1990). Swelling equilibria for positively ionized polyacrylamide hydrogels. *Macromolecules*, *23*(4), 1096–1104.

Chen, J., Park, H., & Park, K. (1999). Synthesis of superporous hydrogels: Hydrogels with fast swelling and superabsorbent properties. *Journal of Biomedical Materials Research: An Official Journal of The Society for Biomaterials, 44*(1), 53-62.]

Inomata, H., Goto, S., Otake, K., & Saito, S. (1992). Effect of additives on phase transition of N-isopropylacrylamide gels. *Langmuir*, *8*(2), 687–690.

Seida, Y., & Nakano, Y. (1996). Effect of salt on the property of adsorption in thermosensitive polymer hydrogel. *Journal of Chemical Engineering of Japan*, *29*(5), 767–772.

Hu, Y., Horie, K., Ushiki, H., Yamashita, T., & Tsunomori, F. (1993). Fluorescence studies of volume phase transition in polyacrylamide gels with a pyrenyl probe in acetone/water mixed solvent. *Macromolecules*, *26*(7), 1761–1766.

Sawahata, K., Hara, M., Yasunaga, H., & Osada, Y. (1990). Electrically controlled drug delivery system using polyelectrolyte gels. *Journal of Controlled Release*, *14*(3), 253–262.

Tsujii, K., Hayakawa, M., Onda, T., & Tanaka, T. (1997). A novel hybrid material of polymer gels and bilayer membranes. *Macromolecules*, *30*(24), 7397–7402.

Peppas, N. A. (2010). *Biomedical applications of hydrogels handbook*. Springer Science & Business Media.

Marchetti, M., Prager, S., & Cussler, E. L. (1990). Thermodynamic predictions of volume changes in temperature-sensitive gels. 2. Experiments. *Macromolecules*, *23*(14), 3445–3450.

Park, K., & Robinson, J. R. (1984). Bioadhesive polymers as platforms for oral-controlled drug delivery: Method to study bioadhesion. *International Journal of Pharmaceutics*, *19*(2), 107–127.

Liu, Y., Huglin, M. B., & Davis, T. P. (1994). Preparation and characterization of some linear copolymers as precursors to thermoplastic hydrogels. *European Polymer Journal*, *30*(4), 457–463.

Park, H., & Park, K. (1996). *Hydrogels in bioapplications.* ACS Publications.

Budianto, E., & Amalia, A. (2020). Swelling behavior and mechanical properties of Chitosan-Poly (N-vinyl-pyrrolidone) hydrogels. *Journal of Polymer Engineering*, *40*(7), 551–560.

Park, H., & Park, K. (1994). Honey, I blew up the hydrogels! *Proceedings of the International Symposium on Controlled Release of Bioactive Materials*, *21*, 21–22.

Chirila, T. VConstable, I. JCrawford, G. JVijayasekaran, SThompson, D. EChen, Y. CGriffin, B. J. (1993). Poly (2-hydroxyethyl methacrylate) sponges as implant materials: In vivo and in vitro evaluation of cellular invasion. *Biomaterials*, *14*(1), 26–38.

Oxley, H. R., Corkhill, P. H., Fitton, J. H., & Tighe, B. J. (1993). Macroporous hydrogels for biomedical applications: Methodology and morphology. *Biomaterials*, *14*(14), 1064–1072.

Chen, J. U. N., & Park, K. (1999). Superporous hydrogels: Fast responsive hydrogel systems. *Journal of Macromolecular Science— Pure and Applied Chemistry*, *36*(7-8), 917–930.

Obaidat, A. A., & Park, K. (1997). Characterization of protein release through glucose-sensitive hydrogel membranes. *Biomaterials*, *18*(11), 801–806.

Miyata, T., Asami, N., & Uragami, T. (1999). A reversibly antigen-responsive hydrogel. *Nature*, *399*(6738), 766–769.

Wen, X. x., Zhang, D. q., Liao, Y. c., Jia, Z. k., & Ji, S. q. (2012). Effects of water-collecting and -retaining techniques on photosynthetic rates, yield, and water use efficiency of millet grown in a semiarid region. *Journal of Integrative Agriculture*, *11*, 1119–1128.

Abobatta, W. (2018). Impact of hydrogel polymer in agricultural sector. *Adv. Agric. Environ. Sci. Open Access*, *1*(2), 59–64.

Lin, X., Guo, L., Shaghaleh, H., Hamoud, Y. A., Xu, X., & Liu, H. (2021). A TEMPO-oxidized cellulose nanofibers/MOFs hydrogel with temperature and pH responsiveness for fertilizers slow-release. *International Journal of Biological Macromolecules*, *191*, 483–491.

Maitra, J., & Shukla, V. K. (2014). Cross-linking in hydrogels-a review. *American Journal of Political Science*, *4*(2), 25–31.

Elsaeed, S. M., Zaki, E. G., Ibrahim, T. M., Ibrahim Talha, N., Saad, H. A., Gobouri, A. A., ... Mohamed El-Kousy, S. (2021). Biochar grafted on cmc-terpolymer by green microwave route for sustainable agriculture. *Agriculture*, *11*(4), 350.

Essawy, H. A., Ghazy, M. B., Abd El-Hai, F., & Mohamed, M. F. (2016). Superabsorbent hydrogels via graft polymerization of acrylic acid from chitosan-cellulose hybrid and their potential in controlled release of soil nutrients. *International Journal of Biological Macromolecules*, *89*, 144–151.

Xiao, X., Yu, L., Xie, F., Bao, X., Liu, H., Ji, Z., & Chen, L. (2017). One-step method to prepare starch-based superabsorbent polymer for slow release of fertilizer. *Chemical Engineering Journal*, *309*, 607–616.

Thakur, S., Sharma, B., Verma, A., Chaudhary, J., Tamulevicius, S., & Thakur, V. K. (2018). Recent progress in sodium alginate based sustainable hydrogels for environmental applications. *Journal of Cleaner Production*, *198*, 143–159.

Thombare, N., Mishra, S., Siddiqui, M. Z., Jha, U., Singh, D., & Mahajan, G. R. (2018). Design and development of guar gum based novel, superabsorbent and moisture retaining hydrogels for agricultural applications. *Carbohydrate Polymers*, *185*, 169–178.

Song, B., Liang, H., Sun, R., Peng, P., Jiang, Y., & She, D. (2020). Hydrogel synthesis based on lignin/sodium alginate and application in agriculture. *International Journal of Biological Macromolecules*, *144*, 219–230.

Fidelia, N., & Chris, B. (2011). Environmentally friendly superabsorbent polymers for water conservation in agricultural lands. *Journal of Soil Science and Environmental Management*, *2*(7), 206–211.

DeRosa, M. C., Monreal, C., Schnitzer, M., Walsh, R., & Sultan, Y. (2010). Nanotechnology in fertilizers. *Nature Nanotechnology*, *5*(2), 91–91.

Salmawi, K. M. E., El-Naggar, A. A., & Ibrahim, S. M. (2018). Gamma irradiation synthesis of carboxymethyl cellulose/acrylic acid/clay superabsorbent hydrogel. *Advances in Polymer Technology*, *37*(2), 515–512.

Fang, S., Wang, G., Li, P., Xing, R., Liu, S., Qin, Y., ... Li, K. (2018). Synthesis of chitosan derivative graft acrylic acid superabsorbent polymers and its application as water retaining agent. *International Journal of Biological Macromolecules*, *115*, 754–761.

Hietala, S., Holmberg, S., Karjalainen, M., Näsman, J., Paronen, M., Serimaa, R., ... Vahvaselkä, S. (1997). Structural investigation of radiation grafted and sulfonated poly (vinylidene fluoride), PVDF, membranes. *Journal of Materials Chemistry*, *7*(5), 721–726.

Ghobashy, M. M., Abd El-Wahab, H., Ismail, M. A., Naser, A. M., Abdelhai, F., El-Damhougy, B. K., ... Alkhursani, S. A. (2020). Characterization of Starch-based three components of gamma-ray cross-linked hydrogels to be used as a soil conditioner. *Materials Science and Engineering B*, *260*, 114645.

Elbarbary, A. M., Abd El-Rehim, H. A., El-Sawy, N. M., Hegazy, E. S. A., & Soliman, E. S. A. (2017). Radiation induced crosslinking of polyacrylamide incorporated low molecular weights natural polymers for possible use in the agricultural applications. *Carbohydrate Polymers*, *176*, 19–28.

Gharekhani, H., Olad, A., Mirmohseni, A., & Bybordi, A. (2017). Superabsorbent hydrogel made of NaAlg-g-poly (AA-co-AAm) and rice husk ash: Synthesis, characterization, and swelling kinetic studies. *Carbohydrate Polymers*, *168*, 1–13.

Ahmad, N. N. R., Fernando, W. J. N., & Uzir, M. H. (2015). Parametric evaluation using mechanistic model for release rate of phosphate ions from chitosan-coated phosphorus fertiliser pellets. *Biosystems Engineering*, *129*, 78–86.

Narayanan, A., Kartik, R., Sangeetha, E., & Dhamodharan, R. (2018). Super water absorbing polymeric gel from chitosan, citric acid and urea: Synthesis and mechanism of water absorption. *Carbohydrate Polymers*, *191*, 152–160.

Alharbi, K., Ghoneim, A., Ebid, A., El-Hamshary, H., & El-Newehy, M. H. (2018). Controlled release of phosphorous fertilizer bound to carboxymethyl starch-g-polyacrylamide and maintaining a hydration level for the plant. *International Journal of Biological Macromolecules*, *116*, 224–231.

Raafat, A. I., Eid, M., & El-Arnaouty, M. B. (2012). Radiation synthesis of superabsorbent CMC based hydrogels for agriculture applications. *Nuclear Instruments & Methods in Physics Research. Section B, Beam Interactions with Materials and Atoms*, *283*, 71–76.

Jyothi, A. N., Pillai, S. S., Aravind, M., Salim, S. A., & Kuzhivilayil, S. J. (2018). Cassava starch-graft-poly (acrylonitrile)-coated urea fertilizer with sustained release and water retention properties. *Advances in Polymer Technology*, *37*(7), 2687–2694.

Kaith, B. S., Sharma, R., Kalia, S., & Bhatti, M. S. (2014). Response surface methodology and optimized synthesis of guar gum-based hydrogels with enhanced swelling capacity. *RSC Advances*, *4*(76), 40339–40344.

Jiang, C., Sun, G., Zhou, Z., Bao, Z., Lang, X., Pang, J., ... Chen, X. (2019). Optimization of the preparation conditions of thermo-sensitive chitosan hydrogel in heterogeneous reaction using response surface methodology. *International Journal of Biological Macromolecules*, *121*, 293–300.

Kademani, B. S., Kumar, V., Sagar, A., & Kumar, A. (2006). World literature on thorium research: A scientometric study based on Science Citation Index. *Scientometrics*, *69*(2), 347–364.

Heidari, S., Esmaeilzadeh, F., Mowla, D., Jokar, H., Cortés, F. B., Nassar, N. N., & Franco Ariza, C. A. (n.d.). *Acrylamide-Acrylic Acid as Copolymer Gel for Water Shut-Off at High Pressure and Temperature Conditions*. Available at SSRN 4127813.

Sundholm, F., Serimaa, R., & Sundholm, G. (1996). Polyelectrolytes and Electrochemically Active Membranes: Synthesis, Characterisation and Applications. *National Programme on Materials and Structure Research, 267*.

Abahmane, L. (2011). Date palm micropropagation via organogenesis. In *Date palm biotechnology* (pp. 69–90). Springer.

Abahmane, L. (2013). Recent achievements in date palm (*Phoenix dactylifera* L.) micropropagation from inflorescence tissues. *Emirates Journal of Food and Agriculture*, *25*(11), 863–874. doi:10.9755/ejfa.v25i11.16659

Abbas, M. S., Dobeie Amani, M., & Azzam Clara, R. (2021). Identification of salt tolerant Genotypes among Egyptian and Nigerian Peanut (Arachis hypogaea L.) Using Biochemical and Molecular Tools. In *Mitigating Environmental Stresses for Agricultural Sustainability in Egypt*. Springer International Publishing AG., doi:10.1007/978-3-030-64323-2_16437

Abd-Alla, M. H., & El-Enany, A. W. E. (2012). Production of acetone-butanol-ethanol from spoilage date palm (Phoenix dactylifera L.) fruits by mixed culture of Clostridium acetobutylicum and Bacillus subtilis. *Biomass and Bioenergy*, *42*, 172–178. doi:10.1016/j.biombioe.2012.03.006

Abd-El Hamed, Rasmia, & Zayed. (2017). Evaluation Physical and Chemical Characteristics of Some Seedlings Date Palm Fruits (Maghal) in the North Delta Egypt. *International Journal of Advances in Agricultural Science and Technology*, *4*(7), 13–32.

Abdelal, A. F., Mahmoud, H. M., & El-Agamy, S. Z. (1983). The effect of pollen source on fruit characteristics of Zaghloul dates (Phoenix dactylifera, L.). *Assuit J Agr Sci*, *14*, 347–355.

Abd-ElBaky, M. A. (2012). Using morphological and Anatomical features as taxonomical evidences to differentiate between some soft and semidry Egyptian cultivars of date palm. *Journal of Horticultural Science & Ornamental Plants*, *4*(2), 195–200.

Abdel-Hamid, N. (2000). Effect of time, rate and patterns of thinning, leaf/bunch ratio and male type on" Zaghloul" date yield and quality. *Arab Universities Journal of Agricultural Sciences*, *8*(1), 305–317.

Abd-ElHamied (2014). Effect of Some Agro-Management Systems on Growth and Production of Date Palm Off-Shoots under North Sinai Conditions. *IOSR Journal of Agriculture and Veterinary Science, 7*(11), 41.

Abdel-Monaim, M. F., Khalil, M. S. M., & Sahar, H. A. (2018). Application of date palm leaves compost (DPLC) and plant growth promoting rhizobacteria(PGPR) for controlling faba bean root rot disease in New Valley, Egypt. *Journal of Phytopathology and Pest Management*, *5*(1), 88–100.

Abdel-Raheem & Abdel-Salam. (2016). *Citrus Pests Bionomics and Control in Egypt*. Lambert Academic Publishing.

Abdel-Raheem, M., Alghamdi, H. A., & Reyad, N. F. (2020b). Nano Essential oils against the red palm weevil, *Rhynchophorus ferrugineus* Olivier (Coleoptera: Curculionidae). *Entomological Research*, *50*(5), 215–220. doi:10.1111/1748-5967.12428

Abdel-Raheem, M., Alghamdi, H., & Reyad, N. (2018). *Botanical Pesticide and Insect Pests*. Lambert Academic Publishing.

Abdel-Raheem, M., Dimetry, N. Z., & Amin, A. E.-R. (2020a). *Nano-Preparations from Botanical Products for Controlling Insect pests*. Lambert Academic Publishing.

Abdi, G. H., & Hedayat, M. (2010). Yield and fruit physiochemical characteristics of 'Kabkab' date palm as affected by methods of potassium fertilization. *The Free Library Science and Technology*. www.google.com

Abdrabou, Fergany, Azzam, & Nahid. (2017). Devolopment of Some Canola Genotypes to Salinity Tolerance using Tissue Culture Technique. *Egypt. J. Agron., 39*(3), 431–448. doi:10.21608/agro.2017.1949.1083

Abobatta, W.F. (2022). *Impact of shading practice on growth of Murcott mandarin seedlings*. Unpublished work.

Abobatta, W. F. (2018). Challenges for Citrus Production in Egypt. *Acta Scientific Agriculture*, *2*(8), 40–41.

Abobatta, W. F. (2019). Overview of Role of Magnetizing Treated Water in Agricultural Sector Development. *Adv Agri Tech Plant Sciences*, *2*(1), 180023.

Abobatta, W. F. (2020). Citriculture and Climate Change. *Modern Concepts & Developments in Agronomy*, *6*(3), 649–650. doi:10.31031/MCDA.2020.06.000639

Abobatta, W. F. (2021a). Managing citrus orchards under climate change. *MOJ Eco Environ Sci*, *6*(2), 43–44.

Abobatta, W. F. (2021b). Fruit orchards under climate change conditions: Adaptation strategies and management. *J Appl Biotechnol Bioeng*, *8*(3), 99–102. doi:10.15406/jabb.2021.08.00260

Abo-Doma & Azzam. (2007). Hunting of some differentially expressed genes under salt stress in wheat. Egypt. *J. Plant Breed*, *11*(3), 233–244.

Abo-Doma, A., & Azzam Clara, R. (2007). Molecular genetic relationships among some bread wheat cultivars (Triticum aestivum L.). *Egyptian Journal of Genetics and Cytology*, *36*(2), 387–400.

Aboelnaga, Abodoma, Lamyaa, & Azzam. (2020). Assessment of biodiversity among some sesame genotypes using ISSR and SRAP markers. *Arab Univ. J. Agric. Sci.*, *82*(3), 1–15.

Abo-Rekab, Darwesh, & Hassan. (2010). Effect of arbuscular mycorrhizal fungi, NPK complete fertilizers on growth and concentration nutrients of acclimatized date palm plantlets. *Mesopotomia J. of Agric*, *38*(1), 1–11.

Abou-Sreea, Azzam, Al-Taweel, Abdel-Aziz, & Belal. (2021). Natural Biostimulant Attenuates Salinity Stress Effects in Chili Pepper by Remodeling Antioxidant, Ion and Phytohormone Balances and Augments Gene Expression. *Plants, 10*(11), 2316. doi:10.3390/plants10112316

Abul-Soud, M. A., Emam, M. S. A., & Mohammed, Sh. M. (2021). Smart hydroponic greenhouse (Sensing, monitoring and control) prototype based on Arduino and IoT. *International Journal of Plant and Soil Science*, *33*(4), 63–77.

Aceto, N., Bardia, A., Miyamoto, D. T., Donaldson, M. C., Wittner, B. S., Spencer, J. A., Yu, M., Pely, A., Engstrom, A., Zhu, H., Brannigan, B. W., Kapur, R., Stott, S. L., Shioda, T., Ramaswamy, S., Ting, D. T., Lin, C. P., Toner, M., Haber, D. A., & Maheswaran, S. (2014). Circulating Tumor Cell Clusters Are Oligoclonal Precursors of Breast Cancer Metastasis. *Cell*, *158*(5), 1110–1122. doi:10.1016/j.cell.2014.07.013 PMID:25171411

Adams, W. T., Zuo, J., Shimizu, J. Y., & Tappeiner, J. C. (1998). Impact of alternative regeneration methods on genetic diversity in coastal Douglas-fir. *Forest Science*, *44*, 390–396.

Afzal, M., Khan, M. A., Pervez, M. A., & Ahmed, R. (2011). Root induction in the aerial offshoots of date palm (*Phoenix dactylifera* L.) cultivar Hillawi. *Pakistan Journal of Agricultural Sciences*, *48*(1), 11–17.

Agrawal, K., & Kamboj, N. (2019). Smart agriculture using IOT: A futuristic approach. *International Journal of Information Dissemination and Technology*, *9*(4), 186–190.

Agrios, G. N. (2005). *Plant pathology* (5th ed.). Elsevier Academic Press.

Ahmad, N., Rab, A., & Ahmad, N. (2016). Light-induced biochemical variations in secondary metabolite production and antioxidant activity in callus cultures of Stevia rebaudiana (Bert). *Journal of Photochemistry and Photobiology. B, Biology*, *154*, 51–56. doi:10.1016/j.jphotobiol.2015.11.015 PMID:26688290

Ahmed, M. F. A. (2005). *Effect of adding some biocontrol agents on non-target microorganisms in root diseases infecting soybean and broad bean plants* [M.Sc. Thesis]. Faculty of Agriculture Moshtohor, Benha Univ.

Ahmed, M. F. A. (2013). *Studies on non-chemical methods to control some soil borne fungal diseases of bean plants Phaseolus vulgaris L.* [Ph.D. Thesis]. Faculty of Agriculture, Cairo Univ., Giza, Egypt.

Ahmed, M. F. A. (2018). Management of date palm root rot diseases by using some biological control agents under organic farming system. *Novel Research in Microbiology Journal, 2*(2), 37-47.

Ahmed, W., Nawaz, M. A., Saleem, B. A., & Asim, M. (2005). Incidence of mango midge and its control in different mango growing countries of the world. In *International Conference on Mango and date Palm: Culture and Export* (pp. 98-101). Academic Press.

Ahmed, F. A., Hamdy, I., Ibrahim, I. M., & Kamel, M. K. h. (2014). Reducing Inorganic N Partially in Zaghloul Date Palm Orchards by Using Humic Acid and Effective Microorganisms. *World Rural Observ*, *6*(2), 102–110.

Ahmed, M. F. A. (2018). Evaluation of some biocontrol agents to control Thompson seedless grapevine powdery mildew disease. Ahmed Egyptian Journal of Biological Pest Control. *Ahmed Egyptian Journal of Biological Pest Control*, *28*(93), 694–700.

Aira, M., Gómez-Brandón, M., González-Porto, P., & Domínguez, J. (2011). Selective reduction of the pathogenic load of cow manure in an industrial-scale continuous-feeding vermireactor. *Bioresource Technology*, *102*(20), 9633–9637. doi:10.1016/j.biortech.2011.07.115 PMID:21875788

Aizen, M. A., & Harder, L. D. (2009). The global stock of domesticated honey bees is growing slower than agricultural demand for pollination. *Current Biology*, *19*(11), 915–918. doi:10.1016/j.cub.2009.03.071 PMID:19427214

Akhtar, M., & Malik, A. (2000). Role of organic amendments and soil organisms in the biological control of plant parasitic nematodes: A review. *Bioresource Technology*, *74*(1), 35–47. doi:10.1016/S0960-8524(99)00154-6

Al-Aa'reji. (2010). Effect of organic fertilizer, urea and sulphur on vegetative growth and concentration of some nutrient of young peach trees cv. Dixired. *J.Tikrit Univ.Agric.Sci*, *10*(2), 76–86.

Al-Alawi, R. A., Al-Mashiqri, J. H., Al-Nadabi, J. S., Al-Shihi, B. I., & Baqi, Y. (2017). Date palm tree (Phoenix dactylifera L.): Natural products and therapeutic options. *Frontiers in Plant Science*, *8*(845), 1–12. doi:10.3389/fpls.2017.00845 PMID:28588600

Alarcón, J. J., Ortuno, M. F., Nicolás, E., Navarro, A., & Torrecillas, A. (2006). Improving water-use efficiency of young lemon trees by shading with aluminised-plastic nets. *Agricultural Water Management*, *82*(3), 387–398. doi:10.1016/j.agwat.2005.08.003

Al-Busaidi, K. T. S. (2013). Effects of organic and inorganic fertilizers addition on growth and yield of banana (*Musa* AAA cv. Malindi) on a saline and non-saline soil in Oman. *Journal of Horticulture and Forestry*, *5*(9), 146–155.

Al-Farsi, M. A., & Lee, C. Y. (2008). Optimization of phenolics and dietary fibre extraction from date seeds. *Food Chemistry*, *108*(3), 977–985. doi:10.1016/j.foodchem.2007.12.009 PMID:26065761

Al-Farsi, M., Alasalvar, C., Morris, A., Baron, M., & Shahidi, F. (2005). Comparison of antioxidant activity, anthocyanins, carotenoids, and phenolics of three native fresh and sun-dried date (Phoenix dactylifera L.) varieties grown in Oman. *Journal of Agricultural and Food Chemistry*, *53*(19), 7592–7599. doi:10.1021/jf050579q PMID:16159191

Alfieri, S. A., Jr., Langdon, K. R., Kimbrough, J. W., El-Gholl, N. E., & Wehlburg, C. (1994). Diseases and Disorders of Plants in Florida. Florida Department of Agriculture & Consumer Services, Division of Plant Industry.

Alford, D. V. (1984). *A colour atlas of fruit pests their recognition, biology and control*. Wolfe Publishing Ltd.

Al-Harrasi, A., Rehman, N. U., Hussain, J., Khan, A. L., Al-Rawahi, A., Gilani, S. A., Al-Broumi, M., & Ali, L. (2014). Nutritional assessment and antioxidant analysis of 22 date palm (Phoenix dactylifera L.) varieties growing in Sultanate of Oman. *Asian Pacific Journal of Tropical Medicine*, *7*, S591–S598. doi:10.1016/S1995-7645(14)60294-7 PMID:25312188

Al-Harrasi, A., Rehman, N. U., Hussain, J., Khan, A. L., Al-Rawahi, A., Gilani, S. A., ... Ali, L. (2014). Nutritional assessment and antioxidant analysis of 22 date palm (Phoenix dactylifera) varieties growing in Sultanate of Oman. *Asian Pacific Journal of Tropical Medicine*, *7*, S591–S598.

Al-Hooti, S., Sidhu, J. S., & Qabazard, H. (1997). Physicochemical characteristics of five date fruit cultivars grown in the United Arab Emirates. *Plant Foods for Human Nutrition (Dordrecht, Netherlands)*, *50*(2), 101–113. doi:10.1007/BF02436030 PMID:9201745

Ali-Dinar, H. M., Alkhateeb, A. A., Al-Abdulhadi, I., Alkhateeb, A., Abugulia, K. A., & Abdulla, G. R. (2002). Bunch thinning improves yield and fruit quality of date palm (Phoenix dactylifera L.). Egypt. *J. Applied Sci*, *17*(11), 228–238.

Alikamanoğlu, S., Yaycılı, O., Atak, Ç., & Rzakoulieva, A. (2007). Effect of Magnetic Field and Gamma Radiation on Paulowinia Tomentosa Tissue Culture. *Biotechnology, Biotechnological Equipment*, *21*(1), 49–53. doi:10.1080/13102818.2007.10817412

Alikhani-Koupaei, M., & Aghdam, M. S. (2021). Defining date palm leaf pruning line in bearing status by tracking physiological markers and expression of senescence-related genes. *Plant Physiology and Biochemistry*, *167*, 550–560. doi:10.1016/j.plaphy.2021.08.035 PMID:34454314

Alikhani-Koupaei, M., Aghdam, M. S., & Faghih, S. (2020). Physiological aspects of date palm loading and alternate bearing under regulated deficit irrigation compared to cutting back of bunch. *Agricultural Water Management*, *232*, 106035. doi:10.1016/j.agwat.2020.106035

Ali, Y. S. S. (2008). Use of Date Palm Leaves Compost as A Substitution to Peatmoss. *American Journal of Plant Physiology*, *3*(4), 131–136.

Al-Jubouri, A.A.A. & Hamza, J.H. (2006). *Magnetically water treatment technology and its impact in the agricultural field.* Baghdad University - Faculty of Agriculture – Dept. of Field Crop Sci. Crop Pro.

Al-Khayri, J. M. (2005). Date palm Phoenix dactylifera L. In S. M. Jain & P. K. Gupta (Eds.), *Protocols of somatic embryogenesis in woody plants* (pp. 309–319). Springer, Netherlands. doi:10.1007/1-4020-2985-3_25

Al-Khayri, J. M. (2007). Date palm Phoenix dactylifera L. micropropagation. In S. M. Jain & H. Häggman (Eds.), *Protocols for micropropagation of woody trees and fruits* (pp. 509–526). Springer, Netherlands. doi:10.1007/978-1-4020-6352-7_46

Allam, H. M., Hussein, F., & Abdalla, K. M. (1973). Seasonal requirements of water by Sakkoti dates grown at Asswan. *Moshtohor Fac Res Bull, 39*, 1–10.

Allard, R.W. (1960). *Principles of plant breeding*. Wiley.

Al-Meer, Ahmed, Al-Najim, & Yaseen. (2008). Effect of treatment by polyethelene glyol on acclimatization of date palm (*Phoenix dactylifera* L.) cv. Barhee propagated in vitro. *El-Basra Journal for Date Palm Research, 7*(1), 1-9.

Al-Mssallem, M. Q., Alqurashi, R. M., & Al-Khayri, J. M. (2020). Bioactive compounds of date palm (Phoenix dactylifera L.). *Bioactive Compounds in Underutilized Fruits and Nuts*, 91-105.

Al-Mssallem, M. Q., Elmulthum, N. A., & Elzaki, R. M. (2019). Nutritional security of date palm fruit: An empirical analysis for Al-Ahsa region in Saudi Arabia. *Scientific Journal of King Faisal University, 20*, 47–54.

Al-Najm, A.R.A. (2009). Effect of Indole butyric acid and Indole- 3-acetic acid on root initiation data palm (*Phoenix dactylifera*)c.v Hillawi. *Basrah Journal for Date Palm Research, 8*(1), 13–20.

Al-Qurashi, A. D., & Awad, M. A. (2011). 5-aminolevulinic acid increases tree yield and improves fruit quality of (Rabia) and (Sukkariat- Yanbo) date palm cultivars under hot arid climate. *Scientia Horticulturae, 129*, 441–448.

Al-Redhaiman, K. N. (2005). Modified atmosphere extends storage period and maintains quality of (Barhi) date fruits. *Acta Horticulturae*, (682), 979–986. doi:10.17660/ActaHortic.2005.682.127

Al-Shahib, W., & Marshall, R. J. (2003). The fruit of the date palm: Its possible use as the best food for the future? *International Journal of Food Sciences and Nutrition, 54*(4), 247–259. doi:10.1080/09637480120091982 PMID:12850886

Al-Taha, H.A. (2013). Effect of Different Treatments on Regeneration, Growth and acclimatization of Ananas Plantlets Produced By Plant Tissue Culture 2-Application Effect of Culture Media and Urea Fertilizer on Growth and Acclimatization of In vitro Propagated Pineapple (*Ananas Comosus* L. Meer. cv.Del Mont) Plantlet. *Tekret Journal, 13*(1), 230–23.

Al-Temimi. (2012). Effect the addition of equivalent ratio of mechanical firtilizarers on growth of date palms (*Phoenix dectylifera* L.) Barhi cultivar. *Basra Science Journal, 38*(4), 60–73.

Al-Turki, S., Shahba, M. A., & Stushnoff, C. (2010). Diversity of antioxidant properties and phenolic content of date palm (Phoenix dactylifera L.) fruits as affected by cultivar and location. *Journal of Food Agriculture and Environment, 8*(1), 253–260.

Aluko, O. B., & Seig, D. A. (2000). An experimental investigation of the characteristics of and conditions for brittle fracture in two-dimensional soil cutting. *Soil & Tillage Research, 57*(3), 143–157. doi:10.1016/S0167-1987(00)00156-2

Aml, R. M. Y., Hala, S. E., & Saleh, M. M. S. (2011). Olive seedlings growth as affected by humic, amino acids, macro and trace elements applications. *Agriculture and Biology Journal of North America, 2*(7), 1101–1107.

Amro, El- Sayed, & El Gammal. (2014). Effect of Effective Microorganisms (EM) and Potassium Sulphate on Productivity and Fruit Quality of "Hayany" Date Palm Grown Under Salinity Stress IOSR. *Journal of Agriculture and Veterinary Science (IOSR-JAVS), 7*(6), 90-99.

Anderson, D. L., Sedgley, H., Shortant, T. R. T., & Allwood, A. J. (1982). Insect pollination of mango in nort-hen Australia. *Aust. Agric. Res., 33*(3), 541–548. doi:10.1071/AR9820541

Andersson, G. K. S., Rundlöf, M., & Smith, H. G. (2012). Organic farming improves pollination success in Strawberries. *PLoS One, 7*(2), 1–4. doi:10.1371/journal.pone.0031599 PMID:22355380

Angeli, G., Anfora, G., Baldessari, M., Germinara, G. S., Rama, F., De Cristofaro, A., & Ioriatti, C. (2007). Mating disruption of codling moth *Cydia pomonella* with high densities of Ecodian sex pheromone dispensers. *Journal of Applied Entomology*, *131*(5), 311–318. doi:10.1111/j.1439-0418.2007.01172.x

Ansebo, L., Coracini, M. D. A., Bengtsson, M., Liblikas, I., Ramírez, M., Borg-Karlson, A.-K., Tasin, M., & Witzgall, P. (2004, August 1). Antennal and behavioural response of codling moth Cydia pomonella to plant volatiles. *Journal of Applied Entomology*, *128*(7), 488–493. doi:10.1111/j.1439-0418.2004.00878.x

Apolo-Apolo,, O.E., & Perez-Ruiz,, M., Martinez-Guanter, & Valente, J. (2020). A-Cloud based environment for generating yield estimation maps from apple orchard using UAV imagery and deep learning technique. *Frontiers in Plant Science*, *11*(1086), 15.

Aracama, C. V., Michael, E. K., Sandra, W. B., & Nancy, P. L. (2008). Comparative growth, morphology, and anatomy of easy- and difficult-to-acclimatize Sea Oats (*Uniola paniculata*) Genotypes During *in vitro* Culture and Ex Vitro acclimatization. *Amer Journal Soc. Hort. Sci.*, *133*(6), 830–843. doi:10.21273/JASHS.133.6.830

Araj, S. E., & Wratten, S. D. (2015). Comparing existing weeds and commonly used insectary plants as floral resources for a parasitoid. *Biological Control*, *81*, 15–20. doi:10.1016/j.biocontrol.2014.11.003

Arancon, N. Q., Edwards, C. A., & Lee, S. 2002. Management of plant parasitic nematode populations by use of vermicomposts. *Proc Brighton Crop Prot Conf Pests Dis 8B-2,* 705–716.

Araujo, J. A., Borralho, N. M. G., & Dehon, G. (2012). The importance and type of non-additive genetic effects for growth in *Eucalyptus globulus. Tree Genetics & Genomes*, *8*(2), 327–337. doi:10.100711295-011-0443-x

Arditti, J., & Emst, R. (1993). *Micropropagation of Orchid.* Wiley.

Arneson, P. A., & Losey, J. E. (1997). *Integrated pest mangement-lecture notes.* Cornell University.

Arshad, M., & Frankenberger, W. T. Jr. (1993). Microbial production of plant growth regulators. In F. B. Metting Jr., (Ed.), *Soil microbial ecology: applications in agricultural and environmental management* (pp. 307–347). Marcell Dekker.

Arunkumar, J. (2000). Effect of vermicomposted sludge on growth of Amaranths dubius. *Ecotoxicol Environ Monit*, *14*(2), 157–160.

Arvidsson, J., Keller, T., & Gustafsson, K. (2004). Specific draught for mouldboard plough, chisel plough and disc harrow at different water contents. *Soil & Tillage Research*, *79*(2), 221–231. doi:10.1016/j.still.2004.07.010

Ashraf, Z., & Hamidi-Esfahani, Z. (2011). Date and date processing: A review. *Food Reviews International*, *27*(2), 101–133.

Asselin, C. (2010). *Choisir le bon pneu pour diminuer la compaction du sol.* La référence en nouvelles technologies agricoles au Québec.

Atiyeh, R. M., Arancon, N. Q., Edwards, C. A., & Metzger, J. D. (2000). Infl uence of earthworm- processed pig manure on the growth and yield of greenhouse tomatoes. *Bioresource Technology*, *75*(3), 175–180. doi:10.1016/S0960-8524(00)00064-X

Atiyeh, R. M., Edwards, C. A., Subler, S., & Metzger, J. D. (2001). Pig manures vermicompost as a component of a horticultural bedding plant medium: Effects on physiochemical properties and plant growth. *Bioresource Technology*, *78*(1), 11–20. doi:10.1016/S0960-8524(00)00172-3 PMID:11265782

Attalla, A. M., Haggag, M. N., Attia, M. M., & Ibrahim, A. M. F. (1988). Fruit and leaf mineral content of four date palm varieties grown in Alexandria. *J Agr Res Tanta Univ*, *14*, 319–331.

Atwood, J. T. (1984). The relationships of the slipper orchids (subfamily cypripedioideae, orchidaceae). *Selbyana*, *7*(2/4).

Awad, M. A. (2006, May). Accelerating ripening of date palm fruit (Phoenix dactylifera L.) cv." Helaly" by some pre and post-harvest treatments. In *International Conference on Date Palm Production & Processing Technology* (*Vol. 13*). Academic Press.

Awad, M. A. (2008). Promotive effects of a 5- aminolevulinic acid- based fertilizer on growth of tissue culture- derived date palm plants (*Phoenix dactylifera* L.) during acclimatization. *Scientia Horticulturae*, *118*, 48–52.

Awad, M. A. (2010). Pollination of date palm (Phoenix dactylifera L.) cv. Khenazy by pollen grain-water suspension spray. *Journal of Food Agriculture and Environment*, *8*, 313–317.

Awad, R. M., El-Sayed, H. A., & El-Razk, A. (2020). Vegetative Growth and Quality of Washington Navel Orange as Affected by Shading Nets and Potassium Silicate Spraying. *Egyptian Academic Journal of Biological Sciences. H, Botany*, *11*(1), 81–90. doi:10.21608/eajbsh.2020.119321

Awasthi, O. P., Singh, I. S., & More, T. A. (2009). Performance of intercrops during establishment phase of aonla (*Emblica officinalis*) orchard. *Indian Journal of Agricultural Sciences*, *79*(8), 587–591.

Axelrod, D. (1960). The evolution of flowering plants. In S. Tax (Ed.), *Evolution after Darwin.1* (pp. 227–305). University of Chicago Press.

Azzam & Abo-Doma. (2007). Genetic relationships among some canola cultivars (*Brassica napus* L.) based on ISSR and RAPD-analyses. *Egyptian Journal of Genetics and Cytology*, *36*(2), 355–367.

Azzam & Mahrous. (2010). Performance and genetic relationships among ten Egyptian wheat cultivars as revealed by RAPD-PCR analysis. *Egypt. J. Plant Breed.*, *14*(3), 87-102.

Azzam, Al-Taweel, Abdel-Aziz, Rabe, & Abou-Sreea. (2021). Salinity Effects on Gene Expression, Morphological, and Physio-Biochemical Responses of Stevia rebaudiana Bertoni In Vitro. *Plants*, *10*(4), 820. doi:10.3390/plants10040820

Azzam, Azer, Khalifa, & Abol-Ela. (2007). Characterization of peanut mutants and molecular markers associated with resistance to pod rot diseases and aflatoxin contamination by RAPD and ISSR. *Arab Journal of Biotechnology*, *10*(2), 301–320.

Azzam, C. R. (2004). Gibberellin 20-oxidase isolation and transformation its anti-sense by Agrobacterium tumfaciens to produce dwarf sunflower plants. In *The 2nd International Conf. of Biotechnology*. El-Baath University.

Azzam, El-Rahman, & Eman. (2011). Evaluation of Genetic Relationships of some Barley Cultivars based on Phenotypic, Seed Quality and Molecular analysis. Egypt. *J. Plant Breed.*, *15*(4), 1–25.

Azzam, C. R., Abd El Naby, Z., & Mohamed, N. (2019). Salt Tolerance Associated With Molecular Markers In Alfalfa. *Journal of Bioscience and Applied Research*, *5*(4), 416–428. doi:10.21608/jbaar.2019.110864

Azzam, C. R., Zeinab, A.-E., & Kh, S. A. (2012). Influence of Agro-Ecological Conditions on Gene Expression, Yield and Yield Components of the Mono-Cut (Fahl) Type of Berseem. Egypt. *J. Plant Breed.*, *16*(2), 135–159. doi:10.12816/0003961

Azzam, C. R., Zein, S. N., & Abbas, S. M. (2007). Biochemical genetic markers for levels of resistance to Cowpea Aphid Borne Mosaic Potyvirus (CABMV) in sesame (*Sesamum indicum* L.) irradiated with gamma ray. *J. Plant Breed.*, *11*(2), 861–885.

Babu, C. (2010). *Use of magnetic water and polymer in agriculture*. Ph.D., Agronomy D.W.S.R. Centre.

Bale, J. S., Van Lenteren, J. C., & Bigler, F. (2008). Biological control and sustainable food production. *Philosophical Transactions of the Royal Society of London. Series B, Biological Sciences*, *363*(1492), 761–776. doi:10.1098/rstb.2007.2182 PMID:17827110

Baliga, M. S., Baliga, B. R. V., Kandathil, S. M., Bhat, H. P., & Vayalil, P. K. (2011). A review of the chemistry and pharmacology of the date fruits (Phoenix dactylifera L.). *Food Research International*, *44*(7), 1812–1822. doi:10.1016/j.foodres.2010.07.004

Balilashaki, K., Moradi, S., Vahedi, M., & Khoddamzadeh, A. A. (2020). A molecular perspective on orchid development. *The Journal of Horticultural Science & Biotechnology*, *95*(5), 542–552. Advance online publication. doi:10.1080/14620316.2020.1727782

Ballard, J., Ellis, D. J., & Payne, C. C. (2000, October 01). Uptake of Granulovirus from the Surface of Apples and Leaves by First Instar Larvae of the Codling Moth Cydia pomonella L. (Lepidoptera: Olethreutidae). *Biocontrol Science and Technology*, *10*(5), 617–625. doi:10.1080/095831500750016415

Ball, R. D. (2005). Experimental designs for reliable detection of linkage disequilibrium in unstructured random population association studies. *Genetics*, *170*(2), 859–873. doi:10.1534/genetics.103.024752 PMID:15781715

Banejad, H., & Abdosalehi, E. (2009). The effect of magnetic field on water hardness reducing. In *Thirteenth international water technology conference, IWTC* (pp. 117-128). Academic Press.

Bangels, E., Alhmedi, A., Akkermans, W., Bylemans, D., & Belien, T. (2021). Towards a Knowledge-Based Decision Support System for Integrated Control of Woolly Apple Aphid, Eriosoma lanigerum, with Maximal Biological Suppression by the Parasitoid *Aphelinus mali. Insects*, *12*(6), 479. doi:10.3390/insects12060479 PMID:34063971

Baniulis, D., Gelvonauskienė, D., Rugienius, R., Sasnauskas, A., & Stanienė, G. (2013). Orchard Plant Breeding, Genetics, And Biotechnology Research At The Institute Of Horticulture, LRCAF. *Sodininkyste ir Darzininkyste*, *32*, 3–4.

Barber, K. N., & Kaupp, W. J. (1993). Specificity testing of the nuclear polyhedrosis virus of the gypsy moth, Lymantria dispar (L.) (Lepidoptera: Lymantriidae). Can Entomol, 125, 1055–1066.

Barik, T., Gulati, J. M. L., Garnayak, L. M., & Bastia, D. K. (2011). Production of vermicompost from agricultural wastes. *Agricultural Reviews (Karnal)*, *31*(3), 172–183.

Barnes, B. N. (1990). *FFTRI monitoring manual for orchard research institute.* Fruit and Fruit Technology Research Institute.

Barnes, B. N., & Blomefield, T. L. (1997). Goading growers towards mating disruption: the South African experience with Grapholita molesta and Cydia pomonella (Lepidoptera, Tortricidae). *Goading growers towards mating disruption: the South African experience with Grapholita molesta and Cydia pomonella (Lepidoptera, Tortricidae), 20*(1), 45-56.

Barnes, B. N. (1999). *Fruit fly baiting on deciduous fruit. Doing the right things right.* Deciduous Fruit Grower.

Bastías, R. M., Manfrini, L., & Grappadelli, L. C. (2012). Exploring the potential use of photo-selective nets for fruit growth regulation in apple. *Chilean Journal of Agricultural Research*, *72*(2), 224–231. doi:10.4067/S0718-58392012000200010

Bauer, C., & Zapp, P. (2005). Towards generic factors for land use and water consumption. In A. Dubreuil (Ed.), *Life cycle assessment of metals: issues and research directions.* SETAC—USA.

Bautista, N., Morales, O., Carrillo, J., & Bravo, H. (1998). Mortalidad de Phyllocnistis citrella con un aceite mineral y nim. *Manejo Integrado de Plagas*, *50*, 29–33.

Beers, E. H., Hull, L. A., & Jones, V. P. (2020). Sampling pest and beneficial arthropods of apple. In *Handbook of Sampling Methods for Arthropods in Agriculture* (pp. 383–416). CRC Press. doi:10.1201/9781003067900-18

Bekheet, S. (2013). Direct organogenesis of date palm (Phoenix dactylifera L.) for propagation of true-to-type plants. *Scientia Agrícola*, *4*(3), 85–92.

Bekheet, S. (2013). Direct Organogenesis of Date Palm (*Phoenix dactylifera* L.) for Propagation of True-to-Type Plants. *ScientiaAgriculturae*, *4*(3), 85–92.

Bekheet, S. A., & El-Sharabasy, S. F. (2015). Date palm status and perspective in Egypt. In *Date palm genetic resources and utilization* (pp. 75–123). Springer. doi:10.1007/978-94-017-9694-1_3

Belarmino, M. M., & Mii, M. (2000). *Agrobacterium-mediated* genetic transformation of a Phalaenopsis orchid. *Plant Cell Reports*, *19*(5), 435–442. doi:10.1007002990050752 PMID:30754879

Belickaya, M. N. (2011). Izmenenie radial'nogo prirosta duba pri defoliacii kron listogryzushchimi vreditelyami. In Bolezni i vrediteli v lesah Rossii: vek XXI. Materialy Vserossijskoj konferencii s mezhdunarodnym uchastiem i V ezhegodnyh chtenij pamyati O.A. Kataeva, IL SO RAN. Academic Press.

Bellusci, F., Pellegrino, G., & Musacchio, A. (2009). Different levels of inbreeding depression between outcrossing and selfing Serapias species. *Biologia Plantarum*, *53*(1), 175–178. doi:10.100710535-009-0029-8

Ben Amor, H., Elaoud, A., Ben Hassen, H., Ben Amor, T., Ben Salah, N., Stuerga, D., & Elmoueddeb, K. (2021). Experimental study and data analysis of the effects of iIons in water on evaporation under static magnetic conditions. *Arabian Journal for Science and Engineering*. Advance online publication. doi:10.100713369-021-05519-5

Ben Amor, H., Elaoud, A., Ben Salah, N., & Elmoueddeb, K. (2017). Effect of Magnetic Treatment on Surface Tension and Water Evaporation. *International Journal of Advance Industrial Engineering.*, *5*(3), 119–124. Doi.Org/10.14741/Ijae/5.3.4

Ben Salah, M., & Hellali, R. (1998). Metaxenic effects of nine pollinators on three palm date varieties (Phoenix dactylifera, L.) growing in Tunisia coastal oasis. In *First international conference on date palms*. United Arab Emirates Univ.

Bentley, J. W., & O'Neil, R. J. (1997). On the ethics of biological control of insect pests. *Agriculture and Human Values*, *14*(3), 283–289. doi:10.1023/A:1007477300339

Berling, M., Blachere-Lopez, C., Soubabere, O., Lery, X., Bonhomme, A., Sauphanor, B., & Lopez-Ferber, M. (2009, February 15). Cydia pomonella granulovirus Genotypes Overcome Virus Resistance in the Codling Moth and Improve Virus Efficiency by Selection against Resistant Hosts". *Applied and Environmental Microbiology*, *75*(4), 925–930. doi:10.1128/AEM.01998-08 PMID:19114533

Bhattacharjee, S. K. (2006): *Advances in Ornamental Horticulture.* Pointer.

Bhawalkar, V. U. (1993). The living soil. *Extended Abstracts of congress on Traditional Science & Technologies of India*, 10.5-10.8.

Biglari, F. (2009). *Assessment of antioxidant potential of date (phoenix dactylifera) fruits from iran, effect of cold storage and addition to minced chicken meat* (Unpublished Master thesis). University Sains Malaysia.

Bilir, N., Kang, K. S., Zang, D., & Lindgren, D. (2004). Fertility variation and status number between a base population and a seed orchard of *Pinus brutia* Ten. *Silvae Genetica*, *53*(1-6), 161–163. doi:10.1515g-2004-0029

Billany, D. J., & Brown, R. M. (1980). The web1spinning larch sawfly Cephalcia lariciphila Wachtl. (Hy1menoptera: Pamphiliidae) a new pest of Larix in En1gland and Wales. *Forestry*, *53*(1), 71–80. doi:10.1093/forestry/53.1.71

Bisbis, M. B., Gruda, N. S., & Blanke, M. M. (2019). Securing horticulture in a changing climate- A mini review. *Horticulturae, 5*(3), 56. doi:10.3390/horticulturae5030056

Bitar, Abu-Qaoud, & Al-Said. (2019). Studies on Date Palm Propagation by Offshoots. *Palestinian Journal of Technology & Applied Sciences,* (2), 62-68.

Blank, R. H., & Gill, G. S. C. (1997). Thrips (Thysanoptera: Terebrantia) on flowers and fruit of Citrus in New Zeland. *New Zealand Journal of Crop and Horticultural Science*, *25*(4), 319–332. doi:10.1080/01140671.1997.9514023

Blomefield, T. L. (1997). Managing resistance of codling moth, Cydia pomonella (L.) in South African pome fruit orchards. Combating Resistance, IARC-Rothamsted, Harpenden, 14-16.

Boavida, C., Neuenschwander, P., & Schulthess, F. (1992). Spatial distribution of *Rastrococcus invadens* Williams (Hom., Pseudococcidae) in mango trees. *Journal of Applied Entomology*, *114*(1-5), 381–391. doi:10.1111/j.1439-0418.1992.tb01141.x

Bohlen, P. J., & Edwards, C. A. (1995). Earthworm effects on N dynamics and soil respiration in microcosms receiving organic and inorganic nutrients. *Soil Biology & Biochemistry*, *27*(3), 341–348. doi:10.1016/0038-0717(94)00184-3

Bonetta, L. (2008). Epigenomics: Detailed analysis. *Nature*, *454*(7205), 795–798. doi:10.1038/454795a PMID:18685708

Borah, P. K., Kataky, M., Bhagabati, K. N., Pathak, J. J., & Bora, L. C. (2002). A new bacterial disease of orchid from India. *J Agri Sci Soc North-East India*, *15*, 1–4.

Bosančić, B., Mićić, N., Blanke, M., & Pecina, M. (2018). A main effects meta principal components analysis of netting effects on fruit: Using apple as a model crop. *Plant Growth Regulation*, *86*(3), 455–464. doi:10.100710725-018-0443-z

Bosily, M. A., Noaman, M. M., El-Banna, M. N., Azzam Clara, R., & Nassar, M. A. (2018). Breeding for barley resistance to leaf rust disease using marker-assisted selection. *Proceeding of The 7th Field Crops Research Institute Conference*, 397-437.

Bostanian, N. J., & Racette, G. (2001). Attract and kill, an effective technique to manage apple maggot, Rhagoletis pomonella [Diptera: Tephritidae] in high density Quebec apple orchards. *Phytoprotection*, *82*(1), 25–34. doi:10.7202/706212ar

Botes, J. (2018). *Impact of shade netting on internal and external quality of 'Nadorcott' mandarin fruit* (Doctoral dissertation). Stellenbosch University.

BP. (2007). *What is a Carbon Footprint?* http://www.bp.com/liveassets/bp_internet/globalbp/STAGING/global_assets/downloads/A/ABP_ADV_ what_on_earth_is_a_carbon_footprint.pdf

Brittain, C., Bommarco, R., Vighi, M., Settele, J., & Potts, S. G. (2010). Organic farming in isolated landscapes does not benefit flower-visiting insects and pollination. *Biological Conservation*, *143*(8), 1860–1867. doi:10.1016/j.biocon.2010.04.029

Brunt, A. A., Crabtree, K., Dallwitz, M. J., Gibbs, A. J., Watson, L., & Zurcher, E. J. (Eds.). (1997). *Plant viruses online: Descriptions and lists from the VIDE database*. http://biology.anu.edu.au/Groups/MES/vide/

Bulpitt, C. (2005). The uses and misuses of orchids in medicine. *QJM*, *98*(9), 625–631. doi:10.1093/qjmed/hci094 PMID:16006500

Burdon, R. D., & Wilcox, P. L. (2011). Integration of molecular markers in breeding. In *Genetics, genomics and breeding of conifers* (pp. 276–322). CRC Press.

Burnett, H.C. (1965). *Orchid diseases*. Fla Dept Agric Consumer Serv Div Plant Ind.

Burnett, H.C. (1974). *Orchid diseases*. Fla. Dept. Agr. and Consumer Serv., Div. of Plant Industry.

Burnett, H. C. (1957). Orchid diseases. *Proceedings of the Annual Meeting of the Florida State Horticultural Society*, *70*, 380–385. Retrieved September 20, 2021, from http://fshs.org/proceedings-o/1957-vol-70/380385%20%28BURNETT%29.pdf

Bush, G. L. (1966). The taxonomy, cytology, and evolution of the genus Rhagoletis in North America (Diptera, Tephritidae). *Bulletin of the Museum of Comparative Zoology*, *134*, 431–562.

Bush, G. L., & Smith, J. J. (1998). The genetics and ecology of sympatric speciation: A case study. *Researches on Population Ecology*, *40*(2), 175–187. doi:10.1007/BF02763403

Busov, V. B., Strauss, S. H., & Pilate, G. (2010). *Transformation as a tool for genetic analysis in Populus. Genetics and genomics of Populus*. Springer.

Butani, D. K., & Butani, D. C. (1979). Insects and fruits (No. 634 B8). Delhi, India: Periodical Export Book Agency.

Bǎders, & Purin, Lıbiete, Nartišs, & Jansons. (2014). Fragment ǎcijas ilgtermin, a dinamika meža ainav ǎbez cilv¯eka saimniecisk ǎs darb¯ıbas ietekmes [Long-term fragmentation dynamics in semi-natural forestlandscape]. *Mezzinatne*, *28*, 91–107.

Cahenzli, F., Sigsgaard, L., Daniel, C., Herz, A., Jamar, L., Kelderer, M., Jacobsen, S. K., Kruczyńska, D., Matray, S., Porcel, M., Sekrecka, M., Świergiel, W., Tasin, M., Telfser, J., & Pfiffner, L. (2019). Perennial flower strips for pest control in organic apple orchards-A pan-European study. *Agriculture, Ecosystems & Environment*, *278*, 43–53. doi:10.1016/j.agee.2019.03.011

Cai, R., Yang, H., He, J., & Zhu, W. (2009). The effects of magnetic fields on water molecular hydrogen bonds. *Journal of Molecular Structure*, *938*(1-3), 15–19. doi:10.1016/j.molstruc.2009.08.037

Campbell, A. J., Biesmeijer, J. C., Varma, V., & Wäckers, F. L. (2012). Realising multiple ecosystem services based on the response of three beneficial insect groups to floral traits and trait diversity. *Basic and Applied Ecology*, *13*(4), 363–370. doi:10.1016/j.baae.2012.04.003

Campillo, C., Fortes, R., Prieto, M. D. H., & Babatunde, E. B. (2012). Solar radiation effect on crop production.pp.167-194 *Solar Radiation*, *1*, 494.] Available from: https://www.intechopen.com/books/solar-radiation/solar-radiation-effect-on-crop-production

Cardé, R. T. (2007). *Using pheromones to disrupt mating of moth pests. In Perspectives in ecological theory and integrated pest management*. Cambridge University Press.

Carreño, E., Inocencio, C., Alcaraz, F., Ríos, S., Palazón, J., Vázquez, L., . . . Laguna, E. (2007). Morphological systematics of date-palm diversity (Phoenix, Arecaceae) in Western Europe and some preliminary molecular results. *V International Symposium on the Taxonomy of Cultivated Plants, 799*.

Carvalho, L. C., Osorio, M. L., Chaves, M. M., & Amancio, S. (2001). Chlorophyll inflorescences as an indicator of photosynthetic functioning of in vitro grapevine and chestnut plantlets under ex vitro acclimatization. *Plant Cell, Tissue and Organ Culture*, *67*(3), 271–280. doi:10.1023/A:1012722112406

Castrignano, A., Buttafuoco, G., Khosla, R., Mouazen, A. M., Moshou, D., & Naud, O. 2020. Agricultural internet of things and decision support for precision smart farming. Academic Press.

Cating, R. A., Hong, J. C., Palmateer, A. J., Stiles, C. M., & Dickstein, E. R. (2008). First report of bacterial soft rot on Vanda orchids caused by Dickeya chrysanthemi (Erwinia chrysanthemi) in the United States. *Plant Disease*, *92*(6), 977–977. doi:10.1094/PDIS-92-6-0977A PMID:30769750

Cating, R. A., & Palmateer, A. J. (2011). *Bacterial soft rot of Oncidium orchids caused by a Dickeya sp. (Pectobacterium chrysanthemi) in Florida*. Academic Press.

Cating, R. A., Palmateer, A. J., McMillan, R. T., & Dickstein, E. R. (2009b). First report of a bacterial Soft rot on Tolumnia orchids caused by a Dickeya sp. in the United States. *Plant Disease*, *93*(12), 1354. doi:10.1094/PDIS-93-12-1354B PMID:30759519

Cating, R. A., Palmateer, A. J., & McMillan, R. T. Jr. (2009a). First report of Sclerotium rolfsii on Ascocentrum and Ascocenda orchids in Florida. *Plant Disease*, *93*(9), 963. doi:10.1094/PDIS-93-9-0963B PMID:30754542

Cating, R. A., Palmateer, A. J., Stiles, C. M., Rayside, P. A., & Daviso, D. A. (2013). *Black rot of orchids caused by Phytophthora palmivora and Phytophthora cactorum.* Plant Pathology Department, Florida Cooperative Extension Service, Institute of Food and Agricultural Sciences, University of Florida.

Chandra, S., Bandopadhyay, R., Kumar, V., & Chandra, R. (2010). Acclimatization of tissue cultured plants: From laboratory to land. *Biotechnology Letters*, *32*(9), 1199–1205. doi:10.100710529-010-0290-0 PMID:20455074

Chandrasekaran, M., & Bahkali, A. H. (2013). Valorization of date palm (Phoenix dactylifera) fruit processing by-products and wastes using bioprocess technology–Review. *Saudi Journal of Biological Sciences*, *20*(2), 105–120. doi:10.1016/j.sjbs.2012.12.004 PMID:23961227

Chandrasekar, B. R., Ambrose, G., & Jayabalan, N. (2005). Influence of biofertilizers and nitrogen source level on the growth and yield of *Echinochloa frumentacea* (Roxb.). *Link. J. Agric. Technol.*, *1*, 223–234.

Chang, W. H., Yang, S. Y., Lin, C. L., Wang, C.-H., Li, P.-C., Chen, T.-Y., Jan, F.-J., & Lee, G.-B. (2013). Detection of viruses directly from the fresh leaves of a Phalaenopsis orchid using a microfluidic system. *Nanomedicine; Nanotechnology, Biology, and Medicine*, *9*(8), 1274–1282. doi:10.1016/j.nano.2013.05.016 PMID:23751373

Chan, Y.-L., Lin, K.-H., & Liao, S. (2005). Gene stacking in Phalaenopsis orchid enhances dual tolerance to pathogen attack. *Transgenic Research*, *14*(3), 279–288. doi:10.100711248-005-0106-5 PMID:16145836

Chao, C. T., & Krueger, R. R. (2007). The date palm (Phoenix dactylifera L.): Overview of biology, uses, and cultivation. *HortScience*, *42*(5), 1077–1082.

Chao, C. T., & Krueger, R. R. (2007). the date palm (Phoenix dactylifera L.): Overview of biology, uses, and cultivation. *HortScience*, *42*(5), 1077–1082. doi:10.21273/HORTSCI.42.5.1077

Chao, Y., Li, L., Girodat, D., Förstner, K. U., Said, N., Corcoran, C., Śmiga, M., Papenfort, K., Reinhardt, R., Wieden, H.-J., Luisi, B. F., & Vogel, J. (2017). *In Vivo* Cleavage Map Illuminates the Central Role of RNase E in Coding and Non-coding RNA Pathways. *Molecular Cell*, *65*(1), 39–51. doi:10.1016/j.molcel.2016.11.002 PMID:28061332

Charati, S. N., Pokharkar, D. S., & Ghorpade, S. A. (2003). Abundance of spiralling whitefly, a newly introduced pest in Maharashtra State. *Journal of Maharashtra Agricultural Universities (India).*

Charmillot, Gourmelon, Fabre, & Pasquier. (2001). Ovicidal and larvicidal effectiveness of several insect growth inhibitors and regulators on the codling moth Cydia pomonella L. (Lep., Tortricidae). *Journal of Applied Entomology*, *125*(3), 147–153.

Chehaibi, S., Hannachi, C., Pieters, J. G., & Verschoore, R. A. (2008). Impacts de la vitesse d'avancement du tracteur sur la structure du sol et le rendement d'une culture de pomme de terre. *Trpiculture*, *26*(3), 195–199.

Chellachamy, V., & Dinakaran, S. (2015). A comparative study on ver- micomposting of epicarp of fruits (pomegranate and Sathukudi) using earthworm *Eisenia foetida*. *International Journal of Recent Scientific Research*, *6*(3), 3125-3129.

Chen, C., Chang, S., Hou, R. F., & Cheng, L. (1996). Deterrent effect of the chinaberry extract on oviposition of the diamondback moth, Plutella xylostella (L.) (Lep., Yponomeutidae). *J. Appl. Ent.*, *120*(1-5), 165–169. doi:10.1111/j.1439-0418.1996.tb01585.x

Cheng, Q. Q. (2011). *Studies on leaf culture and induction of octoploid of Phalaenopsis cultivars* [M. D. Dissertation]. Shantou University.

Chen, J. S., & Hsieh, S. P. Y. (1978). Phytophthora black rot of Phalaenopsis in Taiwan. *Plant Prot Bull, 20*, 161–170.

Chen, P., Taylor, N. J., Dueker, K. G., Keifer, I. S., Wilson, A. K., McGuffy, C. L., Novitsky, C. G., Spears, A. J., & Holbrook, W. S. (2016). pSIN: A scalable, Parallel algorithm for Seismic INterferometry of large-N ambient-noise data. *Computers & Geosciences, 93*, 88–95. doi:10.1016/j.cageo.2016.05.003

Chen, W. H., Hsu, C. Y., Cheng, H. Y., Chang, H., Chen, H. H., & Ger, M. J. (2011). Downregulation of putative UDP-glucose: Flavonoid 3-O-glucosyltransferase gene alters flower coloring in Phalaenopsis. *Plant Cell Reports, 30*(6), 1007–1017. doi:10.100700299-011-1006-1 PMID:21274540

Childers, C. C., & Nakahara, S. (2006). Thysanoptera(thrips) within citrus orchards in Florida: Speciesdistribution, relative and seasonal abundance within trees, and species on vines and ground cover plants. *Journal of Insect Science, 6*(45), 3–7. doi:10.1673/031.006.4501 PMID:20233100

Chinsamy, M., Kulkarni, M. G., & Staden, J. V. (2013). Garden-waste-vermicompost leachate alleviates salinity stress in tomatoseedlings by mobilizing salt tolerance mechanisms. *Plant Growth Regulation, 71*(1), 41–47. doi:10.100710725-013-9807-6

Cho, Y. I., Lee, S., Kim, W., & Suh, S. (2003). *Physical water treatment for the mitigation of mineral fouling in cooling-tower water applications*. Engineering Conferences International, ECI Digital Archives. https://dc.engconfintl.org/heatexchanger/4

Chocarro-Ruiz, B., Fernández-Gavela, A., Herranz, S., & Lechuga, L. M. (2017). Nanophotonic label-free biosensors for environmental monitoring. *Current Opinion in Biotechnology, 45*, 175–183. doi:10.1016/j.copbio.2017.03.016 PMID:28458110

Chuenchitt, S., Dhirabhava, W., Karnjanarat, S., Buangsuwon, D., & Uematsu, T. (1983). A new bacterial disease on orchids Dendrobium sp. caused by Pseudomonas gladioli. *Witthayasan Kasetsat Witthayasat, 17*, 26–36.

Chupp, C. (1953). *A monograph of the genus Cercospora*. Cornell University Press.

Ciglar, I. (1998). *Integrirana zaštita voćaka i vinove loze. Zrinski*. Čakovec.

Clapperton, M. J., Lee, N. O., Binet, F., & Conner, R. L. (2001). Earthworms indirectly reduce the effect of take-all (Gaeumannomyces graminis var. tritici) on soft white spring wheat (Triticium aestivum cv. Fielder). *Soil Biology & Biochemistry, 33*(11), 1531–1538. doi:10.1016/S0038-0717(01)00071-2

Cocco, A., Deliperi, S., & Delrio, G. (2013). Control of Tuta absoluta (Meyrick)(Lepidoptera: Gelechiidae) in greenhouse tomato crops using the mating disruption technique. *Journal of Applied Entomology, 137*(1-2), 16–28. doi:10.1111/j.1439-0418.2012.01735.x

Cochran, W. G. (2007). *Sampling techniques*. John Wiley & Sons.

Cohen, S., & Fuchs, M. (1987). The distribution of leaf area, radiation, photosynthesis and transpiration in a Shamouti orange hedgerow orchard. Part I. Leaf area and radiation. *Agricultural and Forest Meteorology, 40*(2), 123-144. doi:10.1016/0168-1923(87)90002-5

Cohen, Y., Cohen, A., Hetzroni, A., Alchanatis, V., Broday, D., Gazit, Y., & Timar, D. (2008). Spatial decision support system for Medfly control in citrus. *Computers and Electronics in Agriculture, 62*(2), 107-117.

Colauto Stenzel, N. M., Vieira, J. N. C. S., Marur, C. J., Santos, S. M. B., & Gomes, J. C. (2006). Maturation curves and degree-days accumulation for fruits of 'folha murcha' orange trees. *Scientia Agrícola*, *63*(3), 219–225. doi:10.1590/S0103-90162006000300002

Coleman, G. (2012). Transmission of cymbidium mosaic virus in oncidium orchids by periplaneta australasiae. Faculty of the Graduate School of the University of Maryland, College Park.

Coll, M., & Abd-Rabou, Sh. (1998). Effect of oil emulsion sprays on parasitoids of the black parlatoria, Parlatoria ziziphi, in grapefruit. *BioControl*, *43*(1), 29–37. doi:10.1023/A:1009974330554

Committee on the Status of Pollinators in North America NRC. (2007). *Status of pollinators in North America*. National Academies Press.

Conti, F. (2001). Monitoring of *Pezothrips kellyanus* on citrus in eastern Sicily. *Proceedings of the 7th International Symposium on Thysanoptera*.

Cook, J. P., McMullen, M. D., Holland, J. B., Tian, F., Bradbury, P., Ross-Ibarra, J., Buckler, E. S., & Flint-Garcia, S. A. (2012). Genetic architecture of maize kernel composition in the nested association mapping and inbred association panels. *Plant Physiology*, *158*(2), 824–834. doi:10.1104/pp.111.185033 PMID:22135431

Cotxarrera, L., Trillas-Gayl, M. I., Steinberg, C., & Alabouvette, C. (2002). Use of sewage sludge compost and Trichoderma asperellum isolates to suppress Fusarium wilt of tomato. *Soil Biology & Biochemistry*, *34*(4), 467–476. doi:10.1016/S0038-0717(01)00205-X

Couture, I. (2004). *Analyse d'eau pour fin d'irrigation*. MAPAQ Montérégie-Est. https://www.agrirescau.net/petitsfruits/documents/61548/analyse-d_eau-pour-fin-d_irrigation

Couture, I. (2006). *Principaux critères pour évaluer la qualité de l'eau en micro-irrigation Colloque sur l'irrigation l'eau, source de qualité et de rendement*. Boucherville.

Craig, A. M. (2009). *Investigation of the quality of water treated by magnetic fields; Bachelor of Engineering (Environmental)*. University of Southern, Queensland Faculty of Engineering and Surveying.

Cramer, W. A., Zakharov, S. D., Hasan, S. S., Zhang, H., Baniulis, D., & Zhalnina, M. V. (2011). Membrane proteins in four acts: Function precedes structure determination. *Methods (San Diego, Calif.)*, *55*(4), 415–420. doi:10.1016/j.ymeth.2011.11.001 PMID:22079407

Cronjé, P. J. R., Botes, J., Prins, D. M., Brown, R., North, J., Stander, O. P. J., Hoffman, E. W., Zacarìas, L., & Barry, G. H. (2020). The influence of 20% white shade nets on fruit quality of 'Nadorcott' mandarin. *Acta Horticulturae*, (1268), 279–284. doi:10.17660/ActaHortic.2020.1268.36

Crosbie, T. M., Eathington, S. R., Johnson, G. R., Edwards, M., & Reiter, R. (2003). Plant breeding: past, present, and future. In *Plant breeding: the Arnel R. Hallauer International Symposium*. Blackwell.

Čuček, L., Kravanja, Z., & Klemeš, J. J. (2012). A review of footprint analysis tools for monitoring impacts on sustainability. *Journal of Cleaner Production*, *34*, 9–20. doi:10.1016/j.jclepro.2012.02.036

Cunningham, I. C. (1984, November). Mango insect pests. In *Proceedings of the First Australian Mango Research Workshop* (pp. 211-224). Academic Press.

Dag, A., Eisenstin, D., & Gazit, S. (2000). Effect of temperature regime on pollen and the effective pollination of kent mango in Israel. *Scientia Horticulturae*, *86*(1), 1–11. doi:10.1016/S0304-4238(99)00134-X

Dag, A., & Gazit, S. (2000). Mango pollinators in Israel. *Journal of Applied Horticulture*, *2*(1), 39–43. doi:10.37855/jah.2000.v02i01.12

Daly, A., Condé, B., & Duff, J. (2013). Orchid disease in the Northern territory. Northern Territory Government.

Darwesh Rasmia, S. S., Abeer Abd-El Kareim, H. E., & Mona, H. M. (2014). Effect of Foliar Spraying With 5- Aminolevulinic Acid and Different Types Amino Acids on Growth of Date Palm of Plantlets afte Acclimatization in the Green House. *International Journal of Plant and Soil Science*, *3*(10), 1317–1332.

Darwesh Rasmia, S. S., Eman Zayed, M., & Hala Farag, M. A. (2021). Influence of different carbohydrate source on improving in vitro rooting, growth and ex vitro survival of date palm Plantlet. *Plant Cell Biotechnology and Molecular Biology*, *22*(29&30), 109–122.

Darwesh, El-Banna, & Zayed. (2016). Enhancing growth of date palm plantlets by different fertilizer treatments in the acclimatization greenhouse. *International Journal of Scientific Research & Engineering Trends*, *2*(1), 42–49.

Darwesh. (2015). Morphology, physiology and anatomy in vitro affected acclimatization ex vitro date palm plantlets: A Review. International Journal of Chemical. *Environmental & Biological Sciences*, *3*(2), 175–182.

Darwesh, R. S. S., & Mohamed, H. f. (2009). The adverse effect of GA3 (Gibberellic acid) on salinity of of *Phoenix dactylifera* L. Plantlets in vitro rooting stage. *4th conference on Recent Technology in Agriculture*, 1-8.

Darwesh, R. S. S., Zaid, Z. E., & Sidky, R. A. (2011). Effect of Ammonium Nitrate and GA_3 on Growth and Development of Date Palm Plantlets in Vitro and Acclimatization Stage. *Research Journal of Agriculture and Biological Sciences*, *7*(1), 17–22.

Das, A. K., Rahman, M. A., Saha, S. R., Sarmin, N. S., Hoque, M. A., & Bhuiyan, F. (2020). Transforming Malta orchard into agroforestry system with different crops for improving productivity, profitability and land uses. *Ann. Bangladesh Agric., 24*(1), 113-125. www.doi.org/10.3329/aba.v24i1.51940

Das, G., Patra, J. K., & Baek, K. H. (2017). Insight into MAS: A molecular tool for the development of stress-resistant and quality rice through gene stacking. *Front. Plant Sci.*, *8*, 1321. doi:10.3389/fpls.2017.01321 PMID:28775736

daSilva, J. A. T., Chin, D. P., Van, P. T., & Mii, M. (2011). Transgenic orchids. *Scientia Horticulturae*, *130*(4), 673–680. doi:10.1016/j.scienta.2011.08.025

daSilva, J. A. T., Dobranszki, J., Cardoso, J. C., Chandler, S. F., & Zeng, S. (2016). Methods for genetic transformation in Dendrobium. *Plant Cell Reports*, *35*(3), 483–504. doi:10.100700299-015-1917-3 PMID:26724929

Daum, T. (2018). *ICT Applications in Agriculture. Encyclopedia of Food Security and Sustainability*. Elsevier.

Davies, T. J., Barraclough, T. G., Chase, M. W., Soltis, P. S., Soltis, D. E., & Savolainen, V. (2004). Darwin's abominable mystery: Insights from a supertree of the angiosperms. *Proceedings of the National Academy of Sciences of the United States of America*, *101*(7), 1904–1909. doi:10.1073/pnas.0308127100 PMID:14766971

de Chandra, L., Pathak, P., Rao, A. N., & Rajeevan, P. K. (2019). Breeding approaches for improved genotypes. Commercial Orchids.

de Paula, Figueiredo, & de Paula. (2014). Physiological changes in eucalyptus hybrids under different irrigation regimes. *Revista Ciencia Agronomica*, *45*(4), 805–814. doi:10.1590/S1806-66902014000400019

De Souza, A., Garcí, D., Sueiro, L., Gilart, F., Porras, E., & Licea, L. (2006). Pre-sowing magnetic treatments of tomato seeds increase the growth and yield of plants. *Bioelectromagnetics: Journal of the Bioelectromagnetics Society, The Society for Physical Regulation in Biology and Medicine. The European Bioelectromagnetics Association*, *27*(4), 247–257. doi:10.1002/bem.20206

DeBach, P. (1964). *Biological control of insect pests and weeds*. Academic Press.

Debbarma, T., & Hath, T. K. (2021). Pest Complex of Lemon CV. Assam Lemon (*Citrus limon* L. Burm) in Terai Region of West Bengal, India. *International Journal of Current Microbiology and Applied Sciences*, *10*(04), 736–744. doi:10.20546/ijcmas.2021.1004.075

De, L. C., Pathak, P., Rao, A. N., & Rajeevan, P. K. (2014). *Commercial Orchids*. Walter de Gruyter GmbH & Co KG.

Department of Climate Change and Energy Efficiency (DCCEE). (2011). National greenhouse gas account factors. Commonwealth of Australia.

Depommier, D. (2003). The tree behind the forest: Ecological and economic importance of traditional agroforestry systems and multiple uses of trees in India. *Tropical Ecology*, *44*(1), 63–71.

Dhanapal, R., Kumar, D. V. S. R., Lakshmipathy, R., Rani, C. S., & Kumar, R. M. (2019). Pathogenicity testing of indigenous isolates of entomopathogenic fungus, *Lecanicillium lecanii* against tobacco caterpillar, *Spodoptera litura. Journal of Experimental Zoology India*, *22*(2), 753–756.

Diatta, I. L. D., Kane, A., Agbangba, C. E., Sagna, M., Diouf, D., Aberlenc-Bertossi, F., Duval, Y., Borgel, A., & Sane, D. (2014). Inoculation with arbuscular mycorrhizal fungi improves seedlings growth of two sahelian date palm cultivars (*Phoenix dactylifera* L., cv. Nakhla hamra and cv. Tijib) under salinity stresses. *Advances in Bioscience and Biotechnology*, *5*, 64–72.

Diengdoh, R. V., Kumaria, S., Tandon, P., & Das, M. C. (2017). Asymbiotic germination and seed storage of Paphiopedilum insigne, an endangered lady's slipper orchid. *SAJB*, *112*, 215–224. doi:10.1016/j.sajb.2017.05.028

Dillard, C. J., & German, J. B. (2000). Phytochemicals: Nutraceuticals and human health. *Journal of the Science of Food and Agriculture*, *80*(12), 1744–1756. doi:10.1002/1097-0010(20000915)80:12<1744::AID-JSFA725>3.0.CO;2-W

Dobeie, Abbas, Soliman, & Azzam. (2017). *In vitro* screening of some Egyptian and Nigerian peanut genotypes for salt tolerance. Egypt. *J. Plant Breed.*, *21*(6), 1035–1050. doi:10.12816/0046384

Dodge, O. B., & Ricket, H. W. (1943). Orchidaceae (orchids). In Diseases and pests of ornamental plants. Jacques Cattell Press.

Doitsidis, L., Fouskitakis, G. N., Varikou, K. N., Rigakis, I. I., Chatzichristofis, S. A., Papafilippaki, A. K., & Birouraki, A. E. (2017). Remote monitoring of the Bactrocera oleae (Gmelin)(Diptera: Tephritidae) population using an automated McPhail trap. *Computers and Electronics in Agriculture*, *137*, 69–78. doi:10.1016/j.compag.2017.03.014

Dominguez, J., & Edwars, C. A. (2004). Vermicomposting organic wastes: A review. In S. H. Shakir Hanna & W. Z. A. Mikhatl (Eds.), Soil Zoology for Sustainable Development in the 21st Century (pp. 369–395). Academic Press.

Donnet, L. (1988). *Les aimants pour votre santé*. L'eau merveilleuse.

Doube, B. M., Stephens, P. M., Davorena, C. W., & Ryderb, M. H. (1994). Interactions between earthworms, benefi cial soil microorganisms and root pathogens. *Applied Soil Ecology*, *1*(1), 3–10. doi:10.1016/0929-1393(94)90018-3

Dovjik, I., Nemera, D. B., Cohen, S., Shahak, Y., Shlizerman, L., Kamara, I., Florentin, A., Ratner, K., McWilliam, S.C., Puddephat, I.J., FitzSimons, T.R., Charuvi, D. & Sadka, A. (2021). Top Photoselective Netting in Combination with Reduced Fertigation Results in Multi-Annual Yield Increase in Valencia Oranges (Citrus sinensis). *Agronomy*, *11*(10), 2034. doi:10.3390/agronomy11102034

Dubey, K. (2010). *Development of Agro-forestry models for Eastern Uttar Pradesh. Project Report*. ICFRE.

Duchovskis, P., Stanys, V., Sasnauskas, A., & Bobinas, C. (2007). Cold resistance of *Prunus domestica* L. and *Prunus cerasifera* Ehrh. in Lithuania. *Acta Horticulturae*, (734), 299–303. doi:10.17660/ActaHortic.2007.734.39

Duff, J. D. (1993). Orchid diseases of the Northern Territory. *Agnote Darwin*, *568*, 3.

Dumroese, R. D., Deborah, S., Dumroese, P., & Brown, R. E. (2011). Allometry, nitrogen status, and carbon stable isotope composition of Pinus ponderosa seedling in two growing media with contrasting nursery irrigation regimes. *Canadian Journal of Forest Research*, *41*, 1091–1101.

Dungey, Yanchuk, & Burdon. (2014). A 'Reality Check' in the Management of Tree Breeding Programs. In Challenges and Opportunities for the World's Forests in the 21st Century. Forestry Sciences. doi:10.1007/978-94-007-7076-8_19

Dvaranauskaitė, A., Venskutonis, P. R., Raynaud, C., Talou, T., Viškelis, P., & Sasnauskas, A. (2009). Variations in the essential oil composition in buds of six blackcurrant (*Ribes nigrum* L.) cultivars at various development phases. *Food Chemistry*, *11*(2), 671–679. doi:10.1016/j.foodchem.2008.10.005

Dwivedi, P., Amin, D., & Sharma, A. (2020). Effect of differential concentration of micronutrient copper and zinc on *in vitro* morphogenesis of *Foeniculum vulgare* Mill. *Plant Physiol. Rep.*, *25*(1), 178–184. doi:10.100740502-019-00478-4

Dwtvedi, S. & Garg, S. (1997). Screening of plant extracts for ovicidal effect on the rice moth, Corcyra cephalonica (Stainton). *Pest Management and Economic Zoology*, *5*(1), 53–55.

Dwyer, M. J., Evernden, D. W., & Mcallister, M. (1976). Handbook of Agricultural Tyre Performance (2nd ed.). NIAE.

Ebadi, H., Raeini-Sarjaz, M., & Gholami Sefidkoohi, M. A. (2018). Vegetative growth and water use efficiency of citrus seedlings in simultaneous application of shading and partial root zone drying. *Iranian Journal of Horticultural Science*, *49*(2), 417-428. doi:10.22059/IJHS.2017.230309.1218

Echegaray, N., Pateiro, M., Gullon, B., Amarowicz, R., Misihairabgwi, J. M., & Lorenzo, J. M. (2020). Phoenix dactylifera products in human health–A review. *Trends in Food Science & Technology*, *105*, 238–250. doi:10.1016/j.tifs.2020.09.017

EC-JRC. (2011). International Reference Life Cycle Data System (ILCD) Handbook- Recommendations for Life Cycle Impact Assessment in the European context. European Commission-Joint Research Centre - Institute for Environment and Sustainability.

Edwards, C. A., Dominguez, J., & Arancon, N. Q. (2004). The infl uence of vermicomposts on pest and diseases. In S. H. Shakir Hanna & W. Z. A. Mikhail (Eds.), Soil zoology for sustainable development in the 21st centuary (pp. 397–418). Academic Press.

Edwards, C. A. (1998). The use of earthworms in processing organic wastes into plant growth media and animal feed protein. In C. A. Edwards (Ed.), *Earthworm ecology* (pp. 327–354). CRC Press.

Edwards, C. A., Arancon, N. Q., Emerson, E., & Pulliam, R. (2007). Supressing plant parasitic nematodes and arthropod pests with vermicompost teas. *BioCycle*, *48*, 38–39.

Edwards, C. A., Arancon, N. Q., & Sherman, R. (2011). *Vermiculture technology*. Taylor and Francis Group, LLC.

Edwards, C. A., & Bohlen, P. J. (1996). *Biology and ecology of earthworms* (3rd ed.). Chapman & Hall.

El Fiorenza, Sharma, Ranjan, & Shashank. (2018). Smart e-agriculture montoring based on Arduino using IOT. *IJSDR, 3*(10).

El Hadrami, A., & Al-Khayri, J. M. (2012). Socioeconomic and traditional importance of date palm. *Emirates Journal of Food and Agriculture, 24*(5), 371–385.

El Hadrami, I., & El Hadrami, A. (2009). Breeding date palm. In *Breeding plantation tree crops: tropical species* (pp. 191–216). Springer.

Elaoud, A., & Chehaibi, S. (2011). Soil compaction due to tractor traffic. *Journal of Failure Analysis and Prevention, 11*(5), 539–545. doi:10.100711668-011-9479-3

El-Bahr, M. K., Ali, Z. A., & Saker, M. M. (2003). A comparative anatomical study of date palm vitro plants. *Arab Journal of Biotechnology, 7*(2), 219–228.

El-Daen. (2019). Effect of fertilization by injection of soil and trunk with NPK on productivity and fruits quality of Sewy date palm. *J. Agric. Res, & Dev., 33*(1).

El-Dengawy, E. F. A., Wanas, A. L. E., & Mervat, H. M. (2017). Improvement of the rooting efficiency and vegetative growth in date palm offshoots by Licorice root extract and auxins mixture applications. J. Plant Production, Mansoura Univ., 8(7), 789 – 796.

El-Gayar, S., Negm, A., & Abdrabbo, M. (2019). Greenhouse Operation and Management in Egypt. Handbook of Environmental Chemistry, 74, 489–560.

El-Geddawy, Azzam, & Khalil. (2008). Somaclonal variation in sugarcane through tissue culture and subsequent screening for molecular polymorphisms. In *The 3rd International Conference "Meeting the Challenges of Sugar Crops & Integrated Industries in Developing Countries."* Sinai University.

El-Ghayaty, S. H. (1983). Effect of different pollinators on fruit setting and some fruit properties of the Siwi and Ahmadi dates. In *Proceedings 1st symposium on date palm.* King Faisal Univ.

El-Hammady, A. M., Khalifa, A. S., & Montasser, A. S. (1991). Effect of potash fertilization on Sewy date palms. II. Effect on yield and fruit quality. *Egyptian Journal of Horticulture, 18*, 199–210.

El-Kassaby, Y. A., Cappa, E. P., Liewlaksaneeyanawin, C., Klápště, J., & Lstiburek, M. (2011). Breeding without breeding: Is a complete pedigree necessary for efficient breeding? *PLoS One, 6*(10), e25737. doi:10.1371/journal.pone.0025737 PMID:21991342

El-Khawaga, A. S. (2013). Effect of anti-salinity agents on growth and fruiting of different date palm cultivars. *Asian Journal of Crop Science, 5*(1), 65–80.

El-Kosary, Bakr, Hussien, & Sheren. (2008). Effect of the irrigation system and soil components on the growth of date palm offshoots in the nursery. *Egyptian Journal of Applied Sciences, 23*(8), 602-620.

Elloumi, M. (2016). *La gouvernance des eaux souterraines en Tunisie*. IWMI Project report. http://gwmena.iwmi.org/wpcontent/uploads/sites/3/2017/04/Rep.7-Groundwater-governance-inTunisia_final_cover.pdf

Elmer, W. H. (2009). Influence of earthworm activity on soil microbes and soilborne diseases of vegetables. *Plant Disease, 93*, 175–179.

El-Salhy, A. M. (2000). Effect of bagging the spathes of Zaghloul date productivity under Assiut conditions. *Assiut J. Agric. Sci., 31*(3), 123–134.

Elsayd, El-Merghany, & El-Dean. (2018). Influence of Potassium Fertilization on Barhee Date Palms Growth, Yield and Fruit Quality Under Heat Stress Conditions. *J. Plant Production. Mansoura Univ.*, *9*(1), 73–80.

El-Sonbaty, A. E. (2021). Effect of organic matter combined with mineral N fertilizers different rates with or without magnetic water on soil fertility and maize productivity. *Plant Cell Biotechnology and Molecular Biology*, *22*(67 &68), 96–112.

El-Zoghbi, M. (1994). Biochemical changes in some tropical fruits during ripening. *Food Chemistry*, *49*(1), 33–37. doi:10.1016/0308-8146(94)90229-1

ENCODE Project Consortium. (2012). An integrated encyclopaedia of DNA elements in the human genome. *Nature*, *489*(7414), 57–74. doi:10.1038/nature11247 PMID:22955616

Energetics. (2007). *The Reality of Carbon Neutrality*. www.energetics.com.au/file?node_id=21228

Engelmann, J., & Hamacher, J. (2008). Plant Virus Diseases: Ornamental Plants. In Encyclopedia of Virology (3rd ed.). Academic Press.

Esmaeilifar, A. (2013). Influence Of Irrigation, Phosphorus And Mycorhiza On Date Palm. *Advances in Environmental Biology*, *7*(1), 123–130.

Esser, R. P., O'Bannon, J. H., & Clark, R. A. (1988). Procedures to detect foliar nematodes from annual nursery or out of state inspections. Nematology Circular No. 160. Fla. Dept. of Consumer Service Division of Plant Industry.

ETAP. (2007). *The Carbon Trust Helps UK Businesses Reduce their Environmental Impact*. Press Release. https://ec.europa.eu/environment/etap/pdfs/jan07_carbon_trust_initiative.pdf

Evans & Goodwin (2011). The role of insect pollinators in avocado (Persea americana) pollination in New Zealand and Australia. *Proceedings VII World Avocado Congress 2011 (Actas VII Congreso Mundial del Aguacate 2011).*

Evans, A., & Jacquemyn, H. (2022). Range Size and Niche Breadth as Predictors of Climate-Induced HabitatChange in Epipactis (Orchidaceae). *Frontiers in Ecology and Evolution*, *10*, 894616. doi:10.3389/fevo.2022.894616

Ewing, B., Goldfinger, S., Wackernagel, M., Stechbart, M., Stechbart, S., Rizk, M., Reed, A. & Kitzes A. (2008). *The Ecological Footprint Atlas*. Global Footprint Network.

Ezz Gadalla, G., Rasmia Darwesh, S., Rehab Sidky, A., & Zeinab Zaid, E. (2008). Growth and Developing of date palm plantlet for pre- acclimatization stage: The effect of growth retardants and vessel types. *Egyptian Journal of Biotechnology*, *29*(6), 173–189.

Fabre, J., & Dereuddre, J. (1990). Encapsulation-dehydration: A new approach to cryopreservation of Solanum shoots tips. *Cryo Letters*, *11*, 413–426.

Fajardo, A. C., Medina, J. R., Opina, O. S., & Cervancia, C. R. (2008). Insect pollinators and floral visitors of mango. *Philippine Agricultural Scientist*, *91*, 372–382.

Falconer, D. S., & Mackay, T. F. C. (1996). *Introduction to quantitative genetics*. Longman.

FAO. (2008). Phocaides, Manuel des techniques d'irrigation sous pression. FAO.

FAO. (2021). Climate-smart agriculture case studies, Projects from around the world. FAO.

Fathy, M. A., Gabr, M. A., & El Shall, S. A. (2010). Effect of Humic Acid Treatments on 'Canino' Apricot Growth, Yield and Fruit Quality. *New York Science Journal*, *3*(12), 109–115.

Fatima, G., Khan, I. A., Jaskani, M. J., & Ul-Ain Rasool, Q. (2010). studies on different cultivars of date palm (Phoenix dactylifera L.) and their comparative root anatomy. *Science International (Lahore)*, *24*(2), 177–180.

Fay, M., & Rankou, H. (2016). *Slipper orchids on the IUCN Red List. In: 2015 Annual Report to the Environment Agency—Abu Dhabi*. Framework Support for Implementing the Strategic Plan of the IUCN Species Survival Commission.

Feder, J. L., Chilcote, C. A., & Bush, G. L. (1988). Genetic differentiation between sympatric host races of Rhagoletis pomonella. *Nature*, *336*, 61–64. doi:10.1038/336061a0

Feng, S., Cokus, S. J., Zhang, X., Chen, P.-Y., Bostick, M., Goll, M. G., Hetzel, J., Jain, J., Strauss, S. H., Halpern, M. E., Ukomadu, C., Sadler, K. C., Pradhan, S., Pellegrini, M., & Jacobsen, S. E. (2010). Conservation and divergence of methylation patterning in plants and animals. *Proceedings of the National Academy of Sciences of the United States of America*, *107*(19), 8689–8694. doi:10.1073/pnas.1002720107 PMID:20395551

Filippo, G. (2017). Molecular-assisted breeding. In P. Roberto & G. Giuseppe (Eds.), *More Food: Road to Survival* (pp. 373–398).

Fillatti, J. J., Sellmer, J., McCown, B., Haissig, B., & Comai, L. (1987). Agrobacterium mediated transformation and regeneration of Populus. *Molecular & General Genetics*, *206*(2), 192–199. doi:10.1007/BF00333574

Firake, D. M., Behere, G. T., Deshmukh, N. A., Firake, P. D., & Thakur, N. A. (2013). Recent scenario of insect-pests of guava in North East India and their eco-friendly management. *Indian Journal of Hill Farming*, *26*(1), 55–57.

Fitzgerald, J. D., & Solomon, M. G. (2004). Can flowering plants enhance numbers of beneficial arthropods in UK apple and pear orchards? *Biocontrol Science and Technology*, *14*(3), 291–300. doi:10.1080/09583150410001665178

Fki, L., Masmoudi, R., Kriaa, W., Mahjoub, A., Sghaier, B., Mzid, R., Mliki, A., Rival, A., & Drira, N. (2011). Date Palm Micropropagation via Somatic Embryogenesis. In Date Palm Biotechnology. Springer.

Fladung, M., Altosaar, I., Bartsch, D., Baucher, M., Boscaleri, F., Gallardo, F., Häggman, H., Hoenicka, H., Nielsen, K., Paffetti, D., Séguin, A., Stotzky, G., & Vettori, C. (2012). European discussion forum on transgenic tree biosafety. *Nature Biotechnology*, *30*(1), 37–38. doi:10.1038/nbt.2078 PMID:22231091

Flanagan, S. P., Forester, B. R., Latch, E. K., Aitken, S. N., & Hoban, S. (2018). Guidelines for planning genomic assessment and monitoring of locally adaptive variation to inform species conservation. *Evolutionary Applications*, *11*(7), 1035–1052. doi:10.1111/eva.12569 PMID:30026796

Fletcher, T. B. (1914). *Some South Indian Insects and Other Animals of Importance: Considered Especially from an Economic Point of View/by T. Bainbrigge Fletcher*. Bishen Singh Mahendra Pal Singh.

Flint, M. L., & Van den Bosch, R. (2012). *Introduction to integrated pest management*. Springer Science & Business Media.

Fonte, A., Garcerá, C., Tena, A., & Chueca, P. (2021). Volume Rate Adjustment for Pesticide Applications against Aonidiella aurantii in Citrus: Validation of CitrusVol in the Growers' Practice. *Agronomy (Basel)*, *11*(7), 1350. doi:10.3390/agronomy11071350

Food and Agriculture Organization of the United Nations (FAO). (2016). *Citrus fruit fresh and processed statistical bulletin*. FAO.

Franck, P., Reyes, M., Olivares, J., & Sauphanor, B. (2007). Genetic archi1tecture in codling moth populations: Comparison between microsatellite and insecticide resistance markers. *Molecular Ecology*, *16*(17), 3554–3564. doi:10.1111/j.1365-294X.2007.03410.x PMID:17845430

Funda, T., Lstiburek, M., Klapšte, J., & El-Kassaby, Y. A. (2012). Optimization of genetic gain and diversity in seed orchard crops considering variation in seed germination. *Scandinavian Journal of Forest Research, 27*(8), 787–793. do i:10.1080/02827581.2012.686627

Gaire, S., Albrecht, U., Batuman, O., Qureshi, J., Zekri, M., & Alferez, F. (2021). Individual protective covers (IPCs) to prevent Asian citrus psyllid and Candidatus Liberibacter asiaticus from establishing in newly planted citrus trees. *Crop Protection,* 105862. doi:10.1016/j.cropro.2021.105862

Gallai, N., Salles, J. M., Settele, J., & Vaissiere, B. E. (2009). Economic valuation of the vulnerability of world agriculture confronted with pollinator decline. *Ecological Economics, 68*(3), 810–821. doi:10.1016/j.ecolecon.2008.06.014

Gantait, S., El-Dawayati, M. M., Panigrahi, J., Labrooy, C., & Verma, S. K. (2018). The retrospect and prospect of the applications of biotechnology in Phoenix dactylifera L. *Applied Microbiology and Biotechnology, 102*(19), 8229–8259.

García, R. R., & Miñarro, M. (2014). Role of floral resources in the conservation of pollinator communities in cider-apple orchards. *Agriculture, Ecosystems & Environment, 183*, 118–126. doi:10.1016/j.agee.2013.10.017

García-Sánchez, F., Simón, I., Lidón, V., Manera, F. J., Simón-Grao, S., Pérez-Pérez, J. G., & Gimeno, V. (2015). Shade screen increases the vegetative growth but not the production in 'Fino 49' lemon trees grafted on Citrus macrophylla and Citrus aurantium L. *Scientia Horticulturae, 194,* 175-180. doi:10.1016/j.scienta.2015.08.005

Gardiner, B., Blennow, K., & Carnus, J.-M. (2010). *Destructive storms in European forests: past and forthcoming impacts.* Final report to EC DG environment. https://ec.europa.eu/environment/forests/fprotection.htm

Garibaldi, L. A., Aizen, M. A., Klein, A. M., Cunningham, S. A., & Harder, L. D. (2011). Global growth and stability of agricultural yield decrease with pollinator dependence. *Proceedings of the National Academy of Sciences of the United States of America, 108*(14), 5909–5914. doi:10.1073/pnas.1012431108 PMID:21422295

Garibaldi, L. A., Steffan-Dewenter, I., Winfree, R., Aizen, M. A., Bommarco, R., Cunningham, S. A., Kremen, C., Carvalheiro, L. G., Harder, L. D., Afik, O., Bartomeus, I., Benjamin, F., Boreux, V., Cariveau, D., Chacoff, N. P., Dudenhoffer, J. H., Freitas, B. M., Ghazoul, J., Greenleaf, S., ... Klein, A. M. (2013). Wild pollinators enhance fruit set of crops regardless of honey bee abundance. *Science, 339*(6127), 1608–1611. doi:10.1126cience.1230200 PMID:23449997

Garratt, M. P. D., Breeze, T. D., Jenner, N., Polce, C., Biesmeijer, J. C., & Potts, S. G. (2014). Avoiding a bad apple: Insect pollination enhances fruit quality and economic value. *Agriculture, Ecosystems & Environment, 184*, 34–40. doi:10.1016/j.agee.2013.10.032 PMID:24748698

Gavrilov, K. (1963). Earthworms, producers of biologically active substances. *Zh Obshch Biol, 24*, 149–154.

Gelvonauskis, B., Shikshnianas, T., Duchovskis, P., & Gelvonauskienė, D. (2004). Investigation of apple genetic resources to select donors resistant to fungal diseases and for their winterhardiness. *Zeszyty problemowe postępów nauk rolniczych, 497*, 659–663.

Germanà, C., Continella, A., & Tribulato, E. (2002, March). Net shading influence on floral induction on citrus trees. *VI International Symposium on Protected Cultivation in Mild Winter Climate: Product and Process Innovation, 614*, 527-533. 10.17660/ActaHortic.2003.614.78

Ge, X., & Weston, P. A. (1995). Ovipositional and feeding deterrent from chinese prickly ash against angoumois grain moth (Lepidoptera: Gelechiidae). *Journal of Economic Entomology, 88*(6), 1771–1775. doi:10.1093/jee/88.6.1771

GFN. (2007). *Ecological Footprint Glossary*. Global Footprint Network. Accessed July 2007 from http://www.footprint-network.org/gfn_sub.php?content=glossary.

Ghazzawy, H. S. (2013). Effects of some applications with growth regulators to improve fruit physical, chemical characteristics and storage ability of Barhee date palm cultivar. *Int. Res. J. Plant Sci, 4*(7), 208–213.

Ghildiyal, M., & Zamore, P. D. (2009). Small silencing RNAs: An expanding universe. *Nature Reviews. Genetics, 10*(2), 94–108. doi:10.1038/nrg2504 PMID:19148191

Godfray, H. C. J., Crute, I. R., Haddad, L., Lawrence, D., Muir, J. F., Nisbett, N., Pretty, J., Robinson, S., Toulmin, C., & Whiteley, R. (2010). The future of the global food system. *Philosophical Transactions of the Royal Society of London. Series B, Biological Sciences, 365*(1554), 2769–2777. doi:10.1098/rstb.2010.0180 PMID:20713383

Godwin, R. J., & Spoor, G. (1977). Soil failure with narrow tines. *Journal of Agricultural Engineering Research, 22*(3), 213–228. doi:10.1016/0021-8634(77)90044-0

Gold, C. S., Nankinga, C., Niere, B., & Godonou, I. (2003). *IPM of banana weevil in Africa with emphasis on microbial control. Biological control in IPM systems in Africa*. CABI Publishing.

Goldson, S. L., Phillips, C. B., & Barlow, N. D. (1994). The value of parasitoids in biologi1cal control. *New Zealand Journal of Zoology, 21*(1), 91–96. doi:10.1080/03014223.1994.9517979

Gontijo, L. M., Beers, E. H., & Snyder, W. E. (2013). Flowers promote aphid suppression in apple orchards. *Biological Control, 66*(1), 8–15. doi:10.1016/j.biocontrol.2013.03.007

Goossens, Y., Annaert, B. J., De Tavernier, J., Mathijs, E., Keulemans, W., & Geeraerd, A. (2017). Life cycle assessment (LCA) for apple orchard production systems including low and high productive years in conventional, integrated and organic farms. *Agricultural Systems, 153*, 81–93. doi:10.1016/j.agsy.2017.01.007

Grant, G. G., & Langevin, D. (1995). Oviposition deterrence, stimulation, and effect on cluth size of Choristoneura (Lepidoptera: Tortricidae) species by extract fractions of host and nonhost foliage. *Environmental Entomology, 24*(6), 1656–1663. doi:10.1093/ee/24.6.1656

Grattapaglia, D., & Resende, M. D. V. (2011). Genomic selection in forest tree breeding. *Tree Genetics & Genomes, 7*(2), 241–255. doi:10.100711295-010-0328-4

Grimaldi, D., & Engel, M. (2005). *Evolution of the insects*. Cambridge University Press.

Grisso, R. D., Yasin, M., & Kocher, M. F. (1994). Tillage implement forces operating in silty clay loam. ASAE paper n° 94-1532, ASAE, St. *Joseph, Michigan, USA, 1994*, 17.

Grogan, J. D., Moris, A., Searcy, S. W., & Stout, B. A. (1987). Microcomputer based tractor performance monitoring and optimisation system. *Journal of Agricultural Engineering Research, 38*(4), 227–243. doi:10.1016/0021-8634(87)90091-6

Gros-Balthazard, M. (2013). Hybridization in the genus Phoenix: A review. *Emirates Journal of Food and Agriculture*, 831–842.

Gross, A., Arusi, R., Fine, P., & Nejidat, A. (2008). Assessment of extraction methods with fowl manure for the production of liquid organic fertilizers. *Bioresource Technology, 99*, 327–334. https://doi.org/10.1016/j.biortech.2006.12.016

Gross, P. (1991). Influence of target pest feeding niche on success rates in classical biological control. *Environmental Entomology, 20*(5), 1217–1227. doi:10.1093/ee/20.5.1217

Growth Promotion of Date Palm Plantlets. (n.d.). *Ex vitro* by Inoculation of Rhizosphere Bacteria. *ARAB PALM Conference*.

Grubb, E., & Ellis, C. (2007). *Meeting the Carbon Challenge: The Role of Commercial Real Estate Owners*. Users & Managers.

Guarnieri, A., Maini, S., Molari, G., & Rondelli, V. (2011). Automatic trap for moth detection in integrated pest management. *Bulletin of Insectology*, *64*(2), 247–251.

Guo, J. L., Cao, W. J., Li, Z. M., Zhang, Y.-H., & Volis, S. (2019). Conservation implications of population genetic structure in a threatened orchid *Cypripedium tibeticum*. *Plant Diversity*, *41*(1), 13–18. doi:10.1016/j.pld.2018.12.002 PMID:30931413

Gupta, M. L., Prasad, A., Ram, M., & Kumar, S. (2002). Effect of the vesicular-arbuscular mycorrhizal (VAM) fungus *Glomus fasiculatum* on the essential oil yield related characters and nutrient acquisition in the crops of different cultivars of menthol mint (*Mentha arvensis*) under field conditions. *Bioresource Technology*, *81*, 77–79.

Gupta, R. K., & Arora, R. K. (2001). Lepidopteran fruit borers on guava in Jammu. *Insect Environment*, *7*(2), 83–84.

Gustavsson, B. A., & Stanys, V. (2000). Field performance of 'Sanna' lingonberry derived by micropropagation vs. stem cuttings. *HortScience*, *35*(4), 742–744. doi:10.21273/HORTSCI.35.4.742

Gutfleisch, O., Willard, M. A., Brück, E., Chen, C. H., Sankar, S. G., & Liu, J. P. (2011). Magnetic materials and devices for the 21st century: Stronger, lighter, and more energy efficient. *Advanced Materials*, *23*(7), 821–842. doi:10.1002/adma.201002180 PMID:21294168

Gutierrez-miceli, Moguel-Zamudio, AbudArchila, Gutierrez-Oliva, & Dendooven. (2008). Sheep manure vermicompost supplemented with a native diazotrophic bacteria and mycorrhizas for maize cultivation. *Bioresource Technology, 99*(15), 7020-7026.

Gutierrez, R. M. P. (2010). Orchids: A review of uses in traditional medicine, its phytochemistry, and pharmacology. *Journal of Medicinal Plants Research*, *4*, 592–638.

Hafez, O. M., Saleh, M. A., & Naguib, M. M. (2012). Quality improvement and storability of some date palm cultivars by safe postharvest treatments. *Australian Journal of Basic and Applied Sciences*, *6*, 542–550.

Häggman, H., Sutela, S., Walter, C., & Fladung, M. (2014). Biosafety Considerations in the Context of Deployment of GE Trees. T. Challenges and Opportunities for the World's Forests in the 21st Century. *Forestry Sciences*, *81*, 491–525. doi:10.1007/978-94-007-7076-8_21

Haji, F. N. P., Barbosa, F. R., Lopes, P. R. C., Moreira, A. N., de Alencar, J. A., & Ferreira, R. C. F. (2002, September). Monitoring mango pests within an integrated production program in Brazil. In *VII International Mango Symposium 645* (pp. 163-165). Academic Press.

Hall, G. (1989). Unusual or interesting records of plant pathogenic oomycetes. *Plant Pathology*, *38*(4), 604–611. doi:10.1111/j.1365-3059.1989.tb01458.x

Hammond, G. (2007). Time to give due weight to the 'carbon footprint' issue. *Nature*, *445*(7125), 256. doi:10.1038/445256b PMID:17230169

Hamrick, J. L., & Godt, M. J. W. (1990). Allozyme diversity in plant species. In Plant population genetics, breeding, and genetic resources. Sinauer Assoc.

Han, K.-S., Park, J.-H., Back, C.-G., & Park, M.-J. (2015). First Report of *Fusarium subglutinans* Causing Leaf Spot Disease on Cymbidium Orchids in Korea. *Mycobiology*, *43*(3), 343–346. doi:10.5941/MYCO.2015.43.3.343 PMID:26539053

Hansen, J. D., & Armstrong, J. W. (1990). The failure of field sanitation to reduce infestation by the mango weevil, *Cryptorhynchus mangiferae* (F.) (Coleoptera: Curculionidae). *International Journal of Pest Management*, *36*(4), 359–361.

Hare, J. D. (2020). Sampling arthropod pests in citrus. In *Handbook of sampling methods for arthropods in agriculture* (pp. 417–431). CRC Press. doi:10.1201/9781003067900-19

Harhash, M. M., & Abdel-Nasser, G. (2007). Impact of potassium fertilization and bunch thining on Zaghloul date palm. In *The Fourth Symposium on Date Palm in Saudi Arabia.* King Faisal University.

Harp, S. J., & Van Cleave, H. W. (1976). New records of natural enemies of the pecan weevil. *Southwestern Entomologist, 1*, 38.

Harrathi, A. (2008). *Le pou noir de l'oranger Parlatoria ziziphi Lucas (Hemiptera: Diaspididae): bioécologie et essai de lutte chimique* [Mémoire du Projet de Fin d'Etudes du Cycle Ingénieur]. Institut National Agronomique de Tunisie, Université de Carthage.

Hasan, M. A., Manna, M., Dutta, P., Bhattacharya, K., Mandal, S., Banerjee, H., Ray, S. K., & Jha, S. (2013). Foliar nutrient content in mango as influenced by organic and inorganic nutrients and their correlative relationship with yield and quality. *Acta Horticulturae*, (992), 201–206.

Haseeb, M. (2005, December). Current status of insect pest problems in guava. In *International Guava Symposium 735* (pp. 453-467). Academic Press.

Hassanen, S. A., & Khalil, R. M. A. (2013). Biotechnological Studies for Improving of Stevia (*Stevia rebaudiana* Bertoni) in vitro Plantlets. *Middle East Journal of Scientific Research, 14*(1), 93–106.

Hassen, A., Belguith, K., Jedidi, N., Cherif, A., Cherif, M., & Boudabous, A. (2001). Microbial characterization during composting of municipal solid waste. *Bioresource Technology, 80*, 217–225.

Hatzopoulos, P., Banilas, G., Giannoulia, K., Gazis, F., Nikoloudakis, N., Milioni, D., & Haralampidis, K. (2002). Breeding, molecular markers and molecular biology of the olive tree. *European Journal of Lipid Science and Technology, 104*(9-10), 574–586. doi:10.1002/1438-9312(200210)104:9/10<574::AID-EJLT574>3.0.CO;2-1

Hawkins, S., Leplé, J., Cornu, D., Jouanin, L., & Pilate, G. (2003). Stability of transgene expression in poplar: A model forest tree species. *Annals of Science, 60*, 427–438. doi:10.1051/forest:2003035

Hawley, G. J., Schaberg, P. G., DeHayes, D. H., & Brissette, J. C. (2005). Silviculture alters the genetic structure of an eastern hemlock forest in Maine,USA. *Canadian Journal of Forest Research, 35*(1), 143–150. doi:10.1139/x04-148

Haynes, R. J., & Naidu, R. (1998). Influence of lime, fertilizer and manure applications on soil organic matter content and soil physical conditions: A review. *Nutrient Cycling in Agroecosystems, 51*(2), 123–137. doi:10.1023/A:1009738307837

Hazarika, B. N. (2003). Acclimatization of tissue cultured plants. *Current Science, 85*, 1704–1712.

He, C., Zeng, Y., Fu, Y., Wu, J., & Liang, Q. (2020). Light quality affects the proliferation of in vitro cultured plantlets of Camellia oleifera Huajin. *PeerJ, 8*, 10016. doi:10.7717/peerj.10016 PMID:33083122

Heffner, E. L., Jannink, J.-L., & Sorrells, M. E. (2011). Genomic selection accuracy using multifamily prediction models in a wheat breeding program. *The Plant Genome, 4*(1), 65–75. doi:10.3835/plantgenome.2010.12.0029

He, H., Shang, Y., Yang, X., Di, Y., Lin, J., Zhu, Y., Zheng, W., Zhao, J., Ji, M., Dong, L., Deng, N., Lei, Y., & Chai, Z. (2019). Constructing an Associative Memory System Using Spiking Neural Network. *Frontiers in Neuroscience, 13*, 650. doi:10.3389/fnins.2019.00650 PMID:31333397

Hennessey, R., Mills, N., & Unruh, T. (1995). *Field release of an exotic parasitic wasp, Mastrus ridibundus (Hymenoptera: Ichneumonidae), for biological control of codling moth, Cydia pomonella (Lepidoptera: Tortricidae), in the United States*. Environmental Assessment.

Hern, A., & Dorn, S. (1999, July 1). Sexual dimorphism in the olfactory orientation of adult Cydia pomonella in response to α-farnesene. *Entomologia Experimentalis et Applicata, 92*(1), 63–72. doi:10.1046/j.1570-7458.1999.00525.x

Hiei, Y., Komari, T., Ishida, Y., & Saito, H. (2000). Development of Agrobacterium-mediated transformation method for monocotyledonous plants. *Ikushugaku Kenkyu, 2*(4), 205–213. doi:10.1270/jsbbr.2.205

Higa, T. (1991). Effective microorganisms: A biotechnology for mankind. *Proc First International Conference on Kyusei Nature Farming*, 8–14.

Hilgeman, R. H. (1954). *The differentiation, development and anatomy of the axillary bud, inflorescence and offshoot in the date palm.* Academic Press.

Hill, C. F. (2004). First report of Phytophthora multivesiculata on cymbidium orchids in New Zealand. *Australasian Plant Pathology, 33*(4), 603–604. doi:10.1071/AP04070

Hill, W. G., Goddard, M. E., & Visscher, P. M. (2008). Data and theory point to mainly additive genetic variance for complex traits. *PLOS Genetics, 4*(2), e1000008. doi:10.1371/journal.pgen.1000008 PMID:18454194

Hindorff, L. A., Sethupathy, P., Junkins, H. A., Ramos, E. M., Mehta, J. P., Collins, F. S., & Manolio, T. A. (2009). Potential etiologic and functional implications of genome-wide association loci for human diseases and traits. *Proceedings of the National Academy of Sciences of the United States of America, 106*(23), 9362–9367. doi:10.1073/pnas.0903103106 PMID:19474294

Hine, R. B. (1962). Pathogenicity of Phytophthora palmivora in the Orchidaceae. *The Plant Disease Reporter, 46*, 643–645.

Hinsley, A., De Boer, H. J., Fay, M. F., Gale, S. W., Gardiner, L. M., Gunasekara, R. S., Kumar, P., Masters, S., Metusala, D., Roberts, D. L., Veldman, S., Wong, S., & Phelps, J. (2018). A review of the trade-in orchids and their implications for conservation. *Botanical Journal of the Linnean Society, 186*(4), 435–455. doi:10.1093/botlinnean/box083

Hirano, T., Ishikawa, K., & Mii, M. (2006). Advances on orchid cryopreservation. In Floriculture Ornamental and Plant Biotechnology: Advances and Topical Issues (Vol. 2). Global Science Book.

Hirano, T., Godo, T., Miyoshi, K., Ishikawa, K., Ishikawa, M., & Mii, M. (2009). Cryopreservation and low-temperature storage of seeds of Phaius tankervilleae. *Plant Biotechnology Reports, 3*(1), 103–109. doi:10.100711816-008-0080-5

Hodel, D. R., Pittenger, D. R., & Downer, A. J. (2003). Effects of leaf removal and tie up on juvenile, transplanted of canary Island date palms (*Phoenix dactylifera* L.) and queen palms (*Syagrus ramanzoffiana*). *Palms, 47*(4), 177–184.

Hoekstra, A. Y. (2003). *Virtual water trade: Proceedings of the International Expert Meeting on Virtual Water Trade.* Value of Water Research Report Series No. 12, UNESCO-IHE Institute for Water Education.

Hoekstra, A. Y. (2008). *Water neutral: Reducing and offsetting the impacts of water footprints.* Value of Water Research Report Series No 28, UNESCO-IHE. www.waterfootprint.org/Reports/Report28-WaterNeutral.pdf

Hoekstra, A. Y., & Hung, P. Q. (2002). *Virtual Water Trade: A Quantification of Virtual Water Flows Between Nations in Relation to International Crop Trade*. Value of Water Research Report Series No 11. IHE.

Holysz, L., Szczes, A., & Chibowski, E. (2007). Effects of a static magnetic field on water and electrolyte solutions. *Journal of Colloid and Interface Science, 316*(2), 996–1002. doi:10.1016/j.jcis.2007.08.026 PMID:17897662

Hossain, M. M., Kant, R., Van, P. T., Winarto, B., Zeng, S., & Teixeira da Silva, J. A. (2013). The application of biotechnology to orchids. *Critical Reviews in Plant Sciences, 32*(2), 69–139. doi:10.1080/07352689.2012.715984

Houghton, J. (2015). *Global warming – the complete briefing* (5th ed.). Cambridge university press. doi:10.1017/CBO9781316134245

Hoy, M. A. (2003). *Insect Molecular Genetics, An Introduction to Principles and Applications* (2nd ed.). Elsevier Science.

Hoyt, S. C., & Burts, E. C. (1974). Integrated control of fruit pests. *Annual Review of Entomology, 19*(1), 231–252. doi:10.1146/annurev.en.19.010174.001311

Hseu, S. H., Shentu, H., & Sung, C. J. (2012). Multiplex PCR-based rapid detection of bacterial diseases on Phalaenopsis orchid. *Plant Pathology Bulletin, 21*, 91–100.

Hsiao, Y. Y., Pan, Z. J., Hsu, C. C., Yang, Y. P., Hsu, Y. C., Chuang, Y. C., Shih, H.-H., Chen, W.-H., Tsai, W.-C., & Chen, H.-H. (2011). Research on orchid biology and biotechnology. *Plant & Cell Physiology, 52*(9), 1467–1486. doi:10.1093/pcp/pcr100 PMID:21791545

Huda, A. K. S., & Luck, J. (2008, February). Early warning of pest/diseases for selected crops using climate information and crop simulation modeling approach under climate change scenarios. In *Proceeding of International Symposium on Agrometeorology and Food Security* (pp. 18-21). Academic Press.

Hull, L. A., Joshi, N. K., & Zaman, F. U. (2009). Large plot reduced risk insecticide study for Lepidopteran pests infesting apples, 2008. *Arthropod Management Tests, 34*(1).

Hussain, M., Farooq, S., Hasan, W., Ul-Allah, S., Tanveer, M., Farooq, M., & Nawaz, A. (2018). Drought stress in sunflower: Physiological effects and its management through breeding and agronomic alternatives. *Agricultural Water Management, 201*, 152-166. doi:10.1016/j.agwat.2018.01.028

Hussain, M. I., Farooq, M., & Syed, Q. A. (2020). Nutritional and biological characteristics of the date palm fruit (Phoenix dactylifera L.)–A review. *Food Bioscience, 34*, 100509. doi:10.1016/j.fbio.2019.100509

Ibrahim, A. E., Mona, H. M., & Darwesh, R. S. S. (2012). Pre- acclimatization stage of date palm (*Phoenix dactylifera*) plantlets cv. Barhee as affected by potassium nitrate and sucrose. *Journal of Biological Chemistry & Environmental Sciences, 7*(3), 1–20.

Ichikawa, K., & Aoki, T. (2000). New leaf spot disease of Cymbidium species caused by *Fusarium subglutinans and Fusarium proliferatum. Journal of General Plant Pathology, 66*(3), 213–218. doi:10.1007/PL00012948

IFICF. (1998). International Food Information Council Foundation. Backgrounder: functional foods. In Food Insight Media Guide. IFICF.

Iglesias, I., & Alegre, S. (2006). The effect of anti-hail nets on fruit protection, radiation, temperature, quality and profitability of 'Mondial Gala' apples. *Journal of Applied Horticulture, 8*(2), 91–100. doi:10.37855/jah.2006.v08i02.22

Ilieva, E., Manin't Veld, W. A., Veenbaas-Rijks, W., & Pieters, R. (1998). Phytophthora multivesiculata, a new species causing rot in Cymbidium. *European Journal of Plant Pathology, 104*(7), 677–684. doi:10.1023/A:1008628402399

Ingvarsson, P. K., Hvidsten, T. R., & Street, N. R. (2016). Towards integration of population and comparative genomics in forest trees. *The New Phytologist, 212*(2), 338–344. doi:10.1111/nph.14153 PMID:27575589

Ioannou, C. S., Papanastasiou, S. A., Zarpas, K. D., Miranda, M. A., Sciarretta, A., Nestel, D., & Papadopoulos, N. T. (2019). Development and Field Testing of a Spatial Decision Support System to Control Populations of the European Cherry Fruit Fly, *Rhagoletis cerasi*, in Commercial Orchards. *Agronomy (Basel), 9*(10), 568. doi:10.3390/agronomy9100568

Ioriatti, C., Anfora, G., Tasin, M., De Cristofaro, A., Witzgall, P., & Lucchi, A. (2011). Chemical ecology and management of *Lobesia botrana* (Lepidoptera: Tortricidae). *Journal of Economic Entomology, 104*(4), 1125–1137. doi:10.1603/EC10443 PMID:21882674

IPCC. (2013). The Physical Science Basis – Anthropogenic and natural radiative forcing. Working group I. *Climatic Change.*

Irawati, A. (2013). Conservation of orchids the gems of the tropics. In M. Normah, H. Chin, & B. Reed (Eds.), *Conservation of Tropical plant species* (pp. 171–187). Springer. doi:10.1007/978-1-4614-3776-5_9

Iriarte, A., Almeida, M. G., & Villalobos, P. (2014). Carbon footprint of premium quality export bananas: Case study in Ecuador, the world's largest exporter. *The Science of the Total Environment, 472*, 1082–1088. doi:10.1016/j.scitotenv.2013.11.072 PMID:24361571

Isajev, V., Ivetić, V., Lučić, A., & Rakonjac, Lj. (2009). Gene pool conservation and tree improvement in Serbia. *Genetika, 41*(3), 309–327. doi:10.2298/GENSR0903309I

Ishaaya, I., Kontsedalov, S., & Horowitz, A. R. (2005). Biorational insecticides: Mechanism and cross-resistance. *Archives of Insect Biochemistry and Physiology: Published in Collaboration with the Entomological Society of America, 58*(4), 192–199. doi:10.1002/arch.20042 PMID:15756702

Isik, Whetten, Zapata-Valenzuela, Ogut, & McKeand. (2011). Genomic selection in loblolly pine–from lab to field. From IUFRO tree biotechnology conference 2011: From genomes to integration and delivery. *BMC, 5*(7).

Ismail, I. A., Abdel-Rahaman, R. S., & Abdel-Raheem, M. A. (2015). Influence of some essential oils, chemical compounds and their mixtures against *Ceroplastes rusci* L. and *Asterolcanium pustolans* Cock on fig trees. *International Journal of Chemtech Research, 8*(9), 187–195.

Ismail, I. A., Abdel-Rahman, R. S., & Abdel-Raheem, M. A. (2016). Utilization of certain plant extracts and entomopathogenic fungi for controlling the black fig fly, *Lonchaea aristella* on fig trees. *International Journal of Chemtech Research, 9*(4), 35–42.

Ismail, I. A., Farag, N. A., Abdel-Rahman, R. S., Abdel-Raheem, M. A., & Radwan, H. M. (2014). Insecticidal activity of some plant extracts rich in coumarin against cowpea beetle, *Closobruchus maculatus* (Fab.) (Coleoptera: Bruchidae). *Egyptian Journal of Biological Pest Control, 24*(2), 465–469.

IUCN. (2017). *The IUCN red list of threatened species*. https://www.iucnredlist.org/

Ivetić, Devetaković, Nonić, Stanković, & Šijačić-Nikolić. (2016). Genetic diversity and forest reproductive material – from seed source selection to planting. *iForest-Biogeosciences and Forestry, 9*(5), 801-812. doi:10.3832/ifor1577-009

Jackson, D. M. (1982, May 15). Searching Behavior and Survival of 1st-Instar Codling Moths. *Annals of the Entomological Society of America, 75*(3), 284–289. doi:10.1093/aesa/75.3.284

Jain, A., & Abhay, K. (2020). Smart Agriculture Monitoring System using IoT. *International Journal for Research in Applied Science and Engineering Technology, 8*(7), 366–372.

Jain, A., Sarsaiya, S., Chen, J., Wu, Q., Lu, Y., & Shi, J. (2021). Changes in global Orchidaceae disease geographical research trends: Recent incidences, distributions, treatment, and challenges. *Bioengineered, 12*(1), 13–29. doi:10.1080/21655979.2020.1853447 PMID:33283604

Jain, S. M. (2012). Date palm biotechnology: Current status and prospective-an overview. *Emirates Journal of Food and Agriculture*, 386–399.

Jalal, J. S., Kumar, P., & Pangtey, Y. (2008). Ethnomedicinal Orchids of Uttarakhand, Western Himalaya. *Ethnobotanical Leaflets., 2008*, 164.

Jamshidi, S., Zand-Parsa, S., Kamgar-Haghighi, A. A., Shahsavar, A. R., & Niyogi, D. (2020). Evapotranspiration, crop coefficients, and physiological responses of citrus trees in semi-arid climatic conditions. *Agricultural Water Management, 227*, 105838. doi:10.1016/j.agwat.2019.105838

Jayawickrama, K. J. S., & Carson, M. J. (2000). A breeding strategy for the New Zealand radiata pine breeding cooperative. *Silvae Genetica, 49*, 82–90.

Jiang, G. L. (2015). Molecular marker-assisted breeding: a plant breeder's review. In Advances in Plant Breeding Strategies: Breed, Biotechnology Molecular Tools. Springer.

Jiang, J. A., Lin, T. S., Yang, E. C., Tseng, C. L., Chen, C. P., Yen, C. W., Zheng, X.-Y., Liu, C.-Y., Liu, R.-H., Chen, Y.-F., Chang, W.-Y., & Chuang, C. L. (2013). Application of a web-based remote agro-ecological monitoring system for observing spatial distribution and dynamics of *Bactrocera dorsalis* in fruit orchards. *Precision Agriculture, 14*(3), 323–342. doi:10.100711119-012-9298-x

Jiao, Y., Lau, O. S., & Deng, X. W. (2007). Light-regulated transcriptional networks in higher plants. *Nature Reviews Genetics, 8*(3), 217-230. doi:10.1038/nrg2049

Jifon, J. L., & Syvertsen, J. P. (2003). Moderate shade can increase net gas exchange and reduce photoinhibition in citrus leaves. *Tree Physiology, 23*(2), 119-127. doi:10.1093/treephys/23.2.119

Johnson, D. (2011). Introduction: date palm biotechnology from theory to practice. In Date palm biotechnology (pp. 1-11). Springer.

Jokar, A., Zare, H., Zakerin, A., & Jahromi, A. A. (2021). The Influence of Photo-selective Netting on Tree Physiology and Fruit Quality of Fig (*Ficus carica* L.) Under Rain-fed Conditions. *International Journal of Fruit Science, 21*(1), 896-910. doi:10.1080/15538362.2021.1936345

Joko, T., Subandi, A., Kusumandari, N., Wibowo, A., & Priyatmojo, A. (2014). Activities of plant cell walldegrading enzymes by bacterial soft rot of orchid. *Archiv für Phytopathologie und Pflanzenschutz, 47*(10), 1239–1250. doi:10.1080/03235408.2013.838374

Jones, S. (2003). Orchids. *Bull Am Orchid Soc.* http://www.aos.org/Default.aspx? id=120

Jones, V. P., Brunner, J. F., Grove, G. G., Petit, B., Tangren, G. V., & Jones, W. E. (2010). A web-based decision support system to enhance IPM programs in Washington tree fruit. *Pest Management Science: Formerly. Pesticide Science, 66*(6), 587–595. doi:10.1002/ps.1913 PMID:20127866

Joseph, J. A., Shukitt-Hale, B., Denisova, N. A., Bielinski, D., Martin, A., McEwen, J. J., & Bickford, P. C. (1999). Reversals of age-related declines in neuronal signal transduction, cognitive, and motor behavioral deficits with blueberry, spinach, or strawberry dietary supplementation. *The Journal of Neuroscience: The Official Journal of the Society for Neuroscience, 19*(18), 8114–8121. doi:10.1523/JNEUROSCI.19-18-08114.1999 PMID:10479711

Joshi, K., & Kamat, P. (1966). Effect of magnetic field on the physical properties of water. *Journal of Evolutionary Biology Research, 2*(1), 7–14.

Joshi, N. K., Hull, L. A., Rajotte, E. G., Krawczyk, G., & Bohnenblust, E. (2011, May 02). Evaluating sex-pheromone- and kairomone-based lures for attracting codling moth adults in mating disruption versus conventionally managed apple orchards in Pennsylvania. *Pest Management Science, 67*(10), 1332–1337. doi:10.1002/ps.2194 PMID:21538805

Joshi, N. K., Leslie, T., Rajotte, E. G., & Biddinger, D. J. (2020). Environmental impacts of reduced-risk and conventional pesticide programs differ in commercial apple orchards, but similarly influence pollinator community. *Chemosphere, 240*, 124926. doi:10.1016/j.chemosphere.2019.124926 PMID:31726586

Jothi, D. B., Tandon, P. L., & Verghese, A. (1994). Hot water immersion as a quarantine treatment for Indian mangoes infested with the oriental fruitfly, Bactrocera dorsalis (Hendel)(Diptera: Tephritidae). *FAO Plant Protection Bulletin*, *42*(3), 158–159.

Junaid, A., & Khan, S. A. (2009). In vitro micropropagation of 'khalas' date palm (*Phoenix dactylifera* l.) an important fruit plant. *Journal of Fruit and Ornamental Plant Research*, *17*(1), 15–27.

Kaiser, C., & Ernst, M. (2016). *Hydroponic Lettuce*. CCDCP-63. Lexington, KY: Center for Crop Diversification, University of Kentucky College of Agriculture, Food and Environment. Available: http://www.uky.edu/ccd/sites/www.uky.edu.ccd/files/hydrolettuce.pdf

Kalcsits, L., Musacchi, S., Layne, D. R., Schmidt, T., Mupambi, G., Serra, S., Mendoza, M., Asteggiano, L., Jaralmasjed, S., Sankaran, S., Khot, L.R. & Espinoza, C. Z. (2017). Above and below-ground environmental changes associated with the use of photoselective protective netting to reduce sunburn in apple. *Agricultural and Forest Meteorology, 237*, 9-17. doi:10.1016/j.agrformet.2017.01.016

Kale, R. D., Bano, K., & Krishnamoorthy, R. V. (1982). Potential of Perionyx excavatus for utilization of organic wastes. *Pedobiologia*, *23*, 419–425.

Kamboj, D. (2020). *Dendrobium and Venda Orchids as Potential Cut Flower in North Indian Market*. GINMA.

Kamjaipai, K. (1983). *Diseases and pest of orchids*. Bangkok Flower Center Co. Ltd.

Kang, K. S., Bila, A. D., Harju, A. M., & Lindgren, D. (2003). Estimation of fertility variation in forest tree populations. *Forestry*, *76*(3), 329–344. doi:10.1093/forestry/76.3.329

Kannangowa, T., Utkhede, R. S., Paul, J. W., & Punja, Z. K. (2000). Effect of mesophilic and thermophilic composts on suppression of Fusarium root and stem rot of greenhouse cucumber. *Canadian Journal of Microbiology*, *46*, 1021–1022.

Kapoor, V. C. (2000). Fruit flies (Diptera: Tephritidae): status, biology and management strategies. *IPM System in Agriculture, 7*.

Karamanis, G., Drosos, C., Papoutsidakis, M., & Tseles, D. (2018). *Implementation of an Automated System for Controlling and Monitoring a Hydroponic Greenhouse. International Journal of Engineering Science Invention*, *7*(10), 27–35.

Kasina, J. M., Mburu, J., Kraemer, M., & Holm-Mueller, K. (2009). Economic benefit of crop pollination by bees: A case of kakamega small-holder farming in western Kenya. *Journal of Economic Entomology*, *102*(2), 467–473. doi:10.1603/029.102.0201 PMID:19449623

Kassem, H. A. (2012). The response of date palm to calcareous soil fertilization. *Journal of Soil Science and Plant Nutrition*, *12*(1), 45–58. doi:10.4067/S0718-95162012000100005

Kassem, H. A., El-Sabrout, M. B., & Attia, M. M. (1997). Effect of nitrogen and potassium fertilization on yield, fruit quality and leaf mineral content in some Egyptian soft varieties. *Alexandria Journal of Agricultural Research*, *42*(1), 137–157.

Kataria, S., & Verma, S. K. (2018). Salinity stress responses and adaptive mechanisms in major glycophytic crops: the story so far. In *Salinity Responses and Tolerance in Plants* (Vol. 1, pp. 1–39). Springer. doi:10.1007/978-3-319-75671-4_1

Kawate, M., & Sewake, K. T. (2014). *Pest management strategic plan for potted orchid production in Hawaii. University of Hawaii at Manoa*.

Kaya, H. K. (1985). Entomogenous nematodes for insect control in IPM systems. In M. A. Hoy & D. C. Herzog (Eds.), *Biological control in agricultural IPM systems* (pp. 282–302). Academic Press. doi:10.1016/B978-0-12-357030-7.50022-3

Kearns, C. A., Inouye, D. W., & Waser, N. M. (1998). Endangered mutualisms: The conservation of plant-pollinator interactions. *Annual Review of Ecology and Systematics*, *29*(1), 83–112. doi:10.1146/annurev.ecolsys.29.1.83

Keel, B. G. (2005). Assisted migration. In M. Allaby (Ed.), *Dictionary of Ecology* (p. 36). Oxford University Press.

Keith, L. M., & Sewake, K. T. (2009). Isolation and characterization of Erwinia spp. causing bacterial soft rot on orchids in Hawaii. *Pac Agr Nat Resour*, *1*, 4–11.

Keith, L. M., Sewake, K. T., & Francis, T. Z. (2005). Isolation and characterization of Burkholderia gladioli from orchids in Hawaii. *Plant Disease*, *89*(12), 1273–1278. doi:10.1094/PD-89-1273 PMID:30791304

Kemper, K. E., Daetwyler, H. D., Visscher, P. M., & Goddard, M. E. (2012). Comparing linkage and association analyses in sheep points to a better way of doing GWAS. *Genetical Research*, *94*(4), 191–203. doi:10.1017/S0016672312000365 PMID:22950900

Kerr, R. J., Dieters, M. J., & Tier, B. (2004). Simulation of the comparative gains from four hybrid tree breeding strategies. *Canadian Journal of Research*, *34*(1), 209–220. doi:10.1139/x03-180

Kerry, B. (1988). Fungal parasites of cyst nematodes. In C. A. Edwards, B. R. Stinner, D. Stinner, & S. Rabatin (Eds.), *Biological interactions in soil* (pp. 293–306). Elsevier.

Keshta, & Hassan Nemat, Azzam, & Hassanin. (2011). Embryogenic callus induction of some sunflower (*Helianthus annuus* L.) genotypes under *in vitro* salt stress. *J. Plant Production. Mansoura Univ.*, *2*(2), 327–333. doi:10.21608/jpp.2011.85524

Kevan, P. G., & Viana, B. F. (2003). The global decline of pollination services. *Biodiversity (Ottawa)*, *4*(4), 3–8. doi:10.1080/14888386.2003.9712703

Khalifa, A. S., Hamdy, Z. H., El-Masry, H. M., Tadros, M. R., & Said, G. (1984). Effect of some growth regulators on thinning" Amhat" date palm. *Agricultural Research Review*, *62*(3A), 255–266.

Khalifa, M. M. A., Azzam Clara, R., & Azer, S. A. (2006). Biochemical markers associated with disease resistance to damping-off and root-rot diseases of peanut mutants and their productivity. *Egyptian J. of Phytopathology*, *34*(2), 53–74.

Khalil, C. A. (2014). The emerging role of epigenetics incardiovascular disease. *Therapeutic Advances in Chronic Disease*, *5*(4), 178–187. doi:10.1177/2040622314529325 PMID:24982752

Khamtham, J., & Akarapisan, A. (2019). *Acidovorax avenae* subsp. *cattleyae* causes bacterial brown spot disease on terrestrial orchid *Habenaria lindleyana* in Thailand. *Journal of Plant Pathology*, *101*(1), 31–37. doi:10.100742161-018-0135-6

Khanal, Kushal, Fulton, Shearer, & Ozkan. (2020). Remote Sensing in Agriculture—Accomplishments, Limitations, and Opportunities. *Remote Sensing*, *12*(3783), 1–29.

Khan, S., & Bi Bi, T. (2012). Direct shoot regeneration system for date palm (*Phoenix dactylifera* L.) cv. Dhakki as a means of micropropagation. *Pakistan Journal of Botany*, *44*(6), 1965–1971.

Khierallah, H. S. M., & Bader, S. M. (2007). Micropropagation of Date Palm (*Phoenix dactylifera* L.) var. Maktoom through Direct Organogenesis. Proc. IIIrd IC on Date Palm. *Acta Horticulturae*, (736), 213–224. doi:10.17660/ActaHortic.2007.736.19

Khosla, P. K., & Khurana, D. K. (1987). *Agroforestry for Rural Need* (Vol. 1). Indian Soc. of Tree Sci.

Kim, S. M., Lee, S., Moon, J. S., & Lee, B. J. (2015). Draft genome sequence of the plant quarantine bacterium Acidovorax cattleyae ATCC33619. Korean Soc Plant Pathol: A. *Bacteriol Bacterial Dis*, *A-521*(2), 108.

Klápště, J., Lstibůrek, M., & El-Kassaby, Y. A. (2014). Estimates of genetic parameters and breeding values from western larch open-pollinated families using marker-based relationship. *Tree Genetics & Genomes*, *10*(2), 241–249. doi:10.100711295-013-0673-1

Klein, A. M., Vaissiere, B. E., Cane, J. H., Steffan-Dewenter, I., Cunningham, S. A., Kremen, C., & Tscharntke, T. (2007). Importance of pollinators in changing landscapes for world crops. *Proceedings of the Royal Society of London*, *274*, 303–313. PMID:17164193

Klein, A.-M., Vaissière, B. E., Cane, J. H., Steffan-Dewenter, I., Cunningham, S. A., Kremen, C., & Tscharntke, T. (2007). Importance of pollinators in changing landscapes for world crops. *Proceedings. Biological Sciences*, *274*(1608), 303–313. doi:10.1098/rspb.2006.3721 PMID:17164193

Knudson, L. (1946). A new nutrient solution for germination of orchid seed. *American Orchid Society Bulletin*, *15*, 14–217.

Kohl, L.M. (2011). Astronauts of the nematode world: an aerial view of foliar nematode biology, epidemiology, and host range. *APSnet Features*. doi:10.1094/APSnetFeature-2011-0111

Kole, C., & Hall, T. C. (2008). *Compendium of transgenic crop plants: transgenic forest tree species*. Wiley. doi:10.1002/9781405181099

Konig, A. O. (2005). Provenance research: evaluating the spatial pattern of genetic variation. In Conservation and management of forest genetic resources in Europe. Arbora Publishers.

Konnert, M., & Hosius, B. (2010). Contribution of forest genetics for a sustainable forest management. *Forstarchiv (Hannover)*, *4*, 170–174.

Koolen, A. J., & Kuipers, H. (1983). *Agricultural Soil Mechanics*. Springer. doi:10.1007/978-3-642-69010-5

Korecký, J., Klápště, J., Lstibůrek, M., Kobliha, J., Nelson, C. D., & El-Kassaby, Y. A. (2013). Comparison of genetic parameters from marker-based relationship, sibship, and combined models in Scots pine multi-site open-pollinated tests. *Tree Genetics & Genomes*, *9*(5), 1227–1235. Advance online publication. doi:10.100711295-013-0630-z

Kostecka, J., Blazej, J.B., & Kolodziej, M. (1996). Investigations on application of vermicompost in potatoes farming in second year of experiment. *Zeszyty Naukowe Akademii Rolniczej W Krakowie, 310*, 69–77.

Kostritsyn, A. K. (1956). *Cutting of a Cohesive Soil Medium with Knives and Cones*. National Institute of Agricultural Engineering.

Koul, O., Jain, M., & Sharma, V. (2000). Growth inhibitory and antifeedant activity of extracts from Melia dubia to Spodoptera litura and Helicoverpa armígera larvae. *Indian Journal of Experimental Biology*, *38*, 63–38. PMID:11233088

Kovačević, Ž. (1952). *Primijenjena entomologija. II. Knjiga Poljoprivredni štetnici*. Sveučilište Zagreb.

Kovanci, O. B. (2015). Co-application of microencapsulated pear ester and codlemone for mating disruption of Cydia pomonella. *Journal of Pest Science*, *88*(2), 311–319. doi:10.100710340-014-0619-x

Kozai, T., Afreen, F., & Zobayed, S. M. A. (2005). *Photoautotrophic (sugar-free medium) micropropagation as a new micropropagation and transplant production system*. Springer. doi:10.1007/1-4020-3126-2

Kreimer, A., Litvin, O., Hao, K., Molony, C., Pe'er, D., & Pe'er, I. (2012). Inference of modules associated to eQTLs. *Nucleic Acids Research*, *40*(13), e98. doi:10.1093/nar/gks269 PMID:22447449

Krishna, K. R. (2018). *Agricultural Drones A Peaceful Pursuit*. Apple Academic Press Inc.

Kui, L., Chen, H., Zhang, W., He, S., Xiong, Z., Zhang, Y., Yan, L., Zhong, C., He, F., Chen, J., Zeng, P., Zhang, G., Yang, S., Dong, Y., Wang, W., & Cai, J. (2017). Building a Genetic Manipulation Tool Box for Orchid Biology: Identification of Constitutive Promoters and Application of CRISPR/Cas9 in the Orchid, Dendrobium officinale. *Front. Plant Sci.*, *7*, 2036. doi:10.3389/fpls.2016.02036 PMID:28127299

Kulkarni, N. S. (2018). Pests of Grapes. In *Pests and Their Management* (pp. 517–557). Springer. doi:10.1007/978-981-10-8687-8_16

Kulkarni, N. S. (2020). Sucking Pests of Grapes. In *Sucking Pests of Crops* (pp. 425–450). Springer. doi:10.1007/978-981-15-6149-8_14

Kulkarni, S., Vijayanand, P., & Shubha, L. (2010). Effect of processing of dates into date juice concentrate and appraisal of its quality characteristics. *Journal of Food Science and Technology*, *47*(2), 157–161.

Kumar, S. & Bhatt, R. (1999). Field evaluation of plant leaf extracts, oil and neem products against mango hopper (Amritodus atikinsoni Lethierry) and thrips (Scirtothrips mangiferae Hood). *Allelopathy Journal*, *6*(2), 271–276.

Kumar, S., Prasad, R., Shukla, A., & Kumar, A. (2018). Management of organic orchard: Effect of vermicompost on fruit yield of ber (Zizyphus mauritiana Lamk.) cv. Seo and soil health in Central India. *Indian J. of Agroforestry, 20*(2), 58-62.

Kumar, S., Joship, C., Pashupati, N., Singh, V. K., & Mansotra, D. K. (2016). Role of insects in pollination of mango trees. *International Research Journal of Biological Sciences*, *5*(1), 64–67.

Kumar, S., Prasad, R., Anil Kumar, A., & Dhyani, S. K. (2019). Integration of fruit trees in agroforestry for sustainability and profitability of farming systems in arid and semi-arid regions. *Indian J. of Agroforestry*, *21*(1), 95–99.

Kurup, S. S., Aly, M. A. M., Lekshmi, G., & Tawfik, N. H. (2014). Rapid in vitro regeneration of date palm (*Phoenix dactylifera* L.) cv. Kheneizi using tender leaf explant. *Emirates Journal of Food and Agriculture*, *26*(6), 539–544. doi:10.9755/ejfa.v26i6.18051

Kurz, W. A., Dymond, C. C., Stinson, G., Rampley, G. J., Neilson, E. T., Carroll, A. L., Ebata, T., & Safranyik, L. (2008). Mountain pine beetle and forest carbon: Feedback to climate change. *Nature*, *454*(7190), 987–990. doi:10.1038/nature06777 PMID:18432244

Labiod, K. (2010). *Etude du caractère incrustant des eaux et son inhibition par des essais chimiques et s'électrodéposition, Master*. University of Constantine.

Lango, A. H. (2010). Hundreds of variants clustered in genomic loci and biological pathways affect human height. *Nature, 467*(7317), 832–838.

Latha, S., Babu, D. V. N., Sathyanarayana, N., Reddy, O. R., & Sharma, R. (1999). Plant parasitic nematodes intercepted from imported ornamental plants. *Indian Phytopathology*, *52*(3), 283–284.

Laza, M., Miter, V., Tripon, F. A., & Badiu, D. E. (2014). The Potential use of vermicompost in orchards. *Bulletin UASVM Horticulture*, *71*(1), 56–58.

Lebert, M., Burger, N., & Horn, R. (1989). Effects of dynamic and static loading on compaction of structured soils. In *Mechanics and related processes in structured agricultural soils* (pp. 73–80). Springer. doi:10.1007/978-94-009-2421-5_7

Lee, T. C., Zhong, P. J., & Chang, P. T. (2015). The effects of preharvest shading and postharvest storage temperatures on the quality of 'Ponkan' (Citrus reticulata Blanco) mandarin fruits. *Scientia Horticulturae*, *188*, 57–65. doi:10.1016/j.scienta.2015.03.016

Lee, Y. A., Chao, C. S., & Jung, C. H. (2013). Combination of a simple differential medium and tox A-specific PCR for isolation and identification of phytopathogenic Burkholderia gladioli. *European Journal of Plant Pathology, 136*(3), 523–533. doi:10.100710658-013-0184-9

Lefebvre, M. P. (2010). *Spatialisation de modèles de fonctionnement hydromécanique des sols appliquée à la prévision des risques de tassement à l'échelle de la France* [Thèse de doctorat]. Université d'Orléans.

Leitch, C. C., Lodh, S., Prieto-Echagüe, V., Badano, J. L., & Zaghloul, N. A. (2014). Basal body proteins regulate Notch signaling through endosomal trafficking. *Journal of Cell Science, 127*, 2407–2419. PMID:24681783

Létourneau, Guillaume, Caron, Anderson, & Cormier. (2015). Matric Potential-Based Irrigation Management of Field-Grown Strawberry: Effects on Yield and Water Use Efficiency. *Agricultural Water Management, 161*, 102–113.

Lewis, T. (1997). Flight and dispersal. In T. Lewis (Ed.), Thrips as crop pests (pp. 175–196). CAB International.

Liao, M. S., Chuang, C. L., Lin, T. S., Chen, C. P., Zheng, X. Y., Chen, P. T., Liao, K.-C., & Jiang, J. A. (2012). Development of an autonomous early warning system for *Bactrocera dorsalis* (Hendel) outbreaks in remote fruit orchards. *Computers and Electronics in Agriculture, 88*, 1–12. doi:10.1016/j.compag.2012.06.008

Li, B., Qiu, W., Fang, Y., & Xie, G. L. (2009). Bacterial stem rot of Oncidium orchid caused by a Dickeya sp. (ex Pectobacterium chrysanthemi) in mainland China. *Plant Disease, 93*(5), 552–552. doi:10.1094/PDIS-93-5-0552B PMID:30764150

Li, D. M., Zhao, C., Liu, X., Liu, X., Lin, Y., Liu, J., Chen, H., & Lv, F. (2015). *De novo* assembly and characterization of the root transcriptome and development of simple sequence repeat markers in Paphiopedilum concolor. *Genetics and Molecular Research, 14*(2), 6189–6201. doi:10.4238/2015.June.9.5 PMID:26125820

Liesebach, H., Liepe, K., & Baucker, C. (2021). Towards new seed orchard designs in Germany – A review. *Silvae Genetica, 70*(1), 84–98. doi:10.2478g-2021-0007

Li, J. W., Ya, J. D., Ye, D. P., Liu, C., Liu, Q., Pan, R., He, Z.-X., Pan, B., Cai, J., Lin, D.-L., & Jin, X.-H. (2021). Taxonomy notes on Vandeae (*Orchidaceae*) from China: Five new species and two new records. *Plant Diversity, 43*(5), 379–389. doi:10.1016/j.pld.2021.01.009 PMID:34816063

Lindley, C. P., Arniputri, R. B., Soliah, L. A., & Cahyono, O. (2017). Effects of organic additives and naphthalene acetic acid (NAA) application on the in vitro growth of black orchid hybrid Coelogyne pandurata Lindley. *Bulgarian Journal of Agricultural Science, 23*(6), 951–957.

Lindner, M., Maroschek, M., Netherer, S., Kremer, A., Barbati, A., Garcia-Gonzalo, J., Seidl, R., Delzon, S., Corona, P., Kolström, M., Lexer, M. J., & Marchetti, M. (2010). Climate change impacts, adaptive capacity, and vulnerability of European forest ecosystems. *Forest Ecology and Management, 259*(4), 698–709. doi:10.1016/j.foreco.2009.09.023

Lin, I., & Yotvat, T. (1990). Exposure of irrigation and drinking water to a magnetic field with controlled power and direction. *Journal of Magnetism and Magnetic Materials, 83*(1-3), 525–526. doi:10.1016/0304-8853(90)90611-S

Lin, W., Wang, J., Xu, X., Wu, Y., Qiu, D., He, B., Sarsaiya, S., Ma, X., & Chen, J. (2020). Rapid propagation in vitro and accumulation of active substances of endangered *Dendrobium cariniferum* Rchb. f. *Bioengineered, 11*(1), 386–396. doi:10.1080/21655979.2020.1739406 PMID:32172675

Lin, Y. H., Lee, P. J., Shie, W. T., Chern, L. L., & Chao, Y. (2015). Pectobacterium chrysanthemi as the dominant causal agent of bacterial soft rot in Oncidium "Grower Ramsey". *European Journal of Plant Pathology, 142*(2), 331–343. doi:10.100710658-015-0618-7

Liobikas, J., Baniulis, D., Stanys, V., Toleikis, A., & Eriksson, O. (2006). Identification and analysis of a novel serine beta-lactamase-like plant protein using a bioinformatics approach. *Biologija (Vilnius, Lithuania), 3*, 126–129.

Li, S. H., Kuoh, C. S., Chen, Y. H., Chen, H. H., & Chen, W. H. (2005). Osmotic sucrose enhancement of single-cell embryogenesis and transformation efficiency in Oncidium. *Plant Cell, Tissue and Organ Culture*, *81*(2), 183–192. doi:10.100711240-004-4955-z

Li, T., Kung, H. J., Mack, P. C., & Gandara, D. R. (2013). Genotyping and genomic profiling of non-small-cell lung cancer: Implications for current and future therapies. *Journal of Clinical Oncology*, *31*(8), 1039–1049. doi:10.1200/JCO.2012.45.3753 PMID:23401433

Liu, J., Chiou, C., Shen, C., Chen, P. J., Liu, Y. C., Jian, C. D., Shen, X. L., Shen, F. Q., & Yeh, K. W. (2014). RNA Interference-Based Gene Silencing of Phytoene Synthase Impairs Growth, Carotenoids, and Plastid Phenotype in Oncidium Hybrid Orchid. 3. *SpringerPlus*, *3*(1), 478. doi:10.1186/2193-1801-3-478 PMID:25221736

Liu, J., Zimmer, K., Rusch, D. B., Paranjape, N., Podicheti, R., Tang, H., & Calvi, B. R. (2015). DNA sequence templates adjacent nucleosome and ORC sites at gene amplification origins in Drosophila. *Nucleic Acids Research*, *43*(18), 8746–8761. doi:10.1093/nar/gkv766 PMID:26227968

Liu, X., & Locasale, J. W. (2017). Metabolomics: A Primer. *Trends in Biochemical Sciences*, *42*(4), 274–284. doi:10.1016/j.tibs.2017.01.004 PMID:28196646

Li, X., Jin, F., Jin, L., Jackson, A., Huang, C., Li, K., & Shu, X. (2014). Development of Cymbidium ensifolium genic SSR markers and their utility in genetic diversity and population structure analysis in cymbidiums. *BMC Genetics*, *15*(1), 124. doi:10.118612863-014-0124-5 PMID:25481640

Li, Y., Klápště, J., Telfer, E., Wilcox, P., Graham, N., Macdonald, L., & Dungey, H. S. (2019). Genomic selection for non-key traits in radiata pine when the documented pedigree is corrected using DNA marker information. *BMC Genomics*, *20*(1), 1026. doi:10.118612864-019-6420-8 PMID:31881838

Lloyd, D. C. (1960). Significance of the type of host plant crop in successful biological control of insect pests. *Nature*, *187*(4735), 430–431. doi:10.1038/187430a0 PMID:14417718

Louis, F., & Schirra, K. J. (2001). Mating disruption of *Lobesia botrana* (Lepidoptera: Tortricidae) in vineyards with very high population densities. *IOBC/WPRS Bulletin*, *24*, 75–79.

Lowe, R., Wu, Y., Tamar, A., Harb, J., Abbeel, O. P., & Mordatch, I. (2017). Multi-agent actor-critic for mixed cooperative-competitive environments. Advances in Neural Information Processing Systems, 6382–6393.

Lu, H. C., Chen, H. H., Tsai, W. C., Chen, W. H., Su, H. J., Chang, C. N., & Yeh, H. H. (2007). Strategies for functional validation of genes involved in reproductive stages of orchids. *Plant Physiology*, *143*(2), 558–569. doi:10.1104/pp.106.092742 PMID:17189336

Lu, J. J., Liu, Y. Y., Xu, J., Mei, Z. W., Shi, Y. J., Liu, P. L., He, J. B., Wang, X. T., Meng, Y. J., Feng, S. G., Shen, C. J., & Wang, H. Z. (2018). High-density genetic map construction and stem total polysaccharide content related QTL exploration for Chinese endemic Dendrobium (*Orchidaceae*). *Front. Plant Sci.*, *9*, 398. doi:10.3389/fpls.2018.00398 PMID:29636767

Lukoševičiūtė, V., Stanienė, G., Blažytė, A., Sasnauskas, A., & Gelvonauskienė, D. (2011). Angliavandenių įtaka kultūrinių kriaušių mikroūglių užsigrūdinimui žemoje teigiamoje temperatūroje. *Sodininkyste ir Darzininkyste*, *30*(3–4), 17–29.

Luo, Y. H., Huang, M. L., & Wu, J. S. (2012). Progress in Oncidium breeding study. *Jiangxi Nongye Daxue Xuebao*, *24*, 15–20.

Luz, Paiva, & Tavares. (2008). Effect of Foliar and Substrate Fertilization on Lady Palm Seedling Growth and Development. *Journal of Plant Nutrition*, *31*, 1311–1318.

Maceljski, M. (2002). *Poljoprivredna entomologija. II. Edition, Zrinski.* Čakovec.

Mahdia, F. G., & Abd-Alla, M. M. (2010). Micro Propagation of *Phoenix dactylifera* L. var karama. *New York Science, 3*(12), 64–69.

Mahmood, A., Hu, Y., Tanny, J., & Asante, E. A. (2018). Effects of shading and insect-proof screens on crop microclimate and production: A review of recent advances. *Scientia Horticulturae, 241*, 241–251. doi:10.1016/j.scienta.2018.06.078

Malausa, T., Fenis, A., Warot, S., Germain, J.-F., Ris, N., Prado, E., Botton, M., Vanlerberghe-Masutti, F., Sforza, R., Cruaud, C., Couloux, A., & Kreiter, P. (2011). DNA markers to disentangle complexes of cryptic taxa in mealybugs (Hemiptera: Pseudococcidae). *Journal of Applied Entomology, 135*(1-2), 142–155. doi:10.1111/j.1439-0418.2009.01495.x

Malik, S. K., Bhandari, D. C., Kumar, S., & Dhariwal, O. P. (2013). Conservation of multipurpose tree species to ensure ecosystem sustainability and farmers livelihood in Indian arid zone. In S. Nautiyal, K. S. Rao, K. Herald, K. V. Raju, & S. Ruediger (Eds.), *Knowledge Systems of Societies for Adaptation and Mitigation of Impacts of Climate Change. Environmental Science and Engineering* (pp. 257–270). Springer. doi:10.1007/978-3-642-36143-2_16

Mallinger, R. E., Gibbs, J., & Gratton, C. (2016). Diverse landscapes have a higher abundance and species richness of spring wild bees by providing complementary floral resources over bees' foraging periods. *Landscape Ecology, 31*(7), 1523–1535. doi:10.100710980-015-0332-z

Mangoud, A. A. H. (2008). Insecticidal effect of some chemical and natural control agents against the black parlatoria scale, Parlatoria ziziphi (Lucas) and associated parasitoids on citrus trees. *Egyptian Journal of Agricultural Research, 86*, 2157–2167.

Mani, M. (1986). *Distribution, Bioecology And Management Of The Grape Mealybug, Maconellicoccus hirsutus (Green) With Special Reference To Its Natural Enemies* (Doctoral Dissertation). University Of Agricultural Sciences, Bangalore.

Mani, M., & Kulkarni, N. S. (2007). Citrus mealybug *Planococcus citri* (Risso) Homoptera; Pseudococcidae)-a major pest of grapes in India. *Entomon, 32*, 235–236.

Mani, M., Kulkarni, N. S., Banerjee, K., & Adsule, P. G. (2008). *Pest management in grapes*. Extension.

Manja, K., & Aoun, M. (2019). The use of nets for tree fruit crops and their impact on the production: A review. *Scientia Horticulturae, 246*, 110–122. doi:10.1016/j.scienta.2018.10.050

Marashi, S., & Mousavi, A. (2006, February). Effects of different methods and degrees of fruit thinning on yield and fruit characteristics of Barhee date cultivar. In *III International Date Palm Conference 736* (pp. 187-192). Academic Press.

Marini, L., Quaranta, M., Fontana, P., Biesmeijer, J. C., & Bommarco, R. (2012). Landscape context and elevation affect pollinator communities in intensive apple orchards. *Basic and Applied Ecology, 13*(8), 681–689. doi:10.1016/j.baae.2012.09.003

Martinez-Ferrer, M. T., Grafton-Cardwell, E. E., & Shorey, H. H. (2003). Disruption of parasitism of the California red scale (Homoptera: Diaspididae) by three ant species (Hymenop1tera: Formicidae). *Biological Control, 26*(3), 279–286. doi:10.1016/S1049-9644(02)00158-5

Martínez-Ferri, E., Muriel-Fernández, J. L., & Díaz, J. R. (2013). Soil water balance modelling using SWAP: An application for irrigation water management and climate change adaptation in citrus. *Outlook on Agriculture, 42*(2), 93–102. doi:10.5367/oa.2013.0125

Martins, K. T., Gonzalez, A., & Lechowicz, M. J. (2015). Pollination services are mediated by bee functional diversity and landscape context. *Agriculture, Ecosystems & Environment, 200*, 12–20. doi:10.1016/j.agee.2014.10.018

Mathews, C. R., Bottrell, D. G., & Brown, M. W. (2004). Habitat manipulation of the apple orchard floor to increase ground-dwelling predators and predation of Cydia pomonella(L.) (Lepidoptera: Tortricidae). *Biological Control, 30*(2), 265–273. doi:10.1016/j.biocontrol.2003.11.006

Matthews, T. J., & Whittaker, R. J. (2015). On the species abundance distribution in applied ecology and biodiversity management. *Journal of Applied Ecology*, *52*(2), 443–454. doi:10.1111/1365-2664.12380

Mazeikiene, I., Bendokas, V., Stanys, V., & Siksnianas, T. (2012). Molecular markers linked to resistance to the gall mite in blackcurrant. *Plant Breeding*, *131*(6), 762–766. doi:10.1111/j.1439-0523.2012.01995.x

McGhee, P. S., Miller, J. R., Thomson, D. R., & Gut, L. J. (2016). Optimizing aerosol dispensers for mating disruption of codling moth, Cydia pomonella L. *Journal of Chemical Ecology*, *42*(7), 612–616. doi:10.100710886-016-0724-9 PMID:27369280

McMillan, R. T. Jr, Palmateer, A., & Cating, R. A. (2009). Problems in controlling Phytophthora cactorum on catteleya orchids. *Proceedings of the Annual Meeting of the Florida State Horticultural Society*, *122*, 426–428.

Mcmillan, R. T., Palmateer, A. J., & Vendrame, W. A. (2008). Cercospora Leaf Spot Caused by *Cercospora dendrobii* on *Dendrobium antennatoum* Lindl. and Its Control. *Proceedings of the Annual Meeting of the Florida State Horticultural Society*, *121*, 353–355.

McMullen. (2009). Genetic properties of the maize nested association mapping population. *Science, 325*(5941), 737–740.

Mditshwa, A., Magwaza, L. S., & Tesfay, S. Z. (2019). Shade netting on subtropical fruit: Effect on environmental conditions, tree physiology and fruit quality. *Scientia Horticulturae, 256*, 108556. doi:10.1016/j.scicnta.2019.108556

Mebdoua, S. (2018). *Pesticide residues in fruits and vegetables.* Bioactive molecules in food. Reference series in phytochemistry. Springer.

Medany, M. (2011). Vermiculture in Egypt: Current development and future potential Food and Agriculture Organization of the United Nations Regional Office for the Near East. Academic Press.

Medany, M. A., Abdrabbo, M. A. A., Awny, A. A., Hassanien, M. K., & Abou-Hadid, A. F. (2009). Growth and productivity of mango grown under greenhouse conditions. *Egyptian Journal of Horticulture*, *36*, 373–382.

Medina, O., Anaya, L. A., Aguilar, A. C., Llaven, A. O., Talavera, T. A., Dendooven, L., Miceli, F. G., & Figueroa, M. S. (2007). *Ex vitro* Survival and Early Growth of *Alpinia purpurata* Plantlets Inoculated with Azotobacter and Azospirillum. *Pakistan Journal of Biological Sciences*, *10*(19), 3454–3457. doi:10.3923/pjbs.2007.3454.3457 PMID:19090169

Meehan, R. R., Dunican, D. S., Ruzov, A., & Pennings, S. (2005). Epigenetic silencing in embryogenesis. *Experimental Cell Research*, *309*(2), 241–249. doi:10.1016/j.yexcr.2005.06.023 PMID:16112110

Mekonnen, M. M., & Hoekstra, A. Y. (2011). The green, blue and grey water footprint of crops and derived crop products. *Hydrology and Earth System Sciences*, *15*(5), 1577–1600. doi:10.5194/hess-15-1577-2011

Melke, A. (2015). The physiology of chilling temperature requirements for dormancy release and bud-break in temperate fruit trees grown at mild winter tropical climate. *Journal of Plant Studies*, *4*(2), 110-156. doi:10.5539/jps.v4n2p110

Merkle, S. A., Andrade, G. M., Nairn, C. J., Powell, W. A., & Maynard, C. A. (2007). Restoration of threatened species: A noble cause for transgenic trees. *Tree Genetics & Genomes*, *3*(2), 111–118. doi:10.100711295-006-0050-4

Meuwissen, T. H. E., Goddard, M. E., & Hayes, B. J. (2001). Prediction of total genetic value using genome-wide dense marker maps. *Genetics*, *157*(4), 1819–1829. doi:10.1093/genetics/157.4.1819 PMID:11290733

Meyers, M., & Hudelson. (2014). *Foliar nematodes University of Wisconsin Extension provides equal opportunities in employment and programming, including Title IX and ADA requirements.* https://pddc.wisc.edu

Michel, J. A., Fornstrom, K. J., & Borrelli, J. (1985). Energy requirements of two tillage systems for irrigated sugarbeets, dry beans and corn. *Transactions of the ASAE. American Society of Agricultural Engineers*, *28*(6), 1731–1735. doi:10.13031/2013.32508

Michener, C. D. (2000). *The bees of the world.* Johns Hopkins University Press.

Mikkelsen, C., & Schluter, M. (2009). *The new EU Regulation for Organic Food and Farming, (EC) No. 834/2007.* International Federation of Organic Agriculture Movements IFOAM EU Group.

Miller, J. R., & Gut, L. J. (2015). Mating disruption for the 21st century: Matching technology with mechanism. *Environmental Entomology*, *44*(3), 427–453. doi:10.1093/ee/nvv052 PMID:26313949

Mills, N. (2005). Selecting effective parasitoids for biological control introductions: Codling moth as a case study. *Biological Control*, *34*(3), 274–282. doi:10.1016/j.biocontrol.2005.02.012

Miranda, M. Á., Barceló, C., Valdés, F., Feliu, J. F., Nestel, D., Papadopoulos, N., Sciarretta, A., Ruiz, M., & Alorda, B. (2019). Developing and Implementation of Decision Support System (DSS) for the Control of Olive Fruit Fly, *Bactrocera Oleae,* in Mediterranean Olive Orchards. *Agronomy (Basel)*, *9*(10), 620. doi:10.3390/agronomy9100620

Misra, C. S. (1920). Index to Indian fruit pests. Report. *Proc. 3rd Ent. Management*, 564-595.

Mohanasundaram, M. (1974). *Occurrence of Kerria lacca (Kerr.) and Aleurocanthus spiniferus (Quain.) on grape vine in Tamil Nadu.* Academic Press.

Mohsen, A.M.A. (2019). Survey of insect mango pollinators and the pollination occurrence of mango trees in relations to fruit yields. *Current Science International, 8*(2), 245-251.

Mondino, P., & González-Andújar, J. L. (2019). Evaluation of a decision support system for crop protection in apple orchards. *Computers in Industry*, *107*, 99–103. doi:10.1016/j.compind.2019.02.005

Moody, S. A., Piearce, T. G., & Dighton, J. (1996). Fate of some fungal spores associated with wheat straw decomposition on passage through the guts of Lumbricus terrestris and Aporrectodea longa. *Soil Biology & Biochemistry*, *28*, 533–537.

Moon, H., Parkb, H., Jeonga, A., Hanb, S.-W., & Parka, C.-J. (2016). Isolation and identification of Burkholderia gladioli on Cymbidium orchids in Korea. *Biotechnology, Biotechnological Equipment.* Advance online publication. doi:10.1080/13102818.2016.1268069

Morgan, D., & Solomon, M. G. (1995). PEST-MAN: A forecasting system for apple and pear pests 1. *Bulletin OEPP*, *23*(4), 601–605. doi:10.1111/j.1365-2338.1993.tb00556.x

Morte, A., & Honrubia, M. (2002). Growth response of Phoenix canariensis to inoculation with Arbuscular mycorrhizal fungi. *Morte & Honrubia, 46*(2), 76- 80.

Mostafa, R. A. A., & El Akkad, M. M. (2011). Effect of fruit thinning rate on yield and fruit quality of Zaghloul and Haiany date palms. *Australian Journal of Basic and Applied Sciences*, *5*(12), 3233–3239.

Mostafa, R. A. A., El-Salhy, A. M., El-Banna, A. A., & Diab, Y. M. (2014). Effect of bunch bagging on yield and fruit quality of Seewy date palm under New Valley conditions (Egypt). Middle East. *Journal of Agricultural Research (Lahore)*, *3*, 517–521.

Mostafazadeh, F. B., Khoshravesh, M., Mousavi, S. F., & Kiani, A. R. (2011). Effects of magnetized water on soil sulfate ions in trickle irrigation. In *Proceedings of the second international conference on environmental engineering and applications.* IACSIT Press.

Mouazen, A. M., Ramon, H., & De Baerdemaeker, J. (2002). SW-Soil and water: Effects of bulk density and moisture content on selected mechanical properties of sandy loam soil. *Biosystems Engineering*, *83*(2), 217–224. doi:10.1006/bioe.2002.0103

Mound, L. A., & Jackman, D. J. (1998). Thrips in the economy and ecology of Australia. *Proceedings of the Sixth Australian Applied Entomological Research Conference*, 472-478.

Moussaid, El Fkihi, & Zennayi. (2019). *Citrus Orchards Monitoring based on Remote Sensing and Artificial Intelligence Techniques: A Review of the Literature.* Science and Technology Publication.

Moustafa, A. A. 2007. Effect of bagging period of spathe after pollination on fruit set, yield and fruit quality of "Seewy" dates under Fayoum Governorate conditions. *Proceeding of the Fourth Symposium on the Date Palm in Saudi Arabia, Al-Hassa*, 123.

Muhsen, K. A., Ibrahim, A., & Jasim, A. M. (2013). Effect of NPK and Hoagland solution on accumulation of date palm plantlets Phoenix dactylifera L. cv. Barhee production in vitro. *Journal of Basrah Researches (Sciences)*, *39*(4B), 32–45.

Mullen, J. D. (1995). *Estimating environmental and human health benefits of reducing pesticide use through integrated pest management programs* (Doctoral dissertation). Virginia Tech.

Munier, P. (1973). The date palm. *Techniques Agricoles et Productions Tropicales,* (24).

Munj, A. Y., Zote, V. K., Rut, R. A., & Salvi, B. R. (2017). Survey and surveillance of pollinators of mango in south konkan coastal region of Maharashtra. *Journal of Entomology and Zoology Studies*, *5*(3), 190–192.

Mupambi, G., Anthony, B., Layne, D. R., Musacchi, S., Serra, S., Schmidt, T., & Kalcsits, L. A. (2018). The influence of protective netting on tree physiology and fruit quality of apple: A review. *Scientia Horticulturae*, *236*, 60–72. doi:10.1016/j.scienta.2018.03.014

Murashige, T., & Skoog, F. (1962). A revised medium for rapid growth and bioassays with tobacco tissue cultures. *Physiologia Plantarum*, *15*(3), 473–497. doi:10.1111/j.1399-3054.1962.tb08052.x

Murugasridevi, K., Jeyarani, S., & Mohankumar, S. (2019). *Occurrence of groundnut Leafminer (GLM), Aproaerema modicella Deventer (Lepidoptera: Gelechiidae) and its parasitoid fauna in various groundnut growing areas of Tamil Nadu.* Academic Press.

Murugasridevi, K., Jeyarani, S., & Kumar, S. M. (2021, December 4). Incidence of Groundnut Leafminer (GLM), *Aproaerema modicella* (Deventer) (Lepidoptera: Gelechiidae) and its Parasitic Fauna on Alternate Leguminous Hosts in Tamil Nadu, India. *Legume Research.* Advance online publication. doi:10.18805/LR-4672

Murugasridevi, K., Jeyarani, S., Nelson, S. J., Kumar, S. M., & Nakkeeran, S. (2021 a). Assessment of Diversity Indices and DNA Barcoding of Parasitic Fauna Associated with Groundnut Leafminer (GLM), *Aproaerema modicella* Deventer (Lepidoptera: Gelechiidae). *Legume Research.* Advance online publication. doi:10.18805/LR-4579

Mustafa, A. B., Harper, D. B., & Johnston, D. E. (1986). Biochemical changes during ripening of some Sudanese date varieties. *Journal of the Science of Food and Agriculture*, *37*(1), 43–53. doi:10.1002/jsfa.2740370107

Mutanal, S. M., Patil, S. J., Shahapurmath, G., & Maheswarappa, V. (2009). Performance of arable crops in a teak based agroforestry system. *Karnataka Journal of Agricultural Sciences*, *22*(4), 854–856.

Muthukrishnan, Benjamin, & Sathishkannan, Kumar, & Rao. (2013). *In vitro* propagation of genus Ceropegia and Retrosynthesis of cerpegin- A Review. *International Journal of Pharmaceutical Sciences and Research, 22*(2), 315–330.

Muthukrishnan, S., Kumar, T. S., & Rao, M. V. (2013). Effects of different media and organic additives on seed germination of *Geodorum densiflorum* (Lam) Schltran endangered orchid. *International Journal of Scientific Research, 2*(8), 2277–8179.

Myburg, A. A., Hussey, S. G., Wang, J. P., Street, N. R., & Mizrachi, E. (2019). Systems and synthetic biology of forest trees: A bioengineering paradigm for Woody biomass feedstocks. *Front. Plant Sci., 10*, 775. doi:10.3389/fpls.2019.00775 PMID:31281326

Myhara, R. M., Karkalas, J., & Taylor, M. S. (1999). The composition of maturing Omani dates. *Journal of the Science of Food and Agriculture, 79*(11), 1345–1350. doi:10.1002/(SICI)1097-0010(199908)79:11<1345::AID-JSFA366>3.0.CO;2-V

Nagavallemma, K. P., Wani, S. P., Lacroix, S., Padmaja, V. V., Vineela, C., Rao, M. B., & Sahrawat, K. L. (2004). *Vermicomposting: Recycling Wastes into Valuable Organic Fertilizer.* Global Theme on Agroecosystems Report no. 8. International Crops Research Institute for the Semi-Arid Tropics, Patancheru.

Nagodawithana, W.T. (1991). *Yeast technology. Universal foods cooperation.* Van Nostrand.

Nalawde & Bhalerao. (2013). Comparative account of Effect of Biofertilizers on the growth and biochemicalparameters of *Vigna mungo* (L.Hepper). *International Journal of Advanced Research in Biological Sciences, 2*(5), 62–66.

Namkoong, G., Kang, H. C., & Brouard, J. S. (1988). Tree breeding: Principles and strategies. Springer, New York, Monograph. *Theoretical and Applied Genetics*, 11.

Nasher, S. H. (2008). The effect of magnetic water on growth of chick-pea seeds. *Engineering and Technology Journal, 26*(9), 4.

Nasir, M. J. (1996). *Effect of IBA on rooted and un-rooted offshoots of date palm (Phoenix dactylifera L.) cv. Hillawii* [M.Sc. Thesis]. Dept. Hort., Univ. Agric., Faisalabad, Pakistan.

Nassef, M. (1999). Juvenile hormone mimic and plant1derived oils as control agents against whitefly, Bemisia tabaci (Genn.), on cotton. *Egypt J. Agrie. Res., 77*(2), 691–699.

Nath, R., & Sikha, D. (2019). Insect pests of citrus and their management. *Int J Pl Pr, 12*(2), 188–196. doi:10.15740/HAS/IJPP/12.2/188-196

Naumann, I. D. (1991). Hymenoptera. In *The Insects of Australia* (2nd ed., Vol. 2, pp. 917–1000). Melbourne University Press.

Navarro, C.C., Aguilar, A. & Marí, F.G. (2008). Pezothrips Kellyanus, Thrips causante de daños en frutos de cítricos. *Levante agrícola: Revista internacional de cítricos, 299*, 298-303.

Navarro, C. C., Aguilar, A., & Marí, F. G. (2011). Population trend and fruit damage of *Pezothrips kellyanus* in Citrus orchards in Valencia (Spain). *Integrated Control in Citrus Fruit Crops IOBC Bulletin SROP, 38*, 204–209.

Nazam El-Din, A. M. M., & Abd El-Hameed, A. K. E. (2001) Study on the storage of Egyptian Siwi date variety (semi-dry date). *The second international conference on date palm.*

Ndiaye, M., Dieng, E. O., & Delhove, G. (2008). Population dynamics and on-farm fruit fly integrated pest management in mango orchards in the natural area of Niayes in Senegal. *Pest Management in Horticultural Ecosystems, 14*(1), 1–8.

Neale, D. B., & Kremer, A. (2011). Forest tree genomics: Growing resources and applications. *Nature Reviews. Genetics, 12*(2), 111–122. doi:10.1038/nrg2931 PMID:21245829

Nebauer, S. G., Renau-Morata, B., Guardiola, J. L., & Molina, R. V. (2011). Photosynthesis down-regulation precedes carbohydrate accumulation under sink limitation in Citrus. *Tree Physiology, 31*(2), 169-177. doi:10.1093/treephys/tpq103

Negi, R. S., Baghel, B. S., & Gautam, U. S. (2013). Effect of method of orchard establishment and propagation on growth and development of aonla (*Emblica officinalis* Gaertn.) plants in wastelands. *Journal of Applied Horticulture, 15*(3), 227–231. doi:10.37855/jah.2013.v15i03.46

Nekar, M. M., Goroji, P. T., & Channakeshava, S. (2007). Growth and yield of greengram (Vigna radiata) near teak tree row. *Journal of Ecobiology, 21*(3), 289–294.

Nestel, D., Cohen, Y., Shaked, B., Alchanatis, V., Nemny-Lavy, E., Miranda, M. A., Sciarretta, A., & Papadopoulos, N. T. (2019). An Integrated Decision Support System for environmentally-friendly management of the Ethiopian fruit fly in greenhouse crops. *Agronomy (Basel), 9*(8), 459. doi:10.3390/agronomy9080459

Neuenschwander, P. (2001). Biological control of the cassava mealybug in Africa: a review. *Biological Control, 21*, 214–229. doi:10.1006/bcon.2001.0937

Newaj, R., & Rai, P. (2005). Aonla-based agroforestry system: A source of higher income under rainfed conditions. *Indian Farming, 55*(9), 24–27.

Ng, T. B., Liu, J., Wong, J. H., Ye, X., Wing Sze, S. C., Tong, Y., & Zhang, K. Y. (2012). Review of research on Dendrobium, a prized folk medicine. *Applied Microbiology and Biotechnology, 93*(5), 1795–1803. doi:10.100700253-011-3829-7 PMID:22322870

Nicolás, E., Barradas, V. L., Ortuño, M. F., Navarro, A., Torrecillas, A., & Alarcón, J. J. (2008). Environmental and stomatal control of transpiration, canopy conductance and decoupling coefficient in young lemon trees under shading net. *Environmental and Experimental Botany, 63*(1-3), 200-206. doi:10.1016/j.envexpbot.2007.11.007

Nimmi, V., & Medhu, G. (2009). Effect of pre-sowing treatment with permanent magnetic field on germination and growth of Chili. *International Agrophysics, 23*, 195–198.

Niu, J. Z., Hull-Sanders, H., Zhang, Y. X., Lin, J. Z., Dou, W., & Wang, J. J. (2014). Biological control of arthropod pests in citrus orchards in China. *Biological Control, 68*, 15–22. doi:10.1016/j.biocontrol.2013.06.005

Nixon, R. W. (1957). *Effect of age and number of leaves on fruit production of the date palm.* Academic Press.

Nixon, R. W., & Carpenter, J. B. (1978). *Growing dates in the United States (No. 207).* Department of Agriculture, Science and Education Administration.

Norikane, Teixeira da Silva, & Tanaka. (2013). Growth of in vitro Oncidesa plantlets cultured under cold cathodefluorescent lamps with super-elevated CO_2 enrichment. *Journal of Plant Sciences, 22*, 1–9.

Norikane, A., Takamura, T., Morokuma, M., & Tanaka, M. (2010). In vitro growth and single-leaf photosynthetic response of Cymbidium plantlets to super-elevated CO2 under cold cathode fluorescent lamps. *Plant Cell Reports, 29*(3), 273–283. doi:10.100700299-010-0820-1 PMID:20094885

OECD. (2013). *OECD guidelines on the production of forest reproductive materials*. Organization for Economic Cooperation and Development Trade and Agriculture Directorate.

Ollerton, J., Winfree, R., & Tarrant, S. (2011). How many flowering plants are pollinated by animals? *Oikos, 120*(3), 321–326. doi:10.1111/j.1600-0706.2010.18644.x

Omar, A. E. D. K., Soliman, S. S., & Ahmed, M. A. (2013). Impact of leaf/bunch ratio and time of application on yield and fruit quality of Barhi date palm trees (Phoenix dactylifera L.) under Saudi Arabian conditions. *Journal of Testing and Evaluation*, *41*(5), 813–817. doi:10.1520/JTE20120340

Onufrieva, K. S., Hickman, A. D., Leonard, D. S., & Tobin, P. C. (2019). Relationship between efficacy of mating disruption and gypsy moth density. *International Journal of Pest Management*, *65*(1), 44–52. doi:10.1080/09670874.2018.1455116

Ordoñez, C., Tejada, M., Benitez, C., & Gonzalez, J. L. (2006). Characterization of a phosphorus–potassium solution obtained during a protein concentrate process from sunflower flour. Application on rye-grass. *Bioresource Technology*, *97*(3), 522–528.

Orlikowski, L. B., & Szkuta, G. (2006). Phythophthora rot of some Orchidaceae – new disease in Poland. *Phytopathologia Polonica*, *40*, 57–61.

Orwa, C., Mutua, A., Kindt, R., Jamnadass, R., & Simons, A. (2009). *Agroforestree Database: A tree reference and selection guide version 4.0.* https://www.worldagroforestry.org/af/treedb/

Osadchuk, V.D., Saranchuk, I.I., Lesyk, O.B., & Olifirovych, V.O. (2020). Selective Breeding in Plant Growing in Bukovina. *Taurian Scientific Herald*, p. 16.

Osman, Moustafa, El-Galil, & Ahmed. (2011). Effect of yeast and Effective Microorganisms (Em1) application on the yield and fruit characteristics of Bartamuda date palm under Aswan conditions. *Assiut J. of Agric. Sci., 42*, 332-349.

Othmani, A., Bayoudh, C., Drira, N., Marrakchi, M., & Trifi, M. (2009). Regeneration and molecular analysis of date palm (Phoenix dactylifera L.) plantlets using RAPD marker. *African Journal of Biotechnology*, *8*, 813–820.

Otsuka, I., & Ozeki, S. (2006). Does Magnetic Treatment of Water Change Its Properties? *The Journal of Physical Chemistry B*, *110*(4), 1509–1512. doi:10.1021/jp056198x PMID:16471705

Pakum, W., Watthana, S., Srimuang, K. O., & Kongbangkerd, A. (2016). Influence of medium composition on *in vitro* propagation of Thai's endangered orchid: *Bulbophyllum nipondhii* Seiden. *Plant Tissue Culture & Biotechnology*, *25*(1), 37–46. doi:10.3329/ptcb.v26i1.29765

Pandey, D. N. (2002). Carbon sequestration in agroforestry systems. *Climate Policy*, *2*(4), 367–377. doi:10.3763/cpol.2002.0240

Pandey, D. N. (2007). Multifunctional agroforestry systems in India. *Current Science, VOL.*, *92*(4), 455–463.

Pandey, V., Patel, M. G., Chaudhari, G. B., Patel, J. R., Bhatt, B. K., Vadodaria, R. P., & Shekh, A. M. (2003). Influence of weather parameters on the population dynamics of mango hopper. *Journal of Agrometeorology*, *5*(1), 51–59. doi:10.54386/jam.v5i1.586

Pan, I. C., Liao, D. C., Wu, F. H., Daniell, H., Singh, N. D., Chang, C., Shih, M. C., Chan, M. T., & Lin, C. S. (2012). Complete chloroplast genome sequence of an orchid model plant candidate: *Erycina pusilla* apply in tropical Oncidium breeding. *PLoS One*, *7*(4), 34738. doi:10.1371/journal.pone.0034738 PMID:22496851

Pant, B. (2013). Medicinal orchids and their uses: Tissue culture a potential alternative for conservation. *African Journal of Plant Science*, *7*(10), 448–467. doi:10.5897/AJPS2013.1031

Pant, B., Shrestha, S., & Pradhan, S. (2011). *In vitro* seed germination and seedling development of *Phaius tancarvilleae* (L'Her.) Blume. *TheScientificWorldJournal*, *9*, 50–52.

Pant, M., Bisht, P., & Gusain, M. P. (2010). *In vitro* propagation through axillary bud culture of *Swertia chirata* Buch-Ham ex Wall: An endangered medicinal herb. *International Journal of Integrative Biology*, *10*(1), 48–53.

Pant, N. (2007). Contribution of agro forestry in rural economy - a case study. *Vegetos*, *20*(1), 89–93.

Pardo, A., & Borges, P. A. V. (2020). Worldwide importance of insect pollination in apple orchards: A review. *Agriculture, Ecosystems & Environment*, *293*, 106839. doi:10.1016/j.agee.2020.106839

Park, Y., & Runkle, E. S. (2018). Far-red radiation and photosynthetic photon flux density independently regulate seedling growth but interactively regulate flowering. *Environmental and Experimental Botany, 155*, 206-216. doi:10.1016/j.envexpbot.2018.06.033

Park, Y.-S. (2014). Conifer Somatic Embryogenesis and Multi- Varietal Forestry. Challenges and Opportunities for the World's Forests in the 21st Century. *Forestry Sciences*, *81*, 425–440. doi:10.1007/978-94-007-7076-8_17

Parsons, S., Wang, B., Judd, J., & Stephenson, T. (1997). Magnetic treatment of calcium carbonate scale-effect of pH control. *Water Research*, *31*(2), 339–342. doi:10.1016/S0043-1354(96)00238-2

Patel, A. T., & Kumar, S. (2020). Chemical control of mango leaf gall midge *Procontarinia matteiana. Ann. Entomol, 38*(1-2), 21-26.

Pathak, R. K. (2003). *Status Report on Genetic Resources of Indian Gooseberry-Aonla (Emblica officinalis Gaertn.) in South and Southeast Asia. IPGRI Office for South Asia National Agriculture Science Centre (NASC).* DPS Marg, Pusa Campus.

Payne, P. J. C. (1956). The relationship between the mechanical properties of soils and the performance of simple cultivation implements. *Journal of Agricultural Engineering Research*, *1*, 23–50.

Payne, P. J. C., & Tanner, D. W. (1959). The relationship between rake angle and the performance of simple cultivation implements. *Journal of Agricultural Engineering Research*, *4*, 312–325.

Pedigo, L. P. (1996). Entomology and pest management (2nd ed.). Prentice-Hall Inc.

Peña, J. E. (2002, September). Integrated pest management and monitoring techniques for mango pests. In *VII International Mango Symposium 645* (pp. 151-161). Academic Press.

Peña, J. E., Sharp, J. L., & Wysoki, M. (2002). *Tropical Fruit Pests and Pollinators Biology, Economic Importance, Natural Enemies and Control.* CABI. doi:10.1079/9780851994345.0000

Perdok, U. D., & Van de Werken, G. (1983, March). Power and Labour requirements in soil tillage. *Soil & Tillage Research, V3*(1), 3–25. doi:10.1016/0167-1987(83)90013-2

Pérez-Escobar, O. A., Chomicki, G., Condamine, F. L., Karremans, A. P., Bogarín, D., Matzke, N. J., Silvestro, D., & Antonelli, A. (2017). Recent origin and rapid speciation of Neotropical orchids in the world's richest plant biodiversity hotspot. *The New Phytologist*, *215*(2), 891–905. doi:10.1111/nph.14629 PMID:28631324

Peters, D. T., & Musunuru, K. (2012). Functional evaluation of genetic variation in complex human traits. *Human Molecular Genetics*, *21*(R1), R18–R23. Advance online publication. doi:10.1093/hmg/dds363 PMID:22936690

Pfalgraz, K.E. (2002). Loss of a Legacy, Fusarium oxysporum in Ornamental Phoenix canariensis. *Principes: Journal of The International Palm Society, 46*(3).

Pintaud, J.-C., Zehdi, S., Couvreur, T., Barrow, S., Henderson, S., Aberlenc-Bertossi, F., . . . Billotte, N. (2010). Species delimitation in the genus Phoenix (Arecaceae) based on SSR markers, with emphasis on the identity of the date palm (Phoenix dactylifera L.). *Diversity, phylogeny, and evolution in the monocotyledons*, 267-286.

Pipfruit N. Z. Inc. (2011) Application to import for release or to release from containment new organisms. Application number 0022269. EPA.

Plant, Zhao, Cun, Chen, & Zhu. (2010). Detection of Acidovorax avenae subsp.cattleyae by real-time fluorescent PCR. Wuli Xuebao, 40, 235–241. https://www.cabdirect.org/abstracts/20103213302.html

Poinar, G. O., Jr. (1986). Entomophagous nematodes. In Biological plant and health protection. New York: Gustav Fi1scher Verlag.

Poinar, G. O. Jr. (1979). *Nematodes for biologi1cal control of insects.* CRC Press.

Poinar, G. O. Jr. (1990). Biology and taxonomy of Steinernematidae and Heterorhabditidae. In R. Gaugler & H. K. Kaya (Eds.), *Entomopatho1genic nematodes in biological control* (pp. 23–61). CRC Press.

Pontikakos, C. M., Tsiligiridis, T. A., Yialouris, C. P., & Kontodimas, D. C. (2012). Pest management control of olive fruit fly (Bactrocera oleae) based on a location-aware agro-environmental system. *Computers and Electronics in Agriculture*, *87*, 39–50. doi:10.1016/j.compag.2012.05.001

Popova, E., Kim, H. H., Saxena, P. K., Engelmann, F., & Pritchard, H. W. (2016). Frozen beauty: The cryo-biotechnology of orchid diversity. *Biotechnology Advances*, *34*(4), 380–403. doi:10.1016/j.biotechadv.2016.01.001 PMID:26792590

Porcel, M., Andersson, G. K. S., Pålsson, J., & Tasin, M. (2018). Organic management in apple orchards: Higher impacts on biological control than on pollination. *Journal of Applied Ecology*, *55*(6), 2779–2789. doi:10.1111/1365-2664.13247

Pospíšilová, J., Synková, H., Haisel, D., Čatský, J., Wilhelmová, N., & Šrámek, F. (1999). Effect of elevated CO_2 concentration on acclimation of tobacco plantlets to ex vitro conditions. -. *Journal of Experimental Botany*, *50*(330), 119–126. doi:10.1093/jxb/50.330.119

POST. (2006). *Carbon footprint of electricity generation.* POST note 268, October 2006, Parliamentary Office of Science and Technology. http://www.parliament.uk/documents/upload/postp n268.pdf

Potting, R. P. J., & Knight, A. L. (2001). Predicting the efficacy of modified modes of action of a pheromone-based attracticide: A bisexual attractant and autosterilisation. *IOBC/WPRS Bulletin.*

Pouya, M. B., Bonzi, M., Gnankambary, Z., Traoré, K., Ouédraogo, J. S., Somé, A. N., & Sédogo, M. P. (2013). Pratiques actuelles de gestion de la fertilité des sols et leurs effets sur la production du cotonnier et sur le sol dans les exploitations cotonnières du Centre et de l'Ouest du Burkina Faso. *Cahiers Agricultures*, *22*(4), 282–292.

Prasada Rao, G. S. L. H. V., & Beevi, S. P. (2008). Forewarning tea mosquito bug *Helopeltis antonii* Sign.(Miridae: Hemiptera) in cashew. In *Proceeding of the International Symposium on Agrometeorology and food security* (pp. 18-21). Academic Press.

Preece, J. E. (2010). *Acclimatization of Plantlets from In vitro to the Ambient Environment. Encyclopedia of Industrial Biotechnology: Bioprocess.* Bioseparation, and Cell Technology. doi:10.1002/9780470054581.eib593

Preti, M., Verheggen, F., & Angeli, S. (2021). Insect pest monitoring with camera-equipped traps: Strengths and limitations. *Journal of Pest Science*, *94*(2), 1–15. doi:10.100710340-020-01309-4

Prins, M. D. T. (2018). *The impact of shade netting on the microclimate of a citrus orchard and the tree's physiology* (Doctoral dissertation). Stellenbosch University.

Prior, R. L., & Cao, G. (2000). Antioxidant phytochemicals in fruits and vegetables: Diet and health implications. *J. Hort. Sci.*, *35*(4), 588–592. doi:10.21273/HORTSCI.35.4.588

Proesmans, W., Smagghe, G., Meeus, I., Bonte, D., & Verheyen, K. (2019). The effect of mass-flowering orchards and semi-natural habitat on bumblebee colony performance. *Landscape Ecology*, *34*(5), 1033–1044. doi:10.100710980-019-00836-5

Pulawska, J., Mikicinski, A., & Orlikowski, L. B. (2013). Acidovorax cattleyae causal agent of bacterial brown spot of Phalaenopsis lueddemanniana in Poland. *Journal of Plant Pathology*, *95*, 407–410.

Qados, A. M. S. A., & Hozayn, M. (2010). Magnetic water technology, a novel tool to increase growth, yield and chemical constituents of lentil (Lens esculenta) under greenhouse condition. *American-Eurasian Journal of Agricultural & Environmental Sciences*, *7*(4), 457–462.

Qin, T. K., & Gullan, P. J. (1998). Systematics as a tool for pest man1agement: case studies using scale insects and mites. *Pest Management—Future Challenges, Vols 1 and 2, Pro1ceedings of the Sixth Australasian Applied Entomological Research Conference*, 479–488.

Qi, Ying, Xin, & Li-N., Shi-Jin, Wen, Juan, Zong, & Ma. (2021). Effects of CO_2 on transplantation of grape plantlets cultured in vitro by promoting photosynthesis. *Scientia Horticulturae*, *287*(20), 110286.

Quaik, S., Embrandiri, A., Rupani, P. F., & Ibrahim, M. H. (2012) Potential of vermicomposting leachate as organic foliar fertilizer and nutrient solution in hydroponic culture: a review. *2nd International Conference on Environment and BioScience IPCBEE*, 43–47.

Quaik, S., & Ibrahim, M. H. (2013). A review on potential of vermicomposting derived liquids in agricultural use. *Int J Sci Res Pub*, *3*(3), 1–6.

Rabeh, M. R. M., & Kassem, H. A. (2003). The effect of bagging the spathes after pollination on yield and quality of Zaghloul and Samany dates. *Zagazig J. Agric. Res.*, *21*(3B), 935–944.

Rader, R., Bartomeus, I., Garibaldi, L. A., Garratt, M. P. D., Howlett, B. G., Winfree, R., Cunningham, S. A., Mayfield, M. M., Arthur, A. D., & Andersson, G. K. S. (2016). Non-bee insects are important contributors to global crop pollination. *Proceedings of the National Academy of Sciences of the United States of America*, *113*, 146–151. doi:10.1073/pnas.1517092112 PMID:26621730

Raffeiner, B., Serek, M., & Winkelmann, T. (2009). *Agrobacterium tumefaciens* mediated transformation of *Oncidium* and *Odontoglossum* orchid species with the ethylene receptor mutant gene etr1-1. *Plant Cell, Tissue and Organ Culture*, *98*(2), 125–134. doi:10.100711240-009-9545-7

Raghavan, V. (2000). Developmental biology of flowering plants. Springer-Verlag GmbH and Co. KG.

Rai, J., Singh, S. P., & Singh, A. K. (2016). Economics of marketing and processing of aonla in district Pratapgarh, Uttar Pradesh. *International Journal of Commerce and Business Management*, *9*(2), 209–213. doi:10.15740/HAS/IJCBM/9.2/209-213

Rajapakse, N. C., & Shahak, Y. (2008). 12 Light-quality manipulation by horticulture industry. Annual Plant Reviews. *Light and Plant Development*, *30*, 290.

Rajeswari, V., & Paliwal, K. (2008). In vitro plant regeneration of red sanders (Pterocarpus santalinus L.f) from cotyledonary nodes. *Indian Journal of Biotechnology*, *7*, 541–546.

Rajmohan, K. (2011). Date palm tissue culture: A pathway to rural development. In S. M. Jain, J. M. Al-Khayri, & D. V. Johnson (Eds.), *Date palm biotechnology*. Springer. doi:10.1007/978-94-007-1318-5_3

Rana, R. R., & Kapoor, P. (2013). Biofertilizersand Their Role in Agriculture. *Popular Kheti*, *1*(1), 56–61.

Raneesha, Halgamuge, Wirasagoda, & Syed. (2019). Adoption of the Internet of Things (IoT) in Agriculture and Smart Farming towards Urban Greening: A Review. *International Journal of Advanced Computer Science and Applications*, *10*(4), 11–28.

Rasmaia Darwesh, S., Essam Madbolly, A., & Ezz Gadalla, G. (2013). Impact of Indole Butyric Acid and Paclobutrazol on Rooting of Date Palm (Phoenix dactylifera L.) Off-Shoots Cultivar Zaghloul. *Journal of Horticultural Science & Ornamental Plants*, *5*(3), 145–150.

Rasmia Darwesh, S. S. (2010). *Phoenix dactylifera* cv. Medjol plantlets as affected by yeast extract and NPK fertilizers. *The Six International Conference Of Sustainable Agricultural Development*, 115-130.

Raveh, E., Cohen, S., Raz, T., Yakir, D., Grava, A., & Goldschmidt, E. E. (2003). Increased growth of young citrus trees under reduced radiation load in a semi-arid climate1. *Journal of Experimental Botany*, *54*(381), 365-373. doi:10.1093/jxb/erg009

Ray. (2016). Remote Sensing in Agriculture. *International Journal of Environment, Agriculture and Biotechnology (IJEAB)*, *1*(3), 362 – 367.

Reardon, R. C., & Podgwaite, J. D. (1994). Summary of efficacy evaluations using aerially applied Gypchek against gypsy moth in the USA. *Journal of Environmental Science and Health. Part B, Pesticides, Food Contaminants, and Agricultural Wastes*, *29*(4), 739–756. doi:10.1080/03601239409372902

Reddy, P. V. R., Varun Rajan, V., Thangam, D., & Chakravarthy, A. K. (2015, April). Stem borers in mango: species diversity and damage patterns. In *4th congress on Insect Science, Punjab Agricultural University, Ludhiana* (pp. 16-17). Academic Press.

Reddy, M. S., Murali, T. S., Suryanarayanan, T. S., Rajulu, M. B. G., & Thirunavukkarasu, N. (2016). Pestalotiopsis species occur as generalist endophytes in trees of Western Ghats forests of southern India. *Fungal Ecology*, *24*, 70–75. doi:10.1016/j.funeco.2016.09.002

Reddy, P. V. R., Gundappa, B., & Chakravarthy, A. K. (2018). Pests of mango. In *Pests and their management* (pp. 415–440). Springer. doi:10.1007/978-981-10-8687-8_12

Redmond. (2017). Controller Design for an Osprey Drone to Support Precision Agriculture Research in Oil Palm Plantations. *ASABE 2017 Annual International Meeting*.

Rees, W. E. (1992). Ecological footprints and appropriated carrying capacity: What urban economics leaves out. *Environment and Urbanization*, *4*(2), 121–130. doi:10.1177/095624789200400212

Reinikka, M. A. (1995). *A history of the Orchid*. Portland Timber Press.

Reis, C. A. F., Gonçalves, F. M. A., Rosse, L. N., Costa, R. R. G. F., & Ramalho, M. A. P. (2011). Correspondence between performance of *Eucalyptus* spp. Trees selected from family and clonal tests. *Genetics and Molecular Research*, *10*(2), 1172–1179. doi:10.4238/vol10-2gmr1078 PMID:21732281

Reis, D., & Vian, B. (2004). Helicoidal pattern in secondary cell walls and possible role of xylans in their construction. *Comptes Rendus Biologies*, *327*(9–10), 785–790. doi:10.1016/j.crvi.2004.04.008 PMID:15587069

Reniprabha, A., Muralidharan, S., & Subramanian, A. (1999). Impact of plant extracts on molting in the larvae of mosquito Culex sp. *Environment and Ecology*, *17*(4), 1022–1024.

Resende, M. F. R. Jr, Muñoz, P., Acosta, J. J., Peter, G. F., Davis, J. M., Grattapaglia, D., Resende, M. D. V., & Kirst, M. (2012). Accelerating the domestication of trees using genomic selection: Accuracy of prediction models across ages and environments. *The New Phytologist*, *193*(3), 617–624. doi:10.1111/j.1469-8137.2011.03895.x PMID:21973055

Reyes, M., Franck, P., Charmillot, P.-J., Ioriatti, C., Olivares, J., Pasqualini, E., & Sauphanor, B. (2007, September 1). Diversity of insecticide resistance mechanisms and spectrum in European populations of the codling moth, Cydia pomonella. *Pest Management Science*, *63*(9), 890–902. doi:10.1002/ps.1421 PMID:17665366

Rezapanah, M., Shojai-Estabragh, S., Huber, J., & Jehle, J. A. (2008, December 1). Molecular and biological characterization of new isolates of Cydia pomonella granulovirus from Iran. *Journal of Pest Science, 81*(4), 187–191. doi:10.100710340-008-0204-2

Ribeiro, C. F., Mizobutsi, E. H., Silva, D. G., Pereira, J. C. R., & Zambolim, L. (1998). Control of Meloidognye javanica on lettuce with organic amendments. *Fitopatologia Brasileira, 23*, 42–44.

Ribeiro, R. V., & Machado, E. C. (2007). Some aspects of citrus ecophysiology in subtropical climates: Re-visiting photosynthesis under natural conditions. *Brazilian Journal of Plant Physiology, 19*(4), 393–411. doi:10.1590/S1677-04202007000400009

Ricci, B., Franck, P., Toubon, J.-F., Bouvier, J.-C., Sauphanor, B., & Lavigne, C. (2009). The influence of landscape on insect pest dynamics: A case study in southeastern France. *Landscape Ecology, 24*(3), 337–349. doi:10.100710980-008-9308-6

Ricketts, T. H., Daily, G. C., Ehrlich, P. R., & Michener, C. D. (2004). Economic value of tropical forest to coffee production. *Proceedings of the National Academy of Sciences of the United States of America, 101*(34), 12579–12582. doi:10.1073/pnas.0405147101 PMID:15306689

Ring, Schroeder, Nyczepir, Payne, & Snow. (1988). Nematodes screened against pecan weevil larvae, 1986. *Insectilcide and Acaricide Tests, 13*(184).

Risch, N., & Merikangas, K. (1996). The future of genetic studies of complex human diseases. *Science, 273*(5281), 1516–1517. doi:10.1126cience.273.5281.1516 PMID:8801636

Rivera-Cruz, M., Trujillo, A., Córdova, G., Kohler, J., Caravaca, F., & Roldán, A. (2008). Poultry Manure and Banana Waste Are Effective Bio-Fertilizer Carriers for Promoting Plant Growth and Soil Sustainability in Banana Crops. *Soil Biology & Biochemistry, 40*, 3092–3095. https://dx.doi.org/10.1016/j.soilbio.2008.09.003

Román, C., Peris, M., Esteve, J., Tejerina, M., Cambray, J., Vilardell, P., & Planas, S. (2022). Pesticide dose adjustment in fruit and grapevine orchards by DOSA3D: Fundamentals of the system and on-farm validation. *The Science of the Total Environment, 808*, 152158. doi:10.1016/j.scitotenv.2021.152158 PMID:34871680

Roubik, D. W. (Ed.). (1995). *Pollination of cultivated plants in the tropics*. FAO.

Roy, S. K., Trehan, S. P., & Sharma, R. C. (2000). Longterm nutrient management in potato-sun" ower- rice system for sustainable productivity. *Intl. Conference on Managing Natural Resources, New Delhi. Extended Summaries, 3*, 920-921.

Rugienius, R., Šikšnianas, T., & Sasnauskas, A. (2013). Žemuogių veislių, hibridų ir atrinktų linijų tyrimai. *Sodininkystė ir daržininkystė, 31*(1–2), 3–13.

Rull, J., Aluja, M., Feder, J., & Berlocher, S. (2006). Distribution and host range of hawthorn1infesting Rhagoletis (Diptera: Tephritidae) in Mexico. *Annals of the Entomological Society of America, 99*(4), 662–672. doi:10.1603/0013-8746(2006)99[662:DAHROH]2.0.CO;2

Rung, Y. W., Shao, Y. C., Ting, F. H., Keng, C. C., Ting, I., Yen, H. L., & Yu, S. C. 2016. Cryopreservation of Orchid Genetic Resources by Desiccation: A Case Study of Bletilla formosana. Cryopreservation in Eukaryotes. doi:10.5772/65302

Sadka, A., Shahak, Y., Cohen, S., Dovjek, I., Nemera, D. B., Wachsmann, Y., Shlizerman, L., Ratner, K., Kamara, I., Morozov, M., & Charuvi, D. (2019). Top netting as a practical tool to mitigate the effect of climate change and induce productivity in citrus: summary of experiments using photo-selective nets. *XI International Symposium on Protected Cultivation in Mild Winter Climates and I International Symposium on Nettings, 1268*, 265-270.

Saez, P. L., Bravo, L. A., Saez, K. L., Sánchez-Olate, M., Latsague, M. I., & Ríos, D. G. (2012). Photosynthetic and leaf anatomical characteristics of Castanea sativa: A comparison between in vitro and nursery plants. *Biologia Plantarum, 56*(1), 15–24. doi:10.100710535-012-0010-9

Saidu, Clarkson, Adamu, Mohammed, & Jibo. (2017). Application of ICT in Agriculture: Opportunities and Challenges in Developing Countries. *International Journal of Computer Science and Mathematical Theory, 3*(1).

Sakai, Keene, Renard, & De Backer. (2016). FBN1: The disease-causing gene for *Marfan syndrome* and other genetic disorders. *Gene, 591*(1), 279-291. doi:10.1016/j.gene.2016.07.033

Sakai, A. (2000). Development of cryopreservation techniques. In F. Engelmann & H. Takagi (Eds.), *Cryopreservation of Tropical Plant Germplasm* (pp. 1–7). IPGRI.

Salampasis, M., & Bournaris, T. (2019). *Information and Communication Technologies in Modern Agricultural Development. 8th International Conference, HAICTA 2017*, Chania, Greece.

Saleh, S. M., Heggi, M. A. M., Abdrabbo, M. A. A. & Farag, A. A. (2017). Heat Waves Investigation During Last Decades in Some Climatic Regions in Egypt. *Egypt. J. Agric. Res., 95*(2).

Saleh, J. (2009). Yield and chemical composition of 'Piarom' Date-Palm *Phoenix dactylifera* L. as affected by nitrogen and phosphorus Levels. *International Journal of Plant Production, 3*(3), 57–64.

Salem, S. A., Abd El-Salam, A. M. E., & Abdel-Raheem, M. A. (2021). Field Evaluation of the Efficacy of *Moringa oleifera* in Controlling Two Main Pests, *Aphis gossypii* and *Bemisia tabaci* Infesting Tomato Plants, Academic. *Journal of Entomology, 14*(1), 24–29.

Salem, S. A., Abd-El Salam, A. M. E., & Ahmed, S. (2017). Evaluate and assess the use of some insecticides of plant origin against *Scritothrips citri* Moulton (*Thysanoptera, Thripidae*) in reducting distortions orange fruits for export. *Bioscience Research, 14*(2), 354–361.

Samietz, J., Graf, B., Höhn, H., Schaub, L., & Höpli, H. U. (2007). Phenology modelling of major insect pests in fruit orchards from biological basics to decision support: The forecasting tool SOPRA. *Bulletin OEPP. EPPO Bulletin. European and Mediterranean Plant Protection Organisation, 37*(2), 255–260. doi:10.1111/j.1365-2338.2007.01121.x

Samietz, J., Hoehn, H., Razavi, E., Schaub, L., & Graf, B. (2015). Decision support for sustainable orchard pest management with the Swiss forecasting system SOPRA. *Acta Horticulturae*, (1099), 383–390. doi:10.17660/ActaHortic.2015.1099.44

Samnegård, U., Alins, G., Boreux, V., Bosch, J., García, D., Happe, A. K., Klein, A. M., Miñarro, M., Mody, K., Porcel, M., Rodrigo, A., Roquer-Beni, L., Tasin, M., & Hambäck, P. A. (2019). Management trade-offs on ecosystem services in apple orchards across Europe: Direct and indirect effects of organic production. *Journal of Applied Ecology, 56*(4), 802–811. doi:10.1111/1365-2664.13292

Sandanayaka, M., Chhagen, A., Page-weir, N. E. M., & Charles, J. G. (2011). Colony optimisation of Mastrus ridens (Hymenoptera: Ichneumonidae), a potential biological control agent of codling moth in New Zealand. *New Zealand Plant Protection, 64*, 227–234. doi:10.30843/nzpp.2011.64.5959

Sarsaiya, S., Shi, J., & Chen, J. (2019). A comprehensive review on fungal endophytes and its dynamics on Orchidaceae plants: Current research, challenges, and future possibilities. *Bioengineered, 10*(1), 316–334. doi:10.1080/21655979.2019.1644854 PMID:31347943

Sasnauskas, A., Rugienius, R., Gelvonauskienė, D., Zalunskaitė, I., & Stanienė, G. (2007). Screening of strawberries with red stele (*Phytophthora Fragariae*) resistance gene *Rpf1* using sequence specific DNA markers. *Acta Horticulturae*, (760), 165–169. doi:10.17660/ActaHortic.2007.760.21

Satyagopal, K., Sushil, S. N., Jeyakumar, P., Shankar, G., Sharma, O. P., Sain, S. K., ... Latha, S. (2014). *AESA based IPM package for banana*. Academic Press.

Satyanarayana, G. (1981). Problems of grape production around Hyderabad. Andhra PRADESH GRAPE Growers Association.

Savolainen, O., Bokma, F., Knurr, T., Karkkainen, K., Pyhajarvi, T., & .... (2007). Adaptation of forest trees to climate change. In *Climate change and forest genetic diversity: Implications for sustainable forest management in Europe* (pp. 19–30). Bioversity International.

Sawaya, W. N., Khalil, J. K., Safi, W. N., & Al-Shalhat, A. (1983). Physical and chemical characterization of three Saudi date cultivars at various stages of development. *Canadian Institute of Food Science and Technology Journal, 16*(2), 87–92. doi:10.1016/S0315-5463(83)72065-1

Sawaya, W. N., Khatchadourian, H. A., Khalil, J. K., Safi, W. M., & Al-Shalhat, A. (1982). Growth and compositional changes during the various developmental stages of some Saudi Arabian date cultivars. *Journal of Food Science, 47*(5), 1489–1492. doi:10.1111/j.1365-2621.1982.tb04967.x

Schaberg, P. G., Dehayes, D. H., Hawley, G. J., & Nijensohn, S. E. (2008). Anthropogenic alterations of genetic diversity within three populations: Implications for forest ecosystem resilience. *Forest Ecology and Management, 256*(5), 855–862. doi:10.1016/j.foreco.2008.06.038

Scherer, L., Svenning, J. C., Huang, J., Seymour, C. L., Sandel, B., Mueller, N., Kummu, M., Bekunda, M., Bruelheide, H., Hochman, Z., Siebert, S., Rueda, O., & van Bodegom, P. M. (2020). Global priorities of environmental issues to combat food insecurity and biodiversity loss. *The Science of the Total Environment, 730*, 139096. doi:10.1016/j.scitotenv.2020.139096 PMID:32388110

Schrock, M., Matheson, D., Blumanhourst, M., & Thompson, J. (1986). A device for aiding gear selection in agricultural tractor. *Transactions of the ASAE. American Society of Agricultural Engineers, 29*(5), 1232–1236. doi:10.13031/2013.30301

Sciarretta, A., Tabilio, M. R., Amore, A., Colacci, M., Miranda, M. Á., Nestel, D., Papadopoulos, N. T., & Trematerra, P. (2019). Defining and evaluating a decision support System (DSS) for the precise pest management of the Mediterranean Fruit Fly, *Ceratitis capitata*, at the farm level. *Agronomy (Basel), 9*(10), 608. doi:10.3390/agronomy9100608

Scortichini, M., D'Ascenzo, D., & Rossi, M. P. (2005). New record of Acidovorax avenae subsp. cattleyae on orchid in Italy. *Journal of Plant Pathology, 87*, 244.

Sedeek, K. E. M., Qi, W., Schauer, M. A., Gupta, A. K., Poveda, L., Xu, S., Liu, Z.-J., Grossniklaus, U., Schiestl, F. P., & Schlüter, P. M. (2013). Transcriptome and Proteome Data Reveal Candidate Genes for Pollinator Attraction in Sexually Deceptive Orchids. *PLoS One, 8*(5), e64621. Advance online publication. doi:10.1371/journal.pone.0064621 PMID:23734209

Seelye, J. F., Burge, G. K., & Morgan, E. R. (2003). Acclimatizing tissue culture plants: Reducing the shock. *Proc. Int. Plant Prop. Soc., 53*, 85–90.

Semiarti, E. (2018). Biotechnology for Indonesian orchid conservation and industry. *Proceeding of Inventing Prosperous Future through Biological Research and Tropical Biodiversity Management, in AIP Conference Proceedings 2002.* 10.1063/1.5050118

Sghir, F., Chliyeh, M., Touati, J., Mouria, B., Ouazzani, A., Touhami, A., Filali-Maltouf, C., El Modafar Moukhli, A., Benkirane, R., & Douira, A. (2014). Effect of a dual inoculation with endomycorrhizae and *Trichoderma harzianum* on the growth of date palm seedlings. *Int. J. Pure App. Biosci., 2*(6), 12–26.

Shafqat, W., Jaskani, M. J., Maqbool, R., Chattha, W. S., Ali, Z., Naqvi, S. A., Haider, M.S., Khan, I., & Vincent, C. I. (2021). Heat shock protein and aquaporin expression enhance water conserving behavior of citrus under water deficits and high temperature conditions. *Environmental and Experimental Botany, 181*, 104270. doi:10.1016/j.envexpbot.2020.104270

Shahak, Y. (2012). Photoselective netting: an overview of the concept, research and development and practical implementation in agriculture. *International CIPA Conference 2012 on Plasticulture for a Green Planet, 1015*, 155-162. 10.17660/ActaHortic.2014.1015.17

Shaheen, M. R. (1986). Pistil receptivity in three cultivar of date palm (Phoenix dactylifera L.). In *Proc 1st Hort Sci, conference*. Tanta Univ.

Shaked, B., Amore, A., Ioannou, C., Valdés, F., Alorda, B., Papanastasiou, S., Goldshtein, E., Shenderey, C., Leza, M., Pontikakos, C., Perdikis, D., Tsiligiridis, T., Tabilio, M. R., Sciarretta, A., Barceló, C., Athanassiou, C., Miranda, M. A., Alchanatis, V., Papadopoulos, N., & Nestel, D. (2018). Electronic traps for detection and population monitoring of adult fruit flies (Diptera: Tephritidae). *Journal of Applied Entomology, 142*(1-2), 43–51. doi:10.1111/jen.12422

Shankar, U. (2017). Integrated pest management in horticultural crops. *Technological Innovations in Integrated Pest Management Biorational and Ecological Perspective*, 307.

Shanmugapriya, P., Rathika, S., Ramesh, T., & Janaki, P. (2019). Applications of Remote Sensing in Agriculture - A Review. *International Journal of Current Microbiology and Applied Sciences, 8*(1), 2270–2283.

Sharma, D. R., & Singh, S. (2006). Management of insect pests of temperate fruits. *Proceedings of advance training course on Emerging Trends in Economic Entomology,* 21-13.

Sharma, A., Verma, P., Mathur, A., & Mathur, A. K. (2018). Overexpression of tryptophan decarboxylase and strictosidine synthase enhanced terpenoid indole alkaloid pathway activity and antineoplastic vinblastine biosynthesis in Catharanthus roseus. *Protoplasma, 255*(5), 1281–1294. doi:10.100700709-018-1233-1 PMID:29508069

Sharma, Kumar, & Malik, Abhilash, & Harender. (2018). Applications of Remote Sensing in Agriculture. Agricultural. *Allied Sciences & Biotech. for Sustainability of Agriculture, Nutrition & Food Security*, 141–146.

Sharma, M., Sood, A., Nagar, P. K., Prakash, O., & Ahuja, P. S. (1999). Direct rooting and hardening of tea microshoots in the field. *Plant Cell, Tissue and Organ Culture, 58*(2), 111–118. doi:10.1023/A:1006374526540

Sharma, S. D., & Kumar, P. (2008). Relationship of Arbuscular Mycorrhizal Fungi and Azotobacter with Plant Growth, Fruit Yield, Soil and Leaf Nutrient Status of Mango Orchards in Northwestern Himalayan Region of India. *Journal of Applied Horticulture, 10*, 172–178.

Shashank, P. R., Doddabasappa, B., Kammar, V., Chakravarthy, A. K., & Honda, H. (2015). Molecular characterization and management of shoot and fruit borer *Conogethes punctiferalis* Guenee (Crambidae: Lepidoptera) populations infesting cardamom, castor and other hosts. In *New horizons in insect science: towards sustainable pest management* (pp. 207–227). Springer. doi:10.1007/978-81-322-2089-3_20

Sheffield, C. S., Kevan, P. G., Pindar, A., & Packer, L. (2013). Bee (Hymenoptera: Apoidea) diversity within apple orchards and old fields in the Annapolis Valley, Nova Scotia, Canada. *Canadian Entomologist, 145*(1), 94–114. doi:10.4039/tce.2012.89

Sheffield, C. S., Kevan, P. G., Westby, S. M., & Smith, R. F. (2008). Diversity of cavity-nesting bees (Hymenoptera: Apoidea) within apple orchards and wild habitats in the Annapolis Valley, Nova Scotia, Canada. *Canadian Entomologist, 140*(2), 235–249. doi:10.4039/n07-058

Shekarriz, P., Kafi, M., Deilamy, S. D., & Mirmasoumi, M. (2014). Coconut water and peptone improve seed germination and protocorm-like body formation of hybrid Phalaenopsis. *Agriculture Science Developments*, *3*(10), 317–322.

Sheppard, W. S. (1989). A history of the introduction of honey bee races into the USA. Part I of a two-part series. *American Bee Journal*, *129*, 617–619.

Sherif, N. A., Benjamin, J. H. F., Kumar, T. S., & Rao, M. V. (2018). Somatic embryogenesis, acclimatization, and genetic homogeneity assessment of regenerated plantlets of *Anoectochilus elatus* Lindl., an endangered terrestrial jewel orchid. *Plant Cell, Tissue and Organ Culture*, *132*(2), 303–316. doi:10.100711240-017-1330-4

Sherwani, A., Mukhtar, M., & Wani, A. A. (2016). *Insect pests of apple and their management. Insect pest management of fruit crops*. Biotech Books.

Shimoda, M., & Honda, K. I. (2013). Insect reactions to light and its applications to pest management. *Applied Entomology and Zoology*, *48*(4), 413–421. doi:10.100713355-013-0219-x

Short, B. D., Khrimian, A., & Leskey, T. C. (2017). Pheromone-based decision support tools for management of Halyomorpha halys in apple orchards: Development of a trap-based treatment threshold. *Journal of Pest Science*, *90*(4), 1191–1204. doi:10.100710340-016-0812-1

Shukla, S. P., & Sharma, A. (2017). *In vitro* seed germination, proliferation, and ISSR marker-based clonal fidelity analysis of *Shorea tumbuggaia* Roxb.: An endangered and high trade medicinal tree of Eastern Ghats. *In Vitro Cellular & Developmental Biology. Plant*, *53*(3), 200–208. doi:10.100711627-017-9818-5

Shulaev, V., Sargent, D. J., Crowhurst, R. N., Mockler, T. C., Folkerts, O., Delcher, A. L., Jaiswal, P., Mockaitis, K., Liston, A., Mane, S. P., Burns, P., Davis, T. M., Slovin, J. P., Bassil, N., Hellens, R. P., Evans, C., Harkins, T., Kodira, C., Desany, B., ... Folta, K. M. (2011). The genome of woodland strawberry (*Fragaria vesca*). *Nature Genetics*, *43*(2), 109–116. doi:10.1038/ng.740 PMID:21186353

Sicher, R. C. (2008). Effects of CO_2 enrichment on soluble amino acids and organic acids in barley primary leaves as a function of age, photoperiod and chlorosis. *Plant Science*, *174*(6), 576–582. doi:10.1016/j.plantsci.2008.03.001

Siddiqui, S. A., Subharwal, M., Gupta, S., & Balkrishan. (1994). Finite range survival model. *Microelectronics and Reliability*, *34*(8), 1377–1380. doi:10.1016/0026-2714(94)90154-6

Siddiqui, Z. A., & Mahmood, I. (1999). Role of bacteria in the management of plant parasitic nematodes: A review. *Bioresource Technology*, *69*, 167–179.

Sidhu, J. S. (2012). Production and Processing of Date Fruits. Handbook of Fruits and Fruit Processing, 629-651] doi:10.1002/9781118352533.ch34

Sidhu, J., Gibbs, R. A., Ho, G. E., & Unkovich, I. (2001). The role of indigenous microorganisms in suppression of Salmonella regrowth in composted biosolids. *Water Research*, *35*, 913–920.

Sikorskaite, S., Rajamäki, M. L., Baniulis, D., Stanys, V., & Valkonen, J. P. T. (2013). Protocol: Optimised methodology for isolation of nuclei from leaves of species in the *Solanaceae* and *Rosaceae* families. *Plant Methods*, *9*(1), 31–39. doi:10.1186/1746-4811-9-31 PMID:23886449

Šikšnianas, T., Stanienė, G., Stanys, V., & Sasnauskas, A. (2008). *Ribes sanguineum* Pursh. as donor of leaf fungal disease resistance in blackcurrant breeding. *Biologija (Vilnius, Lithuania)*, *54*(2), 79–82. doi:10.2478/v10054-008-0015-7

Silva, M., & Pereira, O. L. (2007). First report of Guignardia endophyllicola leaf blight on Cymbidium (Orchidaceae) in Brazil. *Australasian Plant Disease Notes, Australasian Plant Pathology Society*, *2*(1), 31–32. doi:10.1071/DN07015

Silva, M., Pereira, O. L., Braga, I. F., & Lelis, S. M. (2008). Leaf and pseudobulb diseases on Bifrenaria harrisoniae (Orchidaceae) caused by Phyllosticta capitalensis in Brazil. *Australasian Plant Disease Notes, Australasian Plant Pathology Society*, *3*, 53–56.

Simon, S., Bouvier, J. C., Debras, J. F., & Sauphanor, B. (2011). Biodiversity and pest management in orchard systems. *Sustainable Agriculture*, *2*, 693–709. doi:10.1007/978-94-007-0394-0_30

Singh, A. K., Singh, S., Saroj, P. L., Mishra, D. S., Singh, P. P., & Singh, R. K. (2019). Aonla (Emblica officinalis) in India: A review of its improvement, production and diversified uses. *Indian Journal of Agricultural Sciences*, *89*(11), 1773–1781.

Singh, G. (1988). Insect pollinators of mango and their role in fruit setting. *Acta Horticulturae*, (231), 629–632.

Singh, G. (1996). Pollination, pollinators and fruit setting in mango. *Acta Horticulturae*, (455), 116–123.

Singh, L. B. (1968). The mango: Botany, cultivation and utilization. Leonard Hill, London 263pp. Tim, A. Heard, 1999. The role of stingless bees in crop pollination. *Annual Review of Entomology*, *44*, 183–206.

Sinha, Heart, Agarwal, Asadi, & Carretero. (2005). Vermiculture and waste management: study of action of earthworms Eisinia fetida, Eudrilus enginae and Perionyx excavatus on biodegradation of some community wastes in India and Australia. *The Environmentalist, 22*(3), 261 – 268.

Sinha, R. K., Agarwal, S., Chauhan, K., & Valani, D. (2010). The wonders of earthworms & its vermicompost in farm production: Charles Darwin's 'friends of farmers', with potential to replace destructive chemical fertilizers from agriculture. *Agricultural Sciences*, *1*, 76–94.

Siregar, Sari, & Jauhari. (2016). Automation system hydroponic using smart solar solar power plant unit. *JurnalTeknologi (Sciences & Engineering), 78*(5), 55–60.

Śliwiński, T., Kowalczyk, T., Sitarek, P., & Kolanowska, M. (2022). Orchidaceae-Derived Anticancer Agents: A Review. *Cancers (Basel)*, *14*(3), 754. doi:10.3390/cancers14030754 PMID:35159021

Smith, D., Beattie, G. A., & Broadley, R. (1997). *Citrus pests and their natural enemies: integrated pest management in Australia*. Academic Press.

Smith, P., Ashmore, M., Black, H., Burgess, P., Evans, C., Hails, R., Potts, S. G., Quine, T., & Thomson, A. (2011). UK National Ecosystem Assessment – Chapter 14: Regulating Services. UNEP-WCMC.

Smith, C., Coetzee, P. P., & Meyer, J. P. (2003). The effectiveness of a magnetic physical water treatment device on scaling in domestic hot-water storage tanks. *Water S.A.*, *29*(3), 231–236.

Smith, L. A. (1993). Energy requirements for selected crop production implements. *Soil & Tillage Research*, *25*(4), 281–299. doi:10.1016/0167-1987(93)90028-N

Soliman, S. S., & Osman, S. M. (2003). Effect of nitrogen and potassium fertilization on yield, fruit quality and some nutrients content of Samany date palm. *Ann Agr Sci Ain Shams Univ Cairo*, *48*, 283–296.

Soliman, S. S., & Shaaban, S. H. A. (2006). Macronutrient changes of Samany date palm leaflets and fruits as affected by nitrogen and potassium application. *Egypt J Appl Sci*, *21*, 641–660.

Somers, E., Vanderleyden, J., & Srinivasan, M. (2004). Rhizosphere bacterial signaling: A love parade beneath our feet. *Critical Reviews in Microbiology*, *30*, 205–240.

Spencer, C. C., Su, Z., Donnelly, P., & Marchini, J. (2009). Designing genome-wide association studies: Sample size, power, imputation, and the choice of genotyping chip. *PLOS Genetics*, *5*(5), e1000477. doi:10.1371/journal.pgen.1000477 PMID:19492015

Srinidhi, H. V., Chauhan, S. K., & Sharma-S, C. (2007). SWOT analysis of Indian agroforestry. *Indian-Journal-of-Agroforestry*, *9*(1), 1–11.

Srivastava, R. P. (1997). Mango insect pest management. International Book Distributing Co.

Srivastava, S., Kadooka, C., & Uchida, J. Y. (2018). Fusarium species as a pathogen on orchids. *Microbiological Research*, *207*, 188–195. doi:10.1016/j.micres.2017.12.002 PMID:29458853

Stanienė, G., Stanys, V., & Kawecki, Z. (2002). Peculiarities of propagation in vitro of Vaccinium vitis-idaea L. and V. praestans Lamb. *Biologija (Vilnius, Lithuania)*, *48*(1), 84–86.

Stanys, V., Baniulis, D., Morkunaite-Haimi, S., Siksnianiene, J. B., Frercks, B., Gelvonauskiene, D., Stepulaitiene, I., Staniene, G., & Siksnianas, T. (2012). Characterising the genetic diversity of Lithuanian sweet cherry (*Prunus avium* L.) cultivars using SSR markers. *Scientia Horticulturae*, *142*, 136–142. doi:10.1016/j.scienta.2012.05.011

Stanys, V., Weckman, A., Staniene, G., & Duchovskis, P. (2006). *In vitro* induction of polyploidy in japanese quince (*Chaenomeles japonica*). *Plant Cell, Tissue and Organ Culture*, *84*(3), 263–268. doi:10.100711240-005-9029-3

Stefani, F. O. P., Moncalvo, J., Séguin, A., Bérubé, J. A., & Hamelin, R. C. (2009). Impact of an 8-year-old transgenic poplar plantation on the ectomycorrhizal fungal community. *Applied and Environmental Microbiology*, *75*(23), 7527–7536. doi:10.1128/AEM.01120-09 PMID:19801471

Stelinski, L. L., Miller, J. R., Ledebuhr, R., Siegert, P., & Gut, L. J. (2007). Season-long mating disruption of *Grapholita molesta* (Lepidoptera: Tortricidae) by one machine application of pheromone in wax drops (SPLAT-OFM). *Journal of Pest Science*, *80*(2), 109–117. doi:10.100710340-007-0162-0

Stone, A. G., Scheurell, S. J., & Darby, H. M. (2004). Suppression of soilborne diseases in fi eld agricultural systems: organic matter management, cover cropping and other cultural practices. In F. Magdoff & R. Weil (Eds.), *Soil organic matter in sustainable agriculture* (pp. 131–177). CRC Press LLC.

Stovold, G. E., Bradley, J., & Fahy, P. C. (2001). Acidovorax avenae subsp. cattleyae (Pseudomonas cattleyae) causing leafspot and death of Phalaenopsis orchids in New South Wales. *Australasian Plant Pathology*, *30*(1), 73–74. doi:10.1071/AP00066

Stranger, B. E., Stahl, E. A., & Raj, T. (2011). Progress and promise of genome-wide association studies for human complex trait genetics. *Genetics*, *187*(2), 367–383. doi:10.1534/genetics.110.120907 PMID:21115973

Suárez-Jacobo, A., Alcantar-Rosales, V. M., Alonso-Segura, D., Heras-Ramírez, M., Elizarragaz-De La Rosa, D., Lugo-Melchor, O., & Gaspar-Ramirez, O. (2017). Pesticide residues in orange fruit from citrus orchards in Nuevo Leon State, Mexico. *Food Additives & Contaminants: Part B*, *10*(3), 192–199. doi:10.1080/19393210.2017.1315743 PMID:28374639

Subler, S., Edwards, C.A., & Metzger, P.J. (1998). Comparing vermicomposts and composts. *BioCycle, 39*, 63–66.

Sudarsono, S., Elina, J., Giyanto, & Sukma, D. (2018). Pathogen Causing Phalaenopsis Soft Rot Disease – 16S rDNA and Virulence Characterisation. *Plant Protect. Sci., 54*(1), 1–8.

Sudarsono, S., Haristianita, M. D., Handini, A. S., & Sukma, D. (2017). Molecular marker development based on diversity of genes associated with pigment biosynthetic pathways to support breeding for novel colors in Phalaenopsis. *Acta Horticulturae*, (1167), 305–312. doi:10.17660/ActaHortic.2017.1167.44

Sueda, M., Katsuki, A., Nonomura, M., Kobayashi, R., & Tanimoto, Y. (2007). Effects of High Magnetic Field on Water Surface Phenomena. *The Journal of Physical Chemistry C*, *111*(39), 14389–14393. doi:10.1021/jp072713a

Su, J. S., Jiang, J. F., Zhang, F., Liu, Y., Ding, L., Chen, S. M., & Chen, F. D. (2019). Current achievements and prospects in the genetic breeding of Chrysanthemum: A review. *Horticulture Research*, *6*(1), 109. doi:10.103841438-019-0193-8 PMID:31666962

Sumaryono & Riyadi, I. (2011). *Ex vitro* rooting of oil palm (*Elaeis guineensis* Jacq.) plantlets derived from tissue culture. *International Journal of Agricultural Sciences*, *12*(2), 57–62.

Sumbali, G. (2005). *The fungi*. Alpha Science International.

Sung, Ying, Chang, Chen, Chen, & Ho. (2006). Pollinators and their behaviors on mango flowers in Sothern Taiwan. *Formosan Entomol.*, *26*, 161–170.

Surenciski, M. R., Flachsland, E. A., Terada, G., Mroginski, L. A., & Rey, H. Y. (2012). Cryopreservation of *Cyrtopodium hatschbachii* Pabst (*Orchidaceae*) immature seeds by encapsulation dehydration. *Biocell*, *36*(1), 31–36. doi:10.32604/biocell.2012.36.031 PMID:23173302

Surendran, U., Sandeep, O., & Joseph, E. J. (2016). The impacts of magnetic treatment of irrigation water on plant, water and soil characteristics. *Agricultural Water Management*, *178*, 21–29. doi:10.1016/j.agwat.2016.08.016

Syvertsen, J. P., Goñi, C., & Otero, A. (2003). Fruit load and canopy shading affect leaf characteristics and net gas exchange of 'Spring' navel orange trees. *Tree Physiology*, *23*(13), 899-906. doi:10.1093/treephys/23.13.899

Syvertsen, J.P., & Lloyd, D. J. (1994). Citrus. In *Handbook of environmental physiology of fruit crops. Vol. II. Subtropical and tropical crops*. CRC Press.

Szczech, M. M. (1999). Suppressiveness of vermicomposts against fusarium wilt of tomato. *Journal of Phytopathology*, *147*, 155–161.

Szczesniak, M. W., Rosikiewicz, W., & Makalowska, I. (2016). CANTATAdb:A Collection of Plant Long Non-Coding RNAs. *Plant & Cell Physiology*, *57*(1), e8. doi:10.1093/pcp/pcv201 PMID:26657895

Szlachetko, D. L. (2001). Genera et species Orchidalium. *Polish Botanical Journal*, *46*, 11–26.

Tadic, M. (1957). *The Biology of the Codling Moth as the Basis for Its Control*. Univerzitet U Beogradu.

Tai, C., Chang, M., Shieh, R., & Chen, T. (2008). Magnetic effects on crystal growth rate of calcite in a constant-composition environment. *Journal of Crystal Growth*, *310*(15), 3690–3697. doi:10.1016/j.jcrysgro.2008.05.024

Taimoury, H., Oussible, M., Bourarach, E. H., Harif, A., Hassanain, N., Masmoudi, L., & Baamal, L. (2015). Évaluation de la productivité de la pomme de terre (SolanumTuberosum L) sous l'effet de l'irrigation avec une eau traitée magnétiquement dans la région de Chaouia (Maroc). *Revue Marocaine des Sciences Agronomiques et Vétérinaires*, *3*(3), 30–36.

Taiz, L., & Zeiger, E. (1991). *Plant Physiology*. Benjamin Cummings.

Takahashi, Y., Takahashi, K., Watanabe, K., & Kawano, T. (2004). Bacterial black spot caused by Burkholderia andropogonis on Odontoglossum and intergeneric hybrid orchids. *Journal of General Plant Pathology*, *70*(5), 284–287. doi:10.100710327-004-0127-6

Tandon, P. L., & Verghese, A. (1985). *World list of insect, mite and other pests of mango*. Academic Press.

Tang, W., Newton, R., Li, C., & Charles, T. (2007). Enhanced stress tolerance in transgenic pine expressing the pepper *CaPF1* gene is associated with the polyamine biosynthesis. *Plant Cell Reports*, *26*(1), 115–124. doi:10.100700299-006-0228-0 PMID:16937149

Tang, Z. X., Shi, L. E., & Aleid, S. M. (2013). Date fruit: Chemical composition, nutritional and medicinal values, products. *Journal of the Science of Food and Agriculture*, *93*(10), 2351–2361.

Tao, Y. H., Ho, H. H., Wu, Y. X., & He, Y. Q. (2011). Phytophthora nicotianae causing Dendrobium blight in Yunnan Province, China. *International Journal of Plant Pathology*, *2*(4), 177–186. doi:10.3923/ijpp.2011.177.186

Taylor, S. C., Laperriere, G., & Germain, H. (2017). Droplet Digital PCR versus q-PCR for gene expression analysis with low abundant targets: From variable nonsense to publication-quality data. *Scientific Reports*, *7*(1), 2409. doi:10.103841598-017-02217-x PMID:28546538

Tedders, W. L., Weaver, D. J., & Wehunt, E. J. (1973). Pecan weevil: Suppression of larvae with the fungi Metarrhizium anisopliae and Beauveria bassiana and the nematode Neoaplectana dutkyi. *Journal of Economic Entomology*, *66*(3), 723–725. doi:10.1093/jee/66.3.723

Tee, C. S., Maziah, M., Tan, C. S., & Abdullah, M. P. (2003). Evaluation of different promoters Belarmino, driving the GFP reporter gene and selected target tissues for particle bombardment of Dendrobium Sonia 17. *Plant Cell Reports*, *21*(5), 452–458. doi:10.100700299-002-0539-8 PMID:12789448

Tejada, M., Gonzalez, J. L., Hernendez, M. T., & Garcia, C. (2008). Agricultural use of leachates obtained from two different vermicomposting processes. *Bioresource Technology*, *99*(14), 6228–6232. doi:10.1016/j.biortech.2007.12.031 PMID:18215517

Tello, J., Mammerler, R., Čajić, M., & Forneck, A. (2019). Major outbreaks in the nineteenth century shaped grape phylloxera contemporary genetic structure in Europe. *Scientific Reports*, *9*(1), 1–11. doi:10.103841598-019-54122-0 PMID:31772235

Tena, A., Beltra, A., & Soto, A. (2012). Novel defenses of Protop1ulvinaria pyriformis (Hemiptera: Coccidea) against its major parasitoid Metaphycus helvolus (Hymenoptera: Encyrtidae): implications for biological control of soft scales. *Biological Control*, *62*(1), 45–52. doi:10.1016/j.biocontrol.2012.03.005

Tengberg, M. (2012). Beginnings and early history of date palm garden cultivation in the Middle East. *Journal of Arid Environments*, *86*, 139–147.

The Plant List. (2013). *Version 1.1*. http://www.theplantlist.org/1.1/browse/A/Orchidaceae/

Thomas, E., Jalonen, R., Loo, J., Boshier, D., Gallo, L., Cavers, S., Bordács, S., Smith, P., & Bozzano, M. (2014). Genetic considerations in ecosystem restoration using native tree species. *Forest Ecology and Management*, *333*, 66–75. doi:10.1016/j.foreco.2014.07.015

Thorpe, K. W. (2006). A review of the use of mating disruption to manage gypsy moth, *Lymantria dispar* (L.). Academic Press.

Tiwari, P., Sharma, A., Bose, S. K., & Gautam, A. (2022). Biotechnological interventions in Orchids: Recent updates, Translational success, and Commercial outcomes. Research Square. doi:10.21203/rs.3.rs-1382018/v1

Tomati, U., Grappelli, A., & Galli, E. (1983). Fattori difertilita nell'humus di lombrico. *Proceedings of international symposium on agricultural and environmental prospects in earthworm farming*, 49–56.

Tomlinson, P. B. (1990). *The structural biology of palms*. Oxford University Press., www.google.com

TongKe, F. (2013). Smart Agriculture Based on Cloud Computing an d IOT. *Journal of Convergence Information Technology, 8*(2).

Tracewski, K. T., Brunner, J. F., Hoyt, S. C., & Dewey, S. R. (1987). Occurrence of Rhagoletis pomonella (Walsh) in hawthorns, Crataegus, of the Pacific Northwest. *Melanderia, 45*, 19–25.

Traveset, A., Tur, C., Trøjelsgaard, K., Heleno, R., Castro-Urgal, R., Olesen, J. M., & Santos, A. (2016). Global patterns of mainland and insular pollination networks. *Global Ecology and Biogeography, 25*(7), 880–890. doi:10.1111/geb.12362

Trejgell, A., & Tretyn, A. (2011). Shoot multiplication and in vitro rooting of *Carlina onopordifolia* Basser. *Acta Biologica Cracoviensia. Series; Botanica, 53*(2), 68–72. doi:10.2478/v10182-011-0026-z

Tripathi, G., & Bhardwaj, P. (2004). Comparative studies on biomass production, life cycles and composting efficiency of Eisenia fetida (Savigny) and Lampito mauritii (Kinberg). *Bioresource Technology, 92*, 275–283.

Tripathy, B. K., & Anuradha, J. (2018). *Internet of things (IoT). Technologies, Applications, Challenges, and Solutions.* CRC Press Taylor & Francis Group.

Troggio, M., Gleave, A., Salvi, S., Chagne, D., Cestaro, A., Kumar, S., Crowhurst, R. N., & Gardiner, S. E. (2012). Apple, from genome to breeding. *Tree Genetics & Genomes, 8*(3), 509–529. doi:10.100711295-012-0492-9

Trontin, J. F., Walter, C., Klimaszewska, K., Park, Y. S., & Lelu-Walter, M. A. (2007). Recent progress on transformation of four *Pinus* species. Invited review. *Transgenic Plant Journal, 1*(2), 314–329.

Trust, C. (2007). *Carbon Footprint Measurement Methodology, Version 1.1.* Carbon Trust. http://www.carbontrust.co.uk

Tsai, M. S., Lee, T. C., & Chang, P. T. (2013). Comparison of paper bags, calcium carbonate, and shade nets for sunscald protection in 'Murcott'tangor fruit. *Hort Technology, 23*(5), 659-667. doi:10.21273/HORTTECH.23.5.659

Tsai, J., Knutson, B., & Fung, H. H. (2006). Cultural variation in affect valuation. *Journal of Personality and Social Psychology, 90*(2), 288–307. doi:10.1037/0022-3514.90.2.288 PMID:16536652

Tsai, K. H., Chou, C., & Kuo, J. H. (2008). The curvilinear relationships between responsive and proactive market orientations and new product performance: A contingent link. *Industrial Marketing Management, 37*(8), 884–894. doi:10.1016/j.indmarman.2007.03.005

Uchida, J. (1995). Bacterial diseases of Dendrobium, Research extension series, vol 158. Institute of Tropical Agriculture and Human Resources. University of Hawaii.

Uchida, J. Y., & Aragaki, M. (1991a). Phytophthora diseases of orchids in Hawaii. Research extension series, vol 129. College of Tropical Agriculture and Human Resources. University of Hawaii.

Uchida, J. Y., & Aragaki, M. (1991b). Fungal diseases of Dendrobium flowers. Research extension series, vol 133. College of Tropical Agriculture and Human Resources. University of Hawaii.

Uchida, J. Y. (1994). Diseases of orchids in Hawaii. *Plant Disease, 78*(3), 220–224. doi:10.1094/PD-78-0220

Uchida, J. Y., & Aragaki, M. (1980). Nomenclature, pathogenicity, and conidial germination of *Phyllostictina pyriformis. Plant Disease, 64*(8), 786–788. doi:10.1094/PD-64-786

Uddain, J., Zakaria, L., Lynn, C. B., & Subramaniam, S. (2015). Preliminary assessment on Agrobacterium-mediated transformation of Dendrobium Broga Giant orchid's PLBs. *J. Emir. J. Food Agric., 27*(9), 669–677. doi:10.9755/ejfa.2015.05.211

Ünlü, L., Akdemir, B., Ögür, E., & Şahin, İ. (2019). Remote monitoring of European Grapevine Moth, Lobesia botrana (Lepidoptera: Tortricidae) population using camera-based pheromone traps in vineyards. *Turkish Journal of Agriculture-Food Science and Technology*, *7*(4), 652–657. doi:10.24925/turjaf.v7i4.652-657.2382

Unruh, T. R., Knight, A. L., Upton, J., Glenn, D. M., & Puterka, G. J. (2000, June 01). Particle Films for Suppression of the Codling Moth (Lepidoptera: Tortricidae) in Apple and Pear Orchards. *Journal of Economic Entomology*, *93*(3), 737–743. doi:10.1603/0022-0493-93.3.737 PMID:10902324

Ur-Rahman, A. H., James, D. J., Caligari, P. D. S., & Wetten, A. (2007). Difference in competence for in vitro proliferation and ex vitro growth of genetically identical mature and juvenile clones of apomictic Malus species. *Pakistan Journal of Botany*, *39*(4), 1197–1206.

USDA. (1995). *Gypsy moth mangement in the United States: a cooperaive approach. (Draft environmental impact statement)*. USDA Forest Service.

Usman, Gulzar, Wakil, Piñero, Leskey, Nixon, Oliveira-Hofman, Wu, & Shapiro-Ilan. (2020). Potential of entomopathogenic nematodes against the pupal stage of the apple maggot Rhagoletis pomonella (Walsh) (Diptera: Tephritidae). *Journal of Nematology, 52*.

Uthar, S. (2010). Evidence of plant hormone like substances in vermiwash: an ecologically safe option of synthetic chemicals for sustainable farming. *Ecol Eng, 36*, 1089–1092.

Veeresh, G. K. (1985, May). Pest problems in mango—world situation. In *II International Symposium on Mango 231* (pp. 551-565). Academic Press.

Velasco, R., Zharkikh, A., Affourtit, J., Dhingra, A., Cestaro, A., Kalyanaraman, A., Fontana, P., Bhatnagar, S. K., Troggio, M., Pruss, D., Salvi, S., Pindo, M., Baldi, P., Castelletti, S., Cavaiuolo, M., Coppola, G., Costa, F., Cova, V., Dal Ri, A., ... Viola, R. (2010). The genome of the domesticated apple (Malus × Domestica Borkh.). *Nature Genetics*, *42*(10), 833–839. doi:10.1038/ng.654 PMID:20802477

Vendrame, W. A., & Khoddamzadeh, A. A. (2017). Orchid biotechnology. *Horticultural Reviews*, *44*, 173–228.

Verghese, A., & Devi Thangam, S. (2011). Mango hoppers and their management. *Extension folder*, (71-11), 31-11.

Verheij, E. W. M., & Coronel, R. F. (1991). Edible fruits and nuts. Plant resources in South- *East Asia*. Academic Press.

Vessey, J. K. (2003). Plant growth promoting rhizobacteria asbiofertilizers. *Plant and Soil*, *255*, 571–586.

Vining, K. J., Pomraning, K. R., Wilhelm, L. J., Priest, H. D., Pellegrini, M., Mockler, T. C., Freitag, M., & Strauss, S. H. (2012). Dynamic DNA cytosine methylation in the Populus trichocarpa genome: Tissue-level variation and relationship to gene expression. *Biomed Cent Genom*, *13*(1), 27. doi:10.1186/1471-2164-13-27 PMID:22251412

Vithanage, V. (1990). The role of the European honeybee (Apis mellifera L.) in avocado pollination. *Journal of Horticultural Science*, *65*(1), 81–86. doi:10.1080/00221589.1990.11516033

Vitlox, O., Loyen, S. (2002). *Conséquences de la mécanisation sur le compactage du sol et l'infiltration d'eau, actes de la journée d'étude: érosion hydrique et coulées de boue en Wallonie*. Academic Press.

Wachsmann, Y., Zur, N., Shahak, Y., Ratner, K., Giler, Y., Schlizerman, L., Sadka, A., Cohen, S., Garbinshikof, V., Giladi, B., & Faintzak, M. (2012). Photoselective anti-hail netting for improved citrus productivity and quality. *International CIPA Conference 2012 on Plasticulture for a Green Planet, 1015*, 169-176.

Wackernagel, M., & Rees, W. E. (1996). *Our Ecological Footprint: Reducing Human Impact on the Earth*. New Society Publishers.

Wackers, F. L. (2005). Suitability of (extra-) floral nectar, pollen, and honeydew as insect food sources. *Plantprovided Food for Carnivorous Insects*, 17-74.

Wali, V. K., Bakshi, P., Jasrotia, A., Bhushan, B., & Bakshi, M. (2015). *Aonla*. SKUAST-Jammu.

Walsh, B. D. (1867). The apple-worm and the apple maggot. *American Journal of Horticulture*, *2*, 338–343.

Wang, H. M., Tong, C. G., & Jang, S. (2017). Current progress in orchid flowering/flower development research. *Plant Signaling & Behavior*, *12*(5), 5. doi:10.1080/15592324.2017.1322245 PMID:28448202

Wang, J. Y., Liu, Z. J., Zhang, G. Q., Niu, S. C., Zhang, Y. Q., & Peng, C. C. (2020). Evolution of two Ubiquitin-like a system of autophagy in orchids. *Horticultural Plant Journal*, *6*(5), 321–334. doi:10.1016/j.hpj.2020.05.006

Wang, Y., Wei, H., & Li, Z. (2018). Effect of magnetic field on the physical properties of water. *Results in Physics*, *8*, 262–267. doi:10.1016/j.rinp.2017.12.022

Wan, N. F., Ji, X. Y., Gu, X. J., Jiang, J. X., Wu, J. H., & Li, B. (2014). Ecological engineering of ground cover vegetation promotes biocontrol services in peach orchards. *Ecological Engineering*, *64*, 62–65. doi:10.1016/j.ecoleng.2013.12.033

Wan, N. F., Ji, X. Y., & Jiang, J. X. (2014). Testing the enemies hypothesis in peach orchards in two different geographic areas in eastern China: The role of ground cover vegetation. *PLoS One*, *9*(6), e99850. doi:10.1371/journal.pone.0099850 PMID:24963719

Wargovich, M. J. (2000). Anticancer properties of fruits and vegetables. *J. Hort. Sci.*, *35*, 573–575.

Waterhouse, D. F., & Sands, D. P. A. (2001). *Classical biological control of arthropods in Australia. ACIAR Monograph No. 77*. CSIRO Publishing.

Watson, J. C., Wolf, A. T., & Ascher, J. S. (2011). Forested landscapes promote richness and abundance of native bees (hymenoptera: Apoidea: Anthophila) in Wisconsin apple orchards. *Environmental Entomology*, *40*(3), 621–632. doi:10.1603/EN10231 PMID:22251640

Weill, A. (2009). *Les profils de sol agronomiques, un outil de diagnostic de l'état des sols. Centre de références en agriculture et agroalimentaire du Québec*. CRAAQ.

Weissa, M., Jacobb, F., & Duveillerc, G. (2020). Remote sensing for agricultural applications: A meta-review. *Remote Sensing of Environment*, *236*. Advance online publication. doi:10.1016/j.rse.2019.111402

Welter, S., Pickel, C., Millar, J., Cave, F., Van Steenwyk, R., & Dunley, J. (2005). Pheromone mating disruption offers selective management options for key pests. *California Agriculture*, *59*(1), 16–22. doi:10.3733/ca.v059n01p16

Whalon, M. E., & Croft, B. A. (1984). Apple IPM implementation in North America. *Annual Review of Entomology*, *29*(1), 435–470. doi:10.1146/annurev.en.29.010184.002251

Whitehead, V. B., & Rust, D. J. (1972). Control of the fruit-piercing moth Serrodes parfifa (Fabr.)(Lepidoptera: Noctuidae). *Phytophylactica*, *4*(1), 9–12.

White, T. L., Adams, W. T., & Neale, D. B. (2007). *Forest genetics*. CABI. doi:10.1079/9781845932855.0000

Wiedmann, T., & Minx, J. (2008). A Definition of 'Carbon Footprint. In C. C. Pertsova (Ed.), *Ecological Economics Research Trends* (pp. 1–11). Nova Science Publishers. https://www.novapublishers.com/catalog/product_info.php?products_id=5999

Wijayaratne, L. K., & Burks, C. S. (2020). Persistence of mating suppression of the Indian meal moth Plodia interpunctella in the presence and absence of commercial mating disruption dispensers. *Insects*, *11*(10), 701. doi:10.3390/insects11100701 PMID:33066462

Wilcox, P. L., Echt, C. E., & Burdon, R. D. (2007). Gene-assisted selection: applications of association genetics for forest tree breeding. In *Association mapping in plants* (Vol. 278). Springer. doi:10.1007/978-0-387-36011-9_10

Wilkie, J. D., Sedgley, M., & Olesen, T. (2008). Regulation of floral initiation in horticultural trees. *Journal of Experimental Botany*, *59*(12), 3215–3228. doi:10.1093/jxb/ern188 PMID:18653697

Winarto, B., & Setyawati, A. S. (2014). Young shoot nodes derived organogenesis in vitro mass propagation of Ruscus hypophyllum L. *South Western Journal of Horticulture, Biology and Environment*, *5*(2), 63–82.

Winfree, R., Gross, B. J., & Kremen, C. (2011). Valuing pollination services to agriculture. *Ecological Economics*, *71*, 80–88. doi:10.1016/j.ecolecon.2011.08.001

Winkler, K., Helsen, H. H. M., & Wackers, F. (2007). Kader 2. Functionele biodiversiteit in boomgaarden. *Entomologische Berichten*, *67*(6), 236-237.

Wittebolle, L., Marzorati, M., Clement, L., Balloi, A., Daffonchio, D., Heylen, K., De Vos, P., Verstraete, W., & Boon, N. (2009). Initial community evenness favours functionality under selective stress. *Nature*, *458*(7238), 623–626. doi:10.1038/nature07840 PMID:19270679

Witzgall, P., Bäckman, A.-C., Svensson, M., Koch, U., Rama, F., El-Sayed, A., Brauchli, J., Arn, H., Bengtsson, M., & Löfqvist, J. (1999, June 1). Behavioral observations of codling moth, Cydia pomonella, in orchards permeated with synthetic pheromone. *BioControl*, *44*(2), 211–237. doi:10.1023/A:1009976600272

Wu, P., Axmacher, J. C., Li, X., Song, X., Yu, Z., Xu, H., Tscharntke, T., Westphal, C., & Liu, Y. (2019). Contrasting effects of natural shrubland and plantation forests on bee assemblages at neighboring apple orchards in Beijing, China. *Biological Conservation*, *237*, 456–462. doi:10.1016/j.biocon.2019.07.029

Würschum, T., Maurer, H. P., Dreyer, F., & Reif, J. C. (2012). Effect of inter- and intragenic epistasis on the heritability of oil content in rapeseed (*Brassica napus* L.). *Theoretical and Applied Genetics*, *126*(2), 435–441. doi:10.100700122-012-1991-7 PMID:23052025

Wu, W. L., Chung, Y. L., & Kuo, Y. T. (2017). Development of SSR markers in Phalaenopsis orchids, their characterization, cross-transferability, and application for identification. *Orchid Biotechnol.*, *III*, 91–107.

Wysoki, M., Ben-Dov, Y., Swirski, E., & Izhar, Y. (1992, July). The arthropod pests of mango in Israel. In *IV International Mango Symposium 341* (pp. 452-466). Academic Press.

Yang, J., Benyamin, B., McEvoy, B. P., Gordon, S., Henders, A. K., Nyholt, D. R., Madden, P. A., Heath, A. C., Martin, N. G., Montgomery, G. W., Goddard, M. E., & Visscher, P. M. (2010). Common SNPs explain a large proportion of the heritability for human height. *Nature Genetics*, *42*(7), 565–569. doi:10.1038/ng.608 PMID:20562875

Yang, J., Lee, H., Shin, D. H., Oh, S. K., Seon, J. H., Paek, K. Y., & Han, K. (1999). Genetic transformation of Cymbidium orchid by particle bombardment. *Plant Cell Reports*, *18*(12), 978–984. doi:10.1007002990050694

Yano, A., Yoshiaki, O., Tomoyuki, H., & Kazuhiro, F. (2004). Effects of a 60 Hz magnetic field on photosynthetic CO_2 uptake and early growth of radish seedlings. *Bioelectromagnetics*, *25*(8), 572–581. doi:10.1002/bem.20036 PMID:15515039

Yasodha, R., Sumathi, R., & Gurmurthi, K. (2005). Improved micropropagation methods for Teak. *Journal of Tropical Forest Science*, *17*(1), 63–75.

Yee, Klaus, Cha, Linn, Jr., Goughnour, & Feder. (2014). Abundance of apple maggot, Rhagoletis pomonella, across different areas in central Washington, with special reference to black-fruited hawthorns. *Journal of Insect Science, 12.*

Yee, W. L., & Goughnour, R. B. (2008). Host plant use by and new host records of apple maggot, western cherry fruit fly, and other Rhagoletis species (Diptera: Tephritidae) in western Washington state. *The Pan-Pacific Entomologist, 84*(3), 179–193. doi:10.3956/2007-49.1

Yee, W. L., Landolt, P. J., & Darnell, T. J. (2005). Attraction of apple maggot (Diptera: Tephritidae) and nontarget flies to traps baited with ammonium carbonate and fruit volatile lures in Washington and Oregon. *Journal of Agricultural and Urban Entomology*, *22*, 133–149.

Yildirim, M., Dardeniz, A., Kaya, S., & Ali, B. (2016). An automated hydroponics system used in a greenhouse. *Scientific Papers, Series E, Land Reclamation, Earth Observation & Surveying Environmental Engineering, 5*, 63-66.

You, Lü, Zhong, Chen, Li, Liu, & Zhang. (2016). First report of bacterial brown spot in Phalaenopsis spp. caused by Burkholderia gladioli in China. *Plant Dis.*

Yu, H., Yang, S. H., & Goh, C. J. (2000). *DOH1*, a Class 1 *knox* Gene, Is Required for Maintenance ofthe Basic Plant Architecture and Floral Transition in Orchid. *The Plant Cell*, *12*(11), 2143–2159. doi:10.1105/tpc.12.11.2143 PMID:11090215

Yun-Zhu, G., Da-Chuan, Y., Hui-Ling, C., Jian-Yu, S., Chen-Yan, Z., Yong-Ming, L., Huan-Huan, H., Yue, L., Yan, W., Wei-Hong, G., Ai-Rong, Q., & Peng, S. (2012). Evaporation Rate of Water as a Function of a Magnetic Field and Field Gradient. *International Journal of Molecular Sciences*, *13*(12), 16916–16928. doi:10.3390/ijms131216916 PMID:23443127

Yu, X., Kikuchi, A., Matsunaga, E., Morishita, Y., Nanto, K., Sakurai, N., Suzuki, H., Shibata, D., Shimada, T., & Watanabe, K. N. (2009). Establishment of the evaluation system of salt tolerance on transgenic woody plants in the special netted-house. *Plant Biotechnology (Sheffield, England)*, *26*(1), 135–141. doi:10.5511/plantbiotechnology.26.135

Zaid, A., & De Wet, P. (1999). *Climatic requirements of date palm. Date palm cultivation.* Academic Press.

Zaid, A., & de Wet, P. F. (2002b) Climatic requirements of date palm. In Date palm cultivation. Food and Agriculture Organization Plant Production and Protection. Food and Agriculture Organization of the United Nations.

Zaid, A., & De wet, P.F. (2012). *Date Palm Propagation: Offshoot propagation.* www.agrihunt.com, www.google.com

Zaid, A., & de Wet, P. F. (2002). Pollination and bunch management. In A. Zaid (Ed.), *Date palm cultivation* (pp. 145–175). FAO.

Zaid, A., & de Wet, P. F. (2002a). Botanical and systematic description of the date palm. In A. Zaid (Ed.), *Date palm cultivation. Food and agriculture organization plant production and protection* (pp. 1–25). Food and Agriculture Organization of the United Nations.

Zaid, A., El-Korchi, B., & Visser, H. J. (2011). Commercial date palm tissue culture procedures and facility establishment. In S. M. Jain, J. M. Al-Khayri, & D. V. Johnson (Eds.), *Date palm biotechnology*. Springer. doi:10.1007/978-94-007-1318-5_8

Zayed, E. M. M., Rasmia, S. S. D., El-Din, A. F. M. Z., & Farrag, H. M. A. (2014). Impact of different sources of nitrogen fertilizers on performance growth of date palm (*Phoenix dactylifera* L. cv. Bartomouda). *Arab Universities Journal of Agricultural Sciences*, *22*(2), 371–379. doi:10.21608/ajs.2014.14742

Zein, S. N., El-khalik, A., & Khatab, S. (2012). Characterization of Tobacco mosaic Tobamovirus (TMV-S) isolated from sunflower (*Helianthus annuus* L.) in Egypt. *International Journal of Virology*, *8*(1), 27–38. doi:10.3923/ijv.2012.27.38

Zeng, S., Huang, W., Wu, K., Zhang, J., Silva, J. A. T., & Duan, J. (2016). *In vitro* propagation of *Paphiopedilum orchids. Critical Reviews in Biotechnology, 36*(3), 521–534. PMID:25582733

Zhang, D., & Wei, B. (2017). *Robotics and Mechatronics for Agriculture*. CRC Press.

Zhang, F., Maeder, M. L., Unger-Wallace, E., Hoshaw, J. P., Reyon, D., Christian, M., Li, X., Pierick, C. J., Dobbs, D., Peterson, T., Joung, J. K., & Voytas, D. F. (2010). High frequency targeted mutagenesis in Arabidopsis thaliana using zinc finger nucleases. *Proceedings of the National Academy of Sciences of the United States of America, 107*(26), 12028–12033. doi:10.1073/pnas.0914991107 PMID:20508152

Zhang, X., Fan, Y., Jia, Y., Cui, N., Zhao, L., Hu, X., & Gong, D. (2018). Effect of water deficit on photosynthetic characteristics, yield and water use efficiency in Shiranui citrus under drip irrigation. *Nongye Gongcheng Xuebao (Beijing), 34*(3), 143–150.

Zhang, Y. Y., Wu, K. L., Zhang, J. X., Deng, R. F., Duan, J., Silva, J. A. T., Huang, W. C., & Zeng, S. J. (2015). Embryo development in association with symbiotic seed germination in vitro of *Paphiopedilum armeniacum* S.C. Chen et F.Y. Liu. *Scientific Reports, 12*(5), 16356. doi:10.1038rep16356 PMID:26559888

Zhang, Y., Lee, Y. I., Deng, L., & Zhao, S. (2013). Asymbiotic germination of immature seeds and the seedling development of Cypripedium macranthos Sw., an endangered lady's slipper orchid. *Scientia Horticulturae, 164*, 130–136. doi:10.1016/j.scienta.2013.08.006

Zhang, Z. S., He, Q. Y., Fu, X. L., Ou, X. J., Lin, W. Q., & Jiang, J. Y. (2001). Studies on the wide cross of Chinese orchids and the germination of their hybrid seeds. *Journal of South China Agricultural University, 2*, 62–65.

Zhou, J. N., Lin, B. R., Shen, H. F., Pu, X. M., Chen, Z. N., & Feng, J. J. (2012). First report of a soft rot of Phalaenopsis aphrodita caused by Dickeya dieffenbachiae in China. *Plant Disease, 96*(5), 760–760. doi:10.1094/PDIS-11-11-0942 PMID:30727539

Zhou, K., Jerszurki, D., Sadka, A., Shlizerman, L., Rachmilevitch, S., & Ephrath, J. (2018). Effects of photoselective netting on root growth and development of young grafted orange trees under semi-arid climate. *Scientia Horticulturae, 238*, 272–280. doi:10.1016/j.scienta.2018.04.054

Zhu, K., Woodall, C. W., & Clark, J. S. (2012). Failure to migrate: Lack of tree range expansion in response to climate change. *Global Change Biology, 18*(3), 1042–1105. doi:10.1111/j.1365-2486.2011.02571.x

Zirari, A., & Ichir, L. L. (2010). *Effect of exogenous Indole butyric acid (IBA) on rooting and leaf growth of small date palm offshoots (Phoenix dactylifera L.) derived from adult vitro plants of Najda cultivar.* International Society for Horticulture Science. www. Google .com

Zörb, C., Geilfus, C. M., & Dietz, K. J. (2019). Salinity and crop yield. *Plant Biology, 21*(S1), 31–38. doi:10.1111/plb.12884 PMID:30059606

Zuk, O., Hechter, E., Sunyaev, S. R., & Lander, E. S. (2012). The mystery of missing heritability: Genetic interactions create phantom heritability. *Proceedings of the National Academy of Sciences of the United States of America, 109*(4), 1193–1198. doi:10.1073/pnas.1119675109 PMID:22223662

About the Contributors

Waleed Fouad Abobatta, Secretary of Citrus Research Department, Horticulture Research Institute, Agriculture Research Center, is an expert in the International Society of Citriculture, University of California (UC RIVERSIDE). He's Deputy General Secretary of the Arab Federation for the Protection of Wild and Marine Life. He is a member of the Scientific Committee for Greenhouses Plantation {A. R. C} (2014-202), and a member of the Scientific Team for the National Campaign for Navel Orange improvement {H.R.I} (2015-2018). Abobatta is a Scientific Committee Member of international and national conferences. Abobatta is a supervisor of M. Sc. and Ph. D. Thesis in various branches of agricultural science. He published more than 64 scientific papers.

Ahmed Farag is Prof. Dr. at Climate Modification Department, Directore of Central Laboratory for Agricultural Climate, Agriculture Research Center. He's a member of the National Council for Climate Changeand member of the Egyptian Negotiations Committee for Climate Change, Ahmed Awny has published significantly more than 60 research papers with the international journals since 2000. Ahmed Awny has been Certificates of deposit and registration of computer program Information Technology Industry Development Agency Office for the Protection of Intellectual Property Rights for Integrated Management Program for pomegranate Trees and Integrated Management Program for Olive Trees.

* * *

Mahmed Abdrabou Ahmed has a Ph.D. in horticulture crops, June (2005) from the Faculty of Agriculture, Ain Shams University, Cairo, Egypt. He obtained his M.Sc. in Vegetable Crops, May (2001), Faculty of Agriculture, Ain Shams University. He holds several leading positions such as Senior Researcher at Central Laboratory for Agricultural Climate (CLAC), Agricultural Research Center (ARC) (2015), Head of Agrometeorological Research Department Egypt (2012-2016). Deputy of CLAC for research affair from February 2019 till now. He spends 10 years in conducting researches on greenhouse vegetable crops, 10 years in farm irrigation researches, and more than 10 years in climate change researches. Concerning his international and regional activities, Prof. Mohamed Abdrabou has participated in several international, regional conferences, meetings and training courses. He was the National consultant for FAO to make national report about socioeconomic under climate change conditions (2010-2011).

Aula Ali Alwattar does research in organic synthesis, polymer synthesis. The synthesis of non-covalently bound small molecules for the functionalisation of graphene and other 2D materials. Her project is based on the design and synthesis of polycyclic aromatic hydrocarbons (PAHs) to develop a platform of 2D materials and their heterostructures which would allow one to design and fabricate novel functional devices. She had also introduced many new polycyclic heteroarenes. Furthermore, she did work on the formulation of low-cost, hydrogel was used for the removal of some heavy metals and toxic dyes from aqueous solution. Her recent work is evaluating the performance of conductive screen printing inks and body sensors.

Clara R. Azzam is the Head of Cell Department, Field Crops Research Center, Agricultural Research Center, Giza, Egypt. She was born in Budapest, Hungary, on the third of December, 1966. She is married and has a son and a daughter who are dentists and are graduates of the Faculty of Dental Medicine and Surgery, the British University in Egypt (BUE). She is interested in Plant genetics, Transformation, Plant Cell Biology, Plant Tissue Culture, Cytology, Molecular Markers, and Plant improvement via irradiation and mutation breeding. She got a Training course in Germany (13-month) entitled "Utilization of plant genetic resources as a contribution to food security" in the molecular cell biology department of the IPK institute, the work concerned on: in vitro- cell- and tissue culture and establishment of suitable conditions for gene transfer of sunflower (Helianthus annuus L.), learning different techniques in gene transfer of plants, participation in DNA cloning, RNA analysis, and enzyme tests; GUS- assay; analysis of integration and expressions of genes in different kinds of transgenic plants and Learning different PCR techniques, October 1998 – December 1999. She also worked three weeks at the Institute of Crop Plant Research and Plant Breeding of the University of Giessen, Germany, working on the gene transfer of sunflower and canola, August. 1999. Shc participated in some training courses to promote her skills: computer, Experimental system, Statistical analysis, using computer packages in agronomic trials analysis, development of researcher's skills in using World Wide Web (Internet) and information technologies for scientific use, The development of researcher's skills in using World Wide Web (Internet) and information technologies for scientific use, Utilization of VFES and RWES Expert Systems in Agricultural Research and Production, Preparation for TOEFL Test, she has participated in some DAAD Cairo Academy modules related to Self-Marketing, Project Management, Horizon 2020 Workshop, Managing Difficulties by Peaceful Means, How to present scientific paper and Competitive Research Proposal Writing She spent her Postdoctoral fellowship at Washington State University from 1st October 2010 to 1st Jan 2011, which STDF funded She attended some national and international conferences in Egypt or different countries such as Germany, France, Syria as a speaker or Rapporteur of Sessions She is a member of some professional societies as the Egyptian Society of Genetic Engineering (ESGE), the Egyptian Society of Genetics (ESG), the Egyptian Society of Biotechnology and Environmental Science (ESBES), the Egyptian Society of Plant Breeding (ESPB), the Egyptian Society of Nuclear Sciences and Applications (ESNSA). She was also the Focal point of FCRI-ARC for FP7, as well as, Horizon 2020. She was the Principal Investigator of the project "Induce mutations for diseases and economic pests resistance in peanut (Arachis hypogaea L.), which STDF funded from 2010 to 2015. Recently she is the PI of the project "Integrated Chia and Oyster Mushroom System for Sustainable Food Value Chain in Africa (CHIAM)," ERA-NET Cofund on Food Systems and Climate-FOSC Joint Call 2019, FOCS, project ID 43611 (2022-2025). She was a supervisor for some Ph.D. and M.Sc thesis in the field of Biotechnology and Molecular Markers: 13 completed Ph.D. and M.Sc theses and seven ongoing Ph.D. and M.Sc theses in the field of Biotechnology She was awarded the 'Woman of Achievement Award 2005' From ABI; American Biographical Institute, Inc. North Carolina, USA. Also, she was awarded the excellent manuscript year 2007 for a scientific paper entitled "Biochemical markers associated with disease resistance to damping-off and root-rot disease of peanut mutants and their productivity" from

ARC and the Minister of Agriculture. She holds the Shield of the Field Crops Institute and a Certificate of Appreciation from the Field Crops Institute 2022 Recipient of the Agricultural Research Center Shield -2022 and She recipients of the Women's Appreciation Award in Agriculture and Food Science – 2021 (ASRT). She is a reviewer in some scientific Journals like Bentham journals, African Journal of Biochemistry Research, Arabic Agriculture Science Journal, and a member of the Advisory Committee for Research Review, FCRI, ARC. And now, she is a member of the Higher Committee for the Promotion of Professors and Assistant Professors, ARC. She published 74 scientific papers, review articles and and book chapters in national and International Journals and scientific national or international conferences. She holds a Professional Diploma for the Preparation of Arab Women Leaders Internationally and Regionally, July 2020, from the Afro-Asian union for International Law and dispute settlement (AAU) in cooperation with the International Leadership Center (ILC).

K. Dubey is an officer of the Indian Forest Service. Has experience of 35 years of field service in Forestry and forest management. Presently as Principal Chief Conservator of Forest, Research and Training in Uttar Pradesh Forest Department looking after the reseach works caried out in fiel of silviculture and agroforestry.

Anis Elaoud is a professor at the Higher Institute of Environmental Sciences and Technologies, University of Carthage, Tunisia. He is a researcher in Laboratory of Science and Environment Technologies at the University of Carthage, Tunisia. He has more than 14 years of professional experience in the field of higher education and scientific research. Dr. Anis Elaoud is an academic member of the Athens Institute of Research, an expert in the Arab Organization for agricultural Development, and expert consultant at SCP 'CanalSaint Provence'. He has written more than 50 scientific articles in international journals and more than 50 papers in international conferences He conducted research, the most important of which is the study of the effects of magnetic equipment on water and agriculture. Dr. Anis runs a research topic 'Effects of magnetic equipment on water, plants and soil' with an integrated and diverse research team. He presided over and prepared a scientific day on the magnetization of water «Journeìe Magneìtisation de l'Eau JME'2017» and "International Conference on Statistical Modeling and Applications: Agricultural Field and Remote Sensing «ICSMA 2017». He is a member of a project Funded by the Ministry PJEC 'Effet de la magneìtisation de l'eau'; a member of the MAYA project 'ERASMUS+'... Weider is the head of three research partnership agreements between Egypt, Tunisia, and Algeria.

Rehab Y. Ghareeb is currently Associate Professor of Plant Protection and Biomolecular Diagnosis at City of Scientific Research and Technological Application, Borg El-Arab, Alexandria, Egypt. She earned her BSc in Plant Pathology/Alexandria University, Egypt (2003). Her interest in this field of science motivated him for further studies and leads her to finish his MSc in Plant Pathology/Alexandria University, Egypt (2008). Along the way, she has specialized in Plant-pathogenic nematodes, plant-parasitic nematode diseases, nematode-host interaction. As well as the molecular genetics interactions of plant-nematode, gene biosensors in nematode discovery and disaster, gene silencing, and recent tools for nematode control, and got this expertise from City of Scientific Research and Technological Application. In 2015, she got her Ph.D. in Alexandria University, Faculty of Agricultural, Egypt. Her research interest is in Molecular Biology, Green Nanotechnology, and its application in plant pathogenic nematodes management fields. Plant-microbe interactions and the possible ways for discovery the family of genes control that relationship. Plant biotic & abiotic Stress and the philosophy of conditional genes. Defense system in plant against plant pathogen and discovery the micro RNA and Nano RNA control the defense properties. Dr. Rehab has applied his know-how technique in Nematodes Bio-control applications, Eco friendly-Green Nanotechnology; Synthesis, Characterization, and Utilization of Nanoparticles and their Applications as Antimicrobial especially against plant pathogenic nematodes. To date, she has 26 international publications and 2 patents in Green-Technology. With her extensive experience in her area of interest, she has been invited to several speaking engagements both locally and internationally. She is a reviewer in several reputable international journals.

Amira Ibrahim is working in Plant Protection and Biomolecular Diagnosis at City of Scientific Research and Technological Application, Borg El-Arab, Alexandria, Egypt. Dr. Amira Got BSc in Botany and Chemistry, Botany Department, Faculty of Science, Mansoura University, Egypt. She earned her MSc in Genetics /from Mansoura University (2012) and Ph.D. in Genetics /from Mansoura University (2018). Her research interest in Genetics and Molecular Biology, Population genetics, and using Karyotype and Genetic Diversity for Plant conservation. Also her interested in using green synthesized nanoparticles as a biological control for many plant pathogens, studying the cytotoxicity and genotoxicity of many fungicides, pesticides, and biological agents. To date, she published 19 Papers in international journals, 3 chapters, and 1 patent in Molecular Biology.

Rim Jalel is a Dr and a general engineer at AVFA. Dr Rim is an assistant professor at the Tunisian university. She has more than 10 years of professional experience in the field of education. She has written more than 10 scientific articles in international journals and more than 15 papers in international conferences. She works in research area in mechanics, irrigation and agriculture.

Athir Mahmood Haddad, 54 years old, is a Professor working at University of Basrah. He received his Master and Ph.D. degree in Polymer Chemistry in 1994 and in 2002, respectively. During his PhD period, he has focused his research on the Preparation of Hydrogel polymers for soil stabilizer, slow released fertilizers and Polymer drug delivery. He joined the Professor Mauro Angelitti group (University of Camerino, Italy in 2006. From 2007 to 2008, he was focus his research on the Preparation of polymers via ROP, ATRP, RAFT and Click Chemistry with Professor Andrew Whittaker (AIBN, University of Queensland, Australia. At 2011, He worked on the preparation of Janus polymers at University of Michigan/ USA. From October 2015 till 2020, his research activities are mainly centered on the Preparation of poly lactide nano!bers and hydrogels for removal toxic metals and organic dyes (University of Manchester/UK).

Elsayed Zaki received his PhD in physical chemistry from the Department of Chemistry, Faculty of Science, Mansoura University, Egypt. Currently, he is a visiting researcher at the Department of Chemical and Biomolecular Engineering, Iacocca Hall, Lehigh University, USA. He has been a researcher of applied chemistry in the Petroleum Applications Department, Egyptian Petroleum Research Institute from 1 March 2015 to present. He is currently serving as an editorial member of several reputed journals such as: Industrial & Engineering Chemistry Research, American Chemical Society, International Journal of Hydrogen Energy, Archives in Cancer Research, Herald Journal of Agriculture and Food Science Research, on the scientific and technical committee and editorial review board on Chemical and Molecular Engineering (WASET), Open Journal of Applied Sciences, Advances in Chemical Engineering and Science, Journal of the Chemical Society of Pakistan, International Conference on Chemical,

Metallurgy and Environmental Engineering, Universal Researchers in Environmental & Biological Engineering, Asia-Pacific Chemical, Biological & Environmental Engineering Society, International Journal of Chemical and Biomolecular Science in Public Science Framework, Biomedical and Pharmacology Journal, International Academy of Chemical, Civil & Environment Engineering, International Journal of Ambient Energy, he was appointed to editorial board for the International Journal of Materials Science and Applications, he was appointed to editorial board for Advances in Materials, he was appointed to editorial board for International Journal of Science, Technology and Society, Organic & Medicinal Chemistry International Journal (OMCIJ), Universal Journal of Pharmaceutical Research, International Journal of Nano and Material Sciences, he is an advisory board member of World Journal of Pharmacy and Pharmaceutical Sciences, Cogent OA, Taylor & Francis Group, SciFed Journal of Polymerscience. He is a member of the Royal Society of Chemistry (MRSC), Chemicals Development Services Center (CDSC), Egyptian Corrosion Society (ECS), The Arab Society of Material Science, Petroleum and Mineral Resources Society, Egyptian Petroleum Association, The Egyptian Society of Polymer Science and Technology, Egyptian Syndicate of Scientific Professions, Asia-Pacific Chemical, Biological & Environmental Engineering Society (APCBEES), Society of Petroleum Engineers (SPE), World Academy of Science, Engineering and Technology (WASET), The International Association of Engineers (IAENG), Egyptian Association for Science and Engineering (EASE), Sesame User Office (SUO) Synchrotron- Light for Experimental Science and Applications in the Middle East, Sesame Users' Committee (SUC) Synchrotron-Light for Experimental Science and Applications in the Middle East, International Association of Advanced Materials (IAAM), The Society of Digital Information and Wireless Communications (SDIWC), The International Society for Environmental Information Sciences (ISEIS), The Organization for Women in Science for the Developing World (OWSD), and the International Union of Pure and Applied Chemistry.

Index

T

V

W

Y

www.ingramcontent.com/pod-product-compliance
Lightning Source LLC
LaVergne TN
LVHW060008031026
849161LV00025B/2481